T0342341

Remote Sensing Physics

Advanced Textbook Series

Advanced Textbook 3

Remote Sensing Physics

An Introduction to Observing Earth from Space

Rick Chapman
The Johns Hopkins University Applied Physics Laboratory, USA

Richard Gasparovic
The Johns Hopkins University Applied Physics Laboratory (Ret.), USA

This Work is a co-publication of the American Geophysical Union and John Wiley and Sons, Inc.

This edition first published 2022
© 2022 American Geophysical Union

All rights reserved. No part of this publication may be reproduced, stored in a retrieval system, or transmitted, in any form or by any means, electronic, mechanical, photocopying, recording or otherwise, except as permitted by law. Advice on how to obtain permission to reuse material from this title is available at http://www.wiley.com/go/permissions.

Published under the aegis of the AGU Publications Committee

Matthew Giampoala, Vice President, Publications
Carol Frost, Chair, Publications Committee
For details about the American Geophysical Union visit us at www.agu.org.

The right of Rick Chapman and Richard Gasparovic to be identified as the authors of this work has been asserted in accordance with law.

Registered Office
John Wiley & Sons, Inc., 111 River Street, Hoboken, NJ 07030, USA

Editorial Office
111 River Street, Hoboken, NJ 07030, USA

For details of our global editorial offices, customer services, and more information about Wiley products visit us at www.wiley.com.

Wiley also publishes its books in a variety of electronic formats and by print-on-demand. Some content that appears in standard print versions of this book may not be available in other formats.

Limit of Liability/Disclaimer of Warranty
While the publisher and authors have used their best efforts in preparing this work, they make no representations or warranties with respect to the accuracy or completeness of the contents of this work and specifically disclaim all warranties, including without limitation any implied warranties of merchantability or fitness for a particular purpose. No warranty may be created or extended by sales representatives, written sales materials or promotional statements for this work. The fact that an organization, website, or product is referred to in this work as a citation and/or potential source of further information does not mean that the publisher and authors endorse the information or services the organization, website, or product may provide or recommendations it may make. This work is sold with the understanding that the publisher is not engaged in rendering professional services. The advice and strategies contained herein may not be suitable for your situation. You should consult with a specialist where appropriate. Further, readers should be aware that websites listed in this work may have changed or disappeared between when this work was written and when it is read. Neither the publisher nor authors shall be liable for any loss of profit or any other commercial damages, including but not limited to special, incidental, consequential, or other damages.

Library of Congress Cataloging-in-Publication Data

Names: Chapman, Rickey David, 1955- author. | Gasparovic, Richard Francis,
    1941- author.
Title: Remote sensing physics : an introduction to observing earth from
    space / Rick Chapman, The Johns Hopkins University Applied
    Physics Laboratory, USA, Richard Gasparovic, The Johns Hopkins
    University Applied Physics Laboratory (Ret.), USA.
Description: Hoboken, NJ : Wiley-American Geophysical Union, 2022. |
    Series: Advanced textbook series | Includes bibliographical references
    and index.
Identifiers: LCCN 2021038135 (print) | LCCN 2021038136 (ebook) | ISBN
    9781119669074 (paperback) | ISBN 9781119669029 (adobe pdf) | ISBN
    9781119669159 (epub)
Subjects: LCSH: Earth sciences–Remote sensing. | Environmental
    monitoring–Remote sensing. | Artificial satellites.
Classification: LCC QE33.2.R4 C472 2021 (print) | LCC QE33.2.R4 (ebook) |
    DDC 550.28–dc23
LC record available at https://lccn.loc.gov/2021038135
LC ebook record available at https://lccn.loc.gov/2021038136

Cover Design: Wiley
Cover Image: © The National Oceanic and Atmospheric Administration

Set in 9.5/12.5pt STIXTwoText by Straive, Chennai, India

10 9 8 7 6 5 4 3 2 1

Contents

Preface

At no time in history has it been more imperative to understand the present condition of our Earth's environment, especially the impacts of myriad human-induced activities and their portents for the future of our planet. The information provided by Earth observations from space-based remote sensors has become a critical element for this crucial endeavor.

The development of technology for Earth remote sensing has been a true multidisciplinary effort involving a variety of scientific disciplines, sensor and satellite engineering technologies, as well as advances in computer technology. Beginning in the 1960s with early images of cloud patterns observed with various electro-optical sensors, scientists and engineers soon recognized the potential for new information obtainable with instruments operating throughout the electromagnetic spectrum. This quickly led to the launch of infrared sensors for measuring ocean, land, and atmospheric temperatures; optical sensors for ocean color data, vegetation coverage, and atmospheric gases and aerosols; along with microwave instruments for monitoring polar regions, ocean winds and waves, sea level rise, and land surface deformations.

Exploitation of these advances in remote sensing technology would not have been possible without the parallel explosion in computer processing capabilities needed to acquire, store, display and synthesize the massive global data sets transmitted daily from Earth-observing satellites. Now, nearly half

a century later, weather and severe storm forecasts based on satellite observations have become part of daily life, along with rainfall and drought monitoring, and forest fire detection. Moreover, space-based remote sensing technology now affords the only practical means for long-term global monitoring and prediction of such climate variables as sea level rise, ocean temperatures and biological productivity, and atmospheric conditions.

The purpose of this text is to provide an introduction to the physical principles underlying the techniques being used for remote sensing of the Earth. Our focus is on providing the reader with coherent treatments of the basic physics needed to understand exactly what the various instruments are measuring, including how and why the raw signals must be calibrated and corrected for interference or contamination by various environmental factors in order to extract the physical parameters of interest.

We have endeavored also to describe the relevant sensor technologies in sufficient detail for the reader to appreciate the engineering approaches to the acquisition of remotely sensed data. The references cited in each chapter can provide the interested reader with significantly more in-depth information on these engineering aspects. Software systems currently available for processing, display, and extraction of information from remote sensor data is an area we have intentionally omitted. References to some useful starting points on this topic can be found in one of the Appendixes.

An attempt has been made throughout to present the material at a level that should be understandable to upper-class undergraduate science and engineering students, as well as to early-year graduate students in these disciplines.

The book is conceptually divided into three main parts. The first part consists of a brief overview of Earth remote sensing (Chapter 1) and a description of satellite orbits relevant to instruments deployed for Earth observations (Chapter 2). The second major part, comprising Chapters 3 through 6, discusses observations made with passive sensors, i.e., those which make use of natural illumination from the sun and/or thermal radiation from the ocean, land surfaces, and Earth's atmosphere. Observations with active sensors which provide their own illumination, such as radars of various types and lidars, are treated in Chapters 7 through 11 of the final part. The book concludes with a brief overview of two sensing techniques not discussed previously, along with a short summary of future Earth observation missions being planned by NASA and the European Space Agency (Chapter 12).

A compendium of links to a wide variety of remote sensing resources available on the Internet can be found in Appendix F. An extensive bibliography with references to other books, journal publications, and technical reports is included to supplement the material presented in the individual chapters of the book. Our expectation is that these items will prove valuable to students and researchers alike.

A substantial fraction of the material in this book was originally developed for remote sensing courses taught by the authors over many years to Masters Degree students in the Applied Physics curriculum within the Engineering for Professionals Program at the Johns Hopkins Whiting School of Engineering. The authors hereby gratefully acknowledge the numerous and invaluable contributions from these student interactions.

The authors also want to thank the numerous colleagues who reviewed chapters of the book: Dr. Steve Borchardt, Dr. Joshua Broadwater, Dr. Eric Ericson, Dr. David Jansing, Dr. Kevin Kwon, Dr. Carl Lueschen, Mr. Frank Monaldo, Dr. David Porter, Dr. Keith Raney, and Dr. Scott Wunsch. We especially want to thank Dr. Adrienne Criss and Captain James Miller (USN, Ret.) for reviewing the book in its entirety, which was not a small undertaking. We acknowledge the efforts of Mr. Chris Jackson (NOAA) who reprocessed SAR wind data for Figures 10.48-10.52 and Dr. James Churnside (NOAA) who provided lidar data for Figure 11.10. We also appreciate the efforts of numerous editors and reviewers at the AGU and Wiley for their contributions to this book.

Acronyms

ACE	Advanced Composition Explorer
ADCP	Acoustic Doppler Current Profiler
ADEOS	Advanced Earth Observing Satellite
ADM	Atmospheric Dynamics Mission
AIS	Automatic Identification System
ALADIN	Atmospheric LAser Doppler INstrument
ALI	Advanced Land Imager
ALOS	Advanced Land Observing Satellite
ALtiKa	Altimeter Ka-band
AMI	Active Microwave Instrument
AMSR	Advanced Microwave Scanning Radiometer
AMSR-E	Advanced Microwave Scanning Radiometer-EOS
AMSU	Advanced Microwave Sounding Unit
ARM	Atmospheric Radiation Measurement
ASAR	Advanced Synthetic Aperture Radar
ASCAT	Advanced Scatterometer
ASCII	American Standard Code for Information Interchange
ASF	Alaska SAR Facility
ASTER	Advanced Spaceborne Thermal Emission and Reflection Radiometer
AT	Along Track
ATI	Along-Track Interferometry
ATISAR	Along-Track Interferometric SAR
ATLAS	Advanced Topographic Laser Altimeter System
ATM	NASA Airborne Topographic Mapper
ATSR	Along-Track Scanning Radiometer
AVHRR	Advanced Very High Resolution Radiometer
B81	Brown 1981 model
BB	Blackbody
BRDF	Bidirectional Reflectance Distribution Function
BSF	Beam Spread Function
BW	Bandwidth
CAD	Computer Aided Drafting

CALIOP	Cloud-Aerosol Lidar with Orthogonal Polarization
CALIPSO	Cloud-Aerosol Lidar and Infrared Pathfinder Satellite Observation
CAVIS	Clouds, Aerosols, Water Vapor, Ice and Snow instrument
CCD	Charge-Coupled Device, Coherent Change Detection
CCI	Climate Change Initiative
CDOM	Colored Dissolved Organic Material
CFOSAT	China-France Oceanography Satellite
CIE	International Commission on Illumination
CLW	Cloud Liquid Water
CMOD	C-band Geophysical Model Function
CMODIS	Chinese Moderate Resolution Imaging Spectrometer
CMOS	Complementary Metal-Oxide Semiconductor
CNES	French National Centre for Space Studies
CNSA	China National Space Administration
COCTS	Chinese Ocean Color and Temperature Scanner
COSMO-SkyMed	Constellation of Small Satellites for Mediterranean basin Observation
CPI	Coherent Processing Interval
CSA	Canadian Space Agency
CSV	Comma Separated Values format
CYGNSS	Cyclone Global Navigation Satellite System
CZCS	Coastal Zone Color Scanner
CZI	Coastal Zone Imager
D2P	Delay-Doppler Altimeter
DAAC	NASA Distributed Active Archive Center
DC	Direct Current (0 frequency)
DEM	Digital Elevation Model
DIAL	Differential Absorption Lidar
DLR	German Aerospace Center
DMSP	Defense Meterological Satellite Program
DOD	U.S. Department of Defense
DP	Differential Phase Shift
DR	Differential Reflectivity
DSCOVR	Deep Space Climate Observatory
E&M	Electricity and Magnetism
ECOSTRESS	ECOsystem Spaceborne Thermal Radiometer Experiment on Space Station
EGM2008	Earth Gravitational Model 2008
EHF	Extremely High Frequency
ELF	Extremely Low Frequency
EM	Electromagnetic
EnMAP	Environmental Mapping and Analysis Program
ENVISAT	Environmental Satellite
EO	Electro-optic
EOF	Empirical Orthogonal Functions
EOS	Earth Observing System
EOSDIS	Earth Observing System Data and Information System

ERS	European Remote-Sensing Satellite
ESA	European Space Agency
ET	Evapotranspiration
ETM	Enhanced Thematic Mapper
ETM+	Enhanced Thematic Mapper Plus
EUMETSAT	European Organisation for the Exploitation of Meteorological Satellites
EVI	Enhanced Vegetation Index
FAA	U.S. Federal Aviation Administration
FLH	Fluorescence Line Height
FM	Frequency Modulated
FOR	Field of Regard
FOV	Field of View
GAC	Global Area Coverage
GCOM-C	Global Change Observation Mission - Climate
GEDI	Global Ecosystem Dynamics Investigation
GEO-CAPE	GEOstationary Coastal and Air Pollution Events
GEO	Geostationary Earth Orbit
GeoCarb	Geostationary Carbon Observatory
GG	Glazman and Greysukh 1993 model
GISAT	Geo Imaging Satellite
GLAS	Geoscience Laser Altimeter System
GLI	Japanese Global Imager
GLONASS	Global Navigation Satellite System
GmAPD	Geiger-mode Avalanche Photo Diodes
GML	Geiger Mode Lidar
GNSS	Global Navigation Satellite System
GOCI	Geostationary Ocean Color Imager
GOES-3	Geodynamics Experimental Ocean Satellite 3
GOES	Geostationary Operational Environmental Satellites
GOSAT	Greenhouse gases Observing SATellite
GPM	Global Precipitation Measurement
GPO	U.S. Government Publishing Office
GPS	Global Position System
GRACE	Gravity Recovery and Climate Experiment
GRACE-FO	Gravity Recovery and Climate Experiment-Follow On
GSD	Ground Sample Distance
GSFC	Goddard Space Flight Center
HDF	Hierarchical Data Format
HEO	High Earth Orbit
HERA	Hybrid Extinction Retrieval Algorithm
HF	High Frequency
HFSWR	High-Frequency Surface Wave Radar
HH	Horizontal Transmit – Horizontal Receive
HICO	Hyperspectral Imager for the Coastal Ocean
HITRAN	High-resolution Transmission Molecular Absorption Database

HRPT	High-Resolution Picture Transmission
HSI	Hyperspectral Imaging
HV	Horizontal Transmit – Vertical Receive
HV SC	HeaVy-Snow Covered ice
HySI	Hyperspectral Imager
HyspIRI	Hyperspectral Infrared Imager
IF	Intermediate Frequency
IFOV	Instantaneous Field of View
IGS	Information Gathering Satellites
InGaAs	Indium Gallium Arsenide
INSAR	Interferometric SAR
IR	Infrared
ISRO	Indian Space Research Organisation
ISS	International Space Station
JAXA	Japan Aerospace Exploration Agency
JB	Jianbing
JERS	Japanese Earth Resources Satellite
JHU/APL	The Johns Hopkins University Applied Physics Laboratory
JPL	NASA Jet Propulsion Laboratory
JPSS	Joint Polar Satellite System
KARI	Korea Aerospace Research Institute
LAC	Local Area Coverage, LEISA Atmospheric Corrector
LCM	Linear Composite Model
LED	Light Emitting Diode
LEISA	Linear Etalon Imaging Spectrometer Array
LEO	Low Earth Orbit
LF	Low Frequency
LFM	Linear Frequency Modulated
LIS	Lightning Imaging Sensor
LO	Local Oscillator
LOS	Line of Sight
LS	Landsat
LST	Land Surface Temperature, Local Solar Time
LWIR	Long-wave Infrared
MABL	Marine Atmospheric Boundary Layer
MAIA	Multi-Angle Imager for Aerosols
MB	Megabyte, Main Beam
MCW	Modified Chelton and Wentz model
MEO	Mid-Earth Orbit
MERIS	Medium-spectral Resolution, Imaging Spectrometer
MF	Medium Frequency
MFY	Medium First-Year ice
MHS	Microwave Humidity Sounder
MOBY	Marine Optical Buoy
MODIS	Moderate-resolution Imaging Spectrometer

MODTRAN	Moderate Resolution Atmospheric Transmission
MOS	Modular Opto-electric Sensor
MP	Melt Ponds
MSS	Multispectral Scanner
MWIR	Medium-wave Infrared
NASA	National Aeronautics and Space Administration
NASDA	National Space Development Agency of Japan
Nd:YAG	Neodymium-Doped Yttrium Aluminum Garnet
NDBC	U.S. National Data Buoy Center
NDVI	Normalized Difference Vegetation Index
NDWI	Normalized Difference Water Index
NEδL	Noise Equivalent Delta Radiance
NEδT	Noise Equivalent Delta Temperature
NEXRAD	Next-Generation Radar
NF	Noise Figure
NIR	Near-Infrared
NISAR	NASA-ISRO Synthetic Aperture Radar
NISTAR	National Institute of Standards and Technology Advanced Radiometer
NOAA	National Oceanographic and Atmospheric Administration
NOMAD	NASA bio-Optical Marine Algorithm Dataset
NPOESS	National Polar-orbiting Operational Environmental Satellite System
NPP	National Polar-Orbiting Partnership
NRC	U.S. National Research Council
NRCS	Normalized Radar Cross-Section
NRL	U.S. Naval Research Laboratory
NSCAT	NASA Scatterometer
NWP3	Freilich and Dunbar 1993 model
OC3V	Ocean Chlorophyll 3-band VIIRS Algorithm
OC4	Ocean Chlorophyll 4-band Algorithm
OCI	Ocean Color Imager
OCM	Ocean Colour Monitor
OCO	Orbiting Carbon Observatory
OCTS	Ocean Color and Temperature Scanner
OES	Ocean Ecosystem Radiometer
OLCI	Ocean Land Color Imager
OMPS	Ozone Mapping and Profiler Suite
OSC	Orbital Sciences Corporation
OSCAT	OceanSat-2 Scanning Scatterometer
OSMI	Ocean Scanning Multispectral Imager
OSTM	Ocean Surface Topography Mission
PACE	Plankton, Aerosol, Cloud and ocean Ecosystem
PALSAR	Phased Array type L-band Synthetic Aperture Radar
PAR	Photosynthetically Available Radiation
PCR	Pulse Compression Ratio
PD	Power Density

PDF	Probability Density Function
PHyTIR	Prototype HyspIRI Thermal Infrared Radiometer
PNG	Portable Network Graphics format
POES	Polar Operational Environmental Satellites
POLDER	POLarization and Directionality of the Earth's Reflectances
PR	Pressure Ridges
PREFIRE	Polar Radiant Energy in the Far-InfraRed Experiment
PRF	Pulse Repetition Frequency
PRI	Pulse Repetition Interval
PT-JPL	Priestley–Taylor Jet Propulsion Laboratory
QE	Quantum Efficiency
RAM	Random Access Memory
RCM	Radarsat Constellation Mission
RCS	Radar Cross-Section
RF	Radio Frequency
RFSCAT	Rotating Fan Beam SCATterometer
RISAT	Radar Imaging Satellite
RMS	Root Mean Square
ROCSAT	Republic of China Satellite
ROIC	Readout Integrated Circuits
RVI	Radar Vegetation Index
S-NPP	Suomi National Polar-Orbiting Partnership
SABIA-Mar	Satélites Argentino-Brasileño para Información Ambiental del Mar
SAGE-III	Stratospheric Aerosol and Gas Experiment III
SAOCOM	Argentine Microwaves Observation Satellite
SAR	Synthetic Aperture Radar
SARAL	Satellite with ARgos and ALtiKa
SASS	Seasat Scatterometer
SB	Smoothed Brown model
SDP	Simplified Deep Space Perturbations
SDPS	SeaWiFS Data Processing System
SeaBASS	SeaWiFS Bio-optical Archive and Storage System
SeaWiFS	Sea-viewing Wide Field-of-view Sensor
SGLI	Second generation GLobal Imager
SGP	Simplified General Perturbations
SHF	Super High Frequency
SIR-C	Spaceborne Imaging Radar-C band
SLC	Scan Line Corrector
SLF	Super Low Frequency
SMAP	Soil Moisture Active Passive
SMI	Standard Mapped Image
SMMR	Scanning Multichannel Microwave Radiometer
SMOS	Soil Moisture and Ocean Salinity
SNR	Signal-to-Noise Ratio
SOA	Chinese State Ocean Administration

SORCE	Solar Radiation and Climate Experiment
SPL	Single Photon Lidar
SRAL	Sentinel Radar Altimeter
SS MP	SubSurface Melt Ponds
SSA	Small-Slope Approximation
SSM/I	Special Sensor Microwave/Imager
SSMIS	Special Sensor Microwave Imager/Sounder
SST	Sea Surface Temperature
SVD	Singular Value Decomposition
SW	Surface Wave
SWH	Significant Wave Height
SWIR	Short-wave Infrared
SWOT	Surface Water & Ocean Topography
TCI	Temperature Condition Index
TCTE	Total Solar Irradiance Calibration Transfer Experiment
TDI	Time-Delay and Integration
TDRSS	Tracking and Data Relay Satellite System
TDWR	Terminal Doppler Weather Radar
TEMPO	Tropospheric Emissions: Monitoring of Pollution
TES	Temperature Emissivity Separation
TFY	Thick First-Year ice
THF	Tremendously High Frequency
ThFY	Thin First-Year ice
TIRS	Thermal Infrared Sensor
TLE	Two-Line Elements
TM	Thematic Mapper
TMI	TRMM Microwave Imager
TRMM	Tropical Rainfall Measuring Mission
TROPICS	Time-Resolved Observations of Precipitation structure and storm Intensity with a Constellation of Smallsats
TSIS-1	Total and Spectral Solar Irradiance Sensor
TV	Television
UHF	Ultra High Frequency
ULF	Ultra Low Frequency
USGS	U.S. Geological Survey
UTC	Coordinated Universal Time
UV	Ultraviolet
VCI	Vegetation Condition Index
VH	Vertical Transmit – Horizontal Receive
VHF	Very High Frequency
VIIRS	Visible Infrared Imaging Radiometer Suite
VLF	Very Low Frequency
VNIR	Visible and Near-Infrared
VSWIR	Visible to Short Wavelength Infrared
VV	Vertical Transmit – Vertical Receive

WFF	Wallops Flight Facility
WGS	World Geodetic System
WiFS	Wide Field Sensor
WMO	World Meteorological Organization
WS	Wind Speed
WSR	Weather Surveillance Radar
WV	Water Vapor

About the Companion Website

This book is accompanied by a companion website.

www.wiley.com/go/chapman/physicsofearthremotesensing

The website includes:

- Example homework problems
- PDF and Powerpoint files of all figures from the book for downloading
- Latex and Powerpoint files containing all equations used in the text
- Multiple animations for use in the classroom

1

Introduction to Remote Sensing

Remote sensing is commonly defined as the process by which electromagnetic energy is exploited to interrogate some property of the Earth's environment – either its surface or its surrounding atmosphere – by using a sensor system located at some distance from the region of interest.

Yet remote sensing is a bit more general than this. For example, the interior structures of the Earth have been remotely sensed using neutrino detectors. And spacecraft have been deployed to remotely sense the characteristics of other planets, asteroids and comets. Still this definition characterizes most of the remote sensing described in this book, which concentrates on Earth remote sensing using electromagnetic energy.

This image of the Gulf Stream shown in Figure 1.1 is a classic example. The image was created from data obtained by a multiple-wavelength imaging system flown on the NOAA-12 satellite. The data from this camera were transmitted to a ground station located at The Johns Hopkins University Applied Physics Laboratory (JHU/APL), where they were processed using sophisticated algorithms to estimate sea surface temperature. The image you see is a computer-generated false-color map of sea surface temperature.

This book describes the physical basis for such measurements and examines the algorithms used to derive geophysical information from such data.

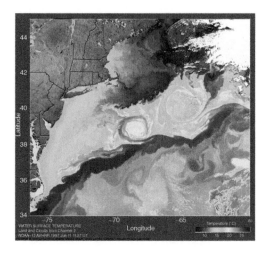

Figure 1.1 Sea surface temperature in Western Atlantic. Source: Courtesy of R. E. Sterner, JHU/APL.

Such calibrated satellite imagery provides a unique source of information on scales that would be otherwise inaccessible to terrestrial sensors. But it is hard today to recall how revolutionary such data really are. For example, in Figure 1.1, the blobs of warm water located just north of the Gulf Stream wall are massive warm core eddies. These eddies are rotating lenses of fluid that are occasionally spun off of the Gulf Stream. The amazing thing is that oceanographers were unaware of the existence of such warm core eddies until the first large-scale black and white images of the ocean were returned from early satellites.

Remote Sensing Physics: An Introduction to Observing Earth from Space, Advanced Textbook 3, First Edition.
Rick Chapman and Richard Gasparovic.
© 2022 American Geophysical Union. Published 2022 by John Wiley & Sons, Inc.
Companion website: www.wiley.com/go/chapman/physicsofearthremotesensing

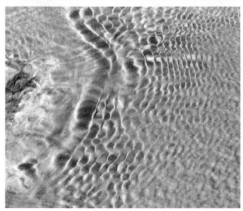

Figure 1.2 Flows near Grand Geyser, Yellowstone National Park.

Remote sensing does not have to occur from so far away. The photographs shown in Figure 1.2 were taken in Yellowstone National Park while awaiting the eruption of Grand Geyser. Water leaking from the geyser flowed down the slight incline towards the raised walkway (top). The middle photograph shows the ripples in the water created by flow past obstacles. The bottom photograph shows a blowup of one set of these ripples.

Surface waves can be created by flow past obstacles, just like a moving boat in still water creates waves. The ripples shown here are capillary waves that propagate slightly upstream of the disturbance. Those waves whose speed is arrested by the flow are stationary and hence can grow. The dispersion relation for surface waves is well known. So wavelength measurements from this single photograph could be used to determine the flow speed!

Without the measurement and a physical model, this is no more than a pretty picture. When combined with quantitative measurements and a physical model this photograph becomes remote sensing data.

This text discusses all aspects of the acquisition, measurement, and physical interpretation of the most common types of remote sensing. While the text primarily concentrates on satellite-based remote sensing of the environment, remote sensors deployed from aircraft and other platforms are also described.

As shown in Figure 1.3, satellites are used for a wide range of applications. They are used for communications, navigation and timing (GPS), as well as military applications such as intelligence collection. Satellites are also used to perform scientific measurements involving space environment, Earth environment, and astronomy (Davis, 2007).

This text concentrates on the Earth environment applications, such as measurements of the atmosphere, land, and oceans. Despite this concentration, the physics of remote sensing are also relevant for other applications.

There are a wide variety of applications for remote sensing data, as shown by the partial list in Box 1.1. These applications span the range from oceans to land to the atmosphere. Many of these applications are discussed in some detail throughout this book.

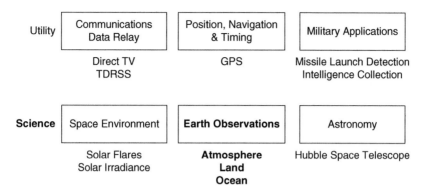

Figure 1.3 Satellite applications.

Box 1.1 Applications for remote sensing data.

- Ocean Observations
  - Sea surface temperature
  - Ocean color
  - Biological productivity
  - Coral bleaching
  - Sea ice concentration and extent
  - Sea level rise and tides
  - Currents, eddies, bathymetry
  - Surface winds
- Atmospheric Observations
  - Weather systems - clouds, storms
  - Temperature and moisture profiles
  - Pollution - dust, volcano plumes
  - CO_2 concentration

- Land Observations
  - Surface temperature
  - Vegetation coverage
  - Snow cover
  - Soil moisture
  - Continental ice sheets
  - Elevation changes
  - Floods
  - Forest fires
  - Urbanization changes
  - Maps
- Earth Radiation Budget
  - Solar insolation
  - Reflected sunlight
  - Emitted thermal radiation

Maybe more important than the specific applications is the ability to regularly make measurements over most or all of the globe and to make those measurements over years or even decades. This makes satellite data particularly useful for radiation budget and climate studies.

Satellites are not ideal – they have a variety of limitations as remote sensing platforms. Most satellites can provide only intermittent observations at any location on the globe. Time intervals between revisits are determined by orbital parameters, sensor swath width and environmental limitations, such as the need for daylight or cloud-free line of sight to the surface. High-resolution sensors typically provide only limited area coverage per orbit. Dwell time per sensed area is typically short, so short duration transient events are seen only by chance. While individual satellites can provide continuous data records of a decade or more, these records are short relative to climate time scales. Time series exceeding the lifetime of an individual sensor require difficult multisensor intercomparisons and intercalibrations.

In addition, interpretation of remote sensing data is not easy. Remote sensors detect properties of electromagnetic radiation emitted,

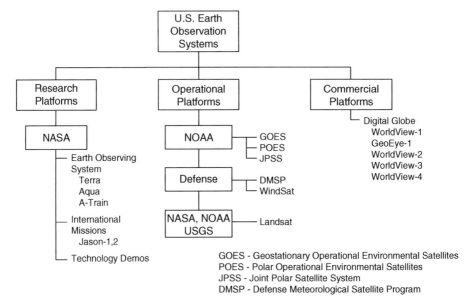

Figure 1.4 Overview of U.S. Earth observation systems.

scattered, or reflected from scenes of interest, but these properties are often affected by multiple geophysical parameters. Models are developed to describe the sensor output as a function of the geophysical parameters. "Inversion" is the process of inferring the geophysical parameters from the sensor output. Inversions are not always unique and validation of geophysical retrieval algorithms is frequently difficult. These are challenges that are discussed throughout this text.

Because this text concentrates on satellite-based environmental remote sensing, it is worth mentioning the organizations responsible for current U.S. Earth observation systems (Figure 1.4). NASA is responsible for research platforms, such as the Terra (Kaufman et al., 1998; Ungar et al., 2003) and Aqua (Parkinson, 2003) Earth observing satellites. They also coordinate with international partners and develop technology demonstrators.

U.S. government satellites that are needed for operational requirements, such as weather prediction, are either run by NOAA or by the military and intelligence communities.

Historically, the earliest Earth observation satellites were developed and launched by U.S. government agencies in the late 1950s and early 1960s to acquire meteorological data (Davis, 2007). The Soviet Union followed in the late 1960s with a similar focus on meteorology. Interest in ocean and land observations came into prominence in the 1970s following major technology developments in optical and microwave sensors. Today, Earth remote sensing has become a truly international endeavor involving tens of countries building and operating scores of both government and commercially funded satellites. In addition, the ready availability of large volumes of digital data from these systems has been the impetus for innovations in processing, display, and dissemination of products for an ever expanding range of applications.

1.1 How Remote Sensing Works

The general process of how remote sensing works is illustrated in Figure 1.5.

A sensor attached to a platform such as a satellite looks down to make measurements. The sensor can typically make measurements of one or more resolution cells within its field

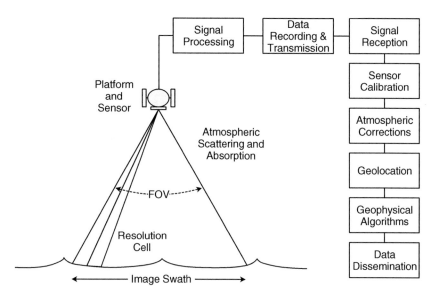

Figure 1.5 Remote sensing process.

of view (FOV). The data from the sensor may be processed on board, and either transmitted to the ground immediately or stored for later transmission. The signals are ultimately received at a ground station, where the main processing begins. Typical processing involves first sensor calibration, followed by corrections for atmospheric effects. The data are then geolocated to ground coordinates by application of coordinate transformations driven by satellite position and attitude information. Geophysical retrieval algorithms are then applied to convert the data from sensor units to geophysical parameters. Finally the data are disseminated.

Almost all remote sensing utilizes electromagnetic energy ranging in wavelength from 100 m to 0.1 μm, a range of nine orders of magnitude! As indicated in Figure 1.6, many remote sensing systems operate at microwave frequencies, in either active configurations such as radar or passive configurations such as radiometers. Another major group of sensors utilizes energy from the infrared wavelengths through the visible and into ultraviolet wavelengths. These are generally referred to as optical systems, because they use optics to receive the radiation instead of antennas.

Note the large gap between the infrared and microwave regions.

The active use of frequency bands is governed by domestic and international agreements in order to avoid interference and conflicts. Some frequency bands are specifically set aside for passive remote sensing by prohibiting emissions in those bands, while virtually all active microwave remote sensors require licensing.

Because of the importance of microwave sensors, Figure 1.6 includes the designators for specifics bands. Originally, the lower RF band names (HF, VHF and so on) were assigned in a sensible system. For historical reasons, microwave frequency band designations were deliberately obscure. To make matters worse, there are two different band designators: the IEEE standard derived from World War II usage, and a newer sequential standard that is used in Europe, NATO, and for some military applications. The IEEE standard is used throughout this text, so when you read about an X-band radar, you'll know that this is a system operating around 10 GHz.

Remote sensors on satellites or even high-altitude aircraft must look through the atmosphere in order to measure radiation from the

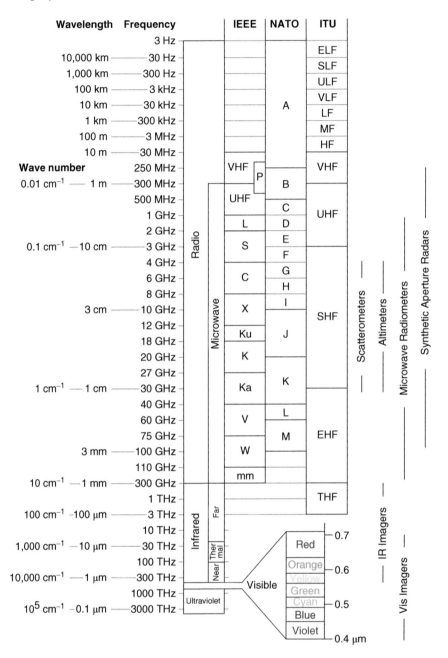

Figure 1.6 Electromagnetic frequencies and wavelengths used in remote sensing, including international band designations. Adapted from Rees (2012).

surface. Figure 1.7 shows the transmittance of the atmosphere along a vertical path from UV to UHF wavelengths. The upper curve is for a midlatitude winter atmosphere and the lower curve is for midlatitude summer atmosphere. The primary difference between the two curves is the amount of water vapor in the atmosphere.

Effective remote sensing of the surface is limited to wavelengths with reasonably high

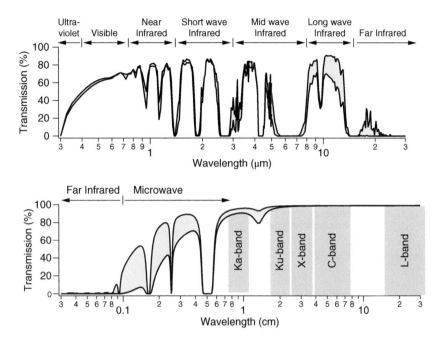

Figure 1.7 Atmospheric transmittance for a dry, midlatitude winter atmosphere (top curve) and midlatitude summer atmosphere (bottom curve) that contains more water vapor. Adapted from Elachi and van Zyle (2006).

transmittance. The atmosphere is relatively transparent in the visible wavelength bands. This is a good thing because otherwise it would be hard to see this book. Moving to the near and mid infrared wavelengths, there are some reasonably transparent windows but also some sizable absorption bands at various wavelengths. Thus, only specific IR wavelengths can be used for surface observations, although absorption band wavelengths are commonly exploited for probing the atmosphere.

At wavelengths between 15 and 1000 μm, the atmosphere is opaque, so there is a gap in remote sensing through the atmosphere in the far infrared. In the microwave region, the atmosphere again becomes mostly transparent with some large absorption bands. At microwave and RF frequencies below X-band, the atmosphere is essentially transparent.

As illustrated in Figure 1.8, there are various types of remote sensors. One basic characteristic is whether the remote sensor is active or passive. Active systems provide their own illumination. Examples are radars or lidars.

Passive systems make use of natural radiation. Examples are microwave radiometers or a camera without a flash.

A second basic characteristic is whether the sensor is imaging (like a camera) or nonimaging (like a radar altimeter that just measures the height of the ground below it). This latter characteristic is a bit fuzzy because you can have hybrid systems, such as a scanning altimeter that measures a single surface height at a time, but scans over all the heights within a given region of interest.

A variety of viewing geometries are commonly used by sensors flown on Earth observation satellites. The three generic types (Figure 1.9) are:

- Nadir view – Data collected only along the satellite ground track – e.g., radar altimeter, laser profiler.
- Area view – Data collected over a swath centered on the ground track. Area covered depends on sensor scan angle – e.g., optical and infrared imagers, atmospheric sounders.

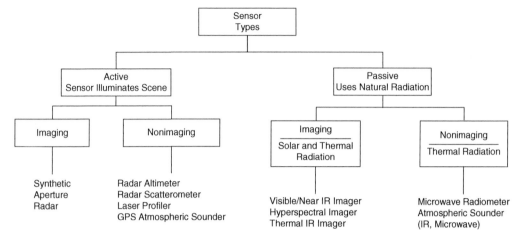

Figure 1.8 Remote sensor systems.

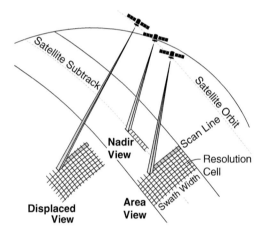

Figure 1.9 Sensor viewing geometries.

- Displaced area view - Data collected over a swath with the center displaced from ground track - e.g., imaging radar.

Independent of the sensor viewing geometry, there are multiple ways to configure an imager. The first configuration illustrated in Figure 1.10 is a linear scanning system. This is essentially a single-pixel camera that scans in the cross-track direction while the satellite flies forward. Early weather satellites worked like this, with the scanning implemented by spinning the entire satellite about the axis oriented along the flight track. Later sensors, and many current ones, use a rotating mirror to scan the detector in a cross-track direction.

The second configuration is a circular-scanning variant of the first. In this case the satellite also spins, but its spin axis is pointed towards nadir. The single sensor is tilted off of nadir, and it scans in a spiral pattern as the satellite orbits. The timing of the spin and the sensor field of view are typically adjusted to minimize coverage gaps. Circular scan systems can provide area coverage while making all measurements at a fixed look angle. This is the approach taken in many passive microwave radiometers.

A third configuration is the push-broom sensor. In the basic configuration a line sensor is oriented cross-track to image a line of pixels. The line sensor is operated at a high frame rate and the motion of the spacecraft allows the sensor to make measurements over an extended area. In more advanced configurations multiple line arrays are placed in the detector plane to image the same spot on the surface multiple times. This can be used to improve sensitivity by what is called time delay integration (or TDI). The individual lines can also have different spectral filters in front of them, an approach often used in multispectral or hyperspectral imagers.

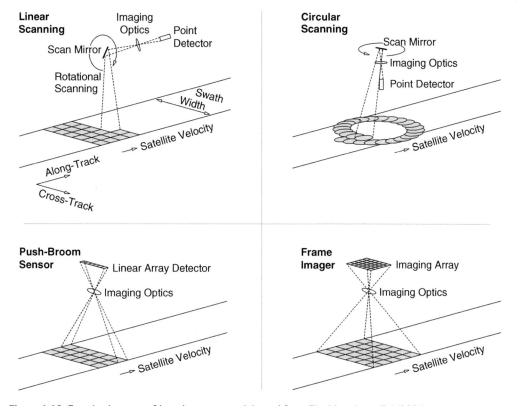

Figure 1.10 Four basic types of imaging systems. Adapted from Elachi and van Zyl (2006).

The final configuration is a framing camera with a two-dimensional array of sensors in the focal plane behind the imaging optics. This is the configuration used in common cameras, and is likely the best approach for acquisition of video rate data.

References

Davis, G. K. (2007). History of the NOAA satellite program. *Journal of Applied Remote Sensing, 1*(1), 012504.

Elachi, C., & van Zyl, J. J. (2006). *Introduction to the physics and techniques of remote sensing.* John Wiley & Sons.

Kaufman, Y. J., Herring, D. D., Ranson, K. J., & Collatz, G. J. (1998). Earth Observing System AM1 mission to Earth. *IEEE Transactions on Geoscience and Remote Sensing, 36*(4), 1045–1055.

National Research Council. (2008). *Earth observations from space: The first 50 years of scientific achievements.* Washington, DC: National Academies Press.

Parkinson, C. L. (2003). Aqua: An Earth-observing satellite mission to examine water and other climate variables. *IEEE Transactions on Geoscience and Remote Sensing, 41*(2), 173–183.

Rees, W. G. (2012). *Physical principles of remote sensing.* Cambridge University Press.

Ungar, S. G., Pearlman, J. S., Mendenhall, J. A., & Reuter, D. (2003). Overview of the Earth Observing One (EO-1) mission. *IEEE Transactions on Geoscience and Remote Sensing, 41*(6), 1149–1159.

2

Satellite Orbits

Satellites have to move in orbit to stay high above the Earth (Figure 2.1). It is important to understand the basics of satellite orbits because the constrained motions of satellites both enable and complicate their use as remote sensing platforms.

A few of the major take aways from this chapter are:

- Orbit type determines the speed of the platform and the range from the sensor to the ground. These in turn drive sensor viewing geometry, ground resolution, and spatial coverage.

- Higher orbits generally provide wider swaths but at lower spatial resolution.
- Orbital altitude determines the field of regard (the region on the ground the satellite can observe) and the temporal observation frequency.
- Low Earth polar orbits allow only 1–2 views of midlatitude areas per day.
- Sun-synchronous orbits allow viewing at the same local solar time each day.
- Geostationary orbits can provide approximately continuous coverage of large areas.

Figure 2.2 shows the height of various types of satellites on a linear scale. Low Earth orbits are those with altitudes of 2000 km or less, where most remote sensing satellites reside. This figure should make it clear that satellites in low Earth orbits can only see a small portion of the Earth at any one time. Geostationary satellites orbit at 36,000 km. Medium Earth orbits are between these two categories, with global navigation satellite systems (e.g., GPS (USA), GLONASS (Russia), Galileo (Europe), BeiDou (China)) at 20,000 km as prime examples.

Figure 2.1 Orbits of some remote sensing satellites. Source: NASA's Goddard Space Flight Center Scientific Visualization Studio, https://svs.gsfc.nasa.gov.

Remote Sensing Physics: An Introduction to Observing Earth from Space, Advanced Textbook 3, First Edition.
Rick Chapman and Richard Gasparovic.
© 2022 American Geophysical Union. Published 2022 by John Wiley & Sons, Inc.
Companion website: www.wiley.com/go/chapman/physicsofearthremotesensing

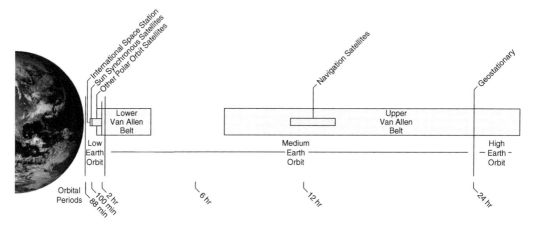

Figure 2.2 Orbital altitudes drawn to scale.

Kepler's law describes satellite orbits around a central mass like the Earth. As illustrated in Figure 2.3, all orbits follow an elliptical path, with the Earth at one focus.

The range r from the center of the Earth to the satellite is then given by:

$$r = \frac{a(1 - e^2)}{1 + e\cos\theta} \tag{2.1}$$

Important definitions in this diagram include:

- length of the semimajor axis (a).
- length of the semiminor axis (b).
- eccentricity $\left(e = \sqrt{1 - \frac{b^2}{a^2}} \right)$.
- apogee, the peak altitude of the satellite = $a(1 + e)$.
- perigee, the minimum altitude of the satellite = $a(1 - e)$.

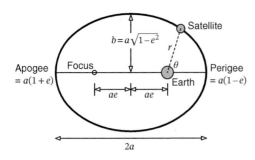

Figure 2.3 Elliptical orbit.

- true anomaly (θ) defined to be the angle to the satellite relative to perigee in an Earth-centered coordinate system.

As illustrated in Figure 2.4, Kepler's law implies that the position of any satellite in elliptical orbit, as measured in a coordinate system relative to the center of the Earth, is specified by six variables called Kepler elements.

In Figure 2.4:

Ω is the longitude of the ascending node of the orbit. The ascending node is the location on the equator that the satellite ground track passes when going from the southern to northern hemisphere.

a is the length of the semimajor axis.

e is the orbit eccentricity.

i is the orbit inclination measured from the equatorial plane.

ω is the argument of periapsis, the angle within the orbital plane between the ascending node and perigee. This defines the orientation of the orbital ellipse.

θ is the true anomaly, the angle within the orbital plane from periapsis to the satellite location.

For a circular orbit, $e = 0$ and since there is no perigee we nominally set $\omega = 0$ to reference the true anomaly to the Equator. Thus only four variables are needed to describe the

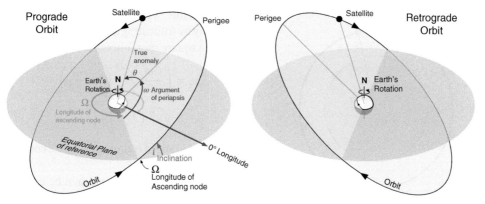

Figure 2.4 Kepler elements defining an elliptical orbit. Adapted from original work of Lasunncty at https://commons.wikimedia.org/wiki/File:Orbit1.svg.

position of a satellite in circular orbit (Ω, a, i, and θ). Although it may be obvious, note that θ increases linearly as a function of time for a circular orbit (2π radians for each orbit) and the ascending node is the start of the orbit at relative time $t = 0$.

Finally, we speak of a prograde orbit when the satellite moves west to east at the ascending node, namely for inclinations less than 90°. Retrograde orbits have inclinations greater than 90° and move east to west, against the rotation of the Earth, at the ascending node.

The subsatellite point is the location where the line from the satellite to the center of the Earth intersects the Earth's surface. The position of the subsatellite point for circular orbits can be expressed analytically, at least assuming a spherical Earth. The latitude and longitude of the subsatellite point on the Earth's surface (in radians) are given by equations (2.2) and (2.3).

$\theta =$ true anomaly (radians).

$T_{orb} =$ orbital period expressed in same units as time t.

$\delta =$ latitude (radians, $-\pi/2$ to $\pi/2$).

$\lambda =$ longitude (radians, 0 to 2π).

$i =$ inclination (radians, 0 to π).

$\lambda_0 =$ longitude (radians) of the ascending node for the particular orbit.

$T_{day} =$ period of one day expressed in same units as time t.

There are several things to note about these equations:

- All angles are expressed in radians.
- The term $2\pi\dfrac{t}{T_{day}}$ makes explicit the dependence of the ascending node on time of day.
- The satellite reaches its maximum latitude when $\theta = \pi/2$. Thus the maximum latitude reached is i for a prograde orbit, and $\pi - i$ for a retrograde orbit.

$$\sin\delta = \sin\theta\sin i \quad \text{where} \quad \theta = 2\pi\frac{t}{T_{orb}} \tag{2.2}$$

$$\lambda = \begin{cases} \lambda_0 + \left[\tan^{-1}(\tan\theta\cos i)\right] - 2\pi\dfrac{t}{T_{day}} & \text{if } \cos\theta \geq 0 \\ \lambda_0 + \left[\tan^{-1}(\tan\theta\cos i) - \pi\right] - 2\pi\dfrac{t}{T_{day}} & \text{if } \cos\theta < 0 \end{cases} \tag{2.3}$$

where:

$t =$ time since the ascending node of the first orbit.

So now let's compute how fast a satellite moves in a circular orbit and how long it takes to complete an orbit, T_{orb}.

Reviewing elementary physics, in a steady orbit the gravitational force is balanced by the

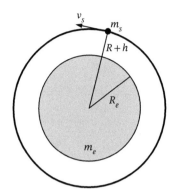

Figure 2.5 Geometry of a circular orbit.

outward centripetal force. The gravitational force (F_g) is given by the usual inverse square law relation, where the distance between the masses is a sum of the radius of the Earth (R) and the height of the satellite (h) (Figure 2.5):

$$F_g = \frac{Gm_e m_s}{(R+h)^2} = \frac{gm_s R^2}{(R+h)^2} \tag{2.4}$$

Likewise, the form of the centripetal force (F_c) should be familiar:

$$F_c = \frac{m_s v_s^2}{R+h} \tag{2.5}$$

where:

m_s = mass of satellite.
m_e = mass of the Earth = 5.976×10^{24} kg.
G = gravitational constant = 6.67×10^{-11} newton m^2 kg^{-2}.
g = acceleration of gravity at Earth's surface = 9.807 m s^{-2} = Gm_e/R^2.
R = mean radius of the Earth = 6371 km.
h = satellite altitude above Earth's surface.
v_s = orbital velocity of satellite.

Setting the two forces equal and solving for the orbital velocity yields equation (2.6):

$$v_s = \left(\frac{Gm_e}{R+h}\right)^{1/2} = \left(\frac{gR^2}{R+h}\right)^{1/2} \tag{2.6}$$

We can approximate equation (2.6) by recognizing that in low Earth orbit the satellite altitude is much less than the radius of the Earth,

$h \ll R$. Thus, we can approximate the satellite velocity by truncating a Taylor series expansion for the square root function:

$$v_s \approx \left(\frac{Gm_e}{R}\right)^{1/2}\left(1 - \frac{h}{2R}\right)$$
$$= \sqrt{gR}\left(1 - \frac{h}{2R}\right) \tag{2.7}$$

The orbital speed is thus $\sqrt{gR}$ with a correction term of order $-h/2R$.

The orbital period T can then be computed from the satellite's angular velocity, ω_s, where:

$$\omega_s = \frac{v_s}{R+h} = \frac{2\pi}{T} \tag{2.8}$$

Substitution of equation (2.6) then yields an expression for the orbital period as a function of satellite altitude:

$$T = \frac{2\pi(R+h)}{v_s} = \frac{2\pi R^{3/2}(1+h/R)^{3/2}}{(Gm_e)^{1/2}}$$
$$= \frac{2\pi(R+h)}{R}\sqrt{\frac{R+h}{g}} \tag{2.9}$$

This formula also has an approximate form appropriate for low Earth orbits:

$$T = 2\pi \frac{R^{3/2}}{(Gm_e)^{1/2}}\left(1 + \frac{3h}{2R}\right)$$
$$= 2\pi\left(\frac{R}{g}\right)^{1/2}\left(1 + \frac{3h}{2R}\right) \tag{2.10}$$

Some sample calculations are shown in Table 2.1.

Figure 2.6 plots the relationships between satellite velocity and orbital period and altitude. The left plot also indicates that the apparent velocity of the subsatellite point is less than the satellite's actual velocity in orbit. It is the subsatellite point's velocity that determines a satellite sensor dwell time at a single point.

It is interesting to note that a satellite orbiting at zero height above the surface (better duck when it passes by) would have a velocity 7.9 km s^{-1} and an orbital period of 84.4 minutes. This turns out to be the exact same

Table 2.1 Orbit parameters for some Low Earth Orbit (LEO) satellites.

Satellite	Altitude (km)	Period (min)	Inclination (deg)
Space Station	370	92	51.6
Landsat-7	705	99	98.2
Radarsat-1	800	101	98.6
NOAA POES	870	102	98.9
Topex/Poseidon	1336	112	66.0

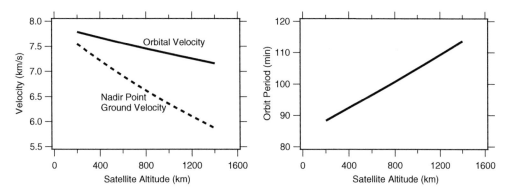

Figure 2.6 Dependence of orbital velocity and period on altitude.

period as that of a pendulum hanging from the surface of the Earth to the center of the Earth.

2.1 Computation of Elliptical Orbits

The computation of elliptical orbits is of course more complicated. We begin by defining an equivalent circular orbit that passes through the perigee and apogee points of the elliptical orbit (the dashed circle in Figure 2.7.) In order to compute the location of the "real" satellite as it travels along its elliptical orbit, we define two fictitious satellites: the "projected" satellite and the "mean" satellite.

The location of the "projected" satellite is the point on the equivalent circular orbit that has

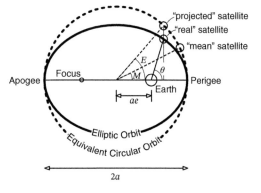

Figure 2.7 Three angular parameters used to compute elliptical orbits: the eccentric anomaly E, true anomaly θ, and mean anomaly M.

the same location along the major axis of the ellipse as the real satellite, and is on the same side of the major axis as the real satellite. So in the diagram, the "projected" satellite would move along the equivalent circular orbit with exactly the same horizontal speed as the "real" satellite.

The location of the fictitious "mean" satellite is equivalent to a satellite in a perfectly constant-speed, circular orbit with the exact same orbital period as the real satellite in its elliptical orbit. So the "mean" satellite also moves around the equivalent circular orbit with the same mean speed as the real satellite.

Then three angular parameters are defined, as illustrated in Figure 2.7:

- The true anomaly θ is the angle between the direction of perigee and the current position of the body, as seen from the main focus of the ellipse (the center of the Earth around which the satellite orbits).
- The eccentric anomaly E is the angle between the direction of perigee and the "projected" satellite.
- The mean anomaly M is the angle between the direction of perigee and the "mean" satellite.

Kepler's equation relates the eccentric anomaly to the mean anomaly:

$$E = M + e \sin E \qquad (2.11)$$

where e is the orbit eccentricity. This equation is useful because the mean anomaly is an angle that advances in time at a constant rate. Kepler's equation is transcendental, so there is no closed form solution – it must be solved numerically. Multiple approaches exist, but an approach that converges quickly utilizes an initial estimate for E:

$$E_0 = M + 0.85\, e\, \text{sign}(\sin M) \qquad (2.12)$$

Next define the auxiliary parameters:

$$f_n = E - e \sin E_n - M$$
$$f_n' = 1 - e \cos E_n$$
$$f_n'' = e \sin E_n \qquad (2.13)$$

A refined estimate of E is then computed from the Laguerre–Conway iteration:

$$E_{n+1} = E_n - \frac{5 f_n}{f_n' + \text{sign}(f_n')\sqrt{|16(f_n')^2 - 20 f_n f_n''|}} \qquad (2.14)$$

Equations (2.13) and (2.14) are then iterated until convergence is reached, which usually takes less than ten iterations.

Given E, the true anomaly can then be computed:

$$r \sin \theta = a\sqrt{1 - e^2}\, \sin E$$
$$r \cos \theta = a(\cos E - e) \qquad (2.15)$$

Note that $r \sin \theta$ and $r \cos \theta$ are essentially the x and y coordinates of the satellite in the orbital plane, as shown in Figure 2.7. These coordinates can then be rotated into the correct orbital plane relative to the Earth at a given time, and then the Earth's rotation can be added to determine the location of the satellite relative to the Earth at any time.

Figure 2.8 shows solutions for Kepler's equation with three different eccentricities.

2.2 Low Earth Orbits

The previous discussion was based on a spherical Earth with a uniform mass distribution, the elementary approach to physics. It turns out that this analysis is just a little wrong, but interestingly, it is wrong in some important ways. In fact the Earth has a bulge at the equator and flattens at the poles. The gravitational potential for this spheroid V can be written in terms of a spherical harmonic expansion that can then be terminated to lowest nontrivial order:

$$V(r) = \frac{-Gm_e}{r}$$
$$\times \left[1 - \frac{J_2}{2}\left(\frac{R_{eq}}{r}\right)^2 (3\sin^2\delta - 1) \right] \qquad (2.16)$$

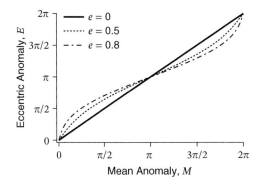 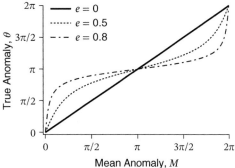

Figure 2.8 Solutions for Kepler's equation with three different eccentricities.

where:

J_2 = quadrapole coefficient $\approx 1.0826 \times 10^{-3}$.
R_{eq} = equatorial radius of the Earth = 6378 km.
δ = latitude.
r = distance from center of Earth to satellite.

This variation in gravitational potential with latitude causes four effects: the satellite velocity varies during the orbit, the orbital period changes, the argument of periapsis changes (where perigee occurs in the orbit), and the orbit precesses. Let's talk about each effect in turn.

The variations in the Earth's gravitational field causes variations in the orbital velocity. The satellite speeds up near the equator where the gravitational potential is larger and slows down near the poles where the gravitational potential is smaller. For example, consider a polar-orbiting satellite that is nominally at an altitude of 800 km. Note that the Earth's equatorial radius is 6378 km, but its polar radius is 6357 km. Thus it appears that the satellite's altitude varies by 21 km as it passes from the pole to the equator. This causes the orbital velocity at the equator to be 13.7 m s^{-1} faster than the velocity near the poles, a result derived from application of equation (2.6). This example shows the variations in orbital speed are typically of the order of 10 m s^{-1} out of a 7 km s^{-1} velocity, or less than 0.2%.

The second variation is in the orbital period. For a satellite in a circular orbit about a non-spherical Earth, the orbit period (known as the nodal period, T_{ns}) becomes:

$$T_{ns} = T_s \left[1 + \frac{3J_2 R_{eq}^2}{2(R+h)^2}(1 - 4\cos^2 i)^2 \right] \quad (2.17)$$

where:

T_s = orbit period for a spherical Earth.
R_{eq} = equatorial radius of the Earth = 6378 km.
R = mean radius of the Earth = 6371 km.
h = satellite altitude.

Notice that for an orbit inclination $i = 60°$, $1 - 4\cos^2 i = 0$ and the orbit period remains unchanged. For $i > 60°$, this factor is positive and the orbit period increases (average orbital velocity slows). For $i < 60°$, the factor is negative and the orbit period decreases (average orbital velocity speeds up). Thus an orbital inclination of 60° is the dividing point between increasing and decreasing orbital period when taking Earth's oblateness into account.

For example: a NOAA/POES satellite may orbit at a mean altitude of $h = 870$ km with an inclination of $i = 98.9°$. The orbital period for a spherical Earth can then be computed from equation (2.9) to be $T_s = 102.25$ min. From equation (2.17) we then obtain, $(T_{ns} - T_s)/T_s = 1.14 \times 10^{-3}$. Hence, the nodal period is about seven seconds longer than the period for a spherical Earth. For our purposes, this variation in orbital period is relatively small. To put this in the proper context, a seven second error in orbital period could lead to a 50 km registration error in a single orbit, but it is a small variation if you are trying to understand the basics of orbits.

The third effect of oblateness is that the argument of periapsis of the satellite will change, affecting where perigee and apogee occur in the orbit. The angular rate of change in the argument of periapsis $\dot{\omega}$ depends on the orbit's inclination, mean motion, semimajor axis and eccentricity:

$$\dot{\omega} = \frac{3nJ_2}{4(1-e^2)^2}\left(\frac{R_{eq}}{a}\right)^2 (5\cos^2 i - 1)$$

(2.18)

where:

$n =$ mean motion of the satellite (angular rate $= \frac{d\theta}{dt}$, typically in degrees per day).
$i =$ inclination.
$R_{eq} =$ equatorial radius of the Earth $= 6378$ km.
$a =$ semimajor axis of orbit.
$e =$ orbit eccentricity.

The fourth and most important effect of oblateness is that the orbital plane of the satellite will rotate (or precess) about the Earth's polar axis, as measured in a coordinate system that is fixed relative to any distant star (Figure 2.9). Thus the orbital plane will no longer be fixed in space. The angular velocity of precession depends on the orbit's inclination, semimajor axis and eccentricity:

$$\omega_p = \frac{-3J_2 \cos i}{2(1-e^2)^2}\left(\frac{R_{eq}}{a}\right)^2 \sqrt{\frac{Gm_s}{a^3}}$$

(2.19)

For a circular orbit, the precession angular velocity simplifies to:

$$\omega_p = \frac{-3}{2}J_2\sqrt{\frac{gR^2}{(R+h)^3}}\left(\frac{R_{eq}}{R+h}\right)^2 \cos i$$

(2.20)

where a positive value indicates a prograde precession.

To illustrate, the NOAA/POES satellites that orbit at an altitude of 870 km with an inclination of 98.9° have an angular precession velocity of 1.99×10^{-7} rad/s, which translates into 359.82 deg/year. Thus this orbit precesses at exactly the same rate as the Earth rotates around the sun. This is referred to as a sun-synchronous orbit.

A sun-synchronous orbit is one where the orbital plane rotates (precesses) about the Earth's axis at the same rate (one revolution per year) that the Earth orbits the Sun. As a consequence, the ground track of the satellite will cross the same latitude (northbound or southbound) at the same local solar time, regardless of the longitude or the date. This type of orbit is particularly desirable for comparing long duration observations of geophysical quantities that depend on the incident solar radiation, e.g., ocean color.

All of this is illustrated in Figure 2.10, which are not-to-scale polar views of the Earth at four times during its yearly orbit around the sun with two orbital planes. The nonprecessing orbit (dashed curves) does not rotate as the Earth rotates about the sun. So the node of the orbit toward the bottom of the page would be after sunset in fall and winter, but in daylight during summer and spring.

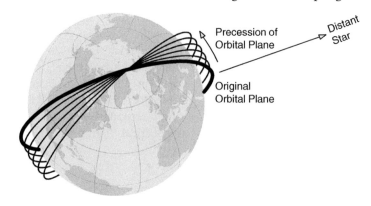

Figure 2.9 Precession of the orbital plane.

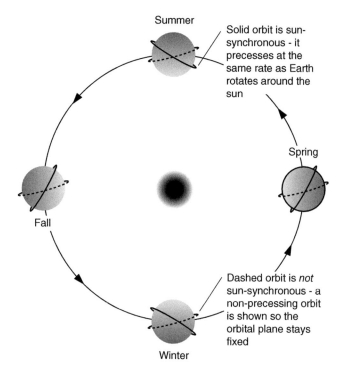

Figure 2.10 Sun synchronous and nonprecessing orbits. Adapted from original work of Brandir XZise / Wikimedia Commons / CC-BY-SA-3.0 / GFDL, https://commons.wikimedia.org/wiki/File:Heliosynchronous_orbit.svg.

The solid curves in Figure 2.10 show a sun-synchronous orbit that rotates as the Earth orbits. In this case, the orbital nodes (corresponding to an equator crossing) always occur at the same time of the solar day. Ignoring the seasonal tilt of the Earth, the nodal times in this illustration might be 8 AM and 8 PM.

Many satellites use sun-synchronous orbits. Optical remote sensors are often designed to keep the sun at certain angles to optimize lighting and viewing geometries. Some radar satellites that use a lot of power are placed in a dusk/dawn sun-synchronous orbit so the solar panels can always be in sunlight.

Figure 2.11 shows the specific periods and inclinations required for a sun-synchronous orbits as a function of altitude. These can be computed directly from equations (2.17) and (2.19).

A sun-synchronous orbit requires an inclination of more than 90°, so it is in a retrograde orbit. At an altitude of 825 km, a satellite in a sun-synchronous orbit will have an inclination of 98.71° and a period of 101.3 minutes. Note that a satellite in this type of orbit will not

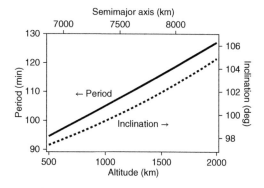

Figure 2.11 Dependence of inclination and nodal period for sun-synchronous orbits.

pass over the poles, leading to gaps in the polar coverage depending on the extent of the ground swaths of the sensors on the satellite, as shown in Figure 2.12.

The ground tracks for three orbits of the sun-synchronous NOAA-20 satellite are illustrated in Figure 2.13. The times are given in UTC, not local solar time. During each 101.3-minute orbital period, the point of solar noon on the Earth rotates 25.5° to the west, so on each orbit the satellite ground track must

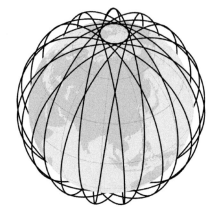

Figure 2.12 Gap in polar coverage for an orbit with an inclination of 98.86°.

move this much to keep at the same solar time. This is exactly what is shown.

Another type of common orbit is a repeating orbit, which is an orbit that exactly repeats its ground track after a certain interval of time. This type of orbit allows data to be collected at the same locations on the Earth with the same viewing geometry many times during the satellite's lifetime. For the track to repeat, the Earth must make an integral number of rotations in the time required for the satellite to make an integral number of orbits. The condition for an orbit that repeats in n_{days} days and n_{orbits} satellite orbits is given by:

$$T_s(\omega_e - \omega_s) = 2\pi \frac{n_{days}}{n_{orbits}} \qquad (2.21)$$

where:

$$T_s = \text{satellite period.}$$
$$\omega_s = \text{satellite angular velocity.}$$
$$\omega_e = \text{Earth's angular velocity of rotation} =$$
$$2\pi/\text{sidereal day.}$$
$$n_{days} = \text{number of Earth rotations (integer).}$$
$$n_{orbits} = \text{number of satellite orbits (integer).}$$

Some sun-synchronous orbits can also be repeating orbits when the sun-synchronous orbit is chosen with the constraint that an integer multiple (n_{orbits}) of the satellite orbital period equals an integer multiple (n_{days}) of the solar day.

Figure 2.14 shows the ground track of an exact repeat orbit with 14 orbits occurring in one day. Such ground track diagrams indicate global coverage (excluding the polar regions) but do not provide a sense of the coverage gaps from one orbit to the next.

Figure 2.15 shows the descending (daylight) imaging locations over about a 30° span of longitude for the Landsat-8 satellite deployed in a 16-day repeat orbit. The dots indicate the location of the image centers, which are by design fixed on the Earth's surface. The edges of four of the imaging areas are shown on one of the central passes. Each pass in this diagram is designated with a path, orbit and day number. The path numbers are assigned sequentially from 1 to 233 around the equator

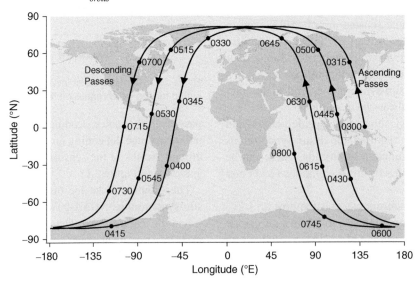

Figure 2.13 Ground track from NOAA-20 on 5 July 2019.

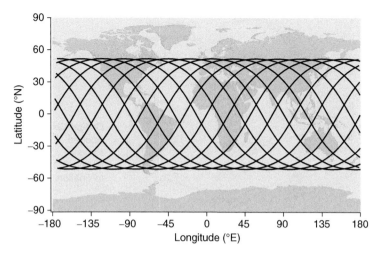

Figure 2.14 Repeat ground track with exactly 14 orbits per day from an altitude of 826 km and an inclination of 51.6°.

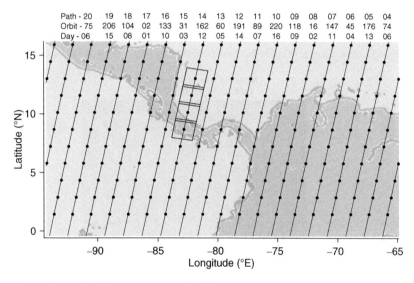

Figure 2.15 Ground track of a 16-day exact repeat orbit.

to make it easier to order data. The orbit numbers indicate the orbit sequence within the 233-orbit repeat cycle. And the day number indicates the day relative to the start of path 1. Notice the first pass in this region crosses the equator at about −89° longitude on day 1. The next descending pass in this region does not occur until 14 orbits later on day 2 and is separated from the initial pass by about 14° of longitude. The coverage pattern eventually fills in over the 16-day repeat cycle.

The NOAA polar orbiting satellites are typically placed in circular, sun-synchronous orbits, as illustrated in Figure 2.16. The orbits are circular to keep the distance to the surface nearly constant. And the orbits are sun-synchronous because the satellites make optical measurements that depend on solar illumination. Note that these satellites do have to carry batteries to allow continuous operation on the dark side of the orbit.

Figure 2.17 illustrates the coverage of the Advanced Very High Resolution Radiometer (AVHRR) instrument on the NOAA-15 satellite as seen from a ground station in Laurel, MD. The green grid illustrates the satellite track and coverage during a single pass.

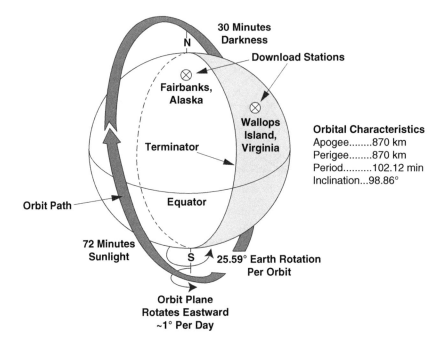

Figure 2.16 NPOESS orbit. Source: *NOAA-N Brochure* published by NOAA & NASA (2004).

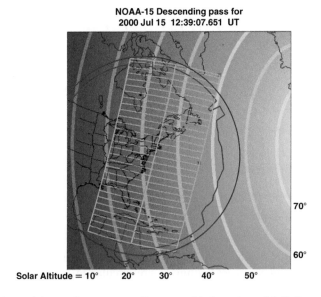

Figure 2.17 Example receiving station coverage. Figure provided courtesy of R. E. Sterner, JHU/APL.

These satellites are designed to continuously broadcast the AVHRR data as they orbit. Ground stations can only receive the broadcast when the satellite appears above the local horizon. The blue circle illustrates the theoretical horizon-to-horizon limits of reception for the Maryland ground station. The ragged red curve inside the blue circle is the actual coverage map when local buildings and terrain are taken into account. This shows that the Maryland site can receive data from this satellite covering the entire east coast of

Canada and the United States, but the coverage is not worldwide.

The overall coloring and the pastel arcs indicate the solar altitude during this pass. Again the instruments are dependent on solar illumination angle, so this is an important parameter.

2.3 Geosynchronous Orbits

A geosynchronous satellite orbits with the same angular velocity as the Earth. That is to say that the satellite's period is one sidereal day. (A sidereal day is 23 h 56 m long. The reason the orbital period is not a full 24 hours is discussed below.)

A geostationary orbit is a geosynchronous orbit with a zero inclination and zero eccentricity. This means the orbital plane of a geostationary satellite coincides with the Earth's equatorial plane. The result is that the satellite will stay above the equator at a fixed longitude.

Keeping a satellite at a fixed location above the Earth is desirable for many reasons. Of course orbit perturbations cause a geostationary satellite to drift from the desired orbit, but periodic maneuvers can be used to correct the orbit.

The altitude of a geosynchronous satellite is computed from the orbital period equation:

$$T = \frac{2\pi(R+h)}{R}\sqrt{\frac{R+h}{g}}$$
$$= 86{,}164 \text{ seconds} = 1 \text{ sidereal day}$$
$$(2.22)$$

The result is that geostationary satellites orbit at an altitude of 35,778 km. Notice that this is 5.6 times the radius of the Earth.

NOAA's GOES weather satellites are in geostationary orbits (Menzel & Purdom, 1994). Communications and TV satellites also use geostationary orbits to maintain constant coverage and to avoid steering the receiving antenna. The penalty paid is increased transmitter power or higher gain antennas because the received power falls off as the inverse square of the transmitter altitude.

A solar day is the time that elapses between the sun reaching its highest point in the sky two consecutive times. Because the Earth is orbiting the sun, this corresponds to the Earth rotating about 361°. The length of the mean solar day is 24 hours.

A sidereal day is the time it takes the Earth to rotate 360° measured relative to a distant star, which is any star other than the sun. The length of the mean sidereal day is 23 hours and 56 minutes.

Refer to Figure 2.18 to see why the sidereal day is slightly shorter than the solar day. The three grey circles show the Earth at three positions as it orbits around the sun. Assume that the leftmost position occurs at solar noon on a Tuesday. Also imagine there is a distant star

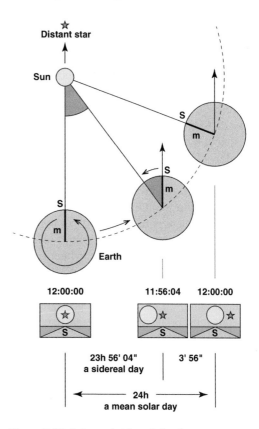

Figure 2.18 Solar and sidereal day. Source: Wikipedia, By Francisco Javier Blanco Gonzalez - 29 May 2009, CC BY-SA 4.0, https://en.wikipedia.org/wiki/Sidereal_time\LY1\textbackslash#/media/File:Sidereal_time.svg.

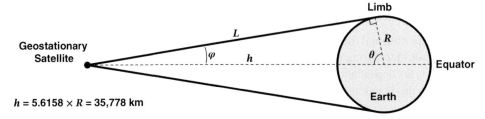

Figure 2.19 Earth coverage from a geostationary satellite.

along the exact same bearing as the sun at this moment of time.

The leftmost colored cartoon illustrates that an observer sitting on the North pole might see the sun and this star aligned due south of her or his position at noon on Tuesday.

The rightmost position corresponds to the position of the Earth a full solar day (24 hours) later, namely noon on Wednesday. During this day, the Earth's position has moved around the sun by an angle of 360° divided by 365.25 days, which is about 1°. Thus to point back to the sun, the Earth actually had to rotate about 361°, assuming we are measuring rotation relative to the stars. So a 24-hour day is how long it takes the Earth to rotate about 361°. Note that if the observer at the North pole has not died of exposure from standing outside for a full day, s/he would see the star appearing 1° to the right of the sun at noon on Wednesday.

The middle position corresponds to the period of time required for the Earth to rotate 360° relative to the stars. This is short of a day by about one 360th of 24 hours, which is four minutes.

All this matters because to keep above the same location on the surface, a geostationary satellite must orbit once per sidereal day.

The Earth-viewing geometry from a geostationary satellite located above the equator is shown in Figure 2.19.

The limb point is defined to be the latitude at which the geostationary satellite would appear to be at the horizon. The latitude θ of the limb point can be easily found from:

$$\sin \theta = \frac{L}{R+h} \quad \text{and}$$

$$L^2 + R^2 = (R+h)^2 = (6.6158R)^2 \quad (2.23)$$

which yields: $L = 6.5398R$ and $\theta = 81.3°$.

Thus, the polar regions above ±81.3° latitude cannot be viewed by a geostationary satellite. In practice, the imaging sensors on these satellites do not provide useful images at latitudes where the interpretation of cloud data is unreliable, which effectively limits the imaging latitude to about ±60° from the equator. To an observer on the ground in this latitude band, the satellite has an elevation angle that is at least 22° above the local horizon.

Figure 2.20 shows the communications coverage, which extends to the northern limb point (i.e., 0° satellite elevation angle at 81° latitude), and the area with useful image data (±60° latitude) for the GOES-East and GOES-West satellites orbiting at 75° and 135° west longitude, respectively.

Next we compute the elevation angle at various latitudes to a geostationary satellite. Figure 2.21 illustrates the geometry of viewing a location on the Earth below the limb from a geostationary satellite.

In this figure φ is the satellite scan angle relative to nadir, θ is the latitude of the observation point, i is the incident angle on the ground, and α is the elevation angle of the satellite relative to the horizon or the grazing angle when viewed from the satellite.

The law of sines yields:

$$\frac{\sin \varphi}{R} = \frac{\sin \theta}{L} = \frac{\sin(\alpha + \pi/2)}{R+h} \quad (2.24)$$

The law of cosines yields either of two formulas for range, L. The first is applicable to a geostationary satellite observing a particular latitude:

$$L^2 = R^2 + (R+h)^2 - 2R(R+h)\cos \theta$$
$$= R^2 + (6.6158R)^2 - 2R(6.6158R)\cos \theta$$
$$= (44.7688 - 13.2316\cos \theta)R^2 \quad (2.25)$$

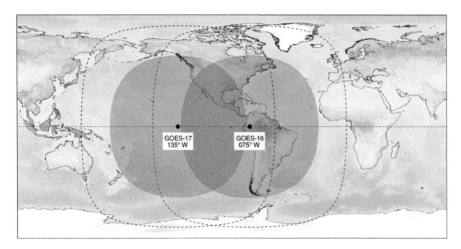

Figure 2.20 GOES-East/West coverage.

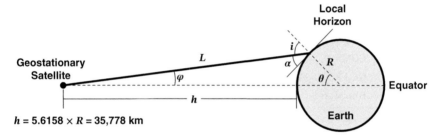

Figure 2.21 Geostationary satellite viewing geometry.

The second is a quadratic formula for range applicable to any satellite based on the satellite scan angle relative to nadir:

$$L^2 - [2(R+h)\cos\varphi]L$$
$$+ [(R+h)^2 - R^2] = 0 \qquad (2.26)$$

These formulas provide simple relationships between the scan angle, latitude, and incident or grazing angle:

$$\sin\varphi = \frac{R}{L}\sin\theta$$

$$\cos\alpha = \frac{R+h}{L}\sin\theta$$

$$\sin i = \frac{R+h}{L}\sin\theta \qquad (2.27)$$

Figure 2.22 plots the scan, incident and grazing/elevation angle for a geostationary satellite. The plot indicates that the grazing angle is greater than 22° for an observer within the imaging area bounded by ±60° latitude.

Table 2.2 lists the geostationary meteorological satellites that are operational as of

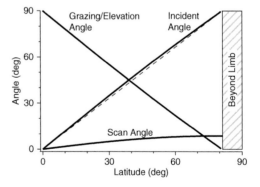

Figure 2.22 Scan, incident, and grazing angles for a geostationary satellite as a function of observation latitude. The dashed line is linear, indicating that incident angle can be approximated by 1.11 × latitude.

December 2020. These satellites are so important for weather prediction that their positions are internationally coordinated.

Geostationary satellites are so important that stand-by satellites are placed in orbit awaiting the failure of one of the operational satellites.

Table 2.2 Operational geostationary meteorological satellites as of December 2020. Adapted from: WMO, https://community.wmo.int/activity-areas/wmo-space-programme-wsp/satellite-status#geocurrent.

Sector	Satellite	Operator	Longitude	Launch
East Pacific	GOES-17 (GOES-West)	US-NOAA	137.2° W	3/2018
	GOES-15	US-NOAA	128.0° W	3/2010
West Atlantic	GOES-16 (GOES-East)	US-NOAA	75.2° W	11/2016
	Electro-L N2	Russia	14.5° W	12/2015
East Atlantic	METEOSAT-11	EUMETSAT	0.0° E	7/2015
	METEOSAT-10	EUMETSAT	9.5° E	7/2012
Indian Ocean	METEOSAT-8	EUMETSAT	41.5° E	9/2016
	Insat-3DR	India	74.0° E	9/2016
	Electro-L N3 (commisioning)	Russia	76.0° E	12/2019
	Feng Yun-2H	China	79.0° E	6/2018
	Insat-3D	India	82.0° E	7/2013
	Feng Yun-2G	China	99.5° E	12/2014
	Feng Yun-4A	China	105.0° E	12/2016
West Pacific	COMS-1	Korea	128.2° E	6/2010
	GEO KOMPSAT-2A	Korea	128.2° E	12/2018
	Himawari-8	Japan	140.7° E	10/2014

Table 2.3 Stand-by geostationary meteorological satellites as of December 2020. Adapted from: WMO, http://www.wmo.int/pages/prog/sat/satellitestatus.php.

Sector	Satellite	Operator	Longitude	Launch
West Atlantic	GOES-14	US-NOAA	105.0° W	6/2009
East Atlantic	METEOSAT-9	EUMETSAT	3.5° W	12/2005
West Pacific	Feng Yun-2F	China	112.0° E	1/2012
	Himawari-9	Japan	140.7° E	11/2016

Table 2.3 lists the stand-by geostationary meteorological satellites as of December 2020. One of these satellites was so new that was still undergoing testing and calibration.

Figure 2.23 is a full disc image obtained from GOES-West, showing the type of image coverage that is possible. The utility for mapping clouds in order to track storms and fronts seems clear.

Figure 2.24 illustrates that the full resolution data can be remapped for more local uses.

The geosynchronous orbit is defined as any orbit with a period of 1 sidereal day. Yet we have spent our time showing examples of the common geostationary orbit, which has a 0° inclination resulting in the orbit lying in the equatorial plane. Figure 2.25 shows the ground track of a satellite in a circular geosynchronous orbit with a 45° inclination. In general the satellite will appear to follow this figure eight pattern above and below the equator. This ground track can be computed directly from the previous equations for circular orbits.

Figure 2.23 GOES-West full disc image. Source: NOAA.

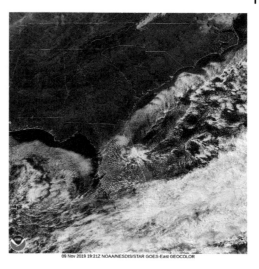

09 Nov 2019 19:21Z NOAA/NESDIS/STAR GOES-East GEOCOLOR

Figure 2.24 GOES-East sector image. Source: NOAA.

The numbers on this orbit indicate hours, so this orbit spends almost four hours over North America. Six satellites located four hours apart in this same orbit could provide nearly continuous coverage looking nearly straight down over the continent, which could prove useful for a number of applications.

Although not a remote sensing application, it is interesting to note that some of the satellites in the Chinese BeiDou navigation system are in geosynchronous orbits with an inclination of 55°.

Figure 2.26 shows the ground track of a satellite in a similar geosynchronous orbit

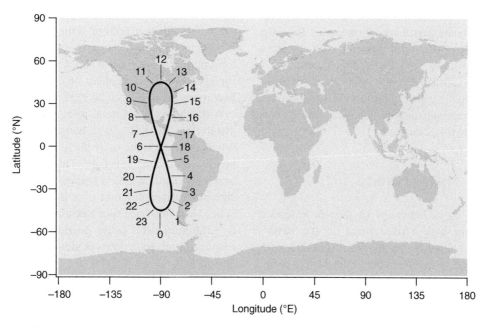

Figure 2.25 Ground track of circular geosynchronous orbit with 45° inclination.

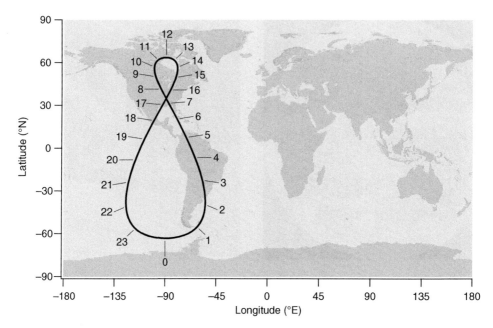

Figure 2.26 Ground track of a Tundra orbit, a geosynchronous orbit with 63.4° inclination and 0.2 eccentricity.

but this time with an inclination of 63.4° and an eccentricity of 0.2. Now the satellite spends a full 10–12 hours over North America, albeit at a higher altitude. The inclination of this orbit was selected to satisfy the equation $5\cos^2 i - 1 = 0$, which is the condition for zero drift in the argument of periapsis. Thus the perigee of this orbit does not drift due to a perturbations from the Earth's oblateness.

This is called a Tundra orbit. NOAA is investigating the potential for just two satellites in a Tundra orbit to provide nearly continuous coverage to the polar region.

2.4 Molniya Orbit

There are other ways to provide extended coverage over a fixed point on the Earth's surface based on highly eccentric orbits with the apogee positioned above the desired point. With a suitable choice of orbit parameters, the satellite can have a long dwell time over the desired point (Figure 2.27).

Again the argument of periapsis should not drift, so the inclination must satisfy the equation $5\cos^2 i - 1 = 0$, which implies an inclination of 63.4°. If the latitude of the apogee is fixed at 63.4° north, and the orbital period is half a sidereal day, the semimajor axis will be about 26,500 km. If the eccentricity is 0.74, the perigee distance is about 6900 km for a minimum altitude of about 500 km, and the apogee distance is about 46,200 km for a maximum altitude about 39,800 km.

With these parameters, the near-apogee dwell time is about eight hours and three satellites in the same orbit plane can provide nearly continuous coverage. This orbit was originally used by the Russians to provide communications coverage over their country. It has been suggested that instruments placed in a Molniya orbit could provide improved monitoring of polar lows (Kidder & Vonder Haar, 1990).

Notice that after 12 hours, the ground track of each satellite advances 180° in longitude. The ground track for a Molniya orbit with an

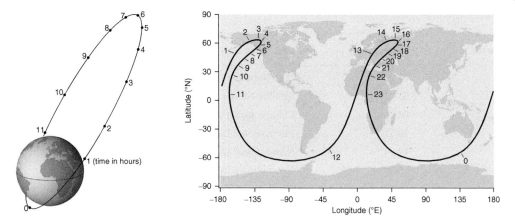

Figure 2.27 Molniya orbit (left panel) and ground track (right panel) with 63.4° inclination, 0.74 eccentricity, 26,600 km semimajor axis, 12-hr orbital period and a −65° argument of perhelion.

inclination of 63.4° and an eccentricity of 0.7 is shown in Figure 2.27. Notice how the satellite at its apogee can spend up to 8 hours over a specific high-latitude location on the globe.

2.5   Satellite Orbit Prediction

The U.S. government tracks all orbiting satellites to produce orbit estimates. These estimates take the form of Two-Line Elements (TLEs) that are available on the internet for most satellites.

TLEs consist of two 80-character lines of ASCII text in a fixed width format, as shown in Box 2.1. Note the satellite name is often in the files, but not part of the TLE. The exact formats are specified in Table 2.4.

The TLEs include mean Kepler elements at a particular time along with a drag term and the first and second derivatives of the angular speed. This allows a predictive code to integrate the solutions forward in time to provide an accurate estimate over a range of times. These TLEs are periodically updated, so only recent TLEs should be used to compute a current satellite position. A good reference for the details of satellite orbit calculations is provided by Capderou (2006).

Box 2.1 Example of a two-line element.

```
NOAA 14
    1 23455U 94089A   97320.90946019  .00000140  00000-0  10191-3 0  2621
    2 23455  99.0090 272.6745 0008546 223.1686 136.8816 14.11711747148495
```

Two components are needed to estimate the position of a satellite: a recent TLE and a compatible propagation model, such as SGP, SGP4, SDP4. TLEs, models, and documentation are available at: www.celestrak.com/NORAD/elements.

2.6   Satellite Orbital Trade-offs

To complete our discussion of satellite orbits, consider the various types of orbits, and what types of observations they would support (Table 2.5).

Table 2.4 Two-line orbital element format.

Line	Column	Description
1	01	Line Number of Element Data
1	03-07	Satellite Number
1	08	Classification (U=Unclassified)
1	10-11	International Designator (Last two digits of launch year)
1	12-14	International Designator (Launch number of the year)
1	15-17	International Designator (Piece of the launch)
1	19-20	Epoch Year (Last two digits of year)
1	21-32	Epoch (Day of the year and fractional portion of the day)
1	34-43	First Time Derivative of the Mean Motion
1	45-52	Second Time Deriv. of Mean Motion (decimal point assumed)
1	54-61	Drag term (decimal point assumed), 'B*' model
1	63	Ephemeris type
1	65-68	Element number
1	69	Checksum (Modulo 10)
2	01	Line Number of Element Data
2	03-07	Satellite Number
2	09-16	i Inclination (Degrees)
2	18-25	Ω Right Ascension of the Ascending Node (Degrees)
2	27-33	e Eccentricity (decimal point assumed)
2	35-42	ω Argument of Perigee (Degrees)
2	44-51	M Mean Anomaly (Degrees)
2	53-63	m Mean Motion (Revolutions per day)
2	64-68	Revolution number at epoch
2	69	Checksum (Modulo 10)

Table 2.5 Orbital trade-offs.

Observation requirement	Orbit
Continuous monitoring	Geostationary orbit
Quasi-continuous monitoring	Constellation in inclined geosynchronous or Molniya orbit
Global coverage including poles	High inclination orbit
Constant sun illumination geometry	Sun-synchronous orbit
Constant illumination on solar cells	Dawn/dusk sun-synchronous orbit
High ground resolution	Low altitude orbit
Minimize atmospheric drag to achieve longer satellite lifetimes	Altitudes above 200 km
Minimize gravity anomaly perturbations	High altitude orbit

References

Capderou, M. (2006). *Satellites: Orbits and missions*. Springer Science & Business Media.

Kidder, S. Q., & Vonder Haar, T. H. (1990). On the use of satellites in Molniya orbits for meteorological observation of middle and high latitudes. *Journal of Atmospheric and Oceanic Technology, 7*(3), 517–522.

Menzel, W. P., & Purdom, J. F. W. (1994). Introducing GOES-I: The first of a new generation of geostationary operational environmental satellites. *Bulletin of the American Meteorological Society, 75*(5), 757–781.

NOAA & NASA (2004). NOAA-N brochure. Retrieved from https://www.nasa.gov/pdf/111742main_noaa_n_booklet.pdf

3

Infrared Sensing

3.1 Introduction

This chapter describes how remote sensing of infrared radiation at wavelengths of 0.7–15 μm can be used to estimate sea surface temperature (SST), land surface temperature (LST), and profiles of atmospheric temperature and humidity.

Sea surface temperature is an important geophysical parameter. At the largest scales, the ocean–atmosphere system is driven by solar radiation that provides excess heat at the equator relative to the poles. The temperature of the sea at the air–water interface is one of the key parameters in heat transfer between the two.

Departures from long-term climatological mean SST (called SST anomalies) are key indicators of changes in environment. Figure 3.1 for example shows the excess temperature signal of an El Niño event emanating from the west coast of South America. The plot shows

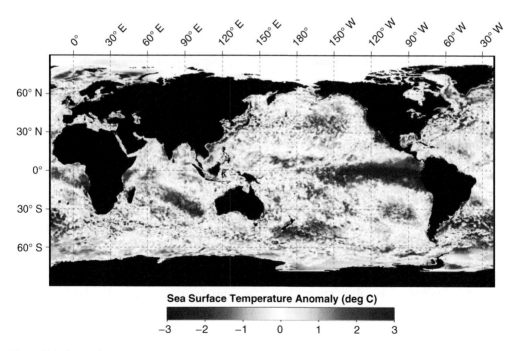

Figure 3.1 Sea surface temperature anomaly map. (Figure based on data from Reynolds et al. (2008).

Remote Sensing Physics: An Introduction to Observing Earth from Space, Advanced Textbook 3, First Edition.
Rick Chapman and Richard Gasparovic.
© 2022 American Geophysical Union. Published 2022 by John Wiley & Sons, Inc.
Companion website: www.wiley.com/go/chapman/physicsofearthremotesensing

the differences between the mean sea surface temperature observed in January 1998 and the climatological average for the years 1971 through 2000.

Overall, long-term infrared remote sensing is both critically important for climate change studies but also provides an ideal starting point for understanding remote sensing physics.

3.2 Radiometry

We begin with a review of radiometry, the measurement, and characterization of electromagnetic radiation, including infrared and visible light. Radiometric techniques in optics characterize the distribution of the radiation's power in space. To begin we must review some definitions.

Radiant energy is the energy of radiation propagating as an electromagnetic wave. It is designated by the letter Q with units of joules (J).

Radiant flux is the rate at which radiant energy is transferred to or from a point or a surface to another surface. It is also the power available for producing a response in a detector. It is designated by the greek letter Φ and equals the time derivative of the energy, dQ/dt with units of watts (W).

Flux density is the radiant flux per unit surface area. It is designated $d\Phi/dA$ with units of watts/m^2 (W m^{-2})

Irradiance is the flux density incident on a receiving surface element of area dA_s. It is designated by the letter $E = d\Phi_{in}/dA_s$ with units of watts/m^2. So irradiance times area is the total power incident on that area.

Power can flow onto an area, but it can also be emitted by an area. **Emittance** is the flux density emitted by a surface element of area dA_s. It is designated by the letter $M = d\Phi_{out}/dA_s$. Emittance is also known as **Radiant Exitance**. As can be seen from

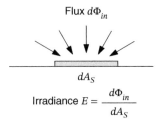

Flux $d\Phi_{in}$

$$dA_S$$

Irradiance $E = \dfrac{d\Phi_{in}}{dA_S}$

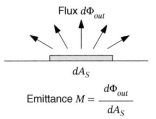

Flux $d\Phi_{out}$

$$dA_S$$

Emittance $M = \dfrac{d\Phi_{out}}{dA_S}$

Figure 3.2 Difference between irradiance and emittance.

Figure 3.2, irradiance and emittance are closely related terms, only the direction of the flow changes.

Note that irradiance and emittance are independent of the direction from which the incident radiation comes, or to which the radiation is emitted.

If you are new to radiometry, these definitions may seem a bit confusing, but the terms, definitions and units are important. Pay special attention to units. It is good practice to use dimensional analysis to ensure the dimensions of the equations are consistent with the units associated with the terms being used. *Thinking about the flow and distribution of energy is the key to understanding radiometry.*

Radiant flux flowing uniformly away from a point will spread evenly across the inside face of a sphere drawn about that point with radius r. The area of that face is $4\pi r^2$. If the radiant flux is flowing unequally, so it has some directionality, then we need a way to account for the angular distribution of the energy. This is the role of solid angles.

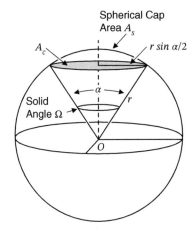

Figure 3.3 Definition of solid angle.

A solid angle is defined to be the ratio of the area on the surface of a sphere to the square of the radius of the sphere. In Figure 3.3, the solid angle Ω is given by the area of the spherical cap A_s/r^2. The solid angle has units of steradians, which is abbreviated sr or ster.

One steradian is the solid angle subtended by an area on the surface of a sphere equal to the square of the radius of the sphere. Maybe easier to see is that the solid angle encompassing the entire sphere is $4\pi r^2/r^2 = 4\pi$. There are 4π steradians about a point, and 2π steradians are subtended by a hemisphere.

Solid angles are the natural extension of two-dimensional angles to three dimensions. Angles are defined as the ratio of the arc length on a circle subtended by the angle, divided by the radius of the circle, with a unit of measurement of radians. Solid angles are the ratio of the area on a sphere subtended by the solid angle divided by the square of radius, with a unit of measurement of steradians.

In many situations of interest here, we will be dealing with remote sensors that have small angular fields of view. So looking at Figure 3.3 we can approximate the solid angle that is subtended by a cone angle of α by replacing the area of the spherical cap with the area

of the disk spanning the cap, A_c. The radius of the disk is $r\sin(\alpha/2)$, hence $A_c = \pi r^2\sin^2(\alpha/2)$. The solid angle is then given by:

$$\Omega = \frac{A_c}{r^2} = \frac{\pi(r\sin\alpha/2)^2}{r^2} = \pi\sin^2\alpha/2$$

$$\approx \frac{\pi\alpha^2}{4} \tag{3.1}$$

We have just computed an approximate relation, but we can also compute an exact relation using simple integration. Referring to Figure 3.4, consider the element of surface area dA_s that extends over ranges of angles $d\theta$ and $d\varphi$. One side of the area dA_s is approximately $r\sin\theta\, d\varphi$, and the other side is $r\, d\theta$, so the area $dA_s = r^2\sin\theta\, d\theta\, d\varphi$, and the solid angle is $\sin\theta\, d\theta\, d\varphi$.

$$d\Omega = \frac{dA_s}{r^2} = \frac{r^2\sin\theta\, d\theta\, d\varphi}{r^2}$$

$$= \sin\theta\, d\theta\, d\varphi \tag{3.2}$$

Now that we have the elemental solid angle, we can compute the solid angle corresponding to a cone of half-angle θ by integration.

$$\Omega = \int_0^\theta \sin\theta\, d\theta \int_0^{2\pi} d\varphi$$

$$= 2\pi(1 - \cos\theta) \tag{3.3}$$

The angle α defined in Figure 3.3 is twice the angle θ so we can derive the exact formula for the solid angle to be:

$$\Omega = 2\pi[1 - \cos(\alpha/2)] = 4\pi\sin^2(\alpha/4) \tag{3.4}$$

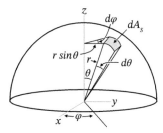

Figure 3.4 Element of surface area on a sphere.

where the right-hand term follows from the trigonometric identity $1 - \cos\varphi = 2\sin^2(\varphi/2)$. For included cone angles of 20° or less, the small angle approximation of equation (3.1) differs from this exact relation by less than 1%.

Now that we have defined solid angles, we will resume with more radiometry definitions.

The **radiant intensity** is the flux Φ per unit solid angle Ω. It is defined by $I = d\Phi/d\Omega$ with units of watts/steradian (W/sr).

For a point source radiating in all directions, the total flux is:

$$\Phi = \int d\Phi = \int I \, d\Omega$$

$$= I \int_0^\pi \sin\theta \, d\theta \int_0^{2\pi} d\varphi = 4\pi I \quad (3.5)$$

An extended source, for example the ocean surface, is characterized by its **radiance** L, which is the flux emitted per unit solid angle, per unit area of the source *projected in the direction of the radiation*, dA_p. Thus:

$$L(\theta, \varphi) = \frac{d\Phi}{d\Omega \, dA_p} = \frac{d\Phi}{d\Omega \, dA_s \cos\theta}$$

$$(3.6)$$

where L has units of $W \cdot m^{-2} \cdot sr^{-1}$.

Note that because the area in the definition is projected into the direction of the radiation, there is a $\cos\theta$ term in the denominator of the definition, where θ is the angle between the normal to the source area and the direction of the radiation (Figure 3.5).

In general, the radiance of a source is a function of both the polar angle θ and the azimuth angle φ.

The emittance from a surface, which has units of W/m^2, is related to the radiance by:

$$M = \int_\Omega L(\theta, \varphi) \cos\theta \, d\Omega \quad (3.7)$$

where the $\cos\theta$ term accounts for the projection of the area into the look direction. Similarly, the total power emitted by a surface of area A_s is:

$$\Phi_{rad} = \int d\Phi = \int_{A_s} \int_\Omega L(\theta, \varphi) \cos\theta \, dA_s \, d\Omega$$

$$(3.8)$$

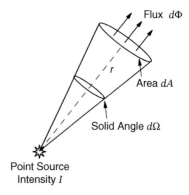

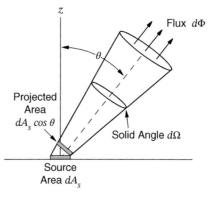

Figure 3.5 Difference between radiant intensity and radiance.

In general, most electromagnetic fields consist of energy with a range of wavelengths. For most applications, we are concerned with how much power a source radiates over some narrow wavelength interval, say from wavelength λ to $\lambda + d\lambda$. We then speak in terms of the **spectral radiance** $L_\lambda = L/d\lambda$. Hence spectral radiance is defined as:

$$L_\lambda = \frac{d\Phi}{d\Omega \, dA_p \, d\lambda} = \frac{d\Phi}{d\Omega \, dA_s \cos\theta \, d\lambda}$$

$$(3.9)$$

where L_λ has units of $W \cdot m^{-2} \cdot sr^{-1} \cdot m^{-1}$, and the total power radiated by a surface of area A_s is:

$$\Phi_{rad} = \int d\Phi = \int_{A_s} \int_\Omega \int_\lambda L_\lambda(\theta, \varphi)$$

$$\times \cos\theta \, dA_s \, d\Omega \, d\lambda \quad (3.10)$$

What we have just reviewed are standard definitions for spectral radiance for visible and infrared radiation. For microwave applications it is common to define spectral radiance in terms of intervals of frequency instead of wavelengths; that is, we speak of the energy radiated over some narrow frequency interval, say from frequency v to $v + dv$. The spectral radiance is then $L_v = L/dv$.

To convert from L_λ to L_v we use the fact that $v\lambda = c$, where c is the speed of light, and include the Jacobian derived from the relationship: $L_\lambda d\lambda = L_v dv$. Hence:

$$L_v = \left|\frac{d\lambda}{dv}\right| L_\lambda = \left(\frac{\lambda^2}{c}\right) L_\lambda \qquad (3.11)$$

where L_v has units of $\text{W·m}^{-2}\text{·sr}^{-1}\text{·Hz}^{-1}$.

Now imagine a surface where the emitted radiation is isotropic, that is, one with a constant radiance for any viewing angle. Such a surface is said to be perfectly diffuse; commonly known also as a Lambertian surface. It follows from equation (3.6) that the radiant intensity I is proportional to $\cos\theta$, where θ is the angle between the surface normal and the viewing direction, a relationship known as **Lambert's cosine law**. As an example, a sheet of white paper is a close approximation to a Lambertian surface.

The total flux emitted into the hemisphere above a Lambertian surface is given by:

$$\Phi_{rad} = \int_\lambda L_\lambda d\lambda \int_{A_s} dA_s \int_\Omega \cos\theta \, d\Omega$$

$$= L A_s \int_0^{2\pi} d\varphi \int_0^{\pi/2} \cos\theta \sin\theta \, d\theta$$

$$= \pi L A_s \qquad (3.12)$$

If you have been paying attention, the factor of π might strike you as odd. The flux is being radiated into a hemisphere, so why is the answer not $2\pi L A_s$? This turns out to be a consequence of Lambert's law – the $\cos\theta$ dependence in Lambert's law limits the flux to half of what might be naively expected.

3.3 Radiometric Sensor Response

3.3.1 Derivation

Now let us apply what we have learned about radiometry to determine the response

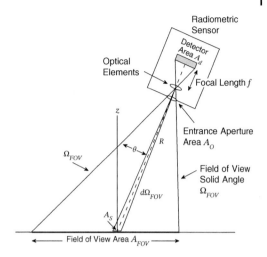

Figure 3.6 Sensor geometry.

of a single-pixel sensor looking down at an extended surface. This sensor is modeled as a set of optical elements (lenses) that project the flux at the entrance aperture of the sensor onto a detector. Referring to Figure 3.6, the important parameters for the sensor are the:

- Entrance aperture area A_o.
- Detector area A_d.
- Solid angle for a single sensing element or pixel Ω_{FOV}.
- Focal length f.
- Wavelength response of the sensor – from wavelength λ_1 to wavelength λ_2.

The geometry has the sensor looking down at an angle θ from nadir, at a range of R from the surface, viewing an area A_{FOV}. We assume that $R \gg f$ and the solid angle is small so $R^2 \gg A_{FOV}$. These conditions imply that R, θ, and the surface radiance of area A_{FOV} do not vary much over the field of view.

Furthermore we will initially ignore any effects due to the atmosphere. Atmospheric effects do turn out to be important, but we will start simple and work our way up to the hard stuff.

Now consider the contribution to the flux $d\Phi_o(\theta)$ at the sensor aperture arising from emissions from a small surface element dA_s. The radiance of the surface is L_s, which we assumed to only be a function of θ. Then,

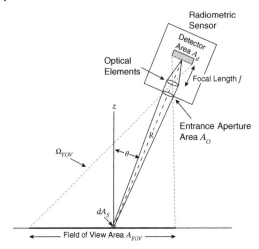

Figure 3.7 Sensor acceptance geometry.

referring to Figure 3.7, the flux at the sensor aperture is:

$$dΦ_o(θ) = L_s(θ) \cos θ \left(\frac{A_o}{R^2}\right) dA_s \quad (3.13)$$

where A_o/R^2 is the solid angle subtended by the entrance aperture as seen from the small surface element dA_s.

The total flux from the surface area within the sensor's field of view is $Φ_o(θ)$, which can now be obtained by integrating over the field of view area:

$$Φ_o(θ) = \int_{A_{FOV}} dΦ_o(θ)$$

$$= \int_{A_{FOV}} L_s(θ) \cos θ \left(\frac{A_o}{R^2}\right) dA_s$$

$$= L_s(θ) \cos θ \left(\frac{A_o}{R^2}\right) A_{FOV} \quad (3.14)$$

The sensor field of view $Ω_{FOV}$ is given by:

$$Ω_{FOV} = \frac{A_{FOV} \cos θ}{R^2} = \frac{A_d}{f^2} \quad (3.15)$$

which is best observed by redrawing the optics as an ideal pinhole camera as in Figure 3.6, and using similar triangles.

Notice that the total flux at the sensor aperture includes the term $A_{FOV} \cos θ/R^2$, which can now be rewritten as A_d/f^2.

So the total flux at the sensor aperture is simply:

$$Φ_o(θ) = A_o Ω_{FOV} L_s(θ) = A_o L_s(θ) \left(\frac{A_d}{f^2}\right) \quad (3.16)$$

If you have ever used a good camera, you will know that lenses are generally characterized by two numbers: the focal length and the f-number, which we will write $f/\#$. The f-number of a lens is defined to be $f/\# = \frac{f}{D}$ where f is the focal length and D is the diameter of the entrance aperture[1]. Noting that $A_o = \frac{π}{4}D^2$, we obtain the term $\frac{A_o}{f^2} = \frac{π}{4(f/\#)^2}$. The flux at the sensor then becomes: $Φ_o = \frac{πA_d}{4(f/\#)^2}L_s$. So more flux is accumulated with lower $f/\#$ or larger detector area.

Some of the flux arriving at the sensor passes all the way through the optics to the detector, but some of it is reflected or scattered away. Let $τ_o(λ)$ be the effective transmittance of the aperture and optical elements in front of the detector. Also assume that one of the optical elements acts as a filter to limit the flux on the detector to a small wavelength interval $Δλ$, where $Δλ = λ_2 - λ_1$. Then the flux at the detector $Φ_{det}$ is given by the expression:

$$Φ_{det}(θ) = \int_{λ_1}^{λ_2} Φ_o(θ)τ_o(λ)dλ$$

$$= A_o Ω_{FOV} \int_{λ_1}^{λ_2} L_s(λ, θ)τ_o(λ)dλ \quad (3.17)$$

where $L_s(λ, θ)$ is the spectral radiance of the surface.

The flux at the detector produces an output voltage V_{det} whose magnitude is determined by the detector responsivity $R_d(λ)$, which has units of volts/watts. Then the output voltage is

---

1 The $f/\#$ is usually expressed in the form f/n where n is the actual $f/\#$ of the lens. So an $f/2$ lens has a focal length that is twice the diameter of the entrance aperture.

given by:

$$V_{det}(\theta) = \int_{\lambda_1}^{\lambda_2} R_d(\lambda)\Phi_{det}(\theta)d\lambda$$

$$= A_o\Omega_{FOV}\int_{\lambda_1}^{\lambda_2} L_s(\lambda, \theta)\tau_o(\lambda)R_d(\lambda)d\lambda$$

$$(3.18)$$

For simplicity, we assume that the wavelength interval $\Delta\lambda$ is small enough that the surface radiance, the optical transmittance, and the detector responsivity are essentially constant over $\Delta\lambda$. We can then replace these quantities in the above integral by their values at the mean wavelength $\bar{\lambda} = (\lambda_1 + \lambda_2)/2$. The detector voltage then becomes:

$$V_{det}(\theta) = A_o\Omega_{FOV}L_s(\bar{\lambda}, \theta)\tau_o(\bar{\lambda})$$
$$\times R_d(\bar{\lambda})(\lambda_2 - \lambda_1) \qquad (3.19)$$

The detector voltage expression, equation (3.19), has a number of important implications:

- The signal from the sensor depends only on the scene radiance and some scale factors that are determined by the characteristics of the sensor design. The sensor factors can be evaluated by calibration procedures.
- The detector signal is independent of the distance between the sensor and the scene being viewed, although this derivation neglected the impact of sensing though the Earth's atmosphere. Atmospheric effects are discussed just below.
- If the scene being viewed is a Lambertian surface, the detector signal does not depend on the viewing direction. In general, however, the radiance, and consequently the detector signal, depends on the viewing direction.
- Any information about the scene must be contained in its radiance. The central problem now becomes one of determining how and what physical properties of a particular type of scene (land, ocean, or atmosphere) are manifested in the scene radiance at the electromagnetic wavelength appropriate to the sensor of interest.

- Also note the larger the aperture, the stronger the signal. In practice, aperture size is highly correlated with spacecraft cost since large apertures often drive satellite size and weight. Thus there are practical limits to the size of optical apertures.

Inclusion of an atmosphere introduces three important effects:

- The flux from the surface arriving at the sensor aperture will be attenuated by an amount determined by the atmospheric transmittance.
- In addition to the intrinsic radiance of the surface, the surface radiance will have a contribution from any downward propagating radiation reflected from the surface into the field of view. This downwelling radiation may originate from sources within the atmosphere (e.g., scattered solar radiation, emitted radiation from atmospheric molecules, and cloud emissions) or from extraterrestrial sources (e.g., direct solar radiation) depending on the wavelength of interest. This contribution is known as the reflected radiance.
- Radiation emitted or scattered by the atmosphere within the field of view between the sensor and the surface will contribute also to the total radiance reaching the sensor aperture. This contribution is known as the path radiance.

The flux at the sensor aperture now becomes:

$$\Phi_o(\theta) = A_o\Omega_{FOV}[\tau_a(\theta)[L_s(\theta)$$
$$+ r_s(\theta)L_d(\theta)] + L_p(\theta)] \qquad (3.20)$$

In this equation the downwelling radiance L_d times the surface reflectance r_s is added to the surface's emitted radiance L_s. This sum represents the radiance leaving the surface. All of that radiance must propagate through the atmosphere, so it is reduced by the atmospheric transmittance, τ_a. Finally, any atmospheric path radiance L_p is added in.

The flux at the detector then becomes this far more complicated looking expression with many more terms:

$$\Phi_{det}(\theta) = A_o \Omega_{FOV} \int_{\lambda_1}^{\lambda_2} \tau_o(\lambda)[\tau_a(\lambda,\theta)[L_s(\lambda,\theta)$$
$$+ r_s(\lambda,\theta)L_d(\lambda,\theta)] + L_p(\lambda,\theta)]d\lambda$$

$$(3.21)$$

and as before, the output voltage from the detector is:

$$V_{det}(\theta) = \int_{\lambda_1}^{\lambda_2} R(\lambda)\Phi_{det}(\theta)d\lambda$$

$$= A_o \Omega_{FOV} \int_{\lambda_1}^{\lambda_2} L_{scene}(\lambda,\theta)\tau_o(\lambda)$$
$$\times R_d(\lambda)d\lambda \qquad (3.22)$$

where the scene radiance is defined to be:

$$L_{scene} = \int_{\lambda_1}^{\lambda_2} [\tau_a(\lambda,\theta)[L_s(\lambda,\theta)$$
$$+ r_s(\lambda,\theta)L_d(\lambda,\theta)] + L_p(\lambda,\theta)]d\lambda$$

$$(3.23)$$

The procedure for extracting information about the surface being viewed now becomes more complex. As before, the physical properties of the surface are still manifested in its radiance L_s, and there may be additional information encoded in the surface reflectance. However, before the surface information can be extracted, corrections must generally be applied to the measurement in order to estimate or eliminate the path radiance and reflected downwelling radiance terms. Finally, the atmospheric transmittance must also be taken into account.

Theoretical and practical techniques for dealing with these atmospheric effects are treated in more detail in subsequent sections discussing optical, infrared, and microwave observations.

3.3.2 Example Sensor Response Calculations

It is instructive to estimate the order of magnitude of the detector voltage from a typical infrared sensor such as the AVHRR instrument on the NOAA POES satellites when viewing the ocean in the nadir direction. We ignore the

atmosphere and assume the temperature of the sea surface is 27 °C (300 K), and that the sensor wavelength response is from 10–11 μm. As we shall see later, the scene radiance is then approximately $10\,\mathrm{W}{\cdot}\mathrm{m}^{-2}{\cdot}\mathrm{sr}^{-1} \cdot \mathrm{\mu m}^{-1}$.

The sensor parameters are:

- Entrance aperture diameter $d_o = 0.2\,\mathrm{m}$ (8-inch aperture).
- Entrance aperture area $A_o = \pi d_o^2/4 = 0.031\,\mathrm{m}^2$.
- Angular field of view $\alpha = 1$ mrad (0.057°).
- Instantaneous field-of-view solid angle $\Omega = \pi\alpha^2/4 = 7.8 \times 10^{-7}$ sr.
- Optical transmittance $\tau_o = 0.8$.
- Detector responsivity $R_d = 10^5$ V/W (typical for HgCdTe detector).

Thus the detector voltage is:

$$V_{det} = A_o \Omega_{FOV} L_s(\overline{\lambda}, 0°)\tau_o(\overline{\lambda}, 0°)$$
$$\times R_d(\overline{\lambda})(\lambda_2 - \lambda_1)$$
$$= \frac{\pi}{4}(0.2)^2 \times \frac{\pi}{4}(10^{-3})^2 \times 10 \times 0.8$$
$$\times 10^5 = 20\,\mathrm{mV} \qquad (3.24)$$

This low-level signal would then be amplified and digitized within the instrument before transmission to the ground.

3.3.3 Response of a Sensor with a Partially-Filled FOV

The sensor response function, equation (3.19), was derived based on an assumption that the scene with mean radiance L_s fills the sensor FOV. It is worth considering how the story changes if the sensor FOV is not filled. Specifically, what is the formula for the response of a sensor when viewing a small surface with radiance L_s and area A_s that lies fully within the sensor FOV, assuming that the surrounding surface that fills the remainder of the sensor FOV has zero emissions? Furthermore, is the sensor response still independent of distance?

The contribution to the flux at the sensor aperture arising from emissions from a small element of the surface with radiance L_s and area A_s ($= dA_s$ in Figure 3.7) is the radiant intensity ($L_s \cos\theta A_s$) times the solid angle subtended by the sensor aperture (A_o/R^2). Thus $\Phi = \frac{L_s \cos\theta A_s A_o}{R^2}$. This form immediately provides an answer to one of the questions: the sensor response is no longer independent of range, but is in fact inversely proportional to R^2.

The above formula is a complete answer to the problem, but some additional insights can be gained by a little further manipulation of this initial result. As illustrated by the short dashed lines in Figure 3.7, the sensor solid angle field of view is given by either $A_{FOV} \cos\theta/R^2$ or A_d/f^2. This fact is best observed by redrawing the optics as an ideal pinhole camera, and using similar triangles. This can be rewritten as $\cos\theta/R^2 = A_d/(A_{FOV}f^2)$. After substitution, the total flux at the sensor aperture then becomes $\Phi = \frac{A_o A_d}{f^2} \frac{A_s}{A_{FOV}} L_s$. Note the first term $\frac{A_o A_d}{f^2}$ only depends on the sensor design, the second term $\frac{A_s}{A_{FOV}}$ is an area dilution factor that equals the ratio of the area of the emitting portion of the surface to the area covered by the sensor, and the final term is the radiance of the emitting portion of the surface.

3.4 Blackbody Radiation

3.4.1 Planck's Radiation Law

A blackbody is an idealized physical body that absorbs all incident electromagnetic radiation, regardless of frequency or angle of incidence. When in equilibrium, such bodies must emit the same amount of radiation as they are receiving, otherwise the body would not be in equilibrium. This is an important concept because with some modification it can be used to model the infrared emissions of many surfaces.

An ideal blackbody:

- Emits energy at a rate determined only by its temperature.

- At a given temperature, it emits the maximum possible energy. Other sources (non-ideal blackbodies) emit less energy.

Planck's Radiation Law specifies the spectral radiance of a blackbody at a given temperature:

$$L_B(\lambda, T) = \frac{2hc^2}{\lambda^5} \left[\frac{1}{\exp\left(\frac{hc}{\lambda k_B T}\right) - 1} \right]$$

(3.25)

where L_B has units of $\text{W} \cdot \text{m}^{-2} \cdot \text{sr}^{-1} \cdot \text{m}^{-1}$, and where:

$h =$ Planck's constant $= 6.6261 \times 10^{-34}$ J·s.
$c =$ velocity of light $= 2.9979 \times 10^8$ m s^{-1}.
$k_B =$ Boltzmann's constant $= 1.38065 \times 10^{-23}$ J K^{-1}.
$T =$ source temperature (K).
$\lambda =$ wavelength (m).

Note the formula only depends on temperature and wavelength. Also note that the formula requires mks units, even though wavelengths are often give in terms of nanometers (nm) or microns (µm). To convert: W·m^{-2}·sr^{-1}·m^{-1} $\times 10^{-6}$ = W·m^{-2}·sr^{-1}·µm^{-1}. Be careful about units!

For example, a blackbody at room temperature of $T = 300$ K has a spectral radiance of $L_B(\lambda, T) = 0.496$ W·m^{-2}·sr^{-1}·µm^{-1} at $\lambda = 3.8$ µm and $L_B(\lambda, T) = 9.58$ W·m^{-2}·sr^{-1}·µm^{-1} at 11.0 µm. These are two common infrared wavelengths used for Earth remote sensing. Values of 0.5–10 W·m^{-2}·sr^{-1}·µm^{-1} are typical for the spectral radiance in the IR at room temperature.

The blackbody hemispherical emittance $M_B(\lambda, T)$ is given by:

$$M_B(\lambda, T) = \pi L_B(\lambda, T)$$

(3.26)

Note the factor π, rather than 2π the solid angle of a hemisphere; a consequence of $\cos\theta$ in the definition of emittance given in equation (3.7).

Figure 3.8 shows the blackbody hemispherical emittance for wavelengths from the UV to 100 µm. A room temperature (300 K) blackbody has a maximum emittance at an infrared

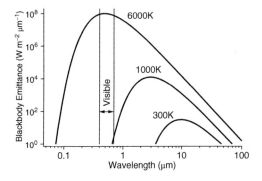

Figure 3.8 Blackbody emittance.

wavelength near 10 μm; at visible wavelengths (near 0.5 μm) the emittance is negligibly small. Even at 1000 K, the emittance at visible wavelengths is about 10^{-5} times smaller than the maximum near 3 μm. Notice that at 6000 K, the peak emittance is at visible wavelengths.

Figure 3.9 shows that the Earth emits approximately as a 290 K blackbody. The smooth curves are spectral radiance from 260, 280 and 300 K blackbodies. The data are from a satellite looking straight down at the Earth. Where there are windows of high transmittance, the Earth's radiance looks like a 290 K blackbody. At other wavelengths where strong atmospheric absorption occurs, the Earth no longer radiates as a blackbody at the same temperature.

While the Earth emits in the infrared, the sun emits energy up into the visible region of the spectrum. The sun emits energy approximately as a 5900 K blackbody. Figure 3.10 shows solar irradiance at the top of the atmosphere compares well with a 5900 K source. The irradiance at sea level is reduced by atmospheric absorption and scattering.

The spectral irradiance at the top of the Earth's atmosphere can be easily computed as the radiance of a 5900 K blackbody multiplied by the solid angle subtended by the sun as observed from the Earth. The peak radiance of a 5900 K blackbody is 2.93×10^7 W·m^{-2}·sr^{-1}·μm^{-1} at $\lambda = 0.491$ μm. Given the sun's radius $R_s = 6.96 \times 10^5$ km and the sun–Earth distance $R_{SE} = 1.5 \times 10^8$ km, the solid angle subtended by the sun as observed from the Earth is $\pi R_s^2 / R_{SE}^2 = 6.76 \times 10^{-5}$ sr. The peak spectral irradiance at the top of the Earth's atmosphere is then 1980 W·m^{-2}·μm^{-1}.

3.4.2 Microwave Blackbody

For microwave sensors, it is more common to use frequency v (Hz) rather than wavelength as

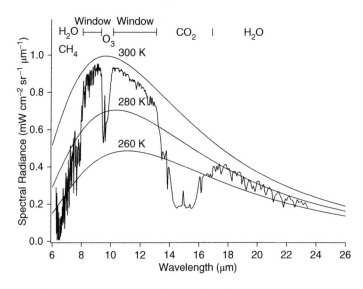

Figure 3.9 Earth's blackbody spectrum as viewed by a nadir looking sensor at the top of the atmosphere.

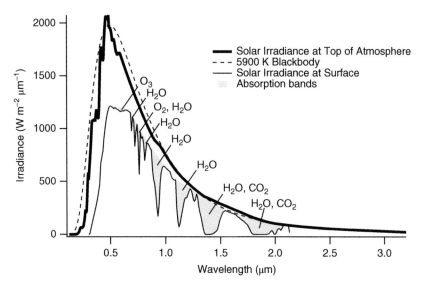

Figure 3.10 Solar blackbody spectrum. Adapted from Wolfe and Zissis (1985).

the independent variable. They are related by: $v\lambda = c$ and so (recall equation (3.11)):

$$L_B(v, T) = L_B(\lambda, T) \left| \frac{d\lambda}{dv} \right| = \frac{\lambda^2}{c} L_B(\lambda, T)$$

(3.27)

Hence:

$$L_B(v, T) = \frac{2hv^3}{c^2} \left[\frac{1}{\exp(\frac{hv}{k_B T}) - 1} \right]$$
$$\times \, \mathrm{W \cdot m^{-2} \cdot sr^{-1} \cdot Hz^{-1}}$$

(3.28)

Thus the spectral radiance of a blackbody can also be expressed in units of radiance per Hertz.

3.4.3 Low-Frequency and High-Frequency Limits

The Rayleigh–Jeans Law is an approximation to Planck's Law that is valid at long wavelengths and low frequencies, such as microwave frequencies. At long wavelengths (low frequencies), the factors in the exponential of Planck's Law are much less than 1: $hc/\lambda k_B T \ll 1$ and $hv/k_B T \ll 1$. In general for small x, we can approximate $\exp(x)$ by $1 + x$. The Rayleigh–Jeans Law then provides these approximate forms for the spectral radiance at microwave frequencies:

$$L_B(\lambda, T) = \frac{2ck_B T}{\lambda^4} \quad \text{or} \quad L_B(v, T) = \frac{2v^2 k_B T}{c^2}$$

(3.29)

Note these expressions are valid for wavelengths much greater than 48 μm or frequencies much less than 6 THz.

There is also an approximation to Planck's Law valid at the high-frequency, short-wavelength limit, where $hc/\lambda k_B T \gg 1$ and $hv/k_B T \gg 1$. In this case, we make the approximation that when $x \gg 1$, $\exp(x) - 1 = \exp(x)$ and so $1/(\exp(x) - 1) \approx \exp(-x)$. The blackbody spectral radiance now becomes:

$$L_B(\lambda, T) = \frac{2hc^2}{\lambda^5} \exp\left(-\frac{hc}{\lambda k_B T} \right) \quad \text{or}$$
$$L_B(v, T) = \frac{2hv^3}{c^2} \exp\left(-\frac{hv}{k_B T} \right)$$

(3.30)

These approximations are valid for mid-wave and long-wave infrared.

3.4.4 Stefan–Boltzmann Law

The Stefan–Boltzmann Law relates the total radiance of a blackbody to its temperature. It can be obtained by integrating the blackbody spectral radiance over frequency or wavelength

$$L_B(T) = \int_0^\infty L_B(\lambda, T)d\lambda = \int_0^\infty L_B(\nu, T)d\nu$$

$$(3.31)$$

Substitute the Planck function into this expression and do the integral to get:

$$L_B(T) = \left(\frac{2\pi^4 k_B^4}{15c^2 h^3}\right) T^4 = \sigma T^4 \quad \text{where}$$

$$\sigma = 1.80498 \times 10^{-8} \text{ W} \cdot \text{m}^{-2} \cdot \text{sr}^{-1} \cdot \text{K}^{-4}$$

$$(3.32)$$

In the end the relationship says the total radiance is proportional to temperature to the fourth power.

Note: The Stefan–Boltzmann constant σ as defined above has units of $\text{W}\cdot\text{m}^{-2}\cdot\text{sr}^{-1}\cdot\text{K}^{-4}$. Many texts define the Stefan–Boltzmann Law in terms of the total emittance M_B of a blackbody, where:

$$M_B(T) = \int_\Omega L_B(T)\cos\theta \; d\Omega = \int_0^{2\pi} d\varphi$$

$$\times \int_0^{\pi/2} L_B(T)\cos\theta \; \sin\theta \; d\theta$$

$$= \pi L_B(T) = \sigma_M T^4 \qquad (3.33)$$

In this case, $\sigma_M = \pi\sigma = 5.6705 \times 10^{-8}$ W·m^{-2}·K^{-4}.

3.4.5 Wein's Displacement Law

Wein's Displacement Law can be used to compute the wavelength of the peak in the blackbody spectrum. Specifically, let λ_{max} be the wavelength at which the Planck function is a maximum. Wien's Displacement Law states that $\lambda_{max}T = 2898$ μm K, a result easily derived by computing the wavelength derivative of the Planck function. It can also be shown that $\nu_{max} = 5.879 \times 10^{10} \, T$ where ν_{max} is the frequency at which the Planck function is a maximum. Note that the product of the two peaks does not equal the speed of light, so $\lambda_{max} \times \nu_{max} \neq c$.

3.4.6 Emissivity

In general, a real material only approximates an ideal blackbody to a varying degree. Let $L_s(\lambda, T, \theta, \varphi)$ be the spectral radiance of a material at the temperature T when viewed from the direction specified by the polar and azimuth angles θ and φ. Let L_B be the spectral radiance of a blackbody under the same conditions.

The emissivity ε of the material is defined by:

$$\varepsilon(\lambda, T, \theta, \varphi) = \frac{L_s(\lambda, T, \theta, \varphi)}{L_B(\lambda, T)} \qquad (3.34)$$

From the definition of an ideal blackbody, it follows that $\varepsilon < 1$ for all real materials. For many surfaces, the emissivity is independent of the azimuth angle φ, so emissivity only depends on λ, T, and θ.

When viewed in the nadir direction, the sea surface emissivity is about 0.98–0.99 at infrared wavelengths (3.8 μm and 11 μm). At microwave frequencies, the value is about 0.4 for the same viewing direction.

3.4.7 Equivalent Blackbody Temperature

Planck's Radiation Law allows us to compute the spectral radiance of a blackbody at temperature T and wavelength λ:

$$L_B(\lambda, T) = \frac{2hc^2}{\lambda^5}\left[\frac{1}{\exp(\frac{hc}{\lambda k_B T}) - 1}\right]$$

$$\times \text{W} \cdot \text{m}^{-2} \cdot \text{sr}^{-1} \cdot \text{m}^{-1} \quad (3.35)$$

This expression can be inverted to derive a formula for the radiometric temperature T for a given spectral radiance L_B and wavelength:

$$T = T_B(\lambda, L) = \frac{hc}{\lambda k_B \ln\left(1 + \frac{2hc^2}{\lambda^5 L_B}\right)}$$

$$(3.36)$$

So we can either compute the spectral radiance at a given wavelength for a blackbody of a given temperature from equation (3.35) or compute the temperature of the blackbody given the spectral radiance at a wavelength from equation (3.36). When doing these calculations, be careful with the units and always use K, not deg C.

For the previously given example, equation (3.35) allows us to compute that a 300 K blackbody will emit a spectral radiance of $9.6\,\text{W·m}^{-2}\text{·sr}^{-1}\text{·}\mu\text{m}^{-1}$ at a wavelength of 11 μm. Equivalently, if we measured a spectral radiance of $9.6\,\text{W·m}^{-2}\text{·sr}^{-1}\text{·}\mu\text{m}^{-1}$ at a wavelength of 11 μm, by replacing L_B in equation (3.36) we could infer that this is equivalent to the emission from a perfect blackbody of temperature 300 K.

This equivalent blackbody temperature is also known as the *radiometric temperature* of the scene. In some texts, the term *brightness temperature* is also used. Notice that for surfaces with a wavelength-dependent emissivity, the radiometric temperature will also vary with wavelength.

3.5 IR Sea Surface Temperature

3.5.1 Contributors to Infrared Measurements

Consider an infrared radiometer viewing the sea surface at an incident angle θ, where θ is measured from the normal to the mean sea surface. The sensor is equipped with a filter that restricts the radiation falling onto the detector to a narrow band centered about the wavelength λ. For simplicity, the sensor is viewing the sea surface either at night, or in the daytime with the viewing geometry restricted to directions where negligible solar radiation is reflected from the surface into the field of view, and there are no clouds in the sky.

As illustrated in Figure 3.11 the radiance of the scene, as viewed at the aperture of the sensor, includes radiant energy from three sources:

- Thermal emissions from the sea surface, which are attenuated by the atmosphere as they propagate to the sensor.

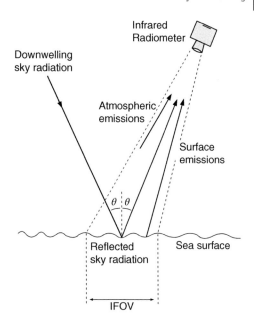

Figure 3.11 Contributions to infrared measurements of the sea surface.

- Downwelling sky radiation (from thermal emissions by atmospheric molecules) that is reflected from the sea surface and then attenuated while propagating to the sensor.
- Emissions from the atmosphere between the sensor and the sea surface (known as path radiance).

Thus the radiance at the sensor = [Surface emission + Reflected sky radiation] attenuated by atmosphere + Path radiance.

Assume we are viewing a flat sea surface with no waves. Then the total scene radiance at the sensor aperture can be written:

$$L_r(T_r, \lambda, \theta) = \tau_a(\lambda, \theta)\varepsilon_w(\lambda, \theta)L_B(\lambda, T_s)$$
$$+ \tau_a(\lambda, \theta)r_w(\lambda, \theta)L_{sky}(T_{sky}(\lambda, \theta))$$
$$+ L_A(\lambda, \theta) \qquad (3.37)$$

In this expression:

L_B is the blackbody radiance at the temperature of the sea surface T_s.

ε_w is the emissivity of the surface.

r_w is the reflectance of the surface.

L_{sky} is the downwelling sky radiance at the sea surface from a zenith angle θ.

T_{sky} is the radiometric temperature of the downwelling sky radiation.

τ_a is the atmospheric transmittance.

L_A is the atmospheric path radiance.

The parameter T_r in equation (3.37) is the radiometric temperature of the scene, that is, the temperature of an ideal blackbody with radiance equal to L_r.

We can simplify the equation by eliminating the explicit references to the dependence of each term on wavelength and viewing angle:

$$L_r = \tau_a \varepsilon_w L_B(T_s) + \tau_a r_w L_{sky}(T_{sky}) + L_A$$

$$(3.38)$$

Thus the total scene radiance is the sum of three terms:

- The blackbody radiance of the sea surface times its emissivity times its atmospheric transmittance.
- The downwelling sky radiance times the reflection coefficient of the sea surface times the atmospheric transmittance.
- The path radiance emitted by the atmosphere towards the sensor.

It is critically important to note that the only sea surface information in equation (3.38) is its surface temperature T_s encoded into the surface's blackbody radiation, and any additional information contained in the surface emissivity, ε_w.

The sensor measures L_r, from which we want to determine the sea surface temperature T_s. It should be clear that the accuracy of our SST estimate depends on our ability to correct for the atmospheric transmittance, the reflected sky radiation, and the path radiance contributions.

L_r is the scene radiance measured by the radiometer. It can be inverted to an equivalent blackbody temperature T_r we call the scene temperature. From this measurement we want to estimate the sea surface radiance, which can be inverted to estimate the sea surface temperature. To do so we need estimates of the surface emissivity (ε_w), atmospheric transmittance (τ_A), the reflected sky radiation ($r_w L_{sky}$), and the path radiance (L_A).

The accuracy of the sea surface temperature estimate will be determined by the accuracy of the various parameters in this equation.

To add context, it is worthwhile to estimate the size of the terms in equation (3.38). In fact, it is informative to make such estimates in two limits: the first where the radiometer is close to the surface, as if in a low flying aircraft, and the second where it is on a satellite looking through the entire atmosphere.

We begin by noting that the general form of Kirchoff's Law states that the sum of the emissivity, reflectance and transmittance must equal one. This is nothing more than a statement that whatever energy goes into a body in equilibrium must either be reemitted from the body, reflected off the body or transmitted through the body. In the case of a deep ocean, there is no transmission through the ocean, so emissivity + reflectance = $\varepsilon_w + r_w = 1$.

Applying Kirchoff's Law allows equation (3.38) to be rewritten:

$$L_r = \tau_a L_B(T_s) - \tau_a r_w L_B(T_s)$$
$$+ \tau_a r_w L_{sky}(T_{sky}) + L_A \qquad (3.39)$$

3.5.2 Correction of Low-Altitude Infrared Measurements

Consider the case where the radiometer is located close enough to the surface that the atmospheric attenuation can be ignored. This is equivalent to saying that $\tau_A(\lambda, \theta) = 1$ and the atmospheric path radiance $L_A(\lambda, \theta) = 0$. It also implies that the incident angle θ is not too large in order to limit the amount of atmosphere within the field of view. Equation (3.39) then can be simplified:

$$L_r = L_B(T_s) - r_w L_B(T_s) + r_w L_{sky}(T_{sky})$$

$$(3.40)$$

This expression indicates that for low-altitude measurements the measured radiance L_r equals the surface radiance $L_B(T_s)$ plus two correction terms: the first accounts for the emissivity of the surface being less than one, and the second accounts for the contributions of reflected downwelling atmospheric radiation.

The relationship between radiance and equivalent blackbody temperature is nonlinear, so to estimate the temperature offsets associated with each these terms we compute:

$$\Delta T_{emissivity} = T_B(L_s - r_w L_s) - T_B(L_s) \quad (3.41)$$

$$\Delta T_{sky} = T_B(L_s - r_w L_s + r_w L_{sky})$$
$$- T_B(L_s) - \Delta T_{emissivity} \quad (3.42)$$

where $T_B(L)$ is the inverse of Planck's Law as defined in equation (3.36). Note these errors are defined so they sum to the total difference between the brightness temperature and the surface temperature.

For a nadir-looking sensor in the 10–11 μm band, we can estimate the error due to emissivity. The surface reflectance at this wavelength, obtained from the Fresnel relations, is of the order 1%. If the surface temperature is 27 °C (300 K), we compute from the Planck function at 10.5 μm that the corresponding radiance is 9.78 W·m^{-2}·sr^{-1}. The result is:

$$\Delta T_{emissivity} = T_B(\varepsilon L_s) - T_B(L_s)$$
$$= T_B(0.99 \times L_s(300\,\text{K})) - 300\,\text{K}$$
$$= T_B(0.99 \times 9.78\,\text{W} \cdot \text{m}^{-2}\text{sr}^{-1}\mu\text{m}^{-1})$$
$$- 300\,\text{K} = -0.65\,\text{K} \quad (3.43)$$

The temperature measured by the radiometer is 0.65 degree cooler than the surface temperature because the surface emissivity is less than one.

To estimate the magnitude of the error due to reflected skylight we will assume a typical value for the near-zenith radiometric temperature of a midlatitude atmosphere in the 10–11 μm band: −40 °C (=233 K). The corresponding sky radiance is L_{sky} = 2.61 W·m^{-2}·sr^{-1}. Substituting these values for the parameters in equation (3.42) yields:

$$\Delta T_{sky} = T_B(L_s - r_w L_s + r_w L_{sky}) - T_B(L_s)$$
$$- \Delta T_{emissivity}$$
$$= T_B(0.99 \times L_s(300\,\text{K}) + 0.01$$
$$\times L_{sky}(233\,\text{K})) - 300\,\text{K} + 0.65\,\text{K}$$
$$= T_B(0.99 \times 9.78 + 0.01 \times 2.61)$$
$$- 300\,\text{K} + 0.65\,\text{K} = +0.17\,\text{K} \quad (3.44)$$

The reflected sky radiance adds an additional 0.17 K to the measurement, partially compensating for the reduced emissivity. Thus we find that the measured radiometric temperature is $T_r = T_s - 0.48$ K, nearly 0.5 K cooler than the actual surface temperature.

Figure 3.12 illustrates the magnitude of these correction terms as a function of the radiometric temperature of the downwelling sky radiation. Notice that the radiometric temperature of the scene gets closer to the surface temperature as the sky temperature increases. Also note that the radiometric temperature is identical to the surface temperature when the downwelling sky radiance is the same

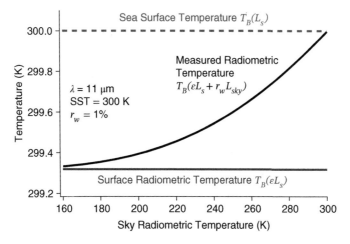

Figure 3.12 Temperature of emissivity and reflected sky terms as a function of radiometric temperature of the downwelling sky radiation.

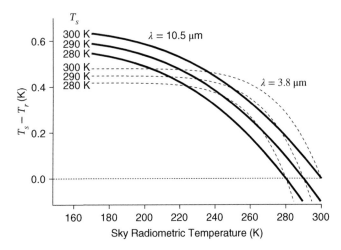

Figure 3.13 Difference between the radiometric and surface temperature as a function of surface temperature for wavelengths of 3.8 μm and 10.5 μm.

as the blackbody radiance from the surface, a situation that is equivalent to measuring the scene radiance inside an ideal blackbody. Normally the sky radiance is cold, but it can be as warm as the surface temperature under foggy conditions where the sky radiance is determined by the temperature of the fog.

Figure 3.13 shows the difference between the radiometric and surface temperature as a function of sky temperature for radiometers with passbands of 10–11 μm and 3.6–4.0 μm. A surface reflectance of 1% was used for the 10–11 μm band, while 2% was used for the 3.6–4.0 μm calculation. For sky temperatures of less than 230 K or so, the difference between the surface and radiometric temperature is smaller at 3.6–4.0 μm than at 10–11 μm, while the opposite holds for warmer sky temperatures.

The important result to be taken from the above analysis is that, at infrared wavelengths, the net effect of the fact that the sea surface is not a true blackbody is to cause the measured radiometric temperature of the surface to differ from the surface temperature by a few tenths of a degree. The exact magnitude of this difference depends on the surface temperature, the sky radiometric temperature, and the wavelength response of the radiometer. The curves in Figure 3.13 do suggest that one approach to correcting for these effects is to leverage the predicted differences in measurements made at multiple wavelengths. The specific algorithms used are discussed later.

3.5.3 Correction of High-Altitude Infrared Measurements

We now consider the more general case where the radiometer is located at a distance away from the sea surface where the effects of the intervening atmosphere cannot be ignored. We start again with equation (3.37), repeated here in simplified form for convenience:

$$L_r = \tau_a \varepsilon_w L_s + \tau_a r_w L_{sky} + L_A \qquad (3.45)$$

Figure 3.14 plots the atmospheric transmittance for a vertical path from space to sea level as a function of wavelength for six different model atmospheres[2]. Notice that the atmosphere's transmittance in the mid-wave (3–5 μm) window and the long-wave (8–13 μm) window varies significantly depending on location and time. The differences are primarily driven by the water content in the atmosphere, with higher transmittance occurring in drier atmospheres.

Because the transmittance in the atmospheric windows can vary by as much as a factor of two, any reasonable attempt at

---

2 The spectral transmittances in this figure were predicted by MODTRAN, a commercially available computer code for predicting atmospheric transmission and radiance. MODTRAN solves the radiative transfer equation (discussed in the next section) for a model atmosphere that takes into account the vertical distribution of the molecular constituents of the atmosphere, and their absorption/emission characteristics as a function of wavelength.

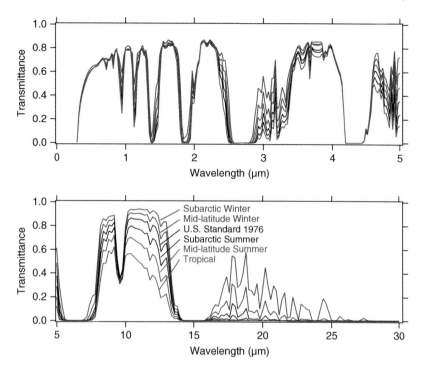

Figure 3.14 Atmospheric transmittance for a vertical path to space for six model atmospheres with 23 km visibility. Source: Modified from Selby and McClatchey (1975).

atmospheric correction will begin by dividing the measured radiance by the estimated atmospheric transmittance:

$$\frac{L_r}{\tau_a} = [\varepsilon_w L_s + r_w L_{sky}] + \frac{L_A}{\tau_a} \qquad (3.46)$$

Note that the term in the square brackets is identical to the radiance measured from a low-altitude sensor, with the same offsets as determined in the last section. Thus, L_r/τ_a is the same radiance as computed in the last section plus an offset (L_A/τ_a) due to the atmospheric path radiance.

We can again use MODTRAN to compute the path radiances that would be observed by a nadir-looking sensor to estimate the radiance correction due to atmospheric path radiance. Application of the inverse blackbody function converts this to a temperature correction.

Figure 3.15 shows the path radiance correction to surface radiometric temperature as a function of wavelength for a sensor at the top of the atmosphere, plotting the temperature difference $T_s - T_r$ from the analysis above. Note that the atmospheric correction to the measured radiometric temperature is 1–10 K

in the few windows where the atmosphere is relatively transparent, more than an order of magnitude larger than the reflected sky radiation term.

3.6 Atmospheric Radiative Transfer

The propagation of radiation through the atmosphere is governed by the theory of radiative transfer, which in its most general form accounts for absorption, emission, and scattering. The MODTRAN calculations used to develop accurate radiometric corrections for estimating sea surface temperature from infrared measurements are solutions to the theory of radiative transfer in the atmosphere at infrared wavelengths. The same theory is also applied to describe the atmosphere's effect on measurements of ocean color. It is thus worthwhile to review some of the basic concepts in radiative transfer theory.

The atmospheric transmittance and path radiance at infrared wavelengths are due to the absorption and emission of radiation by the

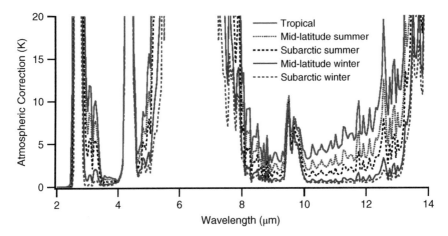

Figure 3.15 Spectral dependence of atmospheric corrections to sea surface temperature for several model atmospheres. Source: Based on Deschamps and Phulpin (1980).

molecular constituents of the atmosphere. The transmittance spectra of five major atmospheric constituents: water vapor, carbon dioxide, methane, nitrous oxide, and ozone, are shown in Figure 3.16. Note the unique wavelength dependence of each constituent.

Scattering of infrared radiation by molecules in the atmosphere is negligible and can be ignored.

Associated with each gas is a volume absorption coefficient κ that determines the

attenuation by that gas over a short segment of path $d\ell$ in the absorbing medium. Integrating the absorption along the path traversed by the radiation gives the total attenuation of the radiation. Every absorber is also an emitter of radiation, and each path segment emits radiation that combines with that being attenuated to contribute to the net radiation observed at the end of the path.

Consider Figure 3.17 to compute the spectral radiance $L_\lambda(h)$ seen by a sensor at a distance

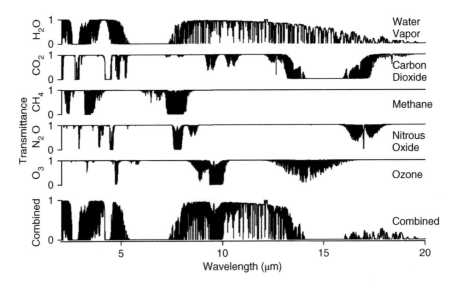

Figure 3.16 Infrared transmittance spectra of five major atmospheric constituents. Data from Earth Observation Data Group, Department of Physics, University of Oxford (2019).

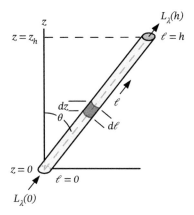

Figure 3.17 Radiance absorption and reemission along a flux tube. Adapted from Stewart (1985).

h from a surface that has a spectral radiance of $L_\lambda(0)$. The tube in the figure indicates the absorbing path between the surface and the sensor. Absorption along this path is characterized by a total absorption coefficient $\kappa(\ell)$, which consists of contributions from water vapor, carbon dioxide, and ozone.

$$\kappa(\ell) = \kappa_{wv}(\ell) + \kappa_{CO_2}(\ell) + \kappa_{O_3}(\ell)$$
(3.47)

Note the wavelength dependence of each coefficient in this expression has been suppressed to simplify the notation.

Consider a small absorbing volume in the path at distance ℓ from the surface, and with infinitesimal length $d\ell$. The radiance L passing through this volume will be attenuated by an amount equal to the absorption coefficient times the radiance entering the volume times the length of the volume:

$$dL_\lambda^a(\ell) = -\kappa(\ell)L_\lambda(\ell)d\ell$$
(3.48)

By Kirchoff's law, the emissivity of the atmosphere in the infinitesimal volume element is:

$$\varepsilon(\ell) = \kappa(\ell)d\ell$$
(3.49)

Hence the emitted radiance from the volume element is given by:

$$dL_\lambda^e(\ell) = \kappa(\ell)L_{BB}(T(\ell))d\ell$$
(3.50)

where $T(\ell)$ is the temperature of the gas at the path distance ℓ above the surface.

The net change in radiance from the volume element is given by:

$$dL_\lambda(\ell) = \kappa(\ell)L_{BB}(T(\ell))d\ell$$
$$- \kappa(\ell)L_\lambda(\ell)d\ell$$
(3.51)

Each volume emits blackbody radiation (the first term) and absorbs some of the incoming radiation (the second term).

Since the absorption depends only on the gas composition, pressure, and temperature along the path, the absorption coefficient is independent of the radiance and equation (3.51) can be written:

$$\frac{dL_\lambda(\ell)}{d\ell} + \kappa(\ell)L_\lambda(\ell) = \kappa(\ell)L_{BB}(T(\ell))$$
(3.52)

Equation (3.52) is the differential form of the radiative transfer equation for an absorbing, nonscattering atmosphere. For a path of length h, the solution of this differential equation is given by:

$$L_\lambda(h) = L_\lambda(0)\exp[-\tau(0,h)] + \int_0^h \kappa(\ell)$$
$$\times L_{BB}(T(\ell))\exp[-\tau(\ell,h)]d\ell$$
(3.53)

where, $\tau(\ell_1,\ell_2) = \int_{\ell_1}^{\ell_2} \kappa(\ell')d\ell'$ is the **optical thickness** or **optical depth** of the layer between ℓ_1 and ℓ_2.

The first term on the right side of equation (3.53) gives the attenuation of the surface radiance $L_\lambda(0)$ by the volume of atmosphere from the surface to $\ell = h$. For convenience we define the atmospheric transmittance τ_A from the surface to $\ell = h$: $\tau_A = \exp[-\tau(0,h)]$.

The second term is the radiance emitted from any point ℓ along the path, attenuated by the atmosphere along the remainder of the path, and then summed over the volume of atmosphere from the surface to $\ell = h$:

$$L_A = \int_0^h \kappa(\ell)L_{BB}(T(\ell))\exp[-\tau(\ell,h)]d\ell$$
(3.54)

This term is identical to what we have previously denoted as the path radiance L_A.

Then equation (3.53) can be written simply as:

$$L_\lambda(h) = \tau_A L_\lambda(0) + L_A \qquad (3.55)$$

which states that the radiance observed by the sensor is just the radiance at the surface times the atmospheric attenuation plus the path radiance.

Note this development has used a coordinate system aligned with the look direction. Now let's shift to the more common coordinates with the vertical axis z. Recalling equation (3.45), we see that the response of a sensor viewing the sea surface depends on the direction of the incident angle to the mean surface θ:

$$L_\lambda(h) = L_r(\lambda, \theta, T_r)$$

$$L_\lambda(0) = \varepsilon_w(\lambda, \theta)L_s(\lambda, T_s)$$

$$+ r_w(\lambda, \theta)L_{sky}(\lambda, T_{sky}(\theta)) \quad (3.56)$$

The solution to the radiative transfer equation can then be expressed in a vertical coordinate system by applying the simple coordinate transformation:

$$z = \ell \cos\theta \quad \text{and} \quad d\ell = \frac{dz}{\cos\theta} \quad (3.57)$$

In this new coordinate system, the optical thickness along the path becomes:

$$\tau(z_1, z_2)\sec\theta = \int_{z_1}^{z_2} \kappa(z')dz' / \cos\theta$$

$$(3.58)$$

because the path lengthens by a factor of $\sec\theta$. The solution of the radiative transfer equation can then be written:

$$L_\lambda(z_h, \theta) = L_\lambda(0)\exp[-\tau(0, z_h)\sec\theta]$$

$$+ \sec\theta \int_0^{z_h} \kappa(z)L_{BB}(T(z))$$

$$\times \exp[-\tau(z, z_h)\sec\theta]dz \quad (3.59)$$

It is usual to define the vertical transmittance of the atmosphere as:

$$\tau_A = \exp[-\tau(0, z_h)] \qquad (3.60)$$

We then make the assumption that the atmosphere consists of homogeneous horizontal layers. The solution can then be rewritten as:

$$L_\lambda(z_h, \theta) = \tau_A^{\sec\theta}L_\lambda(0) + L_A(\theta) \qquad (3.61)$$

where the modified definition of the path radiance is:

$$L_A(\theta) = \sec\theta \int_0^{z_h} \kappa(z)L_{BB}(T(z))$$

$$\times \exp[-\tau(z, z_h)\sec\theta]dz \quad (3.62)$$

To calculate the path radiance, we need to know how the absorption coefficient varies with pressure (equivalent to altitude) and temperature for each gas constituent, and the temperature of the atmosphere as a function of altitude in order to evaluate the blackbody radiance factor in the integrand. These quantities are obtained from laboratory measurements of gas properties and from atmospheric models.

An unrealistic, but simple form of the solution to the radiative transfer equation can be obtained by assuming a satellite sensor looking at the surface through a homogeneous atmosphere of constant temperature T_A. In this case, the absorption coefficient κ is a constant, and it is easy to show that equation (3.59) becomes:

$$L_\lambda(z_h) = L_\lambda(0)\, e^{-\tau\sec\theta}$$

$$+ L_{BB}(T_A)(1 - e^{-\tau\sec\theta}) \quad (3.63)$$

where the optical thickness $\tau = \kappa z_h$.

This takes an even simpler form for nadir-looking geometries where $\sec\theta = 1$. The two terms in the equation for a nadir look through a homogenous atmosphere are plotted in Figure 3.18.

- For $\tau = 0$, the atmosphere is completely transparent, and the sensor sees only the surface radiance: $L_\lambda(z_h) = L_\lambda(0)$.
- For $\tau \gg 1$, the atmosphere is opaque, and the sensor sees only path radiance: $L_\lambda(z_h) = L_A = L_{BB}(T_A)$.
- For $\tau = 1$, the atmosphere is partly transparent, and the sensor response is due to 37% of the surface radiance and 63% of the path radiance: $L_\lambda(z_h) = 0.37L_\lambda(0) + 0.63L_{BB}(T_A)$.

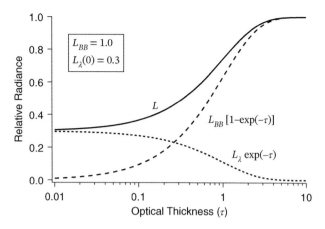

Figure 3.18 Path, surface, and total radiance for a homogeneous atmosphere. Adapted from Stewart (1985).

An expression for the downwelling sky radiance at the sea surface $L_{sky}(\lambda, T_{sky}(\theta))$ (recall equation (3.45)) can be derived following the procedure used to derive equation (3.59):

$$L_{sky}(\lambda, T_{sky}(\theta)) = \sec \theta \int_{z_h}^{0} \kappa(z) L_{BB}(T(z))$$
$$\times \exp[-\tau(z,0) \sec \theta] dz \tag{3.64}$$

Note the change of variables in the integral to reflect the fact that here we are considering radiation starting at the top of the atmosphere and emerging at the surface. For a homogeneous atmosphere, the downwelling sky radiance at the surface is identical to the path radiance, but in general, the two quantities differ because the vertical distribution of the atmospheric gases is not uniform with height.

It is instructive to evaluate equation (3.64) in the limits where the atmosphere is highly absorbing ($\tau \sec \theta \gg 1$) and where the atmosphere is nearly transparent ($\tau \sec \theta \ll 1$). In the first case, where the atmosphere is highly absorbing, the exponential in the integrand falls off quickly away from the ground, essentially making $\kappa(z) L_{BB}(T(z))$ constant over the range of the integral. The term κL_{BB} can thus be moved outside the integral as in equation (3.65a). Next substitute the original definition of the optical thickness along the path from equation (3.58) to obtain equation (3.65b). Then again assume $\kappa(z)$ is large and effectively evaluated only near the ground (equation

(3.65c)). The integral for the optical thickness then reduces to $\kappa(0) \sec \theta z$ as in equation (3.65d). Since the atmosphere is opaque, z_h is essentially ∞, and the integral of the exponential can be evaluated explicitly (equation (3.65e)). The expression then simplifies to the blackbody radiance of the atmosphere at the surface (equation (3.65f)). Thus, infrared sky radiance measurements made within absorption bands are independent of look angle.

$$\lim_{\tau \to \infty} L_{sky}(\lambda, T_{sky}(\theta)) \approx \kappa(0) \sec \theta L_{BB}(T(0))$$
$$\times \int_{z_h}^{0} \exp[-\tau(z,0) \sec \theta] dz \tag{3.65a}$$

$$\approx \kappa(0) \sec \theta L_{BB}(T(0)) \int_{z_h}^{0}$$
$$\times \exp\left[-\int_{0}^{z} \kappa(z') \sec \theta dz'\right] dz \tag{3.65b}$$

$$\approx \kappa(0) \sec \theta L_{BB}(T(0)) \int_{z_h}^{0}$$
$$\times \exp\left[-\kappa(0) \sec \theta \int_{0}^{z} dz'\right] dz \tag{3.65c}$$

$$\approx \kappa(0) \sec \theta L_{BB}(T(0)) \int_{z_h}^{0}$$
$$\times \exp[-\kappa(0) \sec \theta z] dz \tag{3.65d}$$

$$\approx \kappa(0) \sec \theta L_{BB}(T(0)) \frac{1}{\kappa(0) \sec \theta} \tag{3.65e}$$

$$\approx L_{BB}(T(0)) \tag{3.65f}$$

Evaluation of equation (3.64) in the limits where the atmosphere is nearly transparent ($\tau \ll 1$) is easier. In this case the exponential

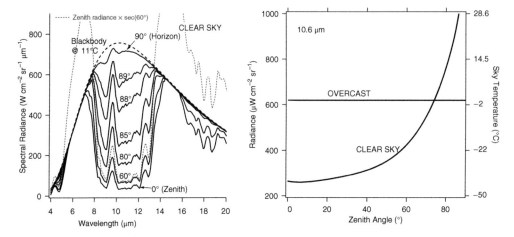

Figure 3.19 Nighttime downwelling sky radiance in the infrared. Left-hand panel similar to Bell et al. (1960) but based on MODTRAN calculations. Right-hand from measurements acquired by one of the authors (rfg).

within the integrand approaches one resulting in equation (3.66a). Now the value of the integral will depend on the vertical distribution of the absorption coefficient and temperature within the atmosphere, but represents the sky radiance for a sensor looking straight up, which we can designate $L_{sky}(0°)$. Thus the angular dependence of the sky radiance in an infrared atmospheric transmission window should be proportional to the sky radiance at $0°$ zenith angle times $\sec\theta$ (equation (3.66b)). This is an approximation only valid in the limit of a transparent path, which can be valid for a wide range of angles about the zenith, but typically begins to fail as the atmospheric path length increases towards the horizon.

$$\lim_{\tau\to 0} L_{sky}(\lambda, T_{sky}(\theta))$$

$$\approx \sec\theta \int_{z_h}^{0} \kappa(z) L_{BB}(T(z)) \, dz \quad (3.66a)$$

$$\approx L_{sky}(0°)\sec\theta \quad\quad\quad (3.66b)$$

Figure 3.19 shows examples of nighttime downwelling sky radiance in the infrared.

The black curves in the left-hand plot show the spectral radiance as a function of wavelength for six different zenith angles. The atmosphere looks like a blackbody when looking right at the horizon ($90°$) or at wavelengths where the atmosphere is highly absorptive, but

deviates from a blackbody when looking up due to transmission windows. The clear sky is coldest when looking straight up ($0°$). The red dashed curve plots the simplest model for the radiance at $60°$ substituting the modeled zenith radiance into equation (3.66b). The agreement between the simple model, and the MODTRAN modeled radiance at $60°$ is quite good in the clearest portions of the atmospheric window but increasingly overpredicts radiance at the wavelengths where the atmosphere is less transparent. At $80°$ and higher zenith angles, the simple model (not shown) disagrees everywhere with the MODTRAN calculations and so is of no value.

The right-hand plot shows the difference in the radiance as a function of zenith angle for a clear and uniformly overcast sky at 10 μm. In general the clouds act like Lambertian surfaces with the temperature of the cloud base, while a clear sky is very cold looking straight up.

3.7 Propagation in Seawater

This section discusses the propagation of electromagnetic radiation in a dielectric medium, such as seawater. This will only be a brief review of material that is readily available in any electromagnetics textbook. Let us begin by defining terms using standard notation and in SI units:

$\vec{E}$ Electric field intensity, V m^{-1}

$\vec{D}$ Electric displacement, C m^{-2}

$\vec{B}$ Magnetic flux density, Weber m^{-2}

$\vec{H}$ Magnetic field intensity, A m^{-1}

$\vec{J}$ Current density, A m^{-2}

$\vec{P}$ Polarization density, C m^{-2}

ρ_t True (free) charge density, C m^{-3}

ϵ_0 Permittivity of free space, F m^{-1}

μ_0 Susceptability of free space

ρ_p Polarization charge density, C m^{-3}

σ Conductivity, $\Omega\cdot$m

χ Electric susceptability

ϵ Permittivity, F m^{-1}

The propagation of electromagnetic energy in seawater is described by Maxwell's equations that relate the vector electric field intensity $\vec{E}$, magnetic flux density $\vec{B}$, electric displacement $\vec{D}$ and magnetic field intensity $\vec{H}$ to the charge density ρ and current density $\vec{J}$:

$$\nabla \cdot \vec{D} = \rho_t \qquad (3.67)$$

$$\nabla \cdot \vec{B} = 0 \qquad (3.68)$$

$$\nabla \times \vec{E} + \frac{\partial \vec{B}}{\partial t} = 0 \qquad (3.69)$$

$$\nabla \times \vec{H} = \vec{J} + \frac{\partial \vec{D}}{\partial t} \qquad (3.70)$$

The electric field intensity can be related to the displacement vector where $\vec{P}$ is the polarization field and ϵ_0 is the permittivity of free space:

$$\vec{D} = \epsilon_0 \vec{E} + \vec{P} \qquad (3.71)$$

The polarization charge density ρ_p can be defined in terms of the gradient of the polarization field:

$$\nabla \cdot \vec{P} = -\rho_p \qquad (3.72)$$

Then equation (3.67) can be rewritten in terms of the total charge density, ρ_{total}:

$$\nabla \cdot \vec{E} = \frac{\rho_t + \rho_p}{\epsilon_0} = \frac{\rho_{total}}{\epsilon_0} \qquad (3.73)$$

Seawater is nonmagnetic, so the magnetic flux density $\vec{B}$ is related to the magnetic field intensity $\vec{H}$:

$$\vec{B} = \mu_0 \vec{H} \qquad (3.74)$$

where μ_0 is the susceptibility of free space.

Seawater is conductive so Ohm's law says the current density $\vec{J}$ equals the electric field intensity $\vec{E}$ times the conductivity of seawater:

$$\vec{J} = \sigma \vec{E} \qquad (3.75)$$

Polarization is also driven by the electric field based on the permittivity ϵ_0 and electric susceptibility χ:

$$\vec{P} = \epsilon_0 \chi \vec{E} \qquad (3.76)$$

Then

$$\vec{D} = \epsilon_0 \vec{E} + \vec{P} = \epsilon_0 (1 + \chi) \vec{E} = \epsilon \vec{E} \qquad (3.77)$$

provides a definition of the permittivity of seawater, ϵ.

The electromagnetic wave equation for a conductive, nonmagnetic, nonmoving medium is now derived by taking the curl of equation (3.69) and using equations (3.70), (3.74), (3.75), and (3.77) to get:

$$\nabla \times (\nabla \times \vec{E}) = -\mu_0 \sigma \frac{\partial \vec{E}}{\partial t} - \mu_0 \epsilon \frac{\partial^2 \vec{E}}{\partial t^2} \qquad (3.78)$$

Next apply the vector identity

$$\nabla \times (\nabla \times \vec{E}) = \nabla(\nabla \cdot \vec{E}) - \nabla^2 \vec{E} \qquad (3.79)$$

and the fact that for a charge-free medium the divergence of the electric field is zero:

$$\nabla \cdot \vec{E} = 0 \qquad (3.80)$$

The result is the standard wave equation:

$$\nabla^2 \vec{E} - \mu_0 \epsilon \frac{\partial^2 \vec{E}}{\partial t^2} - \mu_0 \sigma \frac{\partial \vec{E}}{\partial t} = 0 \qquad (3.81)$$

To solve the wave equation, consider an electromagnetic plane wave propagating in the z-direction in air and incident on an air–water interface at $z = 0$. Note the positive z-axis in this coordinate system is into the water. As usual, we look for solutions in the form of planes waves:

$$\vec{E} = \vec{E}_0 e^{i(kz - \omega t)} \qquad (3.82)$$

Substituting equation (3.82) into the wave equation (3.81) provides the dispersion relation that the wavenumber k must satisfy:

$$k^2 = \mu_0 \epsilon_0 \omega^2 \left[\frac{\epsilon}{\epsilon_0} + \frac{i\sigma}{\epsilon_0 \omega} \right] = \mu_0 \epsilon_0 \omega^2 \tilde{\epsilon} \tag{3.83}$$

where $\tilde{\epsilon}$ is the complex dielectric constant defined by:

$$\tilde{\epsilon} = \epsilon_r + i\epsilon_i = \left(\frac{\epsilon}{\epsilon_0} + \frac{i\sigma}{\epsilon_0 \omega} \right) \tag{3.84}$$

The phase velocity is then:

$$\tilde{v}_p = \frac{\omega}{k} = \frac{1}{\sqrt{\mu_0 \epsilon_0 \tilde{\epsilon}}} \tag{3.85}$$

This expression is general, making it suitable for use in seawater.

In free-space, the dielectric constant becomes 1, so the velocity of an electromagnetic wave in a vacuum is:

$$c = \frac{\omega}{k} = \frac{1}{\sqrt{\mu_0 \epsilon_0}} \tag{3.86}$$

We can now define the complex index of refraction as the ratio of the phase velocity in water and in a vacuum, which according to the previous two equations has the value of the square root of the complex dielectric constant:

$$\tilde{n} = \frac{c}{v_p} = \sqrt{\tilde{\epsilon}} = n + ik_0 \tag{3.87}$$

where the real part n is the index of refraction, and the imaginary part k_0 is the extinction coefficient. Notice that:

$$\tilde{n}^2 = (n + ik_0)^2 = (n^2 - k_0^2) + 2ink_0 = \tilde{\epsilon} \tag{3.88}$$

Equation (3.88) can then be used to derive a relationship for the real and imaginary parts of the complex dielectric constant in terms of the optical constants, n and k_0:

$$\epsilon_r = n^2 - k_0^2$$
$$\epsilon_i = 2nk_0 \tag{3.89}$$

From equation (3.85) we can see that the wavenumber is also complex:

$$k = \frac{\omega}{v_p} = \frac{\omega}{c}\tilde{n} = \frac{\omega n}{c} + i\frac{\omega k_0}{c} \tag{3.90}$$

The plane wave solution then becomes:

$$\vec{E} = \vec{E}_0 \, e^{i\left(\frac{\omega n z}{c} - \omega t \right)} e^{-\frac{k_0 \omega z}{c}} \tag{3.91}$$

Notice that because the wavenumber is complex there are two exponential factors in the plane wave solution. The first is an oscillating factor and the second is a damping factor that exponentially attenuates the electric field amplitude as the wave propagates into the water column. The scale for this attenuation is given by the extinction coefficient. This damping factor describes the absorption of the wave in the water column.

We can then define the electric field skin depth:

$$d_{skin} = \frac{c}{\omega k_0} = \frac{\lambda}{2\pi k_0} \tag{3.92}$$

Now we can rewrite the plane wave solution in terms of the skin depth:

$$\vec{E} = \vec{E}_0 \, e^{i\left(\frac{\omega n z}{c} - \omega t \right)} e^{-\frac{z}{d_{skin}}} \tag{3.93}$$

Radiation flux (power) in the water column is generally expressed in terms of an absorption coefficient α:

$$\Phi(z) = \Phi_0 e^{-\alpha z} \tag{3.94}$$

where α is the Lambert absorption coefficient:

$$\alpha = \frac{2\omega k_0}{c} = \frac{4\pi k_0}{\lambda} \tag{3.95}$$

The flux absorption depth is then given by:

$$\alpha^{-1} = \frac{\lambda}{4\pi k_0} = \frac{d_{skin}}{2} \tag{3.96}$$

These latter equations are nothing more than a result of the flux being proportional to the square of the electric field amplitude. Thus the flux absorption depth is one-half of the electric field skin depth.

The index of refraction n and extinction coefficient k_0 are optical constants that govern the propagation of electromagnetic energy through seawater. The wavelength dependence of these constants for seawater has been measured and tabulated. These can then be used to compute wavelength, phase velocity, attenuation coefficient, or skin depth in seawater.

Table 3.1 Typical values of optical constants for seawater.

Band	Wavelength	n	k_0	α^{-1}
Mid-Wave IR	3.8 μm	1.363	0.00339	89 μm
Long-Wave IR	10.5 μm	1.174	0.0764	11 μm
Visible	500 nm	1.33	10^{-9}	40 m
Microwave	3.0 cm	7.3	2.4	1.0 mm

Table 3.1 shows the index of refraction n, the extinction coefficient k_0, and the flux absorption depth $1/\alpha$ for four wavelengths. At mid-wave infrared wavelengths, the flux absorption depth is about 0.1 mm. The flux absorption depth is even shorter for long-wave IR. This means that seawater absorbs any incident infrared energy in a very thin layer at the air–water interface. This has implications for the thermal boundary layer at the air–water interface that we will discuss. It is also the reason that you should never wear thermal-vision goggles while trying to clean your fish tank.

In contrast, the flux absorption depth at visible wavelengths can be as much as 40 m. Visible light penetrates into the upper portions of the ocean, which is critically important for

all life on earth. Finally, the absorption depth is quite small again at microwave frequencies.

Figure 3.20 plots the flux absorption depth for seawater as a function of wavelengths from the ultraviolet to the infrared. In general, the absorption depth is strongly peaked in the visible.

Figure 3.21 plots the flux absorption depth for seawater as a function of wavelengths from the ultraviolet down to VLF frequencies. At microwave and longer wavelengths, the absorption depth monotonically increases with increasing wavelength, a trend that continues down to VLF wavelengths of 10's of km. So the near-surface layers of the ocean are somewhat transparent only in the visible band and at the very longest radio wavelengths.

Figure 3.21 was created by merging two models and one data source. For wavelengths greater than 1 mm, the dielectric model of Meissner and Wentz (2004) was used. For wavelengths between 100 μm and 0.4 μm the empirical model of Mesenbrink (1996) was used. A linear blend of these two models was used at wavelengths between 100 μm and 1 mm. The measured data from Smith and Baker (1981) were then used at wavelengths below 0.4 μm.

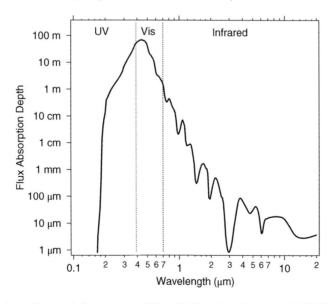

Figure 3.20 Flux absorption depth for seawater (UV to IR). Data from Mesenbrink (1996), Segelstein (1981), and Smith and Baker (1981).

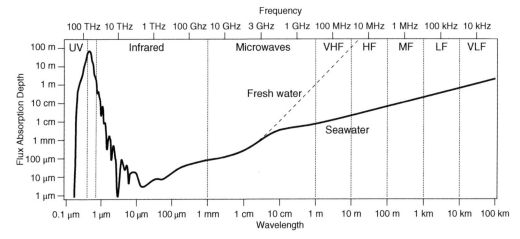

Figure 3.21 Flux absorption depth for seawater (IR to VLF).

3.8 Smooth Surface Reflectance

Consider Figure 3.22 illustrating electromagnetic radiation incident on a dielectric slab of finite thickness. Some of the radiation will be reflected back into the upper hemisphere; some will be absorbed into the dielectric; and some will be transmitted into the lower hemisphere.

We define the terms reflectance, absorptance, and transmittance in terms of the ratios of the outgoing fluxes to the incident flux:

$$\text{Reflectance:} \quad \rho \equiv \frac{\Phi_{reflected}}{\Phi_{incident}} \quad (3.97)$$

$$\text{Absorptance:} \quad \alpha \equiv \frac{\Phi_{absorbed}}{\Phi_{incident}} \quad (3.98)$$

$$\text{Transmittance:} \quad \tau \equiv \frac{\Phi_{transmitted}}{\Phi_{incident}} \quad (3.99)$$

Note these definitions are in terms of flux or power ratios. An amplitude reflectance coefficient can also be defined as a ratio of the reflected versus incoming electric-field strength, with a magnitude given by the square root of ρ.

The conservation of energy requires that the sum of the reflectance, absorptance, and transmittance must equal one: $\rho + \alpha + \tau = 1$. Kirchoff's Law states that matter in local

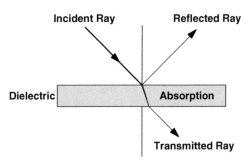

Figure 3.22 Reflection and transmission through a dielectric interface.

equilibrium with its surroundings emits energy at the same rate as it absorbs energy. This requires the emissivity to be equal to the absorptance: $\varepsilon = \alpha$.

If the dielectric slab is the ocean, where the depth is much larger than the absorption depth, then none of the incident flux is transmitted through the ocean: $\tau = 0$. It then follows that the sum of the reflectance and emissivity must equal one: $\rho + \varepsilon = 1$. This is an important result that will be used over and over again.

A word about the terms *reflectance* and *reflectivity*. Reflectance is defined as the ratio of the reflected radiant flux to the incident flux. In contrast, the International Commission on Illumination (CIE) defines reflectivity as the reflectance of a layer of the material that is of sufficient thickness that there is no change

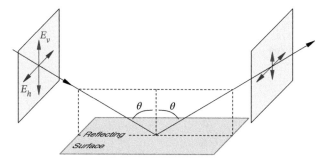

Figure 3.23 Geometry of polarized reflection.

of reflectance with increase in thickness. So the ocean is thick enough that the reflectance and reflectivity should be identical.

Reflection of an electromagnetic wave from a plane boundary depends on the polarization state of the incident radiation and the properties of the reflecting surface. Figure 3.23 illustrates the geometry of polarized reflection. If the electric field of the incident wave is in the plane of incidence, the wave is said to have parallel or vertical polarization. When the electric field of the incident wave is normal to the plane of incidence, the wave is said to have perpendicular or horizontal polarization.

The specular reflection of infrared radiation at a flat plane air–water interface can be treated with the Fresnel formulas for reflection at an interface between a dielectric and a conductive medium. The dielectric (air) is characterized by a real index of refraction (assumed to be unity); the conductive medium (water) is characterized by a complex index of refraction:

$$\tilde{n}(\lambda) = n(\lambda) + ik_0(\lambda) \qquad (3.100)$$

with the index of refraction n and the extinction coefficient k.

The spectral reflectance r_h for the horizontal polarization component of radiation incident on a plane water surface at an angle θ with respect to the surface normal is:

$$r_h(\lambda, \theta) = \frac{(a - \cos \theta)^2 + b^2}{(a + \cos \theta)^2 + b^2} \qquad (3.101)$$

And likewise for vertical polarization:

$$r_v(\lambda, \theta) = r_h(\lambda, \theta) \frac{(a - \sin \theta \tan \theta)^2 + b^2}{(a + \sin \theta \tan \theta)^2 + b^2} \qquad (3.102)$$

where the constants a and b are given by:

$$2a^2 = (n^2 - k_0^2 - \sin^2 \theta) \\ + [(n^2 - k_0^2 - \sin^2 \theta)^2 + 4n^2 k_0^2]^{1/2} \qquad (3.103)$$

$$2b^2 = -(n^2 - k_0^2 - \sin^2 \theta) \\ + [(n^2 - k_0^2 - \sin^2 \theta)^2 + 4n^2 k_0^2]^{1/2} \qquad (3.104)$$

Note that at 0° incident angle, the constant a corresponds to the index of refraction and b corresponds to the extinction coefficient.

In general, thermal radiation is unpolarized, that is, the electric field has a random orientation relative to the plane of incidence, before reflection from a plane surface. The reflection coefficient for unpolarized radiation is simply the average of the vertical and horizontal reflection coefficients:

$$r_u(\lambda, \theta) = [r_v(\lambda, \theta) + r_h(\lambda, \theta)]/2 \quad (3.105)$$

Figure 3.24 shows the reflection coefficients for a wavelength of 10.5 μm. At this wavelength, $n = 1.174$ and $k_0 = 0.0764$. The plot on the left-hand side of the figure is on a linear scale, while the right-hand plot is on a logarithmic scale. The logarithmic plot makes it clear that the reflectance for vertical polarization approaches zero[3] at an incident angle of 49.6°. This is called the Brewster angle. A wave reflected at Brewster's angle will be nearly horizontally polarized if $k_0 \ll n$, but

---

3 The reflectance at 49.6° with $n = 1.174$ and $k_0 = 0.0764$ is actually 9.3×10^{-5}.

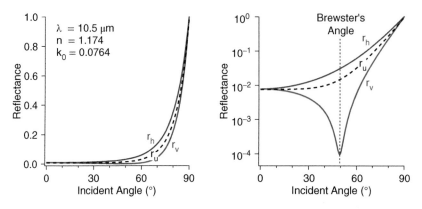

Figure 3.24 Fresnel reflection coefficients for horizontal and vertical polarizations at 10.5 μm.

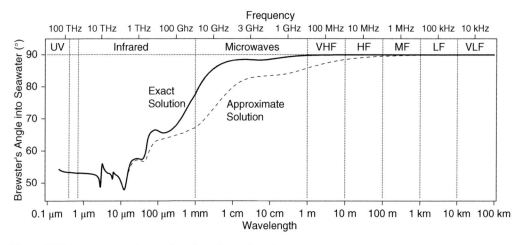

Figure 3.25 Brewster's angle as a function of wavelength.

a mixture of both polarizations if k_0 is too large to be ignored.

Also note that all of the reflectances at 0° incident angles are about 1%. Polarization does not matter when looking straight down.

As just illustrated, Brewster's angle θ_B is the angle where the vertical reflectance goes to zero. There is a simple formula for estimating Brewster's angle valid in the case when $k_0 \ll n$. Then the formula for a and b simplify:

$$a^2 = n^2 - \sin^2\theta \quad \text{and} \quad b^2 = 0 \quad (3.106)$$

Noting that $r_v(\lambda) = 0$ when $a = \sin\theta_B \tan\theta_B$, it can be shown that $\tan\theta_B = n$.

In the more general case, when k_0 cannot be ignored, Brewster's angle is determined from the minima of equation (3.102). Figure 3.25 plots Brewster's angle at the air–sea interface as a function of wavelength from VLF to UV. The solid curve is obtained from the minima of equation (3.102) and the dashed curve is derived from $\tan\theta_B = n$. Note the approximate solution works throughout the UV, visible, and useful portions of the infrared.

3.9 Rough Surface Reflectance

We now consider the effect of surface waves on the radiance reflected or emitted from the sea surface. To do this, we treat the wavy surface as a collection of small specular facets

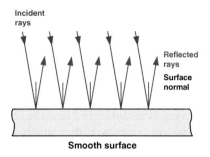 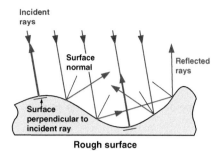

Smooth surface   **Rough surface**

Figure 3.26 Geometry of reflection from a rough surface.

tangent to the local surface elevation. In general, each facet has a tilt, or slope, in two horizontal directions relative to the mean sea surface.

The mean radiance reflected or emitted from the sea surface within the sensor's field of view can now be computed by summing the radiance from each facet, weighted by the probability that the slope of the facet is suitably oriented to direct the radiance toward the sensor (Figure 3.26).

But before showing this formula we need to know the probability distribution of slopes on the sea surface.

For convenience, a coordinate system is chosen with one axis oriented opposite to the direction the wind is blowing, and with the other axis orthogonal to the wind. Each facet then has a slope in the upwind and crosswind direction.

Cox and Munk (1954) used measurements of the sun's glitter pattern (Figure 3.27) to determine that the sea surface slopes are approximately Gaussian distributed with a slope-distribution function p given by:

$$p(\zeta_u, \zeta_c) = \frac{1}{2\pi\sigma^2} \exp\left[-\frac{1}{2}\left(\frac{\zeta_u^2}{\sigma_u^2} + \frac{\zeta_c^2}{\sigma_c^2}\right)\right]$$

(3.107)

where ζ_u and ζ_c are the upwind and crosswind slopes, and σ^2, σ_u^2, σ_c^2, are the total, upwind, and cross wind mean-square slopes, respectively. Thus, the probability P that a wave facet will occur whose slope is within $\pm d\zeta_u/2$ of ζ_u and $\pm d\zeta_c/2$ of ζ_c is: $P = p(\zeta_u, \zeta_c)\, d\zeta_u\, d\zeta_c$

Cox and Munk (1954) also showed that the mean-square upwind and crosswind slopes were linear functions of the wind speed, W,

Figure 3.27 Sun glitter patterns as seen from space. Left-hand panel from International Space Station at 400 km altitude courtesy of ESA/NASA. Right-hand panel from Himawari-8 geostationary satellite on 5 Sep 2018 courtesy JAXA.

with the parameters:

$$\sigma_u^2 = 0.00316 \ W(m/s) \qquad (3.108)$$

$$\sigma_c^2 = 0.003 + 0.00192 \ W(m/s) \qquad (3.109)$$

$$\sigma^2 = \sigma_u^2 + \sigma_c^2 \qquad (3.110)$$

Thus the slope-distribution function is wind dependent[4].

Recall our previous equation for the radiance of a plane sea surface when the atmospheric path radiance can be ignored:

$$L_r(T_r, \lambda, \theta) = L_s(\lambda, T_s)$$
$$- r_w(\lambda, \theta) L_s(\lambda, T_s)$$
$$+ r_w(\lambda, \theta) L_{sky}(T_{sky}(\lambda, \theta))$$
$$(3.111)$$

where θ is the incident angle of the sensor's line of sight relative to the mean sea surface.

This equation can be generalized to a rough surface by applying it to each wave facet where ω is the local incident angle, θ_i is the incident angle of the sensor's line of sight relative to the mean sea surface, and θ_{sky} is the zenith angle of the specular reflection off the surface from sensor into the sky. These three angles are related by Snell's law. The mean radiance from a surface with waves is then given by:

$$L_r(T_r, \lambda, \theta_i) = \iint$$
$$[L_{BB}(\lambda, T_s) - r_w(\lambda, \omega)(L_{BB}(\lambda, T_s)$$
$$- L_{sky}(T_{sky}(\lambda, \theta_{sky})))]$$
$$\times q(\theta_i, \omega) p(\zeta_u, \zeta_c) d\zeta_u d\zeta_c \qquad (3.112)$$

Notice the main term in the integrand is just the radiance of a facet. It is then multiplied by the slope probability density function and integrated over all slopes and incident angles ω satisfying the law of reflection.

Note an additional term $q(\theta_i, \omega)$ in the integrand. This is a geometrical factor that accounts for the horizontal area occupied by all facets with a given slope. See the journal

---

4 Cox and Munk made their measurements using photographs of ocean glitter in 1954! Their approach was to measure the radiance of the glitter pattern, and then invert the mean radiance equation we are going to describe next. This was a clever way of measuring the roughness of the ocean surface.

papers by Cox and Munk (1954, 1956) or Saunders (1968) for more details and one form of the factor q. Zeisse (1995) points out an error in the derivation of q in Cox and Munk (1954), presenting the correct form that should be used in modern-day calculations, especially for calculations performed near the horizon.

Saunders evaluated the equation above for different wind speeds and compared the results with measurements obtained with a pier-mounted radiometer. Figure 3.28 shows representative results from these calculations. Improved calculations of the infrared radiance near the horizon are provided in Zeisse et al. (1999).

Several conclusions are apparent:

- At incident angles less than 50°, the radiance of the sea surface is essentially the same as that of a flat water surface.
- At larger incident angles, the radiance increases with increasing surface roughness, or equivalently, wind speed.
- Near the horizon, the sea surface appears colder than the sky just above the horizon, but it appears warmer as the roughness increases.
- Near the nadir, the radiometric temperature is less than the true sea surface temperature. This is consistent with our previous calculations.

In our initial discussion of the radiance measured by an infrared sensor viewing the sea surface, we explicitly ignored any contribution from reflected solar radiation. The justification for this approach is found in Figure 3.29.

In the atmospheric window region from 10 to 12 μm, the reflected sunlight contribution is negligible at nadir angles of 60° or less where satellite sensors operate. Measurable contributions from reflected sunlight are found at these wavelengths only near the horizon.

In contrast, the sunlight contribution is significant in the 3.8–4.0 μm window. This generally precludes the use of mid-wave measurements for accurate estimates of sea surface temperature during the daytime.

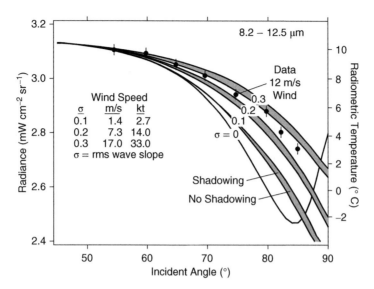

Figure 3.28 Calculation of radiance and radiometric temperature of a wind-roughened sea with an actual sea surface temperature of 13 °C. Figure adapted from data in Saunders (1968) with corrected temperature scale.

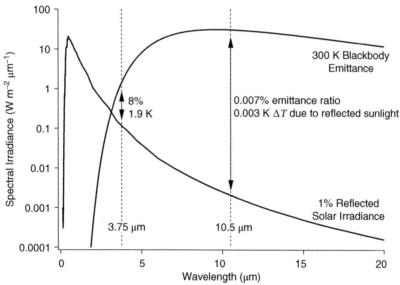

Figure 3.29 Comparison of reflected solar irradiance versus the Earth's blackbody emittance. The reflected solar irradiance curve is based on the top-of-the-atmosphere radiance reflected from a 1% Lambertian surface. The contributions of reflected solar radiance is large in the visible, but small at long-wave infrared wavelengths. Source: Adapted from Stewart (1985).

3.10 Ocean Thermal Boundary Layer

Numerous physical processes at or near the sea surface give rise to an ocean thermal boundary layer. Sunlight is absorbed, heating the upper 5–20 meters of the water column. Wind imparts momentum into the water creating turbulent mixing that can extend to depths of 10–100 meters. There is also a net heat exchange between the surface and the atmosphere. In most situations, there is a net heat loss from the surface due to long-wave IR radiation, evaporation, and conduction, that creates a thin cool skin of less than 1 mm thickness on top of the sea surface.

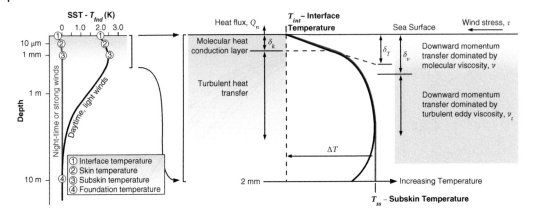

Figure 3.30 Schematic of the sea surface thermal boundary layer. Source: Adapted from Merchant et al. (2019) and Robinson et al. (1984).

Figure 3.30 illustrates some of the complexities of the near-surface boundary layer. The left-hand panel sketches temperature vs depth on a highly nonlinear scale. The shape of the near-surface temperature profile is the result of competing processes. The black curve illustrates how daytime heating can produce a warm layer of water near the surface under conditions of light winds. The red curve illustrates that either convective mixing driven by cooling at the surface, or turbulent mixing driven by momentum imparted into the water by strong winds, can effectively mix the upper layers of the ocean to depths of tens of meters.

Nearly independent of whether the upper tens of meters are thermally stratified or unstratified, a cold skin layer produced mostly by outgoing IR radiation and evaporative cooling is usually present within 1–2 mm of the surface. Taken all together, these processes can produce a complex temperature profile. To be precise, the Group for High Resolution Sea Surface Temperature (www.ghrsst.org) is promoting the use of the following terms:

interface temperature – temperature at the exact air–sea interface (unmeasurable),

skin temperature – temperature measured by an infrared radiometer typically at depths of 10–20 µm,

subskin temperature – temperature at the base of the conductive laminar sublayer typically at depths of 1–2 mm and measurable by passive microwave radiometry,

surface temperature at depth – all measurements of water temperature beneath the subskin depth, and

foundation temperature – shallowest temperature free of diurnal variability.

Four of these terms are labeled in the left-hand panel of Figure 3.30.

Some of the dynamics occurring in this very near surface region are illustrated in the right-hand panel of Figure 3.30. In this plot the stratified and unstratified profiles are plotted with a common interface temperature T_{int}, which is cooler than the subskin temperature T_{ss}. In the absence of any heat exchange across the air–water interface, wind stress on the surface imparts momentum into the water, some of which cascades into near-surface turbulence that effectively mixes water in a layer near the surface with typical extents of 0.1–50 m. It is this mixing that causes the bulk temperature to be constant with depth just below the thermal boundary layer.

Next consider the effect of heat flux through the air–water interface. While the heat flux can be substantial in coastal areas or near major oceanic fronts, such as the Gulf Stream, away from such feature the difference between the air and sea surface temperature is usually no more than a few degrees. Under these conditions, evaporative cooling often dominates over radiative or conductive fluxes. Thus the surface is colder than the bulk temperature, giving up heat to the atmosphere. It is this thermal forcing that creates a conductive laminar sublayer.

The thickness and temperature gradient across the conductive laminar sublayer is estimated by recognizing heat transfer on the water side of the air–water interface is via molecular conduction. From heat transfer analysis, it can be shown that the thermal layer thickness is given by (Saunders, 1967):

$$\delta_T \cong \frac{k\nu}{u_{*w}} \qquad (3.113)$$

where k is a constant of order 7, ν is the kinematic viscosity of water $= 0.01 \, cm^2 \, s^{-1}$ and u_{*w} is the friction velocity in the water. Assuming the frictional stress of the wind is continuous across the interface, it follows that $u_{*w} = u_{*a} \sqrt{\rho_a/\rho_w}$ where u_{*a} is the friction velocity in air, and ρ_a and ρ_w are the air and water densities, respectively.

The friction velocity is a parameter used in boundary layer theory to estimate momentum transfer, in this case from the wind to the sea. The friction velocity in the air can be related to the near surface wind speed. For wind speeds less than about 8 m/s, the friction velocity in air is approximately: $u_{*a} = 0.035 U_{10N}$, where U_{10N} is the wind speed at 10-m height under conditions of neutral atmospheric stability (Edson et al., 2013). The friction velocity in water is about 3% of the friction velocity in air. Thus we can obtain an estimate of thermal layer thickness based on wind speed:

$$\delta_T(cm) = \frac{0.67}{U_{10N} \, (m \, s^{-1})} \qquad (3.114)$$

The thermal conduction equation relates the thermal layer thickness to the net heat flux from the ocean:

$$Q_{net} = k_T \left(\frac{dT}{dz}\right)_{cool \, skin} = k_T \frac{T_{ss} - T_{int}}{\delta_T} \qquad (3.115)$$

It says that the net heat flux from the surface Q equals the temperature gradient (dT/dz) times the thermal conductivity k_T. Note that the thermal conductivity of water is $0.585 \, W \cdot m^2/°C$.

The typical net heat flux from the ocean is a few hundred W m^{-2}. The typical temperature gradient across the cool skin is about $-2 \, °C/cm$ to $-5 \, °C/cm$. Therefore, the typical thickness of the cool skin layer is about 1 mm, and the temperature drop across the cool skin layer is about 0.2–0.5 °C.

Note that cool skin thickness is 10 to 100 times larger than the IR flux skin depth. Hence, IR radiation leaving the sea surface comes from the upper part of the cool skin.

We previously discussed the radiative transfer equation as a way to compute the atmospheric path radiance. Applying this same theory to the thermal boundary layer, it can be shown that the temperature T_s in the emitted radiance term $L_{BB}(\lambda, T_s)$ is identical to the temperature in the cool skin at the depth equal to the flux skin depth, $\alpha(\lambda)^{-1}$. Thus;

$$T_s = T_{int} + \frac{T_{ss} - T_{int}}{\delta_T} \alpha(\lambda)^{-1} \qquad (3.116)$$

which means the "surface temperature" that we have been attempting to measure with an IR sensor is not the interface temperature T_{int}, nor is it the subskin temperature T_{ss}, but somewhere in between. Furthermore the surface temperature (or more precisely skin temperature) depends on wavelength.

The magnitude of this dependence is easy to estimate. Using a cool skin temperature gradient of 3.5 °C cm^{-1}, and recalling that $\alpha(3.8 \, \mu m)^{-1} = 89 \, \mu m$, and $\alpha(10.5 \, \mu m)^{-1} = 11 \, \mu m$, we have that the surface temperature is 31 millidegrees warmer than the interface temperature at 3.8 μm, and 4 millidegrees warmer at 10.5 μm.

The interested reader can find an expanded treatment of the ocean thermal boundary layer and its significance for sea surface temperature measurements in Emery et al. (2001), Robinson et al. (1984), and Robinson (1985).

3.11 Operational SST Measurements

This section describes the satellite sensors providing routine operational SST observations, along with some example algorithms used to estimate sea surface temperature from satellite infrared data.

Beginning in 1978, the National Oceanic and Atmospheric Administration (NOAA) has operated the Polar-orbiting Environmental Satellites (POES) equipped with an instrument for measuring daily global SST data, along with processing and distribution facilities to provide these data to the user community. This instrument, the Advanced Very High Resolution Radiometer (AVHRR), is flown on a pair of satellites, with ascending nodes in the morning and afternoon, respectively. Each satellite provides entire Earth coverage twice per day. Both spacecraft are in circular, sun-synchronous orbits at a nominal altitude of 835 km.

At the present time, the NOAA-19 satellite is the primary afternoon platform. The primary morning platform is the MetOp-B satellite, equipped with an AVHRR instrument provided by NOAA. The MetOp satellites are polar-orbiting meteorological satellites operated by EUMETSAT, the European operational satellite agency. In 1998, NOAA and EUMETSAT agreed to split the morning and afternoon services, along with exchanging data and instruments. NOAA-19, MetOp-B, and MetOp-C are the last spacecraft to carry AVHRR.

The next generation U.S. polar-orbiting operational environmental satellite system is the Joint Polar Satellite System (JPSS). The first JPSS spacecraft, designated NOAA-20, was launched in November 2017. Sea surface temperature measurements are acquired with the Visible Infrared Imaging Radiometer Suite (VIIRS), a 22-band instrument covering wavelengths from 0.41 to 12.5 μm. In addition to SST data, VIIRS provides data for ocean color, clouds, vegetation coverage, snow and ice, and fire.

3.11.1 AVHRR Instrument

The Advanced Very High Resolution Radiometer (AVHRR) makes measurements in six spectral bands with a ground resolution of 1.1 km at nadir (Figure 3.31). The AVHRR scans a single pixel with a field of view of 0.0745° across the surface. The instrument scans in a 360° circle about the axis aligned with the flight direction at a rate of six revolutions per second. Radiometric data from the Earth are acquired over a range of angles ±55° about nadir. Figure 3.32 shows the type of radiometric data that are acquired during each 360° scan. It takes 167 ms to complete each scan. Each scan consists of the following measurements: 2048 samples during the 51 ms spent looking at the Earth, 10 samples during a 0.25 ms interval spent looking at cold space of about 95 K, and 10 samples during a 0.25 ms interval spent looking at a warm target internal to the satellite. The cold space and warm target measurements are used to maintain the radiometric calibration of the sensor.

Figure 3.33 shows one side of the ground footprint of the instrument. The instantaneous field of view (IFOV) of the instrument is 1.1 km

Figure 3.31 Advanced Very High Resolution Radiometer (AVHRR) instrument on the NOAA-M polar orbiting satellite. Source: NASA https://mediaarchive.ksc.nasa.gov/#/Detail/23353.

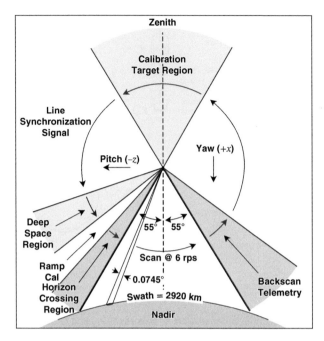

Figure 3.32 Data acquired during each AVHRR scan. Adapted from Robel (2009).

at nadir, growing to an ellipse of 2.4 × 6.5 km at the edges of the swath. Again the scanning is perpendicular to the flight track.

The optical configuration of the sensor is simple in theory but quite complex in practice (Figure 3.34). Light enters through an aperture in the 8-inch primary mirror. The light reflects off the primary onto the secondary mirror back into the sensor through a hole in the primary mirror. This is called a reflective Cassegrain design.

The light initially passes through a beamsplitter that separates the visible light from the IR. Additional beamsplitters and filters are applied to separate the light into various wavelengths, which fall on separate detectors. This design uses reflective optics that are common to all channels in order to assure the complete overlap of the individual footprints for each channel.

The bandpass of the six channels in the AVHRR are listed in Figure 3.34. Channels 1, 2, 3B, 4, and 5 were the original channels present on all versions of the sensor. Channel 3A was added on NOAA-15 and later sensors.

The top panel of Figure 3.35 compares the AVHRR bands with atmospheric transmittance, illustrating that all of the bands are located in

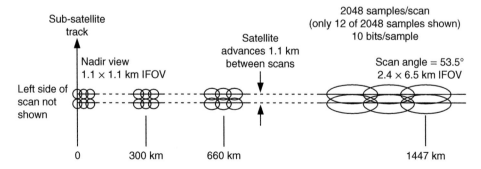

Figure 3.33 AVHRR ground sampling. Adapted from Breaker (1990).

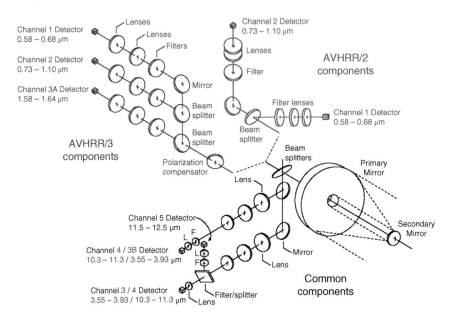

Figure 3.34 AVHRR optical configuration. Adapted from Barnes (1977) and Cracknell (1997).

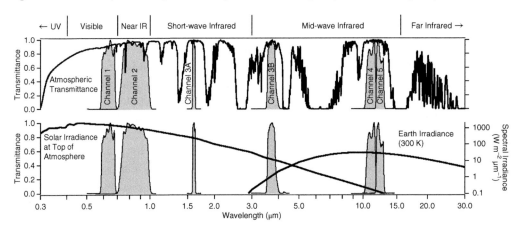

Figure 3.35 AVHRR spectral bands compared to atmospheric transmittance (top panel) and solar and terrestrial radiation (bottom panel).

atmospheric windows. These bands were carefully chosen. For example, Channel 5 is placed near the edge of a window to provide some leverage to measure atmospheric absorption due to water vapor.

The bottom panel of Figure 3.35 compares the AVHRR band locations with the baseline spectral irradiance from the sun and the Earth. Notice that channels 4 and 5 are primarily sensitive to terrestrial radiation, and relatively insensitive to solar irradiance. For this reason the SST algorithms use just channels 4 and 5 during the day, but 3B, 4 and 5 at night. In both cases, channel 4 is the primary sensor, and the other channels are used for atmospheric corrections.

3.11.2 AVHRR Processing

Figure 3.36 illustrates the data processing steps applied to the data to estimate SST. The satellite infrared measurements include

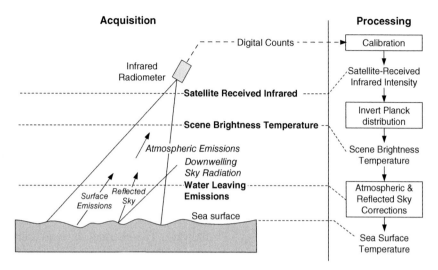

Figure 3.36 Operational SST data processing.

the water-leaving radiance and atmospheric effects. The sensor produces a stream of digital counts that are transmitted to the ground. The first processing step is calibration into radiance units. The second step is the inversion of the Planck distribution to estimate the equivalent blackbody temperature corresponding to the measurement in each channel. The multichannel SST algorithms are then applied to correct for atmospheric effects, resulting in an SST estimate.

Additional steps that are not shown here include geometric registration of the pixels onto the Earth, and identification of cloud free pixels.

As just described, radiometric calibration converts the detector signals in each channel to spectral radiance. The calibration utilizes the warm target and cold space measurements, along with calibration factors determined from laboratory tests.

The brightness or radiometric temperature in each channel is computed by inverting the Planck blackbody function.

$$L(\lambda_1, \lambda_2, T_r) = \int_{\lambda_1}^{\lambda_2} \frac{2hc^2}{\lambda^5} \frac{d\lambda}{\exp(hc/\lambda k T_r) - 1}$$

(3.117)

The idea is to estimate the brightness temperature T_r, given the measured radiance L and the cutoff wavelengths for each channel λ_1 and λ_2.

To compute SST, we must first determine which image pixels contain water, and which are contaminated by land or clouds

Spacecraft tracking data provide orbit parameters used to locate the position of the satellite. The latitude and longitude of each pixel is then computed from the AVHRR viewing geometry. Pixel masks are then derived from coastline maps to separate the land and water pixels.

Cloud contamination of water pixels is determined from pixel brightness temperatures. All clouds other than fog form at altitudes in the atmosphere that are colder than the surface temperature. Thus cloud top temperatures are always less than the SST. This makes it easy to identify cloud-filled pixels. Pixels containing a mix of cloud and water are more difficult to identify (Coakely & Bretherton, 1982). Current algorithms exploit differences in the brightness temperature statistics in small areas to identify partial cloud-filled pixels. In the end, cloud contaminated pixels are flagged and deleted from the SST calculations.

3.11.3 AVHRR SST Algorithms

Multichannel SST algorithms use measurements from two or more channels to correct for atmospheric absorption and path radiance as a function of viewing angle from nadir.

A triple window algorithm is used for nighttime measurements. Specifically, the nighttime algorithm uses T_4, the brightness temperature for channel 4 at 10.3–11.3 μm, for the primary measurement. Channels T_{3B} at 3.6–3.9 microns and T_5 at 11.5–12.5 μm are used for corrections.

$$SST\ (^{\circ}C) = A_0 T_4 + A_1 (T_{3B} - T_5)$$
$$+ A_2 (T_{3B} - T_5)(\sec\theta - 1) + A_3$$
$$(3.118)$$

with:

$A_0 = 1.0291$ (primary measurement)
$A_1 = 2.2754$ (emissivity and sky radiance correction)
$A_2 = 0.7526$ (view angle correction)
$A_3 = -1.1450$ (mean offset correction)

The coefficients A_0 through A_3 are derived from a regression analysis of satellite data with in situ sea surface temperature measurements from buoys and drifters. The values for the coefficients for NOAA-14 are as listed.

The offset characteristics of these channels are shown in Figure 3.37. Note the first coefficient A_0 scales the primary measurement channel T_4 and is thus nearly 1. The second coefficient of about 2.3 scales the difference between T_{3B} and T_5. At low sky temperatures, this difference is negative and the large biases in T_4 are reduced. At high sky temperatures the measurements in T_{3B} and T_5 are more similar and so the correction term is small. The third term corrects for the dependence of the atmospheric contribution to scan angle. And the last term removes a mean bias in the estimate.

The typical SST accuracy obtained from this algorithm is about 0.5 °C, with a bias of about 0.1 °C.

Unfortunately, the nighttime algorithm will not work in the daytime, because the midwave T_{3B} channel is contaminated by reflected skylight. Thus the daytime algorithm is a two-window algorithm utilizing the two long-wave channels, T_4 and T_5. In this case the formula is similar, only the scale factors are changed.

$$SST\ (^{\circ}C) = A_0 T_4 + A_1 (T_4 - T_5)$$
$$+ A_2 (T_4 - T_5)(\sec\theta - 1) + A_3$$
$$(3.119)$$

with

$A_0 = 1.0173$ (primary measurement)
$A_1 = 2.1396$ (emissivity and sky radiance correction)
$A_2 = 0.7797$ (view angle correction)
$A_3 = -0.5430$ (mean offset correction)

The accuracy for this algorithm is slightly worse, about 0.7 °C.

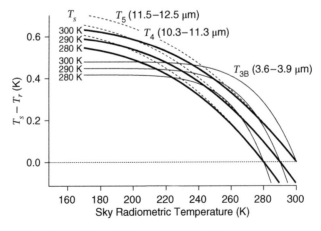

Figure 3.37 Offsets in channels used for multichannel SST algorithms.

3.11.4 Example AVHRR Images

The NOAA polar orbiting satellites orbit the Earth 14 times each day with each pass of the AVHRR instrument providing a 2400 km wide swath. The right-hand panel of Figure 3.38 shows a typical image from one of the visible channels during a single pass down the east coast of the United States; the left-hand panel shows the horizon-to-horizon coverage from the ground station in Maryland (pink circle), the coverage accounting for obstructions near the receiving antenna (inner pink region just inside the horizon), and the swath coverage for the illustrated pass.

Now it should come as no surprise that clouds often obscure the satellite's view of the surface. Figure 3.39 is a typical image from the JHU/APL processing stream where color indicates sea surface temperature and land

and cloud masks have been applied. While clouds obscure half of the ocean surface in this image, approximately half the ocean pixels were clear enough to obtain an estimate of the SST.

It turns out that clouds move much faster than ocean temperature features. So we make use of this fact to substantially reduce the effects of clouds.

Figure 3.40 is a composite of SST estimates from three days of AVHRR passes. The data from each pass is georegistered and then processed for SST along with land and cloud masks. A composite image is then formed by picking the warmest sea surface temperature estimate observed during this three-day window for each and every pixel in the scene. Clouds are typically cold, so cloud contamination is substantially reduced by picking the warmest measurements. And most locations

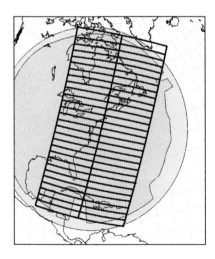

NOAA – 15 AVHRR 2000 Jul 15 12:39 UT CH 2

Figure 3.38 An example AVHRR coverage map. Figure provided courtesy of R. E. Sterner, JHU/APL.

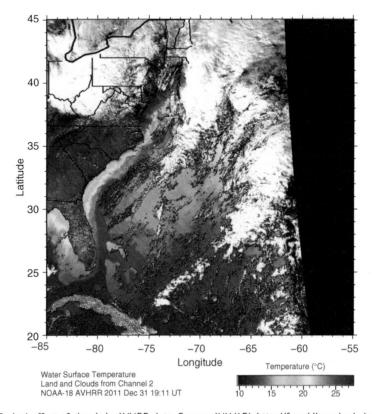

Water Surface Temperature
Land and Clouds from Channel 2
NOAA-18 AVHRR 2011 Dec 31 19:11 UT

Temperature (°C)

Figure 3.39 Typical effect of clouds in AVHRR data. Source: JHU/APL http://fermi.jhuapl.edu/avhrr.

Figure 3.40 A multi-day composite sea surface temperature map. Source: JHU/APL http://fermi .jhuapl.edu/avhrr.

Figure 3.41 Sea surface temperature of Gulf Stream meanders. Source: JHU/APL http://fermi .jhuapl.edu/avhrr.

on the surface are cloud-free at least once in most three-day periods.

In some stormy periods, even a three-day composite is too short, and we have to resort to seven-day composites. But in the end, this approach produces a relatively clear view of major thermal features in the ocean.

Figure 3.41 shows some major meanders of the Gulf Stream, one of the greatest currents in the world's oceans. The Gulf Stream is warm because of the time the water spends absorbing

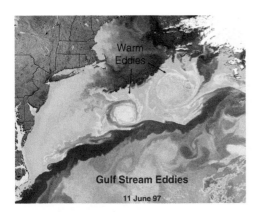

Figure 3.42 Sea surface temperature of Gulf Stream eddies. Source: JHU/APL http://fermi.jhuapl .edu/avhrr.

solar energy in the Gulf of Mexico before making its way up the east coast. A detailed explanation of the physics behind the Gulf Stream is one of the great triumphs of oceanography in the twentieth century.

Sometimes these Gulf Stream meanders pinch off, creating warm core rings on the shelf side, or cold core rings in the southern side of the front. The SST manifestation of some of these rings are shown in Figure 3.42.

While the examples shown so far are images acquired in a single pass, long-term global products are also produced. One product is

the Global Area Coverage data, which maps the AVHRR data into fixed 3×5 km cells (Figure 3.43). This is a reduced resolution product but the grid is uniform and fixed, making it suitable for construction of long-term global time series.

Figure 3.44 shows an example of the Global Area Coverage Data Product.

Figure 3.45 shows results derived from the GAC. In this case multiple GACs were averaged together to estimate the climatological mean sea surface temperature. The differences between the current and mean climatological temperature are then plotted. This is called the SST anomaly and is an excellent way to detect and monitor certain time varying ocean phenomena. In this case, the substantial warming in the eastern equatorial Pacific in the top panel is due to an El Niño, and the cooling in the bottom panel is due to a La Niña, a complex oscillatory interaction between the ocean and the atmosphere.

3.11.5 VIIRS Instrument

The first VIIRS instrument was launched in October 2011 aboard the Suomi National Polar-Orbiting Partnership (S-NPP) spacecraft. Originally planned as a risk-reduction mission to validate new instruments and data

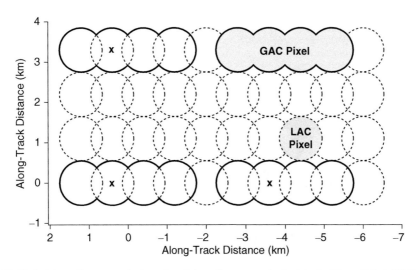

Figure 3.43 Global Area Coverage pixels vs Local Area Coverage pixels. Adapted from Robel (2009). The LAC pixels have a diameter of 1.1 km with a separation of 0.8 km. Symbol "x" indicates the reported location of the GAC pixels.

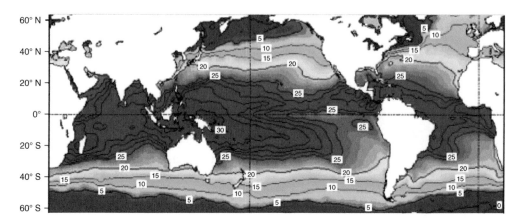

Figure 3.44 An example Global Area Coverage data product. Source: NOAA.

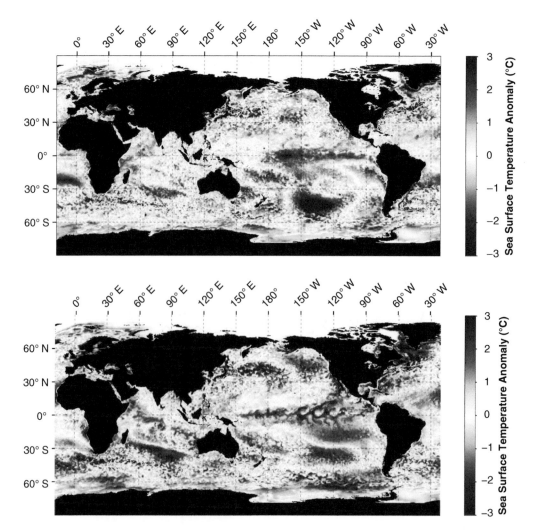

Figure 3.45 Sea surface temperature estimates during an El Niño event in 2010 (top panel) and a La Niña event in 2011 (bottom panel). Figure based on data from Reynolds et al. (2008).

Table 3.2 VIIRS channels for SST processing.

VIIRS band	Central wavelength (μm)	Wavelength range (μm)	SST processing use
M5	0.672	0.662–0.682	Cloud Screening
M7	0.865	0.846 – 0.885	Cloud Screening
M12	3.7	3.61 – 3.79	SST – Night
M15	10.763	10.26 – 11.26	SST – Day, Night
M16	12.013	11.54 – 12.49	SST – Day, Night

processing algorithms, the S-NPP satellite eventually became an operational platform to bridge the gap until the launch of the first JPSS spacecraft (NOAA-20) in 2017.

The design of the VIIRS instrument is based on two legacy research instruments: the Sea-viewing Wide Field-of-view Sensor (SeaWiFS) on the OrbView-2 satellite, and the Moderate-resolution Imaging Spectrometer (MODIS) on NASA's Terra and Aqua spacecraft. Details of these instruments can be found in Chapter 4 (sections 4.9.2 and 4.9.3, respectively).

VIIRS is a whisk-broom scanning radiometer with a field of regard of ±56.28° in the cross-track direction, providing a swath width of 3060 km at a nominal altitude of 830 km. The NOAA-20 satellite executes 14 orbits per day, resulting in full Earth coverage twice per day.

The instrument scans the Earth using a rotating telescope assembly with a rotating half-angle mirror to direct incoming radiation to dichroic beamsplitters and then onto three separate focal plane assemblies: one for visible/near-infrared radiation, a second for short-wave/mid-wave wavelengths, and a third for long-wave infrared radiation. The long-wave IR focal plane assembly used for SST observations is a 16-detector linear array aligned along the flight track.

The scan rate is set to provide adjacent nadir samples on consecutive scans. In the cross-track direction, the pixel sizes increase with scan angle, with varying amounts of overlap from three consecutive scans. VIIRS uses an onboard pixel aggregation scheme to control the pixel growth with increasing scan angle, along with a bow-tie removal scheme to eliminate duplicated pixels at off-nadir angles. Details of the sampling and pixel aggregation scheme can be found in Cao et al. (2017) and Gladkova et al. (2016).

Five of VIIRS 22 spectral bands are used for sea surface temperature observations. These bands have a spatial resolution of 750 m at nadir. Table 3.2 lists the relevant bands and their role in SST processing.

Sea surface temperature estimates are extracted from VIIRS measurement using multichannel algorithms similar to those used with AVHRR (recall equations (3.118) and (3.119)). Nighttime observations use brightness temperatures from three bands, M12, M15, and M16. Reflected sunlight precludes use of the M12 band in the daytime, hence the SST algorithm uses only data from M15 and M16 bands. The coefficients for both algorithms are derived from regression analyses using matchups with in situ measurements from buoys and SST drifters.

3.11.6 SST Accuracy

The simple satellite SST algorithms presented in this chapter have been shown to be accurate to within 0.5–0.7 °C when compared with in situ buoy measurements. More recent algorithms have significantly improved this accuracy, with bias values of less than 0.1 °C and robust standard deviations of 0.2–0.4 °C, depending on the data source.

Figure 3.46 illustrates the statistical differences observed for one of the most modern algorithms using data from AVHRR and ESA's Along-Track Scanning Radiometer (ATSR) compared to buoy measurements. The top panel shows the robust standard deviation of the two measurements, defined to be a factor of 1.48 times the median of the absolute values of the differences between the satellite and buoy SST estimates. The factor of 1.48 is the scaling needed for this statistic to equal the standard deviation of the temperature differences, if those difference were exactly Gaussian distributed. The bottom panel is the median of the temperature differences. These statistics are less influenced by the presence of outliers in the data than the more typical mean and standard deviation.

The bottom panel is labeled "median discrepancy" because at these levels of accuracy it becomes difficult to apportion the errors between the satellite and buoy measurements. It is particularly challenging because the two sensors measure fundamentally different quantities. The satellite measures the ocean skin temperature at depths of less than 1 mm, while in situ buoy measurements usually correspond to measurements at depths between 10 cm and 10 m. In order to achieve these levels of comparison, both the satellite and buoy measurements were projected to a common depth of 20 cm, using an algorithm that fully takes into account an empirically-derived representation of the diurnal cycle in the sea surface temperature profile.

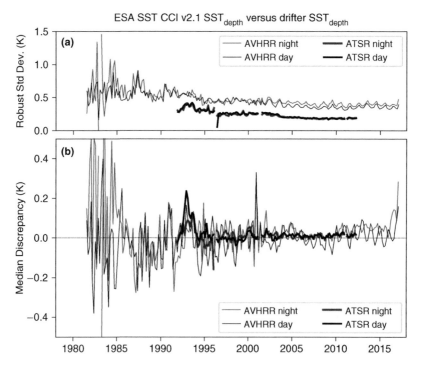

Figure 3.46 Satellite SST accuracies estimated by comparison with buoys. Merchant et al. (2019)/ CC BY 4.0.

The large deviations in the satellite–buoy temperature differences in the 1980s and 1990s are a result of early AVHRR calibration issues and imperfect correction for atmospheric absorption/path radiance. Subsequent sensors had better calibrations. The ATSR had an additional advantage that the system design supports the measurement of brightness temperature at two angles: one near nadir (0° to ~22°) and one forward-looking (~ 53°). Many of the significant variations in these curves lasting one to a few years can be associated with major atmospheric disturbances such as El Niño and La Niña events or major volcanic eruptions.

Merchant et al. (2019) report the ESA CCI SST algorithm has pixel-level uncertainties of 0.18 K with an observational stability relative to drifting buoy measurements of less than 0.003 K yr^{-1}. This level of accuracy is sufficient to meet the requirements of climate change monitoring (Donlon et al., 2002; Minnett et al., 2019).

3.11.7 Applications

While there are a number of important applications of SST measurements, several of which are listed here, there are also fundamental reasons why monitoring the SST is important:

- Hurricane intensity forecasts.
- Storm impact predictions – Nor'Easter snowfall.
- Detection and monitoring of El Niño and La Niña events.
- Tracking and monitoring of major current system features.
- Long-range weather forecasts – North Pacific SST anomalies.
- Ship deployment strategy for commercial fisheries.

The ocean–atmosphere system is driven by solar radiation that produces excess heating at the equator relative to the poles. North–south temperature gradients cause fluid motions in the atmosphere and oceans that redistribute heat toward polar regions. These fluid motions depend on dynamical processes that have long-term mean (climate), seasonal, diurnal, and transient components, such as storms and eddies.

Sea surface temperature is an important factor in air–sea interaction processes. Surface heat flux processes, such as evaporation, conduction, and long-wave radiation, depend on SST. Furthermore ocean circulation is influenced by SST distribution through its effect on atmospheric circulation, stability, and surface winds that drive ocean currents. SST also is important for heat fluxes that drive sea water density which contributes a buoyancy component to ocean circulation.

3.12 Land Temperature – Theory

The determination of land surface temperatures (LST) using thermal infrared measurements from satellites is considerably more difficult than determining sea surface temperatures:

- Land surface areas are much less homogeneous than the sea surface on scales of a few kilometers or less. Hence temperature and emissivity may vary within an image pixel.
- Atmospheric transmission can vary over much smaller spatial scales over land areas than over the ocean.
- Emissivity of a land surface varies with surface composition, which may not be known. Thus each image pixel has two unknowns – temperature and emissivity.
- Emissivity and reflectance must account for surface roughness.
- For a vegetated surface, the canopy temperature is usually less than that of the underlying soil. Shaded soil will be cooler than sun-exposed soil.

- Canopy emissivity will be different from soil emissivity and it will vary with time during the growing season.
- Land temperature exhibits a large diurnal cycle due to solar heating, so the measured LST depends on the time of observation.

Recall the radiance equation for sea surface temperature:

$$L_r(T_r, \theta) = \tau_A(\theta)\varepsilon_w(\theta)L_s(T_s)$$
$$+ \tau_A(\theta)r_w(\theta)L_{sky}(T_{sky}(\theta)) + L_A(\theta)$$
$$(3.120)$$

where the first term is the blackbody radiation of the surface reduced by the atmospheric transmittance, the second term is the reflected skylight, and the third term is the atmospheric path radiance.

For land surface temperature, at long-wave infrared (LWIR) wavelengths where reflected and scattered solar radiation can be ignored, the surface emission and reflection terms are altered to account for emissivity variations within the sensor field of view, and downwelling sky radiance reflection from a rough surface.

The surface emission term then becomes:

$$\tau_A(\theta)\varepsilon(\theta)L_s(T_s) = \tau_A(\theta) \sum_i \varepsilon_i L_s(T_i)a_i \quad (3.121)$$

where there are i different types of land cover within the FOV with emissivities ε_i and temperatures T_i, and a_i is the fractional area of the FOV covered by each land type. Here $\varepsilon(\theta)$ and T_s are the effective emissivity and surface temperature of the surface area within the single-pixel FOV.

If the surface temperature is uniform within the FOV, then the effective emissivity becomes a sum of the different emissivities weighted by the fractional area of the FOV covered by each land type:

$$\varepsilon(\theta) = \sum_i \varepsilon_i a_i \quad (3.122)$$

The reflected sky radiance term becomes more complicated because land surfaces have roughness scales comparable to the

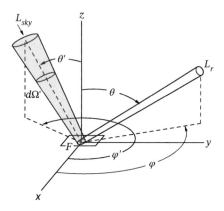

Figure 3.47 Geometry of the Bidirectional Reflectance Distribution Function (BRDF). Adapted from Nicodemus et al. (1977).

wavelengths of infrared radiation, so specular reflection/emission as used with the sea surface does not apply. Instead, the amount of reflected radiation is determined by the Bidirectional Reflectance Distribution Function (BRDF). The BRDF depends on both the incident and scattered radiation angles, as shown in Figure 3.47.

The reflected sky radiance term then becomes:

$$\tau_A(\theta) \int_0^{2\pi} \int_0^{\pi/2} \rho(\theta, \theta', \varphi, \varphi')L_{sky}(\theta', \varphi')$$
$$\times \cos\theta' \sin\theta' d\theta' d\varphi' \quad (3.123)$$

where ρ is the BRDF, θ' and φ' are the incident and azimuth angles of the incident radiation, and θ and φ are the incident and azimuth angles of the observation. Note to compute the radiance observed in a particular direction, the integral has to be performed over all possible angles of the downwelling sky radiance. This is because the rough surface can scatter energy from any direction to any other direction.

The effective emissivity $\varepsilon(\theta)$ is then given by:

$$\varepsilon(\theta) = 1 - \int_0^{2\pi} \int_0^{\pi/2} \rho(\theta, \theta', \varphi')$$
$$\times \cos\theta' \sin\theta' d\theta' d\varphi' \quad (3.124)$$

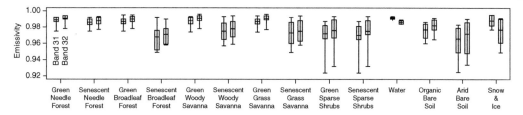

Figure 3.48 Long-wave (10.8–11.3 μm) emissivities of various types of vegetation. The boxes contain 68% of the data, with the horizontal lines indicating mean values. The continuing lines denote the range of the extremes. Modified from Snyder et al. (1998).

For most land cover types, the BRDF is not well known, which makes it difficult to compute the radiance received at the sensor aperture. This problem, along with the difficulty in knowing the surface type composition and emissivities, underlies the justification to search for multi-channel LST algorithms that ignore the reflected sky radiance.

Figure 3.48 shows the emissivities of various types of vegetation with MODIS bands 31 and 32 around 11 μm. There are variations in emissivities of a few percentage for different types of vegetation, but relatively large variations even within a specific type.

For example, arid bare soil has emissivities ranging from 0.93 to 0.98. Thus arid bare soil with a physical temperature of 300 K will exhibit a radiometric temperature that will vary by 3 K, variations far larger than were observed for measurement of the SST.

The four panels in Figure 3.49 show the wavelength dependence of emissivity for two types of forest, water, and ice and snow. The dark curves show the mean and the grey curves

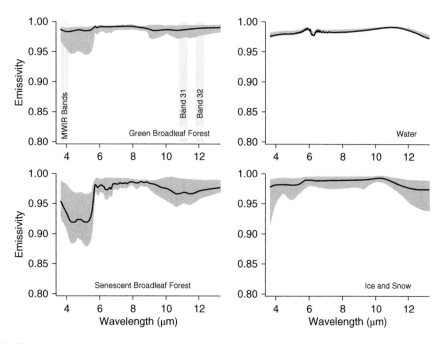

Figure 3.49 Wavelength dependence of emissivity for forest, water, and ice and snow. Modified from Snyder et al. (1998).

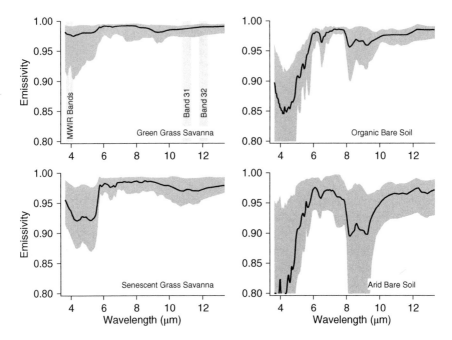

Figure 3.50 Wavelength dependence of emissivity for four other land types. Modified from Snyder et al. (1998).

indicate the extremes. Three MODIS bands are indicated for comparison.

The four panels in Figure 3.50 show the wavelength dependence of emissivity for four other land types. Note the emissivities for all eight types are high and nearly constant for the two LWIR bands, but highly variable at mid-wave bands.

Table 3.3 shows the global yearly coverage fractions for 6 different land types. Shrubs, desert and bare soil is one of the most common types, but suffers from the greatest variability in emissivity.

3.13 Operational Land Temperature

Numerous multichannel LST algorithms have been proposed based on the AVHRR split-window radiometric temperatures T4 (10.3–11.3 μm) and T5 (11.5–12.5 μm), and effective land emissivities in these two long-wave bands.

The general approach is based on the success of the SST split-window algorithms in compensating for atmospheric effects. But LST algorithms are more complex because they have to account for the effective surface emissivity within the field of view. The functional forms of both SST and LST algorithms are based on radiative transfer theory:

- The coefficients for the SST algorithm are based on thousands of colocated, cotemporal comparisons with worldwide in situ data

Table 3.3 Land coverage statistics. Adapted from Snyder et al. (1998).

Land cover type	Global yearly coverage (%)
Ice and Snow	10
Grasslands	17
Shrubs, Desert, and Bare Soil	20
Broadleaf and Mixed Forest	17
Tundra and Cultivated Land	20
Coniferous Forest	16

under a wide range of atmospheric conditions. This is, in fact, an ongoing process.

- No comparable database exists for LST covering all land cover types and a wide range of atmospheric conditions over land.
- Both types of algorithms ignore reflected sky radiance as a small term.

An alternative approach to derive coefficients is to simulate the atmosphere and perform a large number of radiative transfer calculations (including the variability of surface emissivities). This approach is used in some of the most sophisticated estimation approaches, but introduces errors due to model approximations.

In the end it is difficult to validate LST algorithms and determine their absolute accuracies. But it is an important measurement, so researchers have continued to develop and refine LST algorithms.

Sun and Pinker (2005) measured the surface temperature of a test area in Oklahoma consisting of grassland and cropland. They made these measurements with a radiation thermometer and hourly IR images from GOES-8. The goal was to map the diurnal variation of the surface temperature, something that could not be done from one or two satellites in LEO. The GOES imager has two visible and three IR channels. The IR channels are at 3.9, 10.8, and 12 μm.

The daytime GOES radiometric temperatures T_{11} and T_{12} were converted to skin temperature T_s using the following split-window algorithm:

$$T_s(i) = a_0(i) + a_1(i)T_{11} + a_2(i)(T_{11} - T_{12})$$
$$+ a_3(i)(T_{11} - T_{12})^2 + a_4(i)(\sec\theta - 1)$$
$$(3.125)$$

where the index i denotes the surface type, and θ is the incident angle. Notice this is the same functional form as used in the SST algorithms.

At nighttime, the following triple-window algorithm was used:

$$T_s = -16.14 + 1.16T_{11} - 0.72T_{12}$$
$$+ 0.63T_{3.9} + 0.12\frac{(1-\varepsilon_{11})}{\varepsilon_{11}}T_{11}$$
$$- 5.79\frac{(1-\varepsilon_{12})}{\varepsilon_{12}}T_{12} + 2.41\frac{(1-\varepsilon_{3.9})}{\varepsilon_{3.9}}$$
$$+ 288.67(\sec\theta - 1) \qquad (3.126)$$

The explicit emissivities in the nighttime algorithm depend on the surface type, and the coefficients in both algorithms were determined from simulations using the atmospheric radiance and transmittance numerical code MODTRAN 4.0.

Figure 3.51 shows results from Sun and Pinker comparing the LST time variation

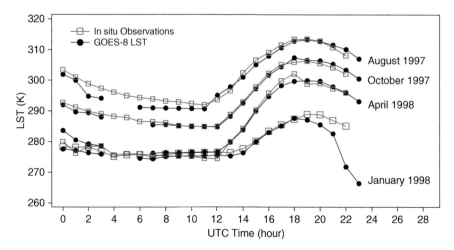

Figure 3.51 Diurnal variations in land surface temperature measurements. Based on data from Sun and Pinker (2005).

measured with the ground radiation thermometer and the GOES-derived measurements.

The GOES nadir footprint is 4 km in diameter. Local time for these plots = UTC - 6 hours. Thus sundown is about 0 hrs UTC, and sunrise is about 12 hrs UTC (varies with season). Thus

either above or below the mean temperature. A portion of the blue dashed sinusoid is used to model the cooling phase, and a portion of the red dashed sinusoid is used to model the midmorning to afternoon warming.

The Sun and Pinker model for the diurnal surface temperature variation is:

$$T_s(t) = \begin{cases} T_0 + T_a \sin(\pi + 2\pi t/T_n) & \text{for} \quad t < t_{sunrise} \\ T_0 + T_a \cos(2\pi(t - t_m)/T_d) & \text{for} \quad t > t_{sunrise} \end{cases} \quad (3.127)$$

the October plot shows the surface cooling at night (0–12 hours), and warming during the first 6–7 hours of the day. Notice the magnitude of the diurnal temperature variation is several degrees – much larger than the diurnal change of SST. Sun et al. (2012) provides a detailed analysis of the accuracies of various LST algorithms applied to GOES data.

Using these results, Sun and Pinker proposed a model for the diurnal variation (shown in Figure 3.52) to be used with AVHRR data, which occur only twice per day. Data from the ARM site in Oklahoma were used to construct the typical pattern of diurnal heating and cooling. The gray vertical bars indicate the times of the morning and afternoon AVHRR passes.

Sun and Pinker reasoned that the magnitude and shape of the curve will vary with land cover type and season. They ended up generalizing this shape to a pair of half sinusoids, each representing the portion of the day that was

where GOES hourly data were used to determine the parameters $t_{sunrise}$, t_m, T_n, and T_d (night and day periods) from monthly mean patterns.

Then two consecutive AVHRR observations are used to determine T_0 and T_a for that day by fitting to the diurnal temperature variation model curve.

The LST accuracies obtained from GOES and AVHRR data are shown in Figure 3.53. This simple algorithm provides LST estimates with an rms error of just over 2° and a bias of a few tenths of a degree.

Other techniques have been developed that improve upon the accuracy of the split-window algorithm by bringing in other sources of information to estimate atmospheric effects. Another approach developed by Coll and Caselles (1997), based on the insight that variations in atmospheric water vapor dominate the atmospheric effects, is discussed below.

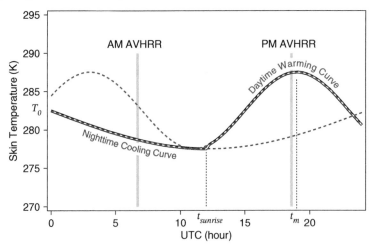

Figure 3.52 Diurnal LST model. Based on data from Sun and Pinker (2005).

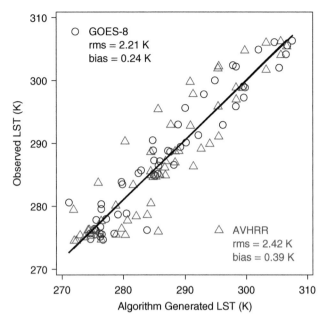

Figure 3.53 Diurnal LST accuracy. Based on data from Sun and Pinker (2005).

The Coll and Caselles multichannel LST algorithm is:

$$T_s = T_4 + a(T_4 - T_5) + b + c(\varepsilon, W)$$

$$c(\varepsilon, W) = \alpha(W)(1 - \varepsilon) - \beta(W)(\varepsilon_4 - \varepsilon_5)$$

$$\text{and} \quad \varepsilon = 0.5(\varepsilon_4 + \varepsilon_5) \qquad (3.128)$$

with W being the total column atmospheric water vapor (gm/cm^2). As before, the subscripts 4 and 5 refer to AVHRR channels 4 and 5, so this is also an LWIR split-window algorithm.

The coefficients are then determined as follows:

- First, determine α and β, which do not depend on ε, from SST data comparisons.
- Second, derive $c(\varepsilon, W)$ from regional emissivity and water vapor data.

The resulting MCLST algorithm is:

$$T_s = T_4 + [1.34 + 0.39(T_4 - T_5)](T_4 - T_5)$$
$$+ \alpha(W)(1 - \varepsilon) - \beta(W)(\varepsilon_4 - \varepsilon_5) + 0.56 \qquad (3.129)$$

Figure 3.54 plots the values of coefficients $\alpha(W)$ and $\beta(W)$ as a function of atmospheric water vapor content W. W is then measured

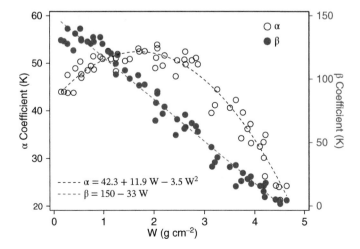

Figure 3.54 MCLST algorithm coefficients α and β as a function of atmospheric water vapor content W (Modified from Coll and Caselles, 1997).

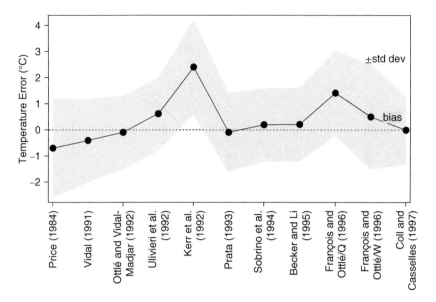

Figure 3.55 AVHRR land surface temperature accuracy. Adapted from Coll and Caselles (1997).

from the POES atmospheric sounder. How this measurement gets made is discussed in Chapter 6.

Figure 3.55 is a plot of the accuracy of the LST estimates obtained from the Coll and Casselles algorithm. The use of the independent sounder data is seen to reduce the rms errors to about 1.5 °C as evaluated against 10 other LST data algorithms.

Multiple satellites and sensors are in use today to make operational estimates of land surface temperature with algorithms similar to the ones described here. One example is the Moderate Resolution Imaging Spectrometer (MODIS) sensor deployed on the NASA Terra and Aqua satellites (Figure 3.56):

- The Terra polar-orbiting satellite was first launched in December 1999.
- The satellite is in a 705-km high, sun-synchronous orbit. The orbit has a mean period of 98.9 minutes with a 16-day repeat cycle. The equatorial crossing time for the descending nodes is 10:30 am.
- The MODIS sensor uses a continuously rotating two-sided mirror to scan across a 110° field of view.

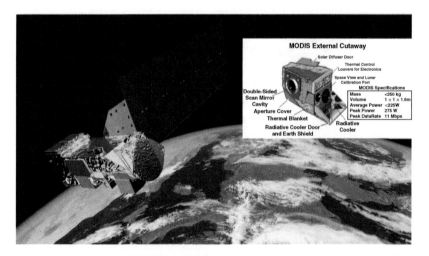

Figure 3.56 Artist's concept of Aqua satellite with MODIS instrument. Source: NASA.

- The system senses the entire equator every two days, with daily full coverage above 30° latitude.
- The sensor has 36 spectral bands: 29 bands have 1.1-km nadir-looking resolution, five have 500-m and two have 250-m resolution.
- A second MODIS sensor was launched on the Aqua Satellite in May 2002. This one has a 1:30 PM equatorial crossing time for the ascending node and is known as the PM satellite.
- In addition to the land surface temperature measurements discussed in this section, MODIS is also used to measure surface reflectance, vegetation indexes, fire area, snow/ice/sea ice, land cover, and ocean color (Justice et al., 1998).

An even more sophisticated split-window algorithm has been developed for global LST estimates from the MODIS sensor:

$$T_S = C + [A_1 + A_2(1 - \varepsilon)/\varepsilon + A_3\Delta\varepsilon/\varepsilon^2]$$

$$\times (T_{31} + T_{32})/2$$
$$+ [B_1 + B_2(1 - \varepsilon)/\varepsilon + B_3\Delta\varepsilon/\varepsilon^2]$$
$$\times (T_{31} - T_{32})/2 \qquad (3.130)$$

where T_S = surface temperature and ε and $\Delta\varepsilon$ are given in terms of the emissivities in the two bands (Snyder & Wan, 1998; Wan, 1999). The two MODIS bands used in the algorithm are channels 31 and 32, at 10.78–11.28 μm and 11.77–12.27 μm, respectively.

The coefficients A_i, B_i and C in this algorithm are obtained from radiative transfer calculations over a wide range of atmospheric conditions, viewing angles, column water vapor values, and atmospheric lower boundary temperatures. The emissivities in this algorithm are obtained from a BRDF and emissivity model for specific land cover types. The estimated rms accuracy of the model is about 1.0 °C.

Figure 3.57 shows seasonal differences in the global land surface temperature for the months of January and July 2013. In January, northern

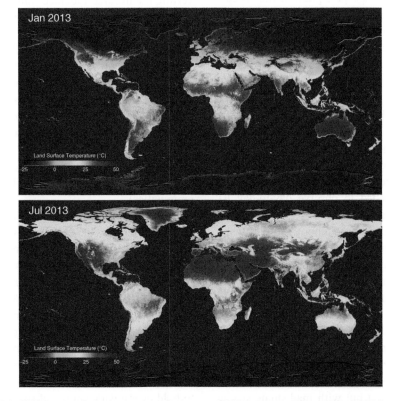

Figure 3.57 MODIS land surface temperature for January 2013 (upper panel) and July 2013 (lower panel). Source: NASA https://neo.sci.gsfc.nasa.gov/view.php?datasetId=MOD11C1_M_LSTDA.

hemisphere regions are snow covered, while the southern hemisphere is experiencing austral summer. The pattern reverses in July with summer heating dominating the northern hemisphere, and cooler winter temperatures prevailing below the equator.

3.14 Terrestrial Evapotranspiration

Estimating and monitoring terrestrial evapotranspiration (ET) is an important application of land surface temperature estimates derived from satellite observations. ET is comprised of three components: vegetation transpiration (65%), soil evaporation (25%), and evaporation from rainfall intercepted by vegetation (10%) (Zhang et al., 2016).

ET is a major component of the Earth's hydrological cycle. For example, evapotranspiration returns to the atmosphere about 67% of the annual average precipitation that occurs over land (Oki & Kanae, 2006). Monitoring ET is an important requirement for agricultural water management and a key variable needed for predicting responses of the terrestrial biosphere to changes in climate due to water availability.

Space-based measurements of land surface temperature (LST) needed for monitoring ET dynamics require sensors with high spatial resolution in order to detect agricultural-scale vegetation plots, along with high temporal resolution needed to observe day-to-day ET variations due to irrigation schedules, as well as variations throughout the day when vegetation may be under stress. None of the current satellite instruments can meet both of these requirements. The Landsat 8 Thermal IR Sensor, TIRS, and the ASTER instrument on the Terra spacecraft have adequate spatial resolution (90–100 m) but inadequate temporal resolution (16-day revisit times). On the other hand, the GOES imager and MODIS/VIIRS can meet the temporal resolution requirement but with inadequate spatial resolution (0.75–1.0 km at best) (Fisher et al., 2016).

NASA's ECOsystem Spaceborne Thermal Radiometer Experiment on Space Station (ECOSTRESS) is a mission launched in July 2018 designed to measure ground temperatures from small plots in order to better understand daily plant temperatures and water use. The space station's low-Earth orbit and inclination (nominal 400 km altitude and 51.6° inclination) result in about 16 orbits a day, with revisits to the same location approximately every 4–5 days at varying times of day.

The ECOSTRESS instrument is a five-band thermal IR radiometer known as PHyTIR – Prototype HyspIRI Thermal Infrared Radiometer. Important characteristics of PHyTIR are listed in Table 3.4. More details of the instrument design including its scanning approach can be found on the ECOSTRESS web site https://ecostress.jpl.nasa.gov/.

The methodology for estimating ET involves a complicated process that begins with converting the infrared radiance values measured at the sensor aperture to land surface temperature estimates. Two important corrections must be implemented to accomplish this: the atmospheric path radiance must be removed, and then the land surface emissivity must be separated from the apparent surface temperature. The latter correction is nontrivial in that for n radiance measurements at a given pixel (one measurement at each of n wavelength bands), there are $n + 1$ variables (one physical temperature common between all bands and n emissivities).

Fortunately for the ECOSTRESS mission, prior experience with analysis of five-band TIR data from the ASTER instrument produced a validated procedure to do this, known as the Temperature Emissivity Separation (TES) algorithm (Gillespie et al., 1998; Payan & Royer, 2004). At its core, this algorithm adds a constraint to the problem based on an empirical relationship between mean emissivity and the spectral contrast between bands (Matsunaga, 1994). This constraint, which appears to hold over a wide variety of surfaces, allows

Table 3.4 Characteristics of PHyTIR radiometer.

Parameter	Value
TIR Spectral Bands	5
Band Centers and FWHM, μm	1 – 8.29, 0.354 2 – 8.78, 0.310 3 – 9.20, 0.396 4 – 10.49, 0.410 5 – 12.09, 0.611 Only Bands 2, 4, and 5 available after 15 May 2019
Nadir Pixel Size, m	69 × 38 2 pixels cross-track, 1 pixel down-track
Swath Width, km	384
Nominal Radiometric Accuracy at 300K, K	0.5 – Varies with wavelength band
Nominal Radiometric Precision at 300K, K	0.5 – Varies with wavelength band

for the iterative solution of the problem. The interested reader is directed to these papers for details of this algorithm. The final output is a surface temperature T_S and a set of emissivities ϵ_S for each pixel in the scene.

The Priestley-Taylor Jet Propulsion Laboratory (PT-JPL) algorithm is used for the ECOSTRESS ET retrieval algorithm (Fischer et al., 2008). Five general data inputs are required to drive the PT-JPL algorithm: net radiation; near-surface air temperature; water vapor pressure; surface reflectance in the red band; and surface reflectance in the near-IR band. Each of these inputs is obtained from satellite observations, with MODIS being the primary source. The approach for obtaining these inputs can be found in the ECOSTRESS Level 3 Algorithm Theoretical Basis Document (Fisher, 2018). The ECOSTRESS measurements of surface temperature T_S and emissivity ϵ_S are used to calculate the upwelling long-wave radiation contribution R_{LU} to the net radiation used in the PT-JPL algorithm:

$$R_{LU} = \sigma \epsilon_S \, T_S^4 \qquad (3.131)$$

The ECOSTRESS mission requires a modification to the (PT-JPL) algorithm to account for diurnal cycling of the net radiation and surface air temperature inputs, which are instantaneous values obtained at the time of the MODIS overpass. Another adjustment has to be made to account for the spatial resolution differences of the various inputs – MODIS inputs at 1-km resolution, Landsat 8 red and NIR reflectances at 30-m resolution, and ECOSTRESS surface temperature and emissivity at 70-m resolution (Fisher, 2018).

Validation of the ECOSTRESS ET product is an ongoing research effort. Preliminary Stage 1 validation against latent heat flux measurements have been carried out at 82 eddy covariance sites around the world. Results have shown the ECOSTRESS ET product performs well against the site measurements, with an overall correlation $r^2 = 0.88$; bias = 8%; and normalized root-mean-square error = 6% (Fisher et al., 2020).

3.15 Geologic Remote Sensing

Land observations with sensors operating at infrared wavelengths ranging from 0.70–15 μm (Visible to LWIR) can be used to extract several types of geological information such as:

- Landform type from the geometry of surface features. This is usually based on georegistered imaging, as well as surface relief measurements from stereo imaging or topographic radars (discussed in Chapter 10).

- Surface type from temporal changes in surface characteristics. These can include the rates at which thermal changes occur for different surfaces, as well as subsidence, landslides or shifts due to earthquakes measured with stereo imagers or topographic radars.
- Mineral content inferred from spectral or hyperspectral imaging, an example of which is discussed below.

Geological information to be extracted from infrared data is encoded in the surface-leaving radiance of the area of interest, much as the surface temperature of a water body is encoded in its water-leaving radiance. It follows that some type of atmospheric correction is often required to obtain quantitative results from the remote sensor data; several approaches to removing atmospheric effects can be found in Liang (2005). The surface-leaving radiance in the infrared varies because of some combination of surface temperature and emissivity. The emissivity of the surface can in turn vary due to mean slope, surface roughness, or the intrinsic emissivity of the surface material.

Data from a Japanese instrument deployed on the NASA Terra satellite, the Advanced Spaceborne Thermal Emission and Reflection Radiometer (ASTER), have provided useful information for geological research, as well as land surface climatology, volcano monitoring, and land cover changes (ASTER, 2021). The ASTER instrument incorporates three optical subsystems: a visible and near-IR subsystem operating in three spectral bands with a spatial resolution of 15 m; a short-wave IR subsystem with six spectral bands with a resolution of 30 m; and a five-band long-wave IR subsystem with a resolution of 90 m

The left-hand panel of Figure 3.58 is a false color image obtained from the ASTER visible and near-IR bands. This scene, centered on Meteor Crater (also known as Barringer Crater) in Arizona, has 15-m resolution and was constructed from three-bands centered at 0.56, 0.66, and 0.82 μm. While vegetation can play a dominant role in determining the color of land scenes, this is a desert area and most of the spectral differences are due to the underlying rocks. The right-hand panel of Figure 3.58 is a portion of a geological map of the area drawn by E. M. Shoemaker in 1958. The delineation between the Kaibab limestone

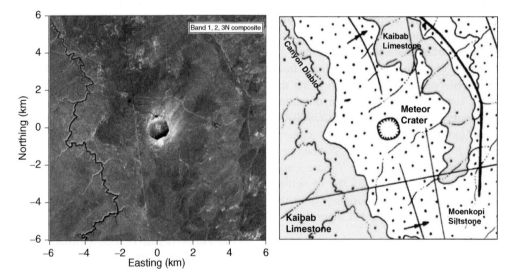

Figure 3.58 Left: ASTER false-color image acquired on 02-21-2014 around Meteor Crater, Arizona. Right: Geological map of the area. Left-hand panel courtesy NASA EOSDIS Land Processes DAAC. Accessed 2020-11-14 from https://doi.org/10.5067/ASTER/AST_09.003. Right-hand panel from Based on Shoemaker (1959).

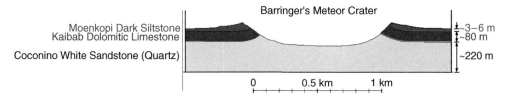

Figure 3.59 Profile of Meteor Crater and the general stratigraphy of the region. Based on Shoemaker (1959).

and the Moenkopi siltstone is easy to see in the false-color image.

There are actually three primary rock types in this image: the white ejecta on the northeast side of the crater is Coconino sandstone, which is nearly pure quartz; the Kaibab dolomitic limestone, which generally lies above the sandstone; and a thin layer of dark Moenkopi siltstone that covers some portions of the image (Shoemaker, 1974).

Figure 3.59 illustrates the profile of Meteor Crater and the general stratigraphy surrounding the region. The figure indicates that even accounting for erosion, the meteorite impact penetrated to Coconino sandstone, and ejected some of that material in the areas adjacent to the crater.

Figure 3.60 plots the emissivity of these three rock types as recorded in the Jet Propulsion Laboratory's ECOSTRESS Spectral Library. This library contains the spectral characteristics of a variety of materials taken from locations around the world. Notice the significant differences in the spectral characteristics of these materials. It is these differences that allows different rock types to be delineated in spectral imagery. The deep dip in emissivity between 8 and 9 µm for pure quartz is a unique signature of quartz.

Figure 3.60 also includes colored bars indicating the wavelengths of ASTER bands 7 and 10–14. Quantitative measurements of the surface-leaving radiance in these bands, as shown in Figure 3.61, can be used to distinguish between the various rock types.

Note that all three rock types have similar emissivities in band 14, but the emissivity of quartz is uniquely low in band 10. An ad hoc measure of the quartz content within each pixel can be constructed by computing the statistic $q_i = \frac{L_{14} - L_{10}}{L_{14} + L_{10}}$, where L_{10} and L_{14} indicates the surface-leaving radiance in bands 10 and 14, respectively. The radiance difference

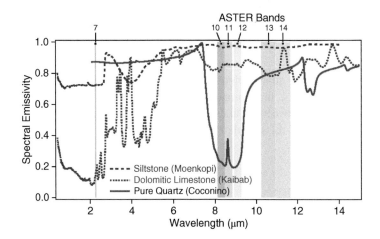

Figure 3.60 Emissivities of quartz, siltstone, and dolomitic limestone. Based on reflectance data from the ECOSTRESS Spectral Library, courtesy of the Jet Propulsion Laboratory, California Institute of Technology, Pasadena, California. Copyright © 2017, California Institute of Technology.

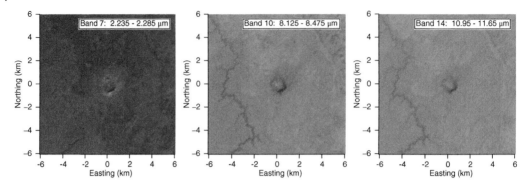

Figure 3.61 ASTER bands 7, 10 and 14. Source: NASA EOSDIS Land Processes DAAC. Accessed 2020-11-14 from https://doi.org/10.5067/ASTER/AST_09.003.

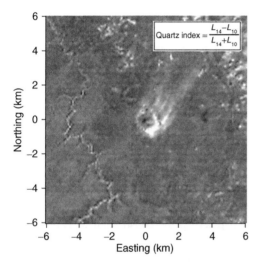

Figure 3.62 Quartz index computed from ASTER images acquired on 02-21-2014 around Meteor Crater, Arizona. Source: NASA EOSDIS Land Processes DAAC. Accessed 2020-11-14 from https://doi.org/10.5067/ASTER/AST_09.003.

in the numerator captures the quartz "signal" while the radiance sum in the denominator normalizes the result to lie between −1 and 1. Figure 3.62 shows the result of this calculation in the vicinity of the crater.

3.15.1 Linear Mixture Theory and Spectral Unmixing

To develop a more quantitative measure of the three rock types in this scene, we have to consider mixing theory. The simplest, linear mixing theory models the surface within each pixel as consisting of a fraction f_j of each of n rock types. The surface-leaving radiance in each pixel at the i-th wavelength is then given by:

$$L_i = \sum_{j=1}^{n} a_{ij} f_j + e_i \qquad i = 1, 2, \dots m \quad (3.132)$$

where a_{ij} is the spectral emissivity of rock type j in spectral band i, and e_i is an error term that allows for differences between the data and the model.

In this example there are three rock types (also called end-members in the literature), so $n = 3$; and we will consider measurements in six spectral bands, so $m = 6$. The emissivity for the three rock types in the six spectral bands is then estimated from Figure 3.60 to be:

$$a = \begin{bmatrix} 0.722 & 0.154 & 0.873 \\ 0.966 & 0.833 & 0.227 \\ 0.968 & 0.854 & 0.247 \\ 0.972 & 0.850 & 0.217 \\ 0.964 & 0.793 & 0.792 \\ 0.962 & 0.863 & 0.820 \end{bmatrix} \quad (3.133)$$

where the columns are siltstone, limestone, and sandstone, and the rows are bands 7, 10, 11, 12, 13, and 14, respectively. In matrix form,

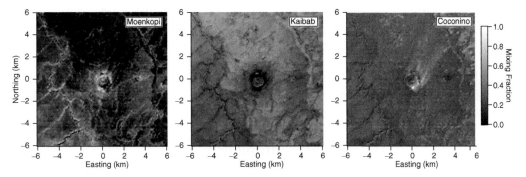

Figure 3.63 Estimation of the mixing fraction of three rock types near Meteor Crater. Source: NASA EOSDIS Land Processes DAAC. Accessed 2020-11-14 from https://doi.org/10.5067/ASTER/AST_09.003.

Equation (3.132) then becomes:

$$\mathbf{L} = \mathbf{A}\mathbf{f} + \mathbf{e} \quad \rightarrow$$

$$\begin{bmatrix} \\ L_i \\ \\ \end{bmatrix} = \begin{bmatrix} a_{11} & & \\ & a_{ij} & \\ & & a_{mn} \end{bmatrix} \times \begin{bmatrix} \\ f_j \\ \end{bmatrix} + \begin{bmatrix} \\ e_i \\ \end{bmatrix}$$

$$(3.134)$$

Theoretically the values of $\mathbf{f}$ that minimize the squared error is given by $\mathbf{f} = \mathbf{A}^{-1}\mathbf{L}$, but this solution is often singular because of correlations in the columns of $\mathbf{A}$. Other approaches such as singular value decomposition (SVD) are often used to avoid such singularities. In SVD, $\mathbf{A}$ is first factored into three matrices: $\mathbf{A} = \mathbf{U}\mathbf{W}\mathbf{V}'$, where $\mathbf{U}$ is an $m \times n$ column-orthogonal matrix, $\mathbf{W}$ is an $n \times n$ matrix of singular values, $\mathbf{V}$ is an $n \times n$ matrix of orthogonal columns, and $\mathbf{V}'$ is the transpose of $\mathbf{V}$. The smallest of the diagonal elements in $\mathbf{W}$ are then zeroed and the inverse is given by $\mathbf{A}^{-1} = \mathbf{V}\mathbf{W}^{-1}\mathbf{U}'$. This leaves a solution:

$$\mathbf{f} = \mathbf{V}\mathbf{W}^{-1}\mathbf{U}'\mathbf{L} \qquad (3.135)$$

If it were only this simple. When equation (3.135) is applied, the results often indicate that the fraction of the surface covered by one or more of the rock types is negative, which is clearly unphysical. So equation (3.134) must be solved with the constraints that $f_i > 0$ and $\sum f_i = 1$. Fortunately, solutions to such constrained least squares problems are provided in many math libraries.

Figure 3.63 shows the results of such calculations, which are often referred to as spectral unmixing.

This type of remote sensing analysis has been conducted on a number of terrestrial features, and the solutions carefully checked by detailed field observations. While there is obvious economic value in detecting and assessing minerals on the surface of the Earth, application of these techniques to other worlds can provide significant scientific value. The analysis of meteor craters and the materials ejected from those impacts provides insights into the characteristics of the stratum below the surface which is of particular value for understanding the geology of other planets such as Mars (Wright & Ramsey, 2006).

3.16 Atmospheric Sounding

While atmospheric absorption and emissions are a complicating factor for the estimation of surface temperature and constituents, the unique characteristics of atmospheric absorption and emissions can be utilized to measure the atmosphere. This process is called atmospheric sounding. This section illustrates

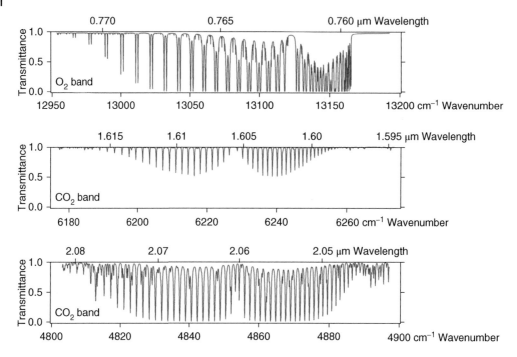

Figure 3.64 O_2 and CO_2 absorption bands. Based on HITRAN calculations from https://spectra.iao.ru/. See Mikhailenko et al. (2005) for more details.

the potential for atmospheric sounding by describing how NASA's Orbiting Carbon Observatory-2 (OCO-2) works.

Molecules in the atmosphere preferentially absorb infrared radiation at a set of specific wavelengths that excite the natural vibrational modes of the molecule. The absorption spectrum for each molecule thus include thousands of generally narrow spectral lines, as can be seen in Figure 3.16. The absorption spectrum of each molecule is unique, so a spectrographic measurement of absorption along a path through the atmosphere can easily be used to assess the relative concentration of various molecules integrated along that path.

The OCO-2 makes high-resolution spectrographic measurements of reflected sunlight in three bands, two CO_2 bands and one O_2 band. The three specific bands used are illustrated in Figure 3.64. We know the solar spectrum at the top of the atmosphere from other measurements. OCO-2 measures the

spectral absorption of the solar spectrum along the path from the sun to the surface and back to the satellite in order to estimate the mean concentrations of CO_2 and O_2 along the path.

OCO-2 can also estimate the vertical distribution of these molecules, hence the term "atmospheric sounding". The leverage for accomplishing this feat arises because the width of the spectral lines for any particular gaseous molecule depends on pressure, and hence height in the atmosphere. Figure 3.65 shows the absorbance[5] spectrum (top panel) and cross-section[6] (bottom panel) for CO_2 at three different pressures (1000 mb, 100 mb,

---

5 Absorbance is the logarithm (base-10) of the transmittance for a sample concentration of one part-per million (ppm) over a 1-m optical path at a temperature of 296K. Thus the units for absorbance are $ppm^{-1} \cdot m^{-1}$.

6 Cross-section is the theoretical absorption cross section of a single CO_2 molecule in units of cm^2 per molecule.

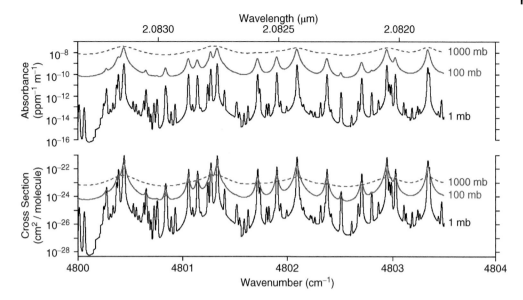

Figure 3.65 Effect of pressure on CO_2 absorption lines. Based on HITRAN calculations from https://spectra.iao.ru/. See Mikhailenko et al. (2005) for more details.

and 1 mb) approximately corresponding to altitudes of ground level, 17 km, and 49 km.

Two primary effects lead to pressure-broadening of spectral absorption lines. Collisional broadening occurs when an emitting molecule collides with another, shortening the effective emission time and hence broadening the emitted spectral line. Of course more impacts occur under higher pressure than lower. The second effect is quasi-static pressure broadening where nearby molecules couple to the emitting molecule, shifting the energy levels in that particle.

OCO-2 gains the leverage to estimate the vertical distribution of CO_2 and O_2 by accurately measuring the total absorption spectrum in the vicinity of each spectral absorption line. Absorption at the peak of a line is indicative of absorption in the upper portions of the atmosphere where there is little pressure broadening, while absorption on the sides of each peak indicate absorption in lower portions of the atmosphere.

Figure 3.66 is a diagram of the OCO-2 optical instrument. The instrument utilizes a common Cassegrain telescope. Light from the telescope is distributed via dichroic beam

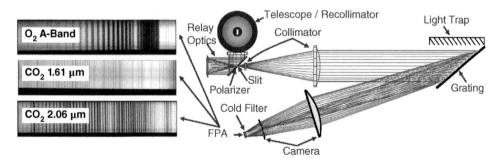

Figure 3.66 OCO-2 instrument optical path (right) and example spectral images recorded by the focal plane array in the three spectral channels (left) (Crisp et al., 2019).

splitters into three similar optical paths, one for each spectral band. The O_2 channel covers 0.758–0.772 µm, and the two CO_2 channels cover 1.594–1.619 µm and 2.042–2.082 µm. The spectrum for each channel is projected onto CCD arrays that produce 1016-element spectra. So ignoring various forms of averaging, each measurement produces approximately 3048 unique measurements corresponding to the 1016 spectral channels in each of three bands.

As illustrated in Figure 3.67, the data processing for OCO-2 is based on an inverse method. At the core of the method is a forward model that can accurately predict what the OCO-2 sensors would measure given inputs including the solar spectrum, the vertical profiles of gases in the atmosphere, some characteristics of aerosols in the atmosphere, the spectral reflectance of the surface, the sensor observation mode (e.g. land or ocean) and viewing geometry, and the instrument characteristics. At the core of this is a full radiative transfer model with a high accuracy model of the line absorption spectra for CO_2, O_2, and some other major atmospheric constituents. This forward model was constructed to be as accurate as it can be given accurate inputs. Some of these inputs are known, such as the solar spectrum or the absorption line characteristics for CO_2 or O_2, but other parameters are initially unknown such as the vertical profile of CO_2, the aerosol contribution to the radiances, or the surface reflectance in the three bands. The goal is to estimate these unknown parameters, which are referred to as the state vector, from the measurements using the forward model.

The inverse method begins by making a guess at the state vector as an initial input to the forward model. The forward model is then run to predict the measurements that would have been observed under the input conditions. The differences between the predicted and measured radiances are then evaluated.

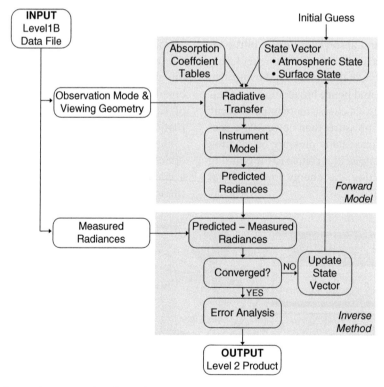

Figure 3.67 General inverse method for estimating parameters from a measurements and a forward model. Adapted from Crisp et al. (2019).

Table 3.5 Atmospheric parameters estimated by OCO-2. (Adapted from Crisp et al. (2019).)

Description	Parameters	Number of elements
Aerosols	5 types × 3 parameters	15
Temperature offset	Scalar	1
Water vapor multiplier	Scalar	1
Surface pressure	Scalar	1
Surface reflectance, mean and spectral slope	2 × 3 bands	6
Surface wind speed (ocean only)	Scalar	1
CO_2	20 levels	20
Spectral dispersion offset	1 shift per band	3
Spectral dispersion slope	1 slope per band	3
Chlorophyll fluorescence (O_2 band) (land only)	Mean + slope	2
Residual EOF amplitudes	3 per band	9

If significant differences remain between the predicted and measured radiances, then the state vector is updated, and the forward model is run again. The updates are done intelligently based on the partial derivatives (Jacobians) of the forward model and the noise characteristics of the measurements so the algorithm converges as quickly as possible. The entire process iterates until the convergence is reached, at which time the computed state vector is recorded as Level-2 data. This approach is computer intensive, but is often close to an optimal approach for estimating multiple parameters from many measurements using extremely complex forward models. For this reason, inverse methods are an increasingly common approach to data estimation in remote sensing.

Table 3.5 lists the 62 parameters in the state vector that are estimated for every OCO-2 measurement. These 62 parameters are derived from the forward model and the 3048 unique spectral measurements reported by OCO-2. You should particularly note that the OCO-2 estimates CO_2 at 20 levels in the atmosphere, surface reflectance in each band, surface wind speed over the ocean and atmospheric pressure at the surface everywhere. We particularly highlight the surface pressure since OCO-2 is one of the first satellite remote sensors capable of accurately estimating surface pressure, which is such a key parameter for numerical weather prediction.

More information on the design of OCO-2 and algorithms used for estimating atmospheric parameters from its measurements can be found in Crisp et al. (2004, 2008, 2019). The similar capabilities of the Japanese Greenhouse gases Observing SATellite (GOSAT) are described in Yoshida et al. (2011).

References

ASTER (Advanced Spaceborne Thermal Emission and Reflection Radiometer). (2021). NASA Jet Propulsion Laboratory. Retrieved from https://asterweb.jpl.nasa.gov/index.asp.

Barnes, W. L. (1977). Visible and infrared imaging radiometers for ocean observations. NASA.

Bell, E. E., Eisner, L., Young, J., & Oetjen, R. A. (1960). Spectral radiance of sky and terrain at wavelengths between 1 and 20 microns. II. Sky measurements. *Journal of the Optical Society of America, 50*(12), 1313–1320.

Breaker, L. C. (1990). Estimating and removing sensor-induced correlation from advanced very high resolution radiometer satellite data. *Journal of Geophysical Research: Oceans, 95*(C6), 9701–9711.

Cao, C., Xiong, X., Wolfe, R., DeLuccia, F., Liu, Q., Blonski, S., et al. (2017). Visible Infrared Imaging Radiometer Suite (VIIRS) Sensor Data Record (SDR) User's Guide Version 1.3. *NOAA Technical Report NESDIS, 142a*, 43.

Coakley, J. A., & Bretherton, F. P. (1982). Cloud cover from high-resolution scanner data: Detecting and allowing for partially filled fields of view. *Journal of Geophysical Research: Oceans, 87*(C7), 4917–4932.

Coll, C., & Caselles, V. (1997). A split-window algorithm for land surface temperature from advanced very high resolution radiometer data: Validation and algorithm comparison. *Journal of Geophysical Research: Atmospheres, 102*(D14), 16697–16713.

Cox, C., & Munk, W. (1954). Measurement of the roughness of the sea surface from photographs of the sun's glitter. *Journal of the Optical Society of America, 44*(11), 838–850.

Cox, C., & Munk, W. (1956). *Slopes of the sea surface deduced from photographs of sun glitter*. University of California Press.

Cracknell, A. P. (1997). *The advanced very high resolution radiometer*. Taylor & Francis.

Crisp, D., Atlas, R. M., Breon, F.-M., Brown, L. R., Burrows, J. P., Ciais, P., et al. (2004). The Orbiting Carbon Observatory (OCO) mission. *Advances in Space Research, 34*(4), 700–709.

Crisp, D., Bösch, H., Brown, L., Castano, R., Christi, M., Connor, B., et al. (2019). Orbiting Carbon Observatory-2 & 3 (OCO- & OCO-3) Level 2 full physics retrieval algorithm theoretical basis. *Jet Propulsion Laboratory, NASA*: Pasadena, CA, USA.

Crisp, D., Miller, C. E., & DeCola, P. L. (2008). NASA Orbiting Carbon Observatory:

Measuring the column averaged carbon dioxide mole fraction from space. *Journal of Applied Remote Sensing, 2*(1), 023508.

Deschamps, P. Y., & Phulpin, T. (1980). Atmospheric correction of infrared measurements of sea surface temperature using channels at 3.7, 11 and 12 μm. *Boundary-Layer Meteorology, 18*(2), 131–143.

Donlon, C. J., Minnett, P. J., Gentemann, C., Nightingale, T. J., Barton, I. J., Ward, B., & Murray, M. J. (2002). Toward improved validation of satellite sea surface skin temperature measurements for climate research. *Journal of Climate, 15*(4), 353–369.

Earth Observation Data Group, Department of Physics, University of Oxford. (2019). Atmospheric infrared spectrum atlas. Spectral data from the RFM line-by-line model with HITRAN 2012 line data and US Standard Atmosphere absorber profiles. Accessed at http://eodg.atm.ox.ac.uk/ATLAS/.

Edson, J. B., Jampana, V., Weller, R. A., Bigorre, S. P., Plueddemann, A. J., Fairall, C. W., et al. (2013). On the exchange of momentum over the open ocean. *Journal of Physical Oceanography, 43*(8), 1589–1610.

Emery, W. J., Castro, S., Wick, G. A., Schluessel, P., & Donlon, C. (2001). Estimating sea surface temperature from infrared satellite and in situ temperature data. *Bulletin of the American Meteorological Society, 82*(12), 2773–2785.

Fisher, J. B. (2018). *ECOSTRESS Level-3 Evapotranspiration L3(ET_PT-JPL) Algorithm Theoretical Basis Document* (tech. rep. No. D-94645). NASA JPL.

Fisher, J. B., Lee, B., Purdy, A. J., Halverson, G. H., Dohlen, M. B., Cawse-Nicholson, K., et al. (2020). ECOSTRESS: NASA's next generation mission to measure evapotranspiration from the International Space Station. *Water Resources Research, 56*(4), e2019WR026058.

Fisher, J. B., Middleton, E., Melton, F., Anderson, M., Hook, S., Hain, C., et al. (2016). Evapotranspiration: A critical variable linking ecosystem functioning, carbon and climate feedbacks, agricultural management, and water resources. In *2016 NASA decadal survey* (9p).

Fisher, J. B., Tu, K. P., & Baldocchi, D. D. (2008). Global estimates of the land–atmosphere water flux based on monthly AVHRR and ISLSCP-II data, validated at 16 FLUXNET sites. *Remote Sensing of Environment, 112*(3), 901–919.

Gillespie, A., Rokugawa, S., Matsunaga, T., Cothern, J. S., Hook, S., & Kahle, A. B. (1998). A temperature and emissivity separation algorithm for Advanced Spaceborne Thermal Emission and Reflection Radiometer (ASTER) images. *IEEE Transactions on Geoscience and Remote Sensing, 36*(4), 1113–1126.

Gladkova, I., Ignatov, A., Shahriar, F., Kihai, Y., Hillger, D., & Petrenko, B. (2016). Improved VIIRS and MODIS SST imagery. *Remote Sensing, 8*(1), 79.

Justice, C. O., Vermote, E., Townshend, J. R. G., Defries, R., Roy, D. P., Hall, D. K., et al. (1998). The Moderate Resolution Imaging Spectrora-diometer (MODIS): Land remote sensing for global change research. *IEEE Transactions on Geoscience and Remote Sensing, 36*(4), 1228–1249.

Liang, S. (2005). *Quantitative remote sensing of land surfaces.* John Wiley & Sons.

Matsunaga, T. (1994). A temperature-emissivity separation method using an empirical relationship between the mean, the maximum, and the minimum of the thermal infrared emissivity spectrum. *Journal of the Remote Sensing Society of Japan, 14*(3), 230–241 (in Japanese with English abstract).

Meissner, T., & Wentz, F. J. (2004). The complex dielectric constant of pure and sea water from microwave satellite observations. *IEEE Transactions on Geoscience and Remote Sensing, 42*(9), 1836–1849.

Merchant, C. J., Embury, O., Bulgin, C. E., Block, T., Corlett, G. K., Fiedler, E., et al. (2019). Satellite-based time-series of sea-surface temperature since 1981 for climate applications. *Scientific Data, 6*(1), 1–18.

Mesenbrink, M. L. (1996). *Complex indices of refraction for water and ice from visible to long wavelengths.* Air Force Institute of Technology, Wright-Patterson AFB, OH.

Mikhailenko, C. N., Babikov, Y. L., & Golovko, V. F. (2005). Information-calculating system spectroscopy of atmospheric gases, the structure and main functions. *Atmospheric and Oceanic Optics, 18*(9), 685–695.

Minnett, P. J., Alvera-Azcárate, A., Chin, T. M., Corlett, G. K., Gentemann, C. L., Karagali, I., et al. (2019). Half a century of satellite remote sensing of sea-surface temperature. *Remote Sensing of Environment, 233*, 111366.

Nicodemus, F. E., Richmond, J. C., Hsia, J. J., Ginsberg, I. W., & Limperis, T. (1977). *Geometrical considerations and nomenclature for reflectance.* US Department of Commerce, National Bureau of Standards.

Oki, T., & Kanae, S. (2006). Global hydrological cycles and world water resources. *Science, 313*(5790), 1068–1072.

Payan, V., & Royer, A. (2004). Analysis of Temperature Emissivity Separation (TES) algorithm applicability and sensitivity. *International Journal of Remote Sensing, 25*(1), 15–37.

Reynolds, R. W., Banzon, V. F., & NOAA CDR Program. (2008). NOAA Optimum Interpolation 1/4 Degree Daily Sea Surface Temperature (OISST) Analysis, Version 2. NOAA National Centers for Environmental Information. doi:10.7289/V5SQ8XB5 (Accessed December 2018).

Robel, J. (2009). NOAA KLM users guide. National Environmental Satellite, Data, and Information Service.

Robinson, I. S., Wells, N. C., & Charnock, H. (1984). The sea surface thermal boundary layer and its relevance to the measurement of sea surface temperature by airborne and spaceborne radiometers. *International Journal of Remote Sensing, 5*(1), 19–45.

Robinson, I. S. (1985). *Satellite oceanography; An introduction for oceanographers and remote-sensing scientists.* Chichester (UK) Horwood.

Saunders, P. M. (1967). The temperature at the ocean-air interface. *Journal of the Atmospheric Sciences, 24*(3), 269–273.

Saunders, P. M. (1968). Radiance of sea and sky in the infrared window 800-1200 cm^{-1}. *Journal of the Optical Society of America, 58*(5), 645–652.

Segelstein, D. J. (1981). *The complex refractive index of water.* Master's thesis, University of Missouri Kansas City.

Selby, J. E. A., & McClatchey, R. A. (1975). Atmospheric transmittance from 0.25 to 28.5 μm: Computer code LOWTRAN 3. Air Force Cambridge Research Laboratories, Air Force Systems Command, United States Air Force.

Shoemaker, E. M. (1959). *Impact mechanics at Meteor Crater, Arizona.* U.S. Geological Survey.

Shoemaker, E. M. (1974). *Guidebook to the geology of Meteor Crater, Arizona.* Center for Meteorite Studies, Arizona State University.

Smith, R. C., & Baker, K. S. (1981). Optical properties of the clearest natural waters (200–800 nm). *Applied Optics, 20*(2), 177–184.

Snyder, W. C., & Wan, Z. (1998). BRDF models to predict spectral reflectance and emissivity in the thermal infrared. *IEEE Transactions on Geoscience and Remote Sensing, 36*(1), 214–225.

Snyder, W. C., Wan, Z., Zhang, Y., & Feng, Y.-Z. (1998). Classification-based emissivity for land surface temperature measurement from space. *International Journal of Remote Sensing, 19*(14), 2753–2774.

Stewart, R. H. (1985). *Methods of satellite oceanography.* University of California Press.

Sun, D., & Pinker, R. T. (2005). Implementation of GOES-based land surface temperature diurnal cycle to AVHRR. *International Journal of Remote Sensing, 26*(18), 3975–3984.

Sun, D., Fang, L., & Yu, Y. Y. (2012). *GOES imager land surface temperature algorithm theoretical basis document.* NOAA NESDIS.

Wan, Z. (1999). *MODIS Land-Surface Temperature Algorithm Theoretical Basis Document (LST ATBD).* University of California, Santa Barbara.

Wolfe, W. L., & Zissis, G. J. (1985). *The infrared handbook.* Arlington: Office of Naval Research, Department of the Navy.

Wright, S. P., & Ramsey, M. S. (2006). Thermal infrared data analyses of Meteor crater, Arizona: Implications for Mars spaceborne data from the Thermal Emission Imaging System. *Journal of Geophysical Research: Planets, 111*(E2).

Yoshida, Y., Ota, Y., Eguchi, N., Kikuchi, N., Nobuta, K., Tran, H., et al. (2011). Retrieval algorithm for CO_2 and CH_4 column abundances from short-wavelength infrared spectral observations by the greenhouse gases observing satellite. *Atmospheric Measurement Techniques, 4*(4), 717–734.

Zeisse, C. R., McGrath, C. P., Littfin, K. M., & Hughes, H. G. (1999). Infrared radiance of the wind-ruffled sea. *Journal of the Optical Society of America A, 16*(6), 1439–1452.

Zeisse, C. R. (1995). Radiance of the ocean horizon. *Journal of the Optical Society of America A, 12*(9), 2022–2030.

Zhang, Y., Peña-Arancibia, J. L., McVicar, T. R., Chiew, F. H., Vaze, J., Liu, C., et al. (2016). Multi-decadal trends in global terrestrial evapotranspiration and its components. *Nature Scientific Reports, 6*, 19124.

4

Optical Sensing – Ocean Color

4.1 Introduction to Ocean Color

Figure 4.1 was created from multispectral data acquired from the Sea-Viewing Wide Field-of-View Sensor (SeaWiFS) aboard the OrbView-2 spacecraft. This instrument collected global ocean and land color data over a 13-year span from 1997 to 2010.

The figure maps chlorophyll concentration, providing a measure of the abundance of the chlorophyll-bearing, photosynthesizing organisms and their responses to changes in solar insolation, temperature, nutrient availability, rainfall, and other environmental factors. These organisms are at the base of the oceanic food chain, so they are important for all life on earth. Photosyntheticorganisms also affect climate – they fix CO_2 and that carbon can fall to the bottom of the ocean when the organisms die. In this chapter we will learn how measurements made at wavelengths of 400–900 nm are used to produce such estimates of chlorophyll concentration and other bio-optical properties. The review article by McClain (2009) and the Ocean Optics Web Book (Mobley et al., 2021) include good reviews of the promise and challenges of satellite ocean color observations.

There are substantial differences with the way visible light interacts with the atmosphere and oceans in comparison with infrared (Table 4.1). For example, most of the signal in the infrared was due to blackbody radiation, which is negligible in the visible. Atmospheric absorption was a significant factor in the infrared, but is negligible in the visible except

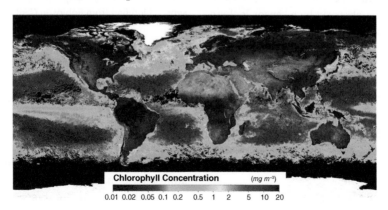

Figure 4.1 Map of chlorophyll obtained from satellite optical data. Based on Summer 2010 data from https://oceancolor.gsfc.nasa.gov.

Remote Sensing Physics: An Introduction to Observing Earth from Space, Advanced Textbook 3, First Edition.
Rick Chapman and Richard Gasparovic.
© 2022 American Geophysical Union. Published 2022 by John Wiley & Sons, Inc.
Companion website: www.wiley.com/go/chapman/physicsofearthremotesensing

Table 4.1 Differences between infrared and visible ocean observations.

	Infrared	Visible
Thermal radiation of ocean or atmosphere	Significant	Negligible, temperature does not matter
Atmospheric Absorption	Significant	Negligible, except for ozone
Atmospheric Scattering	Negligible	Significant
Skylight - Path Radiance	Driven by emissions that are proportional to absorption	Driven by scattering
Water Penetration	Negligible	Significant

for the absorption of blue and UV wavelengths by ozone.

Atmospheric scattering was negligible in the infrared, where both skylight and path radiance were driven by emissions. The story changes at visible wavelengths, where atmospheric radiance and transmittance is dominated by scattering of solar radiation by molecules and aerosols.

Note that ozone absorption in the upper atmosphere is the only significant nonscattering process. This is doubly significant for satellite remote sensing since solar radiation makes two passes through the ozone layer – one incoming and one outgoing.

Finally, as we have seen before, visible light can penetrate tens of meters into the ocean, whereas infrared radiation can penetrate less than 0.1 mm. This penetration is what allows photosynthesis to occur in marine plankton.

Figure 4.2 illustrates the major sources of visible radiation observed by a sensor looking down through the atmosphere at the ocean.

The two paths labeled a contribute to the water-leaving radiance L_w that is initially within the instrument field of view, or IFOV. Note that some of the water-leaving radiance makes it to the sensor (path b) and some is scattered away (path c). Only path b contains direct information regarding what we are trying to measure, namely the ocean color within the IFOV.

Path j leaves the water at an angle that would not have found its way to the sensor, but then

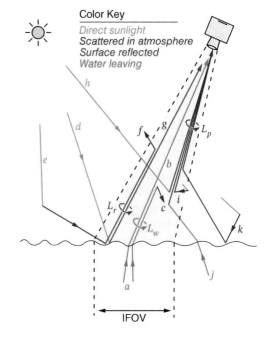

Figure 4.2 Pathways of optical radiation entering a sensor. Modified from Robinson (1983).

is scattered in the atmosphere into the field of view. While it originates within the water and ends up in the sensor, the wavelength dependence of atmospheric scattering affects the color from this path, so it is no longer representative of only the ocean color. It is, therefore, most appropriately considered one of the paths that end up contributing to the atmospheric path radiance L_p. The other paths that contribute are direct sunlight scattered back into the sensor (path h), multiple atmospheric scattering (path *i*), and skylight reflected off the surface outside the field of view and then scattered into the sensor (path *k*).

The other major source of light that makes it to the sensor is radiance L_r, the light reflected off the surface within the field of view back towards the sensor. Examples are reflected sunlight (path *d*) or reflected skylight (path *e*). Some of the reflected light makes it to the sensor (path *g*) and some is scattered away (path *f*).

In the end the total radiance received by the sensor L_T is the sum of three terms: the water-leaving radiance, the reflected solar radiance, and the atmospheric path radiance, with some adjustments made for the amount of the first two terms that are scattered away. As shown in equation (4.1), the first term is the water-leaving radiance L_w times the atmospheric transmittance τ_A. The second term is the sky radiance (including

direct sunlight) L_s times the Fresnel reflection coefficient r_w (L_r in Figure 4.2) times the atmospheric transmittance. And the final term is the atmospheric path radiance L_p.

$$L_T(\lambda, \theta, \varphi) = \tau_A(\lambda, \theta)L_w(\lambda, \theta, \varphi)$$
$$+ \tau_A(\lambda, \theta)r_w(\lambda, \theta, \varphi)L_s(\lambda, \theta, \varphi)$$
$$+ L_p(\lambda, \theta) \tag{4.1}$$

There are two primary challenges in ocean color measurements. First, the water-leaving spectral radiance L_w has to be related to some physical properties of interest, such as chlorophyll concentration, the concentration of dissolved organic matter or the concentration of suspended sediments in coastal waters. The second challenge is to estimate the effects of the atmosphere, both the transmittance and the path radiance. As before, these are not small problems, but they can be dealt with.

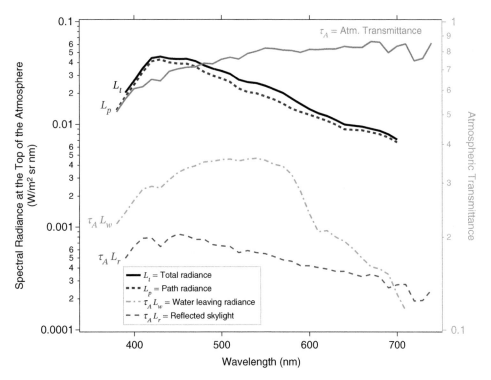

Figure 4.3 Contributions to the total radiance received by a space-based sensor. Adapted from Wilson and Austin (1978), with corrected vertical scale.

Table 4.2 Typical percentage contribution received by an optical sensor at the top of the atmosphere (Robinson, 1985) with permission from John Wiley & Sons.

| Wavelength (nm) | Contribution to signal, % | | | | | |
| | Clear water | | | Turbid water | | |
	τL_w	L_p	τL_r	τL_w	L_p	τL_r
440	14.4	84.4	1.2	18.1	80.8	1.1
520	17.5	81.2	1.3	32.3	66.6	1.1
550	14.5	84.2	1.3	34.9	64.1	1.0
670	2.2	96.3	1.5	16.4	82.4	1.2
750	1.1	97.0	1.9	1.1	97.4	1.5

Figure 4.3 shows why estimation of the path radiance is so important. The figure plots the typical contributions of various terms in the equation for the total radiance received by a space-based sensor looking down through the atmosphere to the ocean.

First, the atmospheric transmittance (τ_A), which is plotted against the right-hand scale, is about 60% in the UV at 350 nm increasing to 80% in the red at 700 nm. The decline of the transmittance towards the UV is due to ozone absorption.

All of the rest of the curves are plotted against the radiance scale on the left-hand side of the plot. The bottom curve L_r is the reflected skylight. L_w is the water-leaving radiance, which is strongly biased to the blue-green. Note that at its peak, the water-leaving radiance is at least 5–6 times the radiance of the reflected skylight. Also note the water-leaving radiance falls to less than the radiance of the reflected skylight above 700 nm. The relatively small contributions from the surface terms to the total radiance at long wavelengths suggests the red channels could be used to estimate path radiance.

The next curve is the path radiance L_p. Notice it is another factor of five brighter than the water-leaving radiance. Thus five out of six photons entering a visible sensor at the top of the atmosphere arise from path radiance due to scattering in the atmosphere, while only one out of six photons has actually interacted with the ocean. The upper curve L_T shows the total radiance.

Table 4.2 further illustrates the relative contributions of the various terms at various wavelengths. Again we see that for clear water 80–85% of the photons at the peak wavelength of 520–550 nm come from path radiance. This increases to greater than 95% at 670–750 nm.

The columns on the right indicate the relative contributions to the radiance when observing turbid waters. In this case the peak of the water-leaving radiance shifts towards green, and the contribution from upwelling light about the peak nearly doubles relative to clear waters. This is because turbid waters contain more particulates, such as suspended sediments, and more dissolved organic matter – the particulates increase scattering and the dissolved organic matter increases absorption in the blue. Of course, turbidity does not affect surface reflection, so those percentages are relatively unchanged.

The processing chain for ocean color measurement is similar to that of infrared sensing (Figure 4.4). The data are acquired then calibrated. Atmospheric corrections are then applied to compute water-leaving radiances

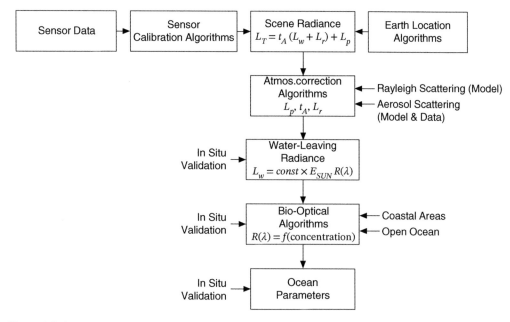

Figure 4.4 Ocean color processing.

at key wavelengths. These water-leaving radiances are then combined in bio-optical algorithms to produce ocean parameters.

4.2 Fresnel Reflection

We begin an investigation of the various terms in equation (4.1) by reviewing the Fresnel reflection coefficients at visible wavelengths r_w. Recall from Chapter 3 that the optical properties of seawater are described by the complex index of refraction, with the real value n being the index of refraction and k_0 being the extinction coefficient (equation (3.100)). For visible wavelengths, the index of refraction for sea water is approximately $n = 1.34$ and essentially independent of wavelength and temperature. Moreover, the extinction coefficient at these wavelengths is $k_0 = 10^{-9}$ implying a flux absorption depth of 40 m.

At visible wavelengths, $k_0 \ll n$, so k_0 can be ignored in the equations for the Fresnel coefficients. The Fresnel reflection coefficients for visible radiant flux incident from the air onto a flat water surface at an incident angle θ are

then given by:

$$r_h(\theta) = \left[\frac{a - \cos\theta}{a + \cos\theta}\right]^2$$

$$r_v(\theta) = \left[\frac{n^2 \cos\theta - a}{n^2 \cos\theta + a}\right]^2 \qquad (4.2)$$

for horizontal and vertical polarization where $a^2 = n^2 - \sin^2\theta$ (equations (3.101)–(3.106)). The reflection coefficient for unpolarized light is the average of the two polarized components: $r = 0.5(r_h + r_v)$.

These expressions can be simplified even further by application of Snell's law. Figure 4.5 shows the incident, reflected and transmitted beams for sun or sky light hitting the flat ocean surface. The air has refractive index $n_a = 1$. The water has refractive index $n_w = 1.34$. The figure defines three angles relative to the normal to the interface: the angle of incidence θ_i, the reflected angle θ_r and the angle of transmittance θ_t. Because the incident angle equals the reflection angle, we define a single angle in air: $\theta_a = \theta_i = \theta_r$. Likewise we define the angle in water $\theta_w = \theta_t$.

Now Snell's law relates the angle of incidence to the angle of transmittance:

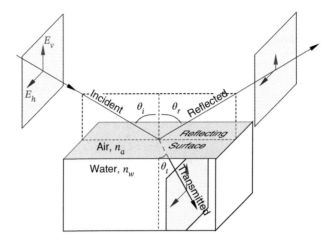

Figure 4.5 Geometry of reflection and transmission at air–water interface.

$n_a \sin \theta_a = n_w \sin \theta_w$. Some algebra can then be used to simplify the equations for the Fresnel coefficients to this more common form:

$$r_\perp = r_h = \frac{\sin^2(\theta_a - \theta_w)}{\sin^2(\theta_a + \theta_w)}$$

$$r_\| = r_v = \frac{\tan^2(\theta_a - \theta_w)}{\tan^2(\theta_a + \theta_w)} \qquad (4.3)$$

where the subscripts $\perp$ and $\|$ refer to the electric field component perpendicular and parallel to the plane of incidence, respectively. Again, the reflection coefficient for unpolarized light is simply the average of the horizontal and vertical reflection coefficients.

Because light penetrates into seawater, we now consider the propagation of upwelling light from beneath the sea surface into the atmosphere. Figure 4.6 shows rays from the water refracting through a flat sea surface into the air. According to Snell's law, a ray propagating vertically in the water, stays vertical in the atmosphere. But a ray propagating at an angle in the water bends toward the horizon in the air. Total internal reflection occurs when:

$$n_w \sin \theta_w \geq \sin \pi/2 = 1 \qquad (4.4)$$

so a ray at an angle of 48.2° bends all the way to the horizon. Rays at angles larger than this are

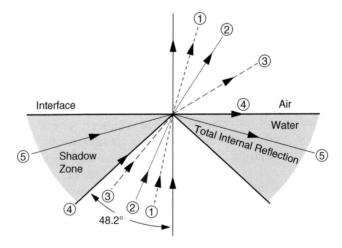

Figure 4.6 Refracting and total internal reflection. Adapted from Martin (2014).

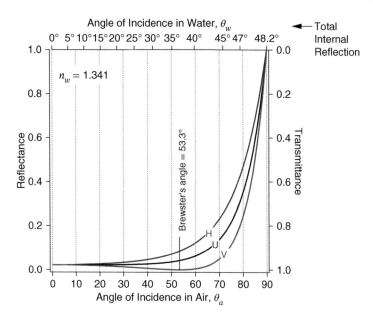

Figure 4.7 Fresnel reflection curves for a flat surface in air and water.

reflected at the air–water interface and do not propagate into the atmosphere.

These rays are said to have undergone a total internal reflection; total because no energy is transmitted into the atmosphere. These paths are reciprocal, so there is a shadow zone looking up from under the surface.

Figure 4.7 plots the Fresnel reflection curves for a flat surface as a function of incident angle in air on the bottom axis or incident angle in water on the top axis. Notice the scale of the incident angle in water only goes up to 48°, the angle of total internal reflection. From top to bottom, the three curves are for horizontal, unpolarized and vertical polarization.

Note the vertical polarization goes through zero at the Brewster's angle of 53.3°, a value slightly different than the 49.6° we saw at infrared wavelengths. All of the vertically-polarized incident energy is transmitted into the water at the Brewster's angle.

The last curves are the reflectance for a flat ocean surface, but the ocean is usually covered with waves, so it is rarely flat. Figure 4.8 plots the reflectance on the left-hand axis

and transmittance on the right-hand axis of a water/air boundary as a function of the angle of observation for several wind speeds. Measurements show that the relationship between wind speed and the variance of surface slopes is nearly linear, so varying wind speeds correspond to different degrees of surface roughness. In all cases the angle of observation is measured relative to the mean surface normal.

These curves are computed from averages of the reflection coefficients weighted by the Gaussian probability density function for surface slopes. The law of reflection and Snell's law are used to relate the mean incident angle and surface slope to the incident, reflected, and transmitted angles for each facet.

The upper left set of blue curves are for radiance coming up from below the surface; the red curves in the lower right corner are for radiance incident from above. These curves are for unpolarized radiation. You can tell it is not vertical polarization because the reflection coefficient from above at zero wind speed has no dip at the Brewster's angle. This is the same as the unpolarized curve in Figure 4.7. As the wind speed increases, the reflection of light

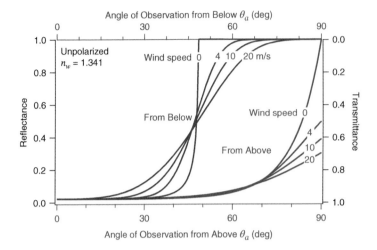

Figure 4.8 Reflection and transmission coefficients for unpolarized radiation incident on a wind-roughened surface. Based on Austin (1974).

from above decreases towards the horizon, but waves make little difference at incident angles less than 60°.

Next consider the upper left set of blue curves for light coming from below. The left portion of these curves indicates for angles within 20° of straight up, about 98% of the light from below is transmitted into the atmosphere and only 2% is reflected back into the water. At zero wind speed, the reflectance increases sharply past 40°, until total internal reflection occurs at 48°. At higher wind speeds, the effects of increased reflection and reduced

transmittance are spread to a wider range of observation angles.

4.3 Skylight

The last few figures covered reflection and transmission through the surface. The next step is to characterize the light coming down onto the surface from the above. Figure 4.9 illustrates the downwelling sky radiance on a clear day. The left-hand panel is a polar plot of the logarithm of the radiance for the sun at

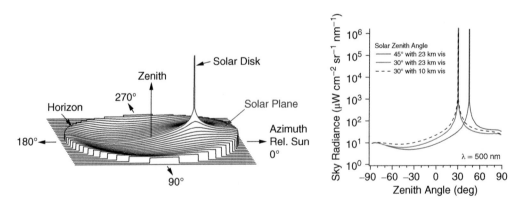

Figure 4.9 Example sky radiance distributions on a clear day. Left-hand panel is a plot of the logarithm of the radiance for the sun at a solar zenith angle $\theta_s = 45°$. The right-hand panel plots sky radiance for three combinations of solar zenith angle and horizontal visibility at the surface. Figure based on MODTRAN calculations.

a solar zenith angle $\theta_s = 45°$. The right-hand panel is a plot of sky radiance as a function of zenith angle, with 0° corresponding to looking straight up[1].

The left-hand panel illustrates that the scattered skylight is at a maximum near the solar disk, and at a minimum for viewing directions corresponding to scattering angles of 90° relative to the sun. This minimum arises because scattering events tend to scatter forward or backwards, and rarely scatter at 90° relative to the source. Sky radiance increases towards the horizon because of the extra path length through the lower portion of the atmosphere that contains most of the aerosol scatterers.

The right-hand panel plots the sky radiance as a function of zenith angle for three combinations of solar zenith angles and horizontal visibility at the surface. These values were computed in the solar plane, which is defined by two vectors: the zenith vector and the vector pointing to the sun. The plots illustrate several effects: the radiance of the solar disk can be a factor of 10,000 larger than the scattered skylight; changing solar angles affect the location of both the maximum and minimum radiances; and an increasing aerosol concentration indicated by decreasing horizontal visibility decreases the solar radiance and increases the radiance away from the sun.

4.4 Water-Leaving Radiance

Water-leaving radiance is precisely the radiance from beneath the surface that has refracted through the air–water interface into the sensor IFOV. Calculation of the water-leaving radiance requires careful treatment of the transition of downwelling radiation from the sun and sky into the water, the scattering of some of that radiation back towards the surface, the underwater absorption of

some of that radiation, and the transition of the upwelling radiation from the water back into the air. These are the topics of this section.

While refraction affects the angle of transmission based on the incident angle and the index of refraction in each medium, refraction also affects radiance. Figure 4.10 illustrates a bundle of rays leaving the water, refracted through the interface into the air. The flux is conserved across the interface, but the solid angle changes, affecting the radiance.

To compute the change to the solid angle, recall the definition of the differential solid angles in air and water:

$$\frac{d\Omega_w}{d\Omega_a} = \frac{\sin\theta_w\, d\theta_w\, d\varphi_w}{\sin\theta_a\, d\theta_a\, d\varphi_a} \quad (4.5)$$

Snell's law states that θ changes, but φ does not:

$$n_w \sin\theta_w = n_a \sin\theta_a \quad (4.6a)$$
$$d\varphi_w = d\varphi_a \quad (4.6b)$$

Differentiate equation (4.6a):

$$n_w \cos\theta_w\, d\theta_w = n_a \cos\theta_a\, d\theta_a \quad (4.7)$$

Then multiply equations (4.6a) and (4.7):

$$n_w^2 \sin\theta_w \cos\theta_w\, d\theta_w = n_a^2 \sin\theta_a \cos\theta_a\, d\theta_a \quad (4.8)$$

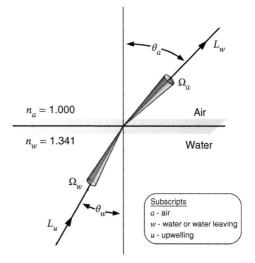

Figure 4.10 Geometry of a single ray bundle at air–water interface. Adapted from Austin (1974).

---

1 Other key terminology for angles include nadir angles that are measured relative to looking straight down, or grazing angles that are measured relative to the horizon.

Set the index of refraction in air $n_a = 1$ and combine equations (4.5) and (4.8), then integrate to get an expression for the change in solid angle across the interface:

$$\Omega_a = n_w^2 \frac{\cos\theta_w}{\cos\theta_a} \Omega_w \qquad (4.9)$$

As illustrated in Figure 4.11, the light can flow in either direction across the interface, and in either case the flux is conserved. The radiance can then be shown to scale as n_w^2. The ray bundles spread leaving the water, so the radiances decrease. The ray bundles compress entering the water, so the radiances increase. Table 4.3 lists the subscripts used throughout this text.

The left-hand panel of Figure 4.12 is an image that was taken looking up from under the water with a 14-mm fisheye lens on a clear day with calm seas. The whole sky,

Table 4.3 Subscripts used in this text.

a	Air	s	Sky
w	Water	i	Incident
d	Downwelling	r	Reflected
u	Upwelling	t	Transmitted

horizon-to-horizon, is visible in this image having been refracted into a cone that is referred to as Snell's window. Rays outside of this cone undergo total internal reflection and hence are dark. The diagram in the right-hand panel illustrates the geometry of Snell's window.

Some example measurements of the directional distribution of light in the ocean on a perfectly clear day are shown in Figure 4.13. The dotted curve is the directional distribution of the relative downwelling radiance measured at a depth of 1.75 m. The solar zenith angle for this measurement was 29.5°, corresponding to a refracted angle in water of 21.7°. The radiance in this case was normalized to 1 at an angle of 90°. The solid curve was measured at a depth of 20.4 m with a solar zenith angle of 75.5°. The peak radiance is reduced at greater depth because of absorption, and the angle of the peak radiance is greater because the sun is lower in the sky. If the sun was on the horizon, the refracted angle in water would be 48.2°. When looking from below, any angles beyond that value suffer total internal reflection, and hence appear dark.

The lower dashed curve in Figure 4.13 is the directional distribution of the relative

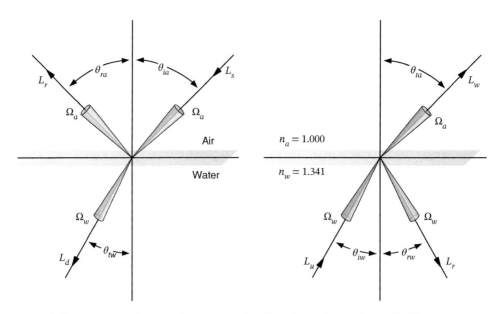

Figure 4.11 Geometry for radiance at the air–water interface. Adapted from Austin (1974).

 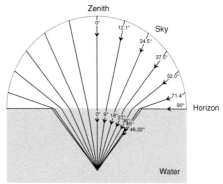

Figure 4.12 Example downwelling light field under the ocean surface. Source: David K. Lynch and Simon Higton from https://epod.usra.edu/blog/2014/06/snells-window.html. With permission from Lynch and Higton.

upwelling radiance measured at a depth of 1.75 m. Notice that the light coming from below is quite diffuse, relative to the downwelling radiance. The upwelling radiance is due to scattering of the downwelling radiance into the upward direction. Most of the upwelling photons are scattered multiple times, with each individual scattering event randomly redirecting the photon into a new direction. Thus multiple scattering throughout the volume effectively diffuses the

directionality of the upwelling radiance. Scattering is also the reason that the downwelling radiance at angles outside of Snell's window are not completely dark.

Figure 4.14 plots the downwelling and upwelling radiance distributions for three different depths: 5, 20, and 53 m. All of the curves have been normalized to 1 at 90° – the absolute radiances decay exponentially with depth. Notice that the downwelling is highly directional at 5 m depth, but becomes less

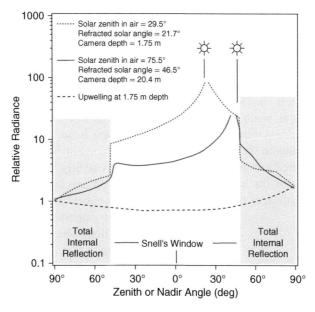

Figure 4.13 Directional distribution of underwater light field measured in the solar plane for two different depths and solar zenith angles. Based on data from Smith (1974).

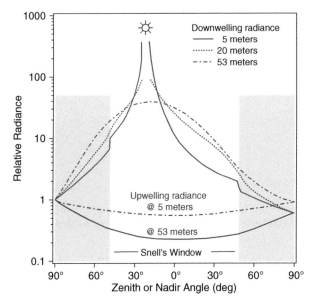

Figure 4.14 Upwelling and downwelling distributions at three depths. Based on data from Smith (1974).

directional with increasing depth. This is also due to scattering in the water column. Furthermore, the distribution of upwelling radiance at 5-m and 20-m depths are approximately the same. In all cases the upwelling light is not highly directional.

4.5 Water Column Reflectance

Figures 4.13 and 4.14 show that underwater scattering is effectively diffuse, so the upwelling radiance is relatively independent of viewing direction and solar incidence angle. This observation motivates us to treat the effects of the water column on the water-leaving radiance in terms of the ratio of the upwelling irradiance to the downwelling irradiance, both measured just below the surface. This factor is called the water column reflectance $R(\lambda)$, also known as the irradiance reflectance:

$$R = \frac{E_u}{E_d} \tag{4.10}$$

where the upwelling and downwelling terms are illustrated in Figure 4.15. Note that R is an effective reflectance arising because of scattering throughout the water column.

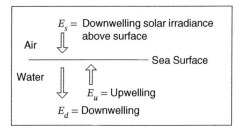

Figure 4.15 Diagram of upwelling and downwelling irradiance.

The water column reflectance is referred to as an apparent optical property. This means the property does depend somewhat on the directional distribution of the incident light fields, and hence will change based on the lighting conditions such as solar angle, atmospheric visibility, and cloud cover. What we need to do next is to relate this reflectance to intrinsic optical properties, that is, the absorption and scattering coefficients of the water column that do not depend on the illumination conditions.

Various relationships between the reflectance and the intrinsic properties of the water column have been derived from radiative transfer models. These relations show the reflectance to be proportional to the beam backscatter coefficient, b_b, divided by the sum of the absorption and backscatter coefficients (Gordon & Morel, 1983):

$$R(\lambda) = f \frac{b_b(\lambda)}{a(\lambda) + b_b(\lambda)} \qquad (4.11)$$

where f is a nondimensional parameter, a is the water absorption coefficient, and b_b is the water beam backscattering coefficient. Equation (4.11) can be interpreted as follows: the reflectance is proportional to the probability that a photon entering the ocean is backscattered to the surface, since the fate of any such photon is either to be absorbed within the medium or backscattered to the surface.

As discussed earlier in Chapter 3, the absorption coefficient is related to the extinction coefficient, k_0, the imaginary part of the index of refraction of water:

$$\alpha = \frac{4\pi k_0(\lambda)}{\lambda} \qquad (4.12)$$

Three assumptions are made in developing an approach to account for scattering due to small particulates in water:

- A plane wave of irradiance E is incident on a small volume element $dA\ dz$.
- The flux scattered from the volume element is rotationally symmetric about the propagation direction. In other words the scattering is only dependent on the angle relative to the original path, not the azimuth about that path.
- There is no wavelength change in the scattered flux – the scattered wavelengths equal the incident wavelengths. This assumption is invalid at wavelengths where fluorescence occurs, a phenomenon discussed in Section 4.10.

Consider the scattering that arises in a specific volume element. The power incident on the volume element is the irradiance times the cross-sectional area of the element:

$$\Phi_0(\lambda) = E(\lambda)\ dA \qquad (4.13)$$

The power $d\Phi_s$ scattered in direction α arising from the scattering volume can then be expressed as the intensity dI times the solid angle $d\Omega$:

$$d\Phi_s(\lambda) = dI(\alpha, \lambda)d\Omega \qquad (4.14)$$

The volume scattering function β is defined to be the ratio of the scattered power per unit length dI/dz to the incident power $E\ dA$, with units of per meter per steradian. So the volume scattering function needs to be integrated over the length of the scattering volume to compute the total power scattered into each solid angle $d\Omega$:

$$\begin{aligned}\beta(\alpha, \lambda) &= \frac{dI(\alpha, \lambda)}{E(\lambda)\ dA\ dz} \\ &= \frac{d\Phi_s(\lambda)}{\Phi_0(\lambda)\ d\Omega\ dz} \quad \text{m}^{-1}\ \text{sr}^{-1}\end{aligned} \qquad (4.15)$$

Next we define the scattering coefficient, b, to be given by the integral over all possible solid angles of the volume scattering function β. Because β only depends on the single scattering angle, this expression can be simplified to an integration over a single dimension.

$$\begin{aligned}b(\lambda) &= \int_{4\pi} \beta(\alpha, \lambda)d\Omega \\ &= 2\pi \int_0^\pi \beta(\alpha, \lambda)\sin\alpha\ d\alpha\end{aligned} \qquad (4.16)$$

If the scattering is isotropic, then we can further simplify: $\beta = $ constant and $b = 4\pi\beta$.

Shifrin (1998) proposed particularly simple models for the volume scattering function β and the scattering coefficient b for pure seawater at 20°C based on fits to data reported by Morel et al. (1974). According to Twardowski et al. (2007), this can be scaled to account for salinity by multiplying by $[1 + 0.3S/37]$, where S is the salinity in practical salinity units. The resultant models are:

$$\beta(\alpha, \lambda) = 9.3 \times 10^{-5} \left(\frac{\lambda_0}{\lambda}\right)^{4.17} (1 + 0.835\cos^2\alpha)$$
$$\times \left(1 + 0.3\frac{S}{37}\right)\ \text{m}^{-1}\ \text{sr}^{-1} \qquad (4.17)$$

$$b(\lambda) = 1.49 \times 10^{-3} \left(\frac{\lambda_0}{\lambda}\right)^{4.17}$$
$$\times \left(1 + 0.3\frac{S}{37}\right)\ \text{m}^{-1} \quad \lambda_0 = 546\ \text{nm} \qquad (4.18)$$

This simple model is likely to be accurate to within a few percent relative to more recent models such as Zhang et al. (2009) and Zhang and Hu (2009).

The backscatter coefficient b_b is defined to be the integral of the volume scattering function over the hemisphere away from the direction of propagation:

$$b_b(\lambda) = 2\pi \int_{\pi/2}^{\pi} \beta(\alpha, \lambda) \sin \alpha \, d\alpha = 0.5 \, b(\lambda)$$

$$(4.19)$$

Note that for either isotropic scattering or scattering by very small particles that scatter symmetrically in the forward or backward directions, the backscatter coefficient is one-half of the scattering coefficient, although this is not generally the case.

The absorption and backscatter coefficients can both be measured in situ using collimated beams of light and sensors situated at the end of the beam to measure attenuation and looking at the beam from off axis to measure backscatter. For many waters, it has been found that a good approximation for the above relation is given by this simpler ratio of backscatter to absorption coefficient:

$$R(\lambda) = G \, \frac{b_b(\lambda)}{a(\lambda)} \qquad (4.20)$$

G in this expression is a constant that depends on the directional distribution of the incident light field and the volume scattering function. For a flat sea surface and the sun at the zenith, $G = 0.33$ (Morel & Prieur, 1977).

4.5.1 Pure Seawater

To compute reflectance, first consider pure seawater. This is water with the same amount of dissolved salts as seawater, but without any foreign particulate matter or dissolved organic matter. The absorption and backscatter coefficients of pure seawater are plotted in Figure 4.16 based on the empirical models presented in the Ocean Optics Web Book (Mobley et al., 2021). The plot shows how scattering decreases towards longer wavelengths,

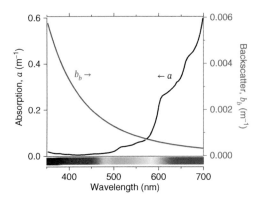

Figure 4.16 Absorption and backscatter coefficients for pure seawater.

while absorption increases. Scattering in pure seawater is due to both water molecules and local fluctuations in the optical density arising from varying sea salt concentration. These scattering elements are physically small, so it should not be surprising that scattering is reduced with increasing wavelength.

The fact that there is little water-leaving radiance at wavelengths above 600 nm is due to strong absorption in this spectral region. Figure 4.17 shows the water column reflectance for pure seawater computed from the model in Mobley et al. (2021).

Again this is a ratio of irradiances and does depend to some extent on the illumination

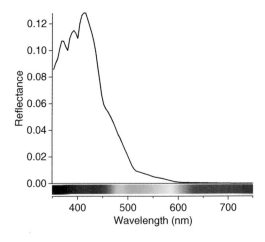

Figure 4.17 Water column reflectance for pure seawater.

geometry. The fact that the reflectance strongly peaks in the blue is why clear ocean waters appear blue. This, in turn, is a result of the increased scattering in ocean water at short wavelengths along with increased absorption in the red.

4.5.2 Case 1 Waters

Historically, optical oceanographers have divided ocean waters into two classes based on the dominant components influencing the optical properties:

Case 1 waters – optical properties are dominated by phytoplankton (single-cell plants) found in the solar-illuminated surface water, and

Case 2 waters – water containing phytoplankton and/or suspended sediments and/or yellow substances ("gelbstoffe") which are dissolved organic materials.

These are not precise categories, and their meaning has shifted over the years. Mobley et al. (2004) suggested it was time to stop using these terms, arguing that there is no way to formulate well-defined and scientifically justified categories for unambiguously classifying a water body according to its optical properties. The arguments made by Mobley are well worth reading, but we are going to ignore those arguments in this text because these categories are still in common use, and provide a simple

if imprecise means of organizing our thinking about the optical properties of different waters.

Deep ocean waters away from coastlines are primarily Case 1 waters. In Case 1 waters, the absorption and backscatter coefficients are modeled with particularly simple forms:

$$a(\lambda) = a_w(\lambda) + a_p(\lambda)c \qquad (4.21)$$

$$b_b(\lambda) = b_{bw}(\lambda) + b_p(\lambda)c \qquad (4.22)$$

Each consists of a sum of two terms: the first based on pure seawater, and the second based on the effects of chlorophyll. The chlorophyll pigment concentration c, measured in mg per cubic meter, is used to scale the second term. It then follows that the reflectance and water-leaving radiance are dependent on chlorophyll concentration, although the relationship is nonlinear and complicated.

The absorption coefficient a, backscatter coefficient b_b, and reflectance R for Case 1 waters are plotted in Figure 4.18 for varying levels of chlorophyll concentration as a function of wavelength. Increasing levels of chlorophyll lead to significant increases in absorption at shorter wavelengths, and additional scattering at all wavelengths. The result is that increasing chlorophyll concentration causes decreasing reflectance at blue wavelengths, whereas the opposite occurs at red wavelengths. The peak in the reflectance indicates why clear water appears blue, while chlorophyll-rich waters appear green.

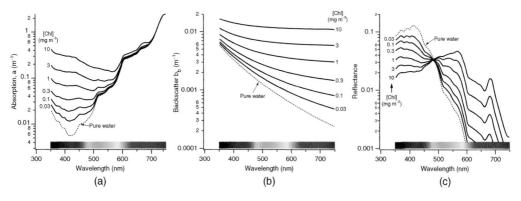

Figure 4.18 Spectral variation of (a) absorption, (b) backscatter, and (c) reflectance for varying chlorophyll concentrations. Figure from models in Mobley et al. (2021)/CC BY 4.0.

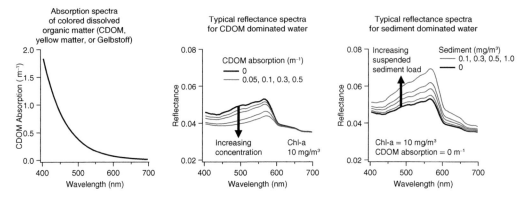

Figure 4.19 Effect of dissolved organic materials and sediments on reflectance. Figure inspired by Robinson (1983).

These results were obtained from the Hydro-Light model, a radiative transfer model that has undergone substantial validation against ocean optics data[2].

4.5.3 Case 2 Waters

Next consider Case 2 waters that may contain sediments and dissolved organic materials in addition to phytoplankton. There are substantial challenges in modeling the contributions from the other optically-active constituents in Case 2 waters:

- Scattering and absorption spectra of these constituents vary widely and the spectra for the different components overlap.
- Suspended sediments can vary in size and reflective properties and frequently have significant reflectance at longer wavelengths.
- Concentrations of each constituent vary seasonally and geographically preventing the development of universal algorithms.
- Case 2 waters are frequently shallow enough that solar illumination can reflect from the bottom making the returns dependent on bottom reflectance.

_____
2 HydroLight utilizes an invariant imbedding technique to solve the radiative transfer equation as described in Mobley (1994). See http://www.oceanopticsbook.info/view/radiative_transfer_theory/level_2/hydrolight for more information on HydroLight.

The assumption of no water-leaving radiance at long wavelengths may not be valid for Case 2 waters, so a different atmospheric correction approach is needed. Developing validated algorithms to retrieve constituent concentrations from Case 2 waters is an active area of current research. The most successful approach has been to acquire extensive in situ data in one location in order to develop and validate site-specific algorithms.

Figure 4.19 illustrate some of the difficulties. The left-hand plot shows the approximate absorption spectrum of colored dissolved organic material, also known as CDOM, gelbstoffe, or yellow substance. The plot makes it clear that dissolved organic material predominantly absorbs in the blue, with little impact in the red. The middle plot shows the typical reflectance spectra for CDOM dominated water, with the arrow indicating increasing concentration of dissolved organic materials. The black line is the clear water spectrum. The right-hand plot shows the typical reflectance spectra for sediment-dominated water. Sediments are small scattering particles and so to the first order they act to increase reflectance at all wavelengths. The arrow indicates increasing suspended sediment load. All of these curves are based on HydroLight calculations.

These additional constituents mix with phytoplankton in unknown proportions, adding many free parameters that need to be estimated.

4.6 Remote Sensing Reflectance

The previous section described how water column constituents, such as chlorophyll concentration, organic material, and suspended sediments, affect the water column reflectance R, a parameter that is readily measured with shipboard instruments. While the use of the water column reflectance was motivated by its relative independence on viewing angles and illumination conditions, some dependency remains. Any dependence on viewing geometry and illumination conditions limits the accuracy of our ability to estimate parameters such as chlorophyll concentrations from measured radiances. This ultimately led to the use of another apparent optical property, the remotely sensed reflectance R_{rs}, for estimating ocean color and chlorophyll concentration.

A simplified treatment of the water-leaving radiance can be developed based on the illustration in Figure 4.20.

The index of refraction in water is 1.34. The transmittance of the surface from below, designated by the symbol t_{w-}, is approximately 0.98 for near-vertical angles. In general the Fresnel equations and optical constants of water can be used to calculate t_{w-}. The water-leaving radiance just above the surface can then be expressed in terms of the upwelling radiance

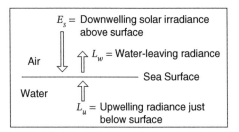

Figure 4.20 Diagram of upwelling and water-leaving radiance.

just below the surface:

$$L_w \cos \theta_a \Omega_a = t_{w-} L_u \cos \theta_w \Omega_w \quad (4.23)$$

Next recall the relationship between the solid angles above and below the surface:

$$\Omega_a \cos \theta_a = n^2 \Omega_w \cos \theta_w \quad (4.24)$$

Applying the relationship yields this simplified expression for the water-leaving radiance:

$$L_w = \left(\frac{t_{w-}}{n^2}\right) L_u \quad (4.25)$$

For sensor viewing geometries within 40° of nadir, the emergent rays arise from the water in a cone with a half angle of 29°. And as illustrated in Figure 4.8, the transmittance from below through a rough surface is approximately constant over a moderate range of near-nadir angles and independent of wind speed. For ocean color measurements made with the sensor pointed within 40° of nadir, a reasonable approximation is to assume the sensor is looking at nadir. Then the water-leaving radiance can be estimated to be about 55% of the upwelling radiance:

$$L_w = \left(\frac{t_{w-}}{n^2}\right) L_u = \left(\frac{0.98}{1.34^2}\right) L_u = 0.55 L_u \quad (4.26)$$

It turns out that L_u is difficult to compute because it is determined by the absorption and scattering of sunlight in the near-surface water column. So we take a heuristic approach based on replacing the upwelling radiance with the upwelling irradiance, so $E_u = f L_u$. If the upwelling radiance was equal in all directions, as if scattered from a Lambertian surface, then $f = \pi$. It turns out that ocean scattering is not quite Lambertian, but ocean data are well approximated by setting f to be between 3.5 and 5, with clearer water being closer to 3.5 (Gordon & Morel, 1983).

Then the water-leaving radiance becomes:

$$L_w = \left(\frac{t_{w-}}{f n^2}\right) E_u \quad (4.27)$$

Next we introduce the concept of the diffuse transmittance t_{w+} from above to below the surface, namely the ratio of the downwelling irradiance just below the surface E_d to the downwelling irradiance just above the surface E_s:

$$E_d = t_{w+}E_s \qquad (4.28)$$

The diffuse transmittance can be computed as a ratio of weighted integrals:

$$t_{w_+} = 1 - \frac{\int_0^{2\pi}\int_0^{\pi/2} r(\theta)L_{sky}(\theta,\varphi)\cos\theta\sin\theta\,d\theta\,d\varphi}{\int_0^{2\pi}\int_0^{\pi/2} L_{sky}(\theta,\varphi)\cos\theta\sin\theta\,d\theta\,d\varphi} \approx 0.94 \qquad (4.29)$$

The ratio of integrals on the right is nothing more than the average reflectance of the surface, and the transmittance is just one minus this reflectance.

Combining terms, the water-leaving radiance can be written:

$$L_w(\lambda) = \frac{t_{w-}}{n^2 f}E_u = \left(\frac{t_{w-}t_{w+}}{n^2 f}\right)R(\lambda)E_s(\lambda) \qquad (4.30)$$

where we have been explicit about the dependence of the water column reflectance and downwelling solar radiation on wavelength.

While equation (4.30) was a component in the original approach for ocean color estimation, the dependence of the water column reflectance on viewing angle and illumination conditions made it less than ideal. This led to the introduction of the remote sensing reflectance R_{rs}, defined as the ratio of the water-leaving radiance to the downwelling solar irradiance at the sea surface:

$$R_{rs}(\lambda) = \frac{L_w(\lambda)}{E_s(\lambda)} = \left(\frac{t_{w-}t_{w+}}{n^2 f}\right)R(\lambda) \qquad (4.31)$$

Notice that R_{rs} has units of sr^{-1}. Applying equation (4.30) and using previously noted values for $t_{w-} = 0.98$, $t_{w+} = 0.94$, $n = 1.34$, and

$f = 3.5$–5, we can obtain a rough estimate for the bounds of R_{rs}:

$$R_{rs}(\lambda) = \left(\frac{t_{w-}t_{w+}}{n^2 f}\right)R(\lambda)$$

$$\approx \begin{cases} 0.15\,R(\lambda) & \text{for pure seawater} \\ 0.10\,R(\lambda) & \text{for Chl} = 10 \text{ mg m}^{-3} \end{cases} \qquad (4.32)$$

While equation (4.31) is an approximate expression for the water-leaving radiance, full radiative transfer codes are normally used for such calculations. Figure 4.21 shows the remote sensing reflectance and the water-leaving spectral radiance for various chlorophyll concentrations appropriate for Case 1 waters based on HydroLight calculations. Also indicated in Figure 4.21 are the spectral bands used for water color measurements in the SeaWiFS satellite sensor, with the specific bands used for chlorophyll estimation shown in black.

The basis of empirical algorithms for estimating chlorophyll concentration is easy to see from this diagram. Notice that the water-leaving radiance is relatively independent of chlorophyll concentration at a wavelength of about 480 nm, while at 550 nm the water-leaving radiance is strongly dependent on chlorophyll concentration. The radiance in any two channels will be affected by other factors such as illumination conditions, but those factors are somewhat wavelength independent. So it seems natural that the algorithms to estimate chlorophyll concentration will utilize ratios of the radiance at multiple wavelengths in the blue and blue-green to the radiance at 550 nm. This approach essentially estimates the varying slopes of these curves that are then related to the pigment concentration.

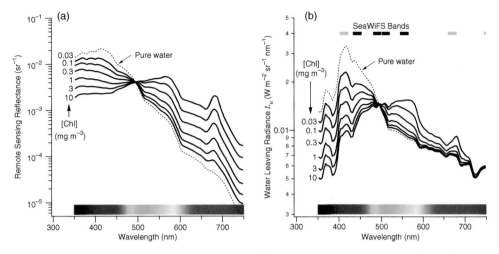

Figure 4.21 Case 1 (a) remote sensing reflectance and (b) water-leaving spectral radiance for varying chlorophyll concentrations.

4.7 Ocean Color Data – Case 1 Water

An empirical approach is used to establish an operationally useful relationship between the water-leaving radiances that can be measured at multiple wavelengths and the chlorophyll pigment concentration. This approach is based on the ratios of the water-leaving radiances at two or more wavelengths, because those ratios depend mostly on water column reflectance ratios that are, in turn, a nonlinear function of the pigment concentration c. The ratio of water-leaving radiances at wavelengths λ_1 and λ_2 can be written:

$$\frac{L_w(\lambda_1)}{L_w(\lambda_2)} = \frac{R_{rs}(\lambda_1)E_s(\lambda_1)}{R_{rs}(\lambda_2)E_s(\lambda_2)} \qquad (4.33)$$

Extensive measurements have been made from ships of the upwelling spectral radiance and coincident chlorophyll concentration. These data can be fit to equations of the form:

$$\log c = A + B\log\left[\frac{L_w(\lambda_1)}{L_w(\lambda_2)}\right] \qquad (4.34)$$

with coefficients A and B derived from in situ data. This equation is then used to estimate chlorophyll pigment concentrations from satellite measurements of water-leaving radiances. The tuning of this equation must be sensor specific. For example, the SeaWiFS measurements are converted to pigment concentration using similar equations, derived from extensive field experiments conducted before the satellite was launched. The key was that these experiments used radiometers with the exact same optical bands as SeaWiFS.

The most challenging part of this entire process is determining the atmospheric correction, which must be applied to estimate the remote sensing reflectances at the wavelengths used to derive the pigment concentrations. The details of this correction are discussed in the next section.

Figure 4.22 plots the chlorophyll concentration at the surface as a function of ratios of the remote sensing reflectance between pairs of bands. The data are plotted on a log-log scale, and these data fall nicely along a set of lines with varying offsets (A) and slopes (B).

The SeaWiFS OC4 algorithm is a refinement of earlier algorithms based on in situ data and comparisons with SeaWiFS images

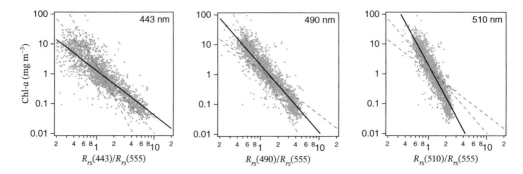

Figure 4.22 Surface chlorophyll as a function of remote sensing reflectance ratios. The dashed curves are the best fits to other wavelength ratios. Data from Werdell and Bailey (2002).

from the first few years on orbit (O'Reilly et al., 2000). The parameters in the OC4 algorithm were readjusted in 2008 based on the NOMAD 2 bio-optical data set. According to Hu et al. (2012), the OC4 algorithm for deriving chlorophyll-a pigment concentration from SeaWiFS data[3] is given by a nonlinear expression in terms of parameter R_4:

$$\log_{10} C_a = 0.3272 - 2.9940\,R_4$$
$$+ 2.7218\,R_4^2 - 1.2259\,R_4^3$$
$$- 0.5683\,R_4^4 \qquad (4.35)$$

The "4" in R_4 and in the algorithm name "OC4" refers to four bands. That is because R_4 is defined to be the maximum value obtained from three remote sensing reflectance ratios: 443, 490, and 510 nm, all relative to 555 nm:

$$R_4 = \max\left(\log_{10}\left[\frac{R_{rs}(443)}{R_{rs}(555)} \right], \right.$$
$$\left. \log_{10}\left[\frac{R_{rs}(490)}{R_{rs}(555)} \right], \ \log_{10}\left[\frac{R_{rs}(510)}{R_{rs}(555)} \right] \right)$$
$$(4.36)$$

The reason this algorithm involves the maximum of these three reflectance ratios can be understood by examining the plot of the water-leaving radiances in Figure 4.21. Specifically note that the peak radiance shifts to the red as the chlorophyll concentration increases. Using the maximum ratio essentially maximizes the signal-to-noise ratio for

these measurements at different levels of chlorophyll concentration.

Figure 4.23 shows log-log comparisons of 2272 in situ measurements of chlorophyll concentration versus the coincident SeaWiFS measurements of the three remote sensing reflectance ratios at 443, 490, and 510 nm. In each panel, the solid line is the OC4 function. The data points show the dependence of the chlorophyll concentration on the R_{rs}-ratios. The black points indicate the reflectance ratios that are maximum among the three bands, while the grey points indicate the remaining reflectance ratios. The figure shows how taking the maximum of the ratios automatically selects different wavelengths depending on the chlorophyll concentration.

Figure 4.24 contains a comparison of 2804 in situ chlorophyll-a observations versus satellite observation using the OC4 algorithm. The dashed lines indicate the 1:5 and 5:1 lines bounding the data set. The overall level of agreement seems remarkable given that the data extend over a range of three orders of magnitude.

4.7.1 Other Uses of Ocean Color

We have concentrated (no pun intended) on the estimation of chlorophyll concentration. This is an important parameter because of its relationship to phytoplankton and the nutrients needed for the phytoplankton to

---

3 Different values of these parameters are used for different satellite sensors.

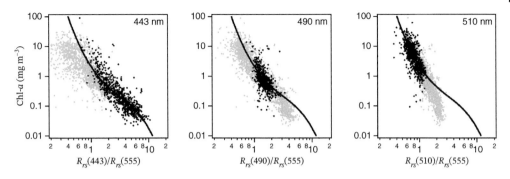

Figure 4.23 Dependence of chlorophyll concentration on R_{rs}-ratios. Data from Werdell and Bailey (2002).

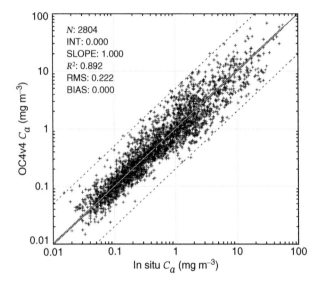

Figure 4.24 OC4 model-data comparisons. O'Reilly et al. (2000).

grow. Other parameters derived from ocean color include:

- Dissolved organic matter arising from decomposing phytoplankton and vegetation.
- Organic and inorganic suspended matter.
- Aquatic primary production which is an indication of the chemical energy contained within an ecosystem as a direct result of photosynthesis. This product is derived from estimates of chlorophyll, dissolved organic carbon, and suspended minerals.
- Ocean biomass, a measure of the organic carbon content of the phytoplankton population, can also be estimated.

4.8 Atmospheric Corrections

As we have seen, geophysical parameters are derived from ocean color data by taking ratios of water-leaving radiances. Estimation of water-leaving radiances from satellite data requires correction for atmospheric effects including path radiance and attenuation. Multiple approaches are used to correct for atmosphere effects in ocean color data (Gordon, 1997).

As before, the total radiance is given by the sum of three terms: the water-leaving radiance times the atmospheric transmittance, the reflected skylight times the atmospheric transmittance and the path radiance:

$$L_T(\lambda, \theta, \varphi) = \tau_A(\lambda, \theta)L_w(\lambda, \theta, \varphi)$$
$$+ \tau_A(\lambda, \theta)r_w(\lambda, \theta, \varphi)L_s(\lambda, \theta, \varphi)$$
$$+ L_p(\lambda, \theta) \qquad (4.37)$$

where $\tau_A(\lambda, \theta)$ is the direct (beam) transmittance of the atmosphere given by:

$$\tau_A(\lambda, \theta) = \exp[-(\tau_r(\lambda) + \tau_{oz}(\lambda) + \tau_a(\lambda)) \sec \theta] \qquad (4.38)$$

with $\tau_r(\lambda)$ = Rayleigh optical depth, $\tau_{oz}(\lambda)$ = ozone optical depth, and $\tau_a(\lambda)$ = aerosol optical depth.

In this formulation the atmospheric transmittance is split into three components: Rayleigh scattering due to molecules in the atmosphere, ozone absorption, and scattering due to aerosols, which are airborne particulates[4]. Each of these constituents has its own optical depth.

Note that the transmittance referred to here is a direct beam transmittance. Thus the transmission losses applied to the water-leaving and reflected radiances arise from both absorption along the path and scattering out of the path. Rays that are scattered into the sensor's path are accounted for in the path radiance term.

To calculate the sky radiance and the path radiance for the equation above, one would need to know the:

- Atmospheric radiance in all directions as a function of altitude.
- Amount and directional properties of scattering at all altitudes.
- Aerosol properties (e.g., size distribution, scattering cross-sections, etc.).
- Parameters for the beam transmittance equation.

It is impractical to obtain all these data for satellite observations, so a semi-empirical

approach is used to estimate the atmospheric correction from multiwavelength ocean color observations.

As illustrated in Figure 4.25, this approach begins by reapportioning the various pathways contributing to the observed radiance. The first term in this reapportioned accounting incorporates all rays that penetrate the surface into the water-leaving radiance term, including those rays that emerge from outside the field of view and are subsequently scattered into the sensor field of view. The second term includes all of the rays reflected from the surface with the path radiance due to sunlight or skylight scattered rays. This second term is then divided into contributions from molecular (Rayleigh) scattering L_R and aerosol (particulate) scattering L_a. The observed radiance is then:

$$L_T(\lambda, \theta) = t_A(\lambda, \theta)L_w(\lambda, \theta) + L_R(\lambda, \theta)$$
$$+ L_a(\lambda, \theta) \qquad (4.39)$$

This reapportionment has the significant impact that the atmospheric transmittance,

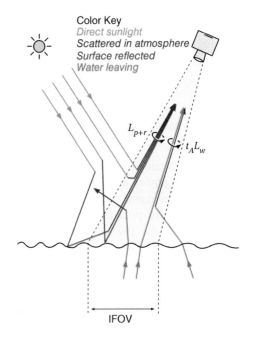

Figure 4.25 Optical pathways for ocean color measurement. Adapted from Robinson (1983).

---

4 Examples of airborne particulates are dust blown from a distant desert, smoke particles from a fire, or salt particles over the ocean.

Beam Attenuation (c)

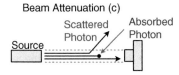

Diffuse Attenuation (K)

All scattered or absorbed
photons are lost to IFOV

All absorbed photons are
lost; as many photons
scatter into IFOV as out

Figure 4.26 Beam versus diffuse attenuation.

which was previously described as a beam transmittance τ_A, now becomes a diffuse transmittance t_A. Figure 4.26 illustrates the differences. As discussed previously, beam attenuation is due to absorption plus scattering – anything scattered out of the path of the beam is considered lost. Diffuse attenuation applies when viewing an extended surface. Photons are still lost in diffuse attenuation to absorption, but just as many photons are scattered back into the sensor field of view as are scattered out.

As illustrated in Figure 4.25, the water-leaving radiance L_w has been redefined to include photons that are emitted from the surface in a direction away from the sensor, but that are scattered back into the sensor field of view. Thus the atmospheric attenuation factor τ_A becomes a diffuse attenuation coefficient. This distinction between beam transmittance/attenuation and diffuse transmittance/attenuation is important and will come up again and again in different contexts.

Next we assume the surface is viewed in a direction where no or little direct sunlight is reflected back to the sensor. This is a reasonable assumption since the sensors used to make ocean color measurements are normally oriented to avoid the glitter pattern.

A strategy for computing the atmospheric correction is now to:

- Calculate the Rayleigh path radiance term $L_R(\lambda, \theta)$ for each wavelength.
- Determine the aerosol path radiance L_a at a particular wavelength λ_0 where the water-leaving radiance is small enough to

be neglected. At this wavelength $L_a(\lambda_0, \theta) = L_T(\lambda_0, \theta) - L_R(\lambda_0, \theta)$.
- Use scaling laws to estimate $L_a(\lambda, \theta)$ at other wavelengths.
- Invert the radiance equation to compute the water-leaving radiance from the difference of the measured radiance and the estimated terms, normalized by the diffuse transmittance.

Specifically, the water-leaving radiance is computed from:

$$L_w(\lambda, \theta) = \frac{L_T(\lambda, \theta) - L_R(\lambda, \theta) - L_a(\lambda, \theta)}{t_A(\lambda, \theta)}$$

(4.40)

According to Gordon and Clark (1981), the diffuse atmospheric transmittance can be well approximated by this function of the ozone optical depth, the Rayleigh optical depth and the aerosol diffuse transmittance:

$$t_A(\lambda, \theta) = \exp\left[-\left(\tau_{oz}(\lambda) + \frac{1}{2}\tau_R(\lambda)\right)\sec\theta\right]$$
$$\times t_a(\lambda, \theta)$$

(4.41)

The factor of $\frac{1}{2}$ in the exponent reflects the difference between the beam and diffuse attenuation for Rayleigh scattering.

From scattering theory it is easy to compute an expression for the Rayleigh optical depth in a standard exponential atmosphere, namely one constrained by gravity:

$$\tau_R(\lambda) = 0.008569 \frac{P}{\lambda^4 P_0}$$

(4.42)

In this expression λ is in microns, P is the surface pressure, and $P_0 = 1013.25$ mbar is the standard atmospheric pressure.

The ozone optical depth τ_{oz} can then be obtained from look-up tables or from satellite measurements. In any case the ozone optical depth $\tau_{oz}(\lambda) \leq 3.5\%$. This means the ozone transmittance $= t_{oz} = \exp(\tau_{oz}) \geq 96.6\%$.

The aerosol diffuse transmittance factor $t_a(\lambda, \theta)$ in the previous equation for the diffuse atmospheric transmittance is approximately 1.0 for $\theta < 60°$. In fact, for correction of the old Coastal Zone Color Scanner data, the correction was done using $t_a(\lambda, \theta) = 1.0$. For data from more modern sensors, such as SeaWiFS and MODIS, it is estimated from the observations.

The Rayleigh path radiance L_R is computed from:

$$L_R(\lambda, \theta) = \frac{F_0'(\lambda, \theta, \theta_0)}{4\pi} \omega_R(\lambda) \tau_R(\lambda)$$
$$\times \sec\theta P_R(\theta, \theta_0) \quad (4.43)$$

where F_0' is the incident solar irradiance, corrected for the round trip passage through the ozone layer. In this expression:

$$F_0'(\lambda, \theta, \theta_0) = F_0(\lambda)\exp[-\tau_{oz}(\lambda)(\sec\theta + \sec\theta_0)] \quad (4.44)$$

where F_0 is the solar irradiance at the top of the Earth's atmosphere, θ_0 is the solar zenith angle, and ω_R is the single-scattering albedo for Rayleigh scattering, usually taken to be $\omega_R = 1$.

The scattering phase function $P_R(\theta, \theta_0)$ is a nondimensional parameter defined to describe the angular distribution of the molecular scattered radiation. In this case, it is defined in terms of a sensor looking down at angle θ into an atmosphere illuminated by the sun at angle of θ_0.

The scattering phase function due to molecules can be written by a sum of the Rayleigh scattering phase function that backscatters light from the sun into the sensor (ψ_-) and the Rayleigh scattering phase function that forward scatters light reflected from the surface into the sensor (ψ_+):

$$P_R(\theta, \theta_0) = P_R(\psi_-) + [r_w(\theta) + r_w(\theta_0)]P_R(\psi_+) \quad (4.45)$$

where the factor $r_w(\theta)$ arises from including the surface reflected rays in the path radiance. The angles $\psi_\pm$ are the scattering angles derived from the solar and viewing angles:

$$\cos\psi_\pm = \pm\cos\theta_0\cos\theta + \sin\theta_0\sin\theta\cos\Delta\varphi \quad (4.46)$$

where φ_0 and φ_{LOS} are the solar and viewing azimuth angles:

$$\Delta\varphi = \varphi_0 - \varphi_{LOS} \quad (4.47)$$

and the Rayleigh scattering phase function is given by:

$$P_R(\psi_\pm) = \frac{3}{4}(1 + \cos^2\psi_\pm) \quad (4.48)$$

This scattering phase function is based on radiation from an idealized dipole, which is a good model for the oscillating "currents" induced in small molecules that cause molecular scattering.

Next, the aerosol path radiance $L_a(\lambda, \theta)$ is determined by assuming that its functional dependence is the same as the Rayleigh radiance. Hence we write that the path radiance equals the solar radiance times the scattering albedo times the aerosol optical thickness times the aerosol scattering phase function:

$$L_a(\lambda, \theta) \approx F_0'(\lambda, \theta, \theta_0)\,\omega_a(\lambda)\,\tau_a(\lambda)\,P_a(\lambda, \theta, \theta_0) \quad (4.49)$$

Let us take each term in turn, but not in order. The solar radiance F_0 is evaluated at the depth of the aerosols in the atmosphere, so one-way atmospheric losses have already been applied. The scattering phase function P_a defines the directionality of the scattered light, while the albedo ω_a defines the overall strength of the scattered light. In other words, the scattering phase function applies to one or a million scatters, whereas the albedo depends on the number of scatterers within a unit volume and the scattering cross-section of each scatterer. The aerosol optical thickness τ_a then accounts for the depth of the scattering layer.

It then follows that the ratio of the path radiance at two different wavelengths can be simplified to the expression below with ε being an effective scattering ratio between wavelengths λ and λ_0:

$$\frac{L_a(\lambda, \theta)}{L_a(\lambda_0, \theta)} = \frac{F_0'(\lambda, \theta, \theta_0)\,\omega_a(\lambda)\,\tau_a(\lambda)P_a(\lambda, \theta, \theta_0)}{F_0'(\lambda_0, \theta, \theta_0)\,\omega_a(\lambda_0)\,\tau_a(\lambda_0)P_a(\lambda_0, \theta, \theta_0)}$$
$$= \varepsilon(\lambda, \lambda_0)\frac{F_0'(\lambda, \theta, \theta_0)}{F_0'(\lambda_0, \theta, \theta_0)} \qquad (4.50)$$

For Coastal Zone Color Scanner (CZCS) data, the assumption was made that the product $\omega_a(\lambda)P_a(\lambda, \theta, \theta_0)$ is independent of wavelength, and that the aerosol optical thickness is inversely proportional to some power of the wavelength. Thus we have the scaling law:

$$\varepsilon(\lambda, \lambda_0) = \left(\frac{\lambda_0}{\lambda}\right)^m \qquad (4.51)$$

where m is known as Angstrom's exponent.

Standard scattering theory can be used to show for spherical particles with a size distribution proportional to the inverse fourth power of the particle radius, $m = 1$. So the CZCS algorithm used $m = 1$, but aerosol measurements can yield differing values for m.

Using these equations, the aerosol path radiance can then be calculated once the wavelength λ_0 is selected. Recalling the earlier plot of water-leaving radiance, the strategy is to select λ_0 to be in the red or near infrared so the water-leaving radiance is small enough to be neglected. For CZCS data, $\lambda_0 = 670$ nm. For SeaWiFS and MODIS, the channels at 750 nm and 865 nm can be used for λ_0.

More recent analyses done for SeaWiFS and MODIS use a different expression for the ratio ε:

$$\varepsilon(\lambda_1, \lambda_2) = exp[c(\lambda_2 - \lambda_1)] \qquad (4.52)$$

In this expression c is a constant that depends on the aerosol type and the relative humidity. The strategy is to use measurements at 750 nm and 865 nm where the water-leaving radiance is approximately zero to estimate $\varepsilon(750, 865)$.

This identifies the aerosol type, and allows calculation of the constant c. Equation (4.52) can then be used to estimate ε at other wavelengths.

In the end, this approach also allows one to model multiple scattering in the atmosphere, including terms involving Rayleigh and aerosol scattering.

The overall approach to estimating ocean parameters from satellite color data is illustrated in Figure 4.27. The process begins by calibrating the sensor data to obtain the scene's spectral radiance at a number of wavelengths. The large path radiance in the optical then needs to be corrected in order to derive the water-leaving radiance L_w from the measured radiances.

The path radiance correction involves:

- Calculating the Rayleigh contribution.
- Measuring the scattered radiance L_p at a wavelength where the water-leaving radiance L_w is small. In this case, the scene radiance is about equal to the path radiance. We can then subtract the Rayleigh term to get the aerosol radiance.
- Applying a scaling law to get the aerosol radiance at other wavelengths.

This approach:

- Does not explicitly include reflected radiation.
- Ignores waves as long as the incident angle is less than 40°, where most of these measurements are made.
- Relies on a relationship between the water-leaving radiance L_w, the upwelling irradiance E_u below surface, and the water column reflectance $R = E_u/E_{sun}$.

The algorithms then use the reflectance ratios at two wavelengths to derive the concentrations of water column constituents, such

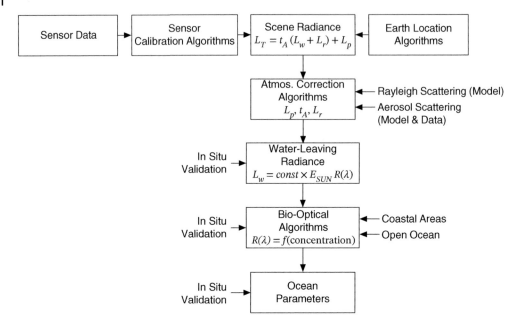

Figure 4.27 Processing steps for satellite color data.

as chlorophyll. The algorithms automatically select the best wavelengths to use.

4.9 Ocean Color Satellite Sensors

4.9.1 General History

Historical Ocean Color Satellites

There is a long history of satellite-based ocean color sensors. Table 4.4 lists 16 ocean color sensors that were used in the past to acquire ocean color data. Most of these sensors had 6–10 bands with resolutions of the order of 1 km. All of the sensors except for HICO were in polar orbits. Although no longer functional, the data acquired by these sensors are part of the growing global database of environmental measurement we have begun to construct.

There are three sensors on this list that are worth noting. The Coastal Zone Color Scanner was the very first sensor with bands selected for ocean color sensing. It operated from 1978 to 1986 and demonstrated the potential for ocean color measurements. The next sensor of

note was SeaWiFS, which was the sole sensor operated from the OrbView-2 spacecraft from 1997 to 2011. This sensor is described in more detail in a later section. And the third sensor of note on this list was MERIS, which provided a decade of ocean color monitoring from one of the first major European environmental satellites, ENVISAT.

Current Ocean Color Satellites

Table 4.5 shows the 14 ocean color sensors that were in operation as of late 2020. Generally the current generation of ocean color sensors have 20–30 bands with spatial resolutions of 250–500 m.

The two MODIS sensors on the Terra and Aqua satellites may be alive, but like the authors of this text, they are very old. The two MODIS sensors have been on orbit since 1999 and 2002, respectively. Not bad for sensors with a design life of five years! The Geostationary Ocean Color Imager (GOCI) was the world's first ocean color sensor placed in geostationary orbit. This is a South Korean sensor, so not surprisingly it acquires data in

Table 4.4 Historical ocean color sensors. (Adapted from: http://www.ioccg.org/, updated December 2020.)

Sensor	Agency	Satellite	Operating dates	Swath (km)	Spatial resolution (m)	# Bands	coverage (nm)	Orbit
CZCS	NASA (USA)	Nimbus-7 (USA)	10/78–06/86	1556	825	6	433–12,500	Polar
CMODIS	CNSA (China)	SZ-3 (China)	03/02–09/02	650–700	400	34	403–12,500	Polar
COCTS CZI	SOA (China)	HY-1A (China)	05/02–04/04	1400/500	1100/250	10/4	402–12,500 420–890	Polar
COCTS CZI	SOA (China)	HY-1B (China)	11/07–02/16	3000/500	1100/250	10/4	402–885 420–890	Polar
GLI	NASDA (Japan)	ADEOS-II (Japan)	12/02–10/03	1600	250/1000	36	375–12,500	Polar
HICO	DOD/NASA (USA)	ISS (USA)	09/09–12/14	50	100	124	380–1000	51.6°
MERIS	ESA (Europe)	ENVISAT (Europe)	03/02–05/12	1150	300/1200	15	412–1050	Polar
MOS	DLR (Germany)	IRS P3 (India)	03/96–05/04	200	500	18	408–1600	Polar
OCI	NEC (Japan)	ROCSAT-1 (Taiwan)	01/99–06/04	690	825	6	433–12,500	Polar
OCM	ISRO (India)	IRS-P4 (India)	05/99–08/10	1420	360/4000	8	402–885	Polar
OCTS	NASDA (Japan)	ADEOS (Japan)	08/96–06/97	1400	700	12	402–12,500	Polar
OSMI	KARI (Korea)	KOMPSAT-1 (Korea)	12/99–01/08	800	850	6	400–900	Polar
POLDER	CNES (France)	ADEOS (Japan)	08/96–06/97	2400	6 km	9	443–910	Polar
POLDER-2	CNES (France)	ADEOS-II (Japan)	12/02–10/03	2400	6000	9	443–910	Polar
POLDER-3	CNES (France)	Parasol (France)	12/04–12/13	2100	6000	9	443–1020	Polar
SeaWiFS	NASA (USA)	OrbView-2 (USA)	08/97–02/11	2806	1100	8	402–885	Polar

the waters around the Korean peninsula. It produces eight daytime and two nighttime images of the region per day. The Visible Infrared Imaging Radiometer Suite (VIIRS) is a sensor deployed on board the Suomi National Polar-orbiting Partnership (Suomi NPP) weather satellite that was launched in 2011, and the NOAA-20 satellite launched in 2018. Finally, the Ocean Land Color Imager (OLCI) deployed on the Sentinel-3A and -3B

Table 4.5 Current ocean color sensors. (Adapted from: http://www.ioccg.org/, updated December 2020.)

Sensor	Agency	Satellite	Launch date	Swath (km)	Spatial resolution (m)	# Bands	coverage (nm)	Orbit
COCTS CZI	CNSA (China)	HY-1C	7-Sep-18	3000 / 950	1100 / 50	10 / 4	402–12,500 / 433–885	Polar
COCTS CZI	CNSA (China)	HY-1D	10-Jun-20	3000 / 950	1100 / 50	10 / 4	402–12,500 / 433–885	Polar
GOCI-II	KARI/KIOST (South Korea)	GeoKompsat-2B	19-Feb-20	1200 × 1500	250/1000	13	412–1240	Geo
GOCI	KARI/KIOST (South Korea)	COMS	26-Jun-10	2500	500	8	400–865	Geo
MODIS-Aqua	NASA (USA)	Aqua (PM)	4-May-02	2330	250/500/1000	36	405–14,385	Polar
MODIS-Terra	NASA (USA)	Terra (AM)	18-Dec-99	2330	250/500/1000	36	405–14,385	Polar
MSI	ESA (Europe)	Sentinel-2A	23-Jun-15	290	10/20/60	13	442–2202	Polar
MSI	ESA (Europe)	Sentinel-2B	7-Mar-17	290	10/20/60	13	442–2186	Polar
OCM-2	ISRO (India)	Oceansat-2	23-Sep-09	1420	360/4000	8	400–900	Polar
OLCI	ESA (Europe)	Sentinel-3A	16-Feb-16	1270	300/1200	21	400–1020	Polar
OLCI	ESA (Europe)	Sentinel-3B	25-Apr-18	1270	300/1200	21	400–1020	Polar
SGLI	JAXA (Japan)	GCOM-C	23-Dec-17	1150–1400	250/1000	19	375–12,500	Polar
VIIRS	NOAA (USA)	Suomi NPP	28-Oct-11	3000	375/750	22	412–12,013	Polar
VIIRS	NOAA/NASA (USA)	NOAA-20	5-Jan-18	3000	370/740	22	412–12,013	Polar

spacecraft is the newest European ocean color sensor.

Future Ocean Color Satellites

Table 4.6 lists the ocean color sensors and satellites that are planned to be launched over the next few years. There are several developments to note on this list. A few more sensors will go into geostationary orbit. A few of the sensors will provide spatial resolutions of better than 100 m, albeit with a reduced coverage area. And a few sensors are moving towards more than 100 spectral bands.

4.9.2 SeaWiFS

This section reviews the capabilities of the Sea-viewing Wide Field-of-view Sensor (Sea-WiFS) in some detail.

Figure 4.28 shows the spectral sensitivities of three spaceborne sensors: CZCS, Sea-WiFS and MODIS. These are compared to the water-leaving radiances at two chlorophyll

Table 4.6 Planned ocean color satellites. (Adapted from: http://www.ioccg.org/, updated December 2020.)

Satellite	Agency	Sensor	Planned launch	Swath (km)	Spatial resolution (m)	# Bands	Spectral coverage (nm)	Orbit
HY-1E	CNSA (China)	CZI	2022	2900 1000	1100 250	10 4	402–12,500 433–885	Polar
EnMAP	DLR (Germany)	HSI	2021	30	30	242	420–2450	Polar
OCEANSAT-3	ISRO (India)	OCM-3	2021	1400	360/1	13	400–1,010	Polar
SABIA-MAR	CONAE (Argentina)	Multi-spectral	2023	200/2200	200/1100	16	380–11,800	Polar
PACE	NASA (USA)	OCI	2022/2023	2000	1000	91	350–2250	Polar
GISAT-1	ISRO (India)	HYSI-VNIR	TBD	250	320	60	400–870	Geo
SBG	NASA (USA)	Hyper-VSWIR TIR-Imager	2026	185 600	30 60–100	>200 8	380–2500	Polar
GLIMR	NASA (USA)	VNIR Imager	>2023	TBD	300	141	340–1040	Geo

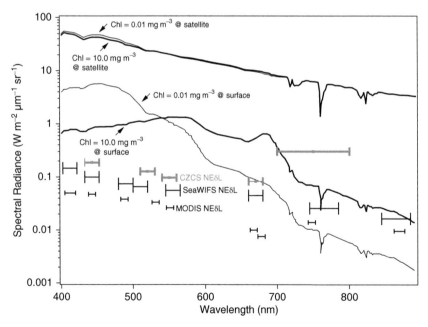

Figure 4.28 Spectral sensitivities of three spaceborne sensors. Adapted from Hooker et al. (1992).

concentrations: an extremely low value of 0.01 mg m^{-3} and an extremely high value of 10 mg m^{-3}. These are the bottom two curves on the plot. The radiance predicted to be measured by the satellite looking down at a cloud-free ocean with the same low and high concentrations of chlorophyll is shown in the top two curves. Note the very small difference in the radiance at the top of the atmosphere for the high and low chlorophyll concentrations, underscoring the need for accurate atmospheric corrections.

The sensor sensitivities on this plot are expressed in terms of the Noise Equivalent delta Radiance (NEδL), which is an indication of the noise due to the instrument after the atmospheric correction processing. Note that the NEδL of the CZCS from 700–800 nm was larger than the expected water-leaving radiances. Thus the atmospheric corrections for CZCS were dominated by sensor noise. The long wavelength sensitivity was improved in the newer sensors, so this is no longer an issue.

Figure 4.29 shows the major processing elements for SeaWiFS. The data from SeaWiFS were broadcast in real time to any receiving station within its line of sight. In addition, data were recorded on board and transmitted to specific sites like the Goddard Space Flight Center, the Wallops Flight Facility, or the download facility operated by Orbital Sciences Corporation (OSC).

All downloaded data were then transferred to the SeaWiFS Data Processing System (SDPS) for processing. Eventually the processed data were transferred to Goddard Distributed Active Archive Center for storage and distribution. All of these data can be downloaded over the web.

The Orbview-2 spacecraft was privately built to host the SeaWiFS sensor (Figure 4.30). The satellite weighed 309 kg. It was actually placed into orbit using a Pegasus rocket that was dropped from an aircraft at 39,000 ft. The SeaWiFS scanner module was a 20-inch cube that weighed 51 kg and used less than 90 W. Size, weight, and power are critical design elements for all spaceborne instruments.

The path of the incoming photons can be traced in the diagram on the right. Notice the tilt axis assembly that allows the field of view to be tilted 20° either fore or aft to avoid viewing sun glint on the ocean.

The key technical specifications for SeaWiFS are summarized in Tables 4.7 and 4.8. The system scanned through an angle of ±58°

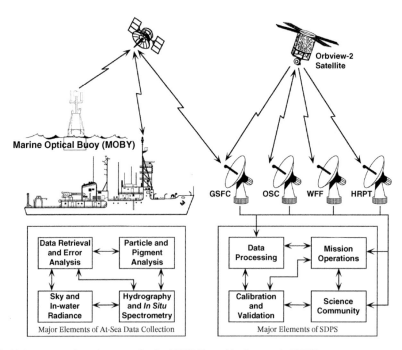

Figure 4.29 Major processing elements for SeaWiFS. From Hooker et al. (1992).

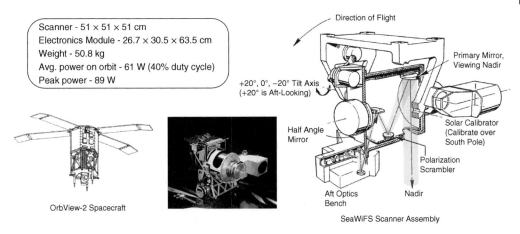

Figure 4.30 Orbview-2 spacecraft and SeaWiFS. From Hooker et al. (1992).

Table 4.7 SeaWiFS technical specifications.

Characteristic	Description
Orbit	Sun-synchronous @ 705 km altitude
Orbit Period	99 minutes
Swath Width	2801 km
Instantaneous FOV	1.6×1.6 milliradians
Spatial Resolution	1.1 km at nadir
Scan	$\pm 58.3°$
Telescope	Off-axis, afocal; 7.6 cm aperture; f/2
Detectors	Silicon
Scan Plane Tilt	$+20°, 0°, -20°$
Polarization Sensitivity	$\leq 1\%$ (worst case)
Quantization	10 bits
Dynamic Range	15,000:1 using bilinear gain
Bright Cloud Recovery	Less than 10 pixels
Weight	≤ 45 kg
Mission Life	5 years

about nadir. It had an f/2 lens with a 7.6 cm aperture and a IFOV of 1.6 milliradians.

The sensor made measurements in the eight spectral bands listed in Table 4.8. You can see that bands 2–5 could be used with the OC4 algorithm, with band 5 acting as the denominator in the radiance ratios (the so-called hinge point). Band 1 was useful for measuring dissolved organic material. All of the long-wave bands were useful for measuring atmospheric aerosols.

Figure 4.31 shows an example chlorophyll concentration map derived from SeaWiFS data overlaid on a true-color image. The chlorophyll concentrations in the northern waters reflect the large phytoplankton populations in these nutrient rich waters. The smaller scale variations in the chlorophyll concentrations

Table 4.8 SeaWiFS bands.

Primary use	Band	Bandwidth (nm)	Req. SNR	Pred. SNR 4:1 TDI[1]
Gelbstoffe Used to separate out the strong blue absorption from gelbstoffe (highly stable humic and fulvic acids from plant decomposition) from chlorophyll pigment-concentration estimates	1	402–422	499	974
Chlorophyll Absorption Used with bands 3 to 5 to determine color boundaries, chlorophyll concentration, and the diffuse attenuation coefficient (k)	2	433–453	674	1040
Pigment Concentration Used with bands 2 and 4 to derive chlorophyll pigment concentrations in coastal (Case-2) waters	3	480–500	667	1100
Chlorophyll Absorption Used with bands 2, 3, and 5 to determine color boundaries, chlorophyll concentration, and the diffuse attenuation coefficient (k)	4	500–520	616	1040
Sediments/Hinge Point Used with bands 2, 3, and 5 to determine color boundaries, chlorophyll concentration, and the diffuse attenuation coefficient (k)	5	545–565	581	967
Atmospheric Aerosols Bands 7 and 8 are better, but Band 6 was included because it is compatible with predecessor (CZCS) ocean-color instrument processing techniques	6	660–680	581	797
Atmospheric Aerosols Used in conjunction with Band 8	7	745–785	455	874
Atmospheric Aerosols Used in conjunction with Band 7	8	845–885	467	770

1 The focal plane has four detectors along the scan direction in each band, allowing time delay and integration (TDI) to increase the SNR by about a factor of two.

reflects the complexities of the underlying currents. The warmer waters of the Gulf Stream to the south have far lower phytoplankton concentrations because the nutrients in those waters have been largely depleted.

4.9.3 MODIS

The Moderate-Resolution Imaging Spectroradiometer (MODIS) is hosted on both the Terra and Aqua satellites (Barnes et al., 1998). MODIS was discussed in Chapter 3 because it has IR bands. It is also included here because it has significant capabilities in the visible bands. Furthermore both MODIS sensors are still in operation.

MODIS is operated in a sun-synchronous polar orbit. It scans a 2300-km swath as it passes overhead. It is a larger instrument than SeaWiFS, occupying more than a cubic meter

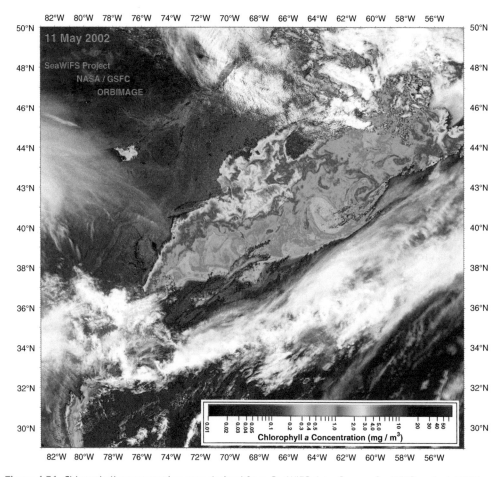

Figure 4.31 Chlorophyll concentration map derived from SeaWiFS data. Source: SeaWiFS project, NASA GSFC and ORBIMAGE.

and weighing more than 250 kg. Its primary aperture is twice the size of SeaWiFS. The basic technical specifications for MODIS are listed in Table 4.9.

Note the along-track scan dimension at nadir of 10 km. This comes about because there are 10 detectors on the focal plane in the along-track direction for each channel. Each detector has a nadir resolution of 1 km. For the scan rate of 20.3 rpm, the two-sided scan mirror rotates 360° in 2.96 seconds. Each rotation gives two scans along-track, one from the front side of the mirror and one from the back side. So about every 1.5 seconds the satellite moves about 10 km along-track, exposing the 10 detectors on the focal plane.

For the channels with 500 m resolution, there are 20 detectors along-track in the focal plane; for 250 m resolution, there are 40 detectors.

MODIS supports the 19 visible and near-IR bands listed in Table 4.10. Channels 1 and 2 provide 250-m resolution at nadir, channels 3–7 provide 500-m resolution, and the rest provide 1-km resolution. The 1-km channels are used for estimating ocean color. Notice the MODIS has larger SNR in the ocean color bands to support atmospheric corrections.

Table 4.11 lists the MODIS IR bands for completeness. All of the IR bands have 1-km resolution at nadir. Note the low NEΔT

Table 4.9 MODIS technical specifications.

Characteristic	Description
Orbit	705 km, sun-synchronous, near-polar, circular orbit with 10:30 am descending node or 1:30 pm ascending node
Scan Rate	20.3 rpm, cross-track
Scan Dimensions	2330 km (cross-track) by 10 km (along-track at nadir)
Telescope	17.78 cm diam. off-axis, afocal (collimated), with intermediate field stop
Size	$1.0 \times 1.6 \times 1.0$ m
Weight	250 kg
Power	225 W (orbital average)
Data Rate	11 Mbps (peak daytime)
Quantization	12 bits
Spatial Resolution	250 m (bands 1–2)
(at nadir)	500 m (bands 3–7)
	1000 m (bands 8–36)
Design Life	5 years

Table 4.10 MODIS visible and near-infrared bands.

Primary Use	Band	Bandwidth (nm)	Spectral radiance ($W \cdot m^{-2} \cdot \mu m^{-1} \cdot sr^{-1}$)	Req. SNR	Nadir res. (m)
Land/Cloud/Aerosols Bounds	1	620–670	21.8	128	250
	2	841–876	24.7	201	
Land/Cloud/Aerosols Properties	3	459–479	35.3	243	500
	4	545–565	29.0	228	
	5	1230–1250	5.4	74	
	6	1628–1652	7.3	275	
	7	2105–2155	1.0	110	
Ocean Color	8	405–420	44.9	880	1000
	9	438–448	41.9	838	
	10	483–493	32.1	802	
	11	526–536	27.9	754	
	12	546–556	21.0	750	
	13	662–672	9.5	910	
	14	673–683	8.7	1087	
	15	743–753	10.2	586	
	16	862–877	6.2	516	
Atmospheric Water Vapor	17	890–920	10.0	167	1000
	18	931–941	3.6	57	
	19	915–961	15.0	250	

Table 4.11 MODIS mid-wave and long-wave IR bands.

Primary Use	Band	Bandwidth (μm)	Spectral radiance (W·m^{-2}·μm^{-1}·sr^{-1})	Req. NEΔT (K)
Surface/Cloud Temperature	20	3.660–3.840	0.45 (300K)	0.05
	21	3.929–3.989	2.38 (335K)	2.00
	22	3.929–3.989	0.67 (300K)	0.07
	23	4.020–4.080	0.79 (300K)	0.07
Atmospheric Temperature	24	4.433–4.498	0.17 (250K)	0.25
	25	4.482–4.549	0.59 (275K)	0.25
Cirrus Clouds/Water Vapor	26	1.360–1.390	6.0	150 (SNR)
	27	6.535–6.895	1.16 (240K)	0.25
	28	7.175–7.475	2.18 (250K)	0.25
Cloud Properties	29	8.400–8.700	9.58 (300K)	0.05
Ozone	30	9.580–9.880	3.69 (250K)	0.25
Surface/Cloud Temperature	31	10.780–11.280	9.55 (300K)	0.05
	32	11.770–12.270	8.94 (300K)	0.05
Cloud Top Altitude	33	13.185–13.485	4.52 (260K)	0.25
	34	13.485–13.785	3.76 (250K)	0.25
	35	13.785–14.085	3.11 (240K)	0.25
	36	14.085–14.385	2.08 (220K)	0.35

requirements for the SST channels (20, 22, 23, 31, and 32).

Figure 4.32 shows an exploded view of the MODIS sensor.

4.9.4 VIIRS

The Visible Infrared Imaging Radiometer Suite (VIIRS) is the newest U.S. sensor capable of making ocean color measurements. It was first launched on the Suomi NPP satellite in 2011. VIIRS is particularly important because it is the sensor of choice for the new generation JPSS satellites, and it will become the only U.S. sensor for ocean color data if the MODIS sensors fail.

Serious questions about the VIIRS program were raised in a National Academy of Sciences Report issued just before launch (National Research Council, 2011). These issues include:

- Serious sensor design issues, such as optical cross-talk between some of the bands and light leakage.
- Insufficient funding for calibration/validation efforts sufficient to meet requirements for climate-quality data.
- Inadequate plans for on-orbit calibration and sensor stability monitoring.
- Program plan was to fund NOAA to develop processing infrastructure and algorithms, but NOAA lacked the expertise to do so. No funding was planned for the NASA Ocean Color Group that has that expertise.

Additional design flaws were identified post-launch, such as progressive decay of sensitivity in some IR channels due to contaminants on the mirrors.

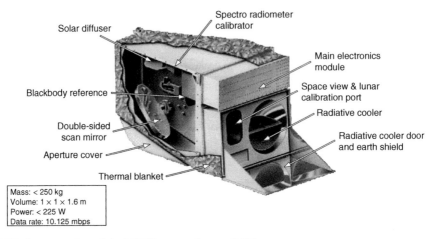

Spectro radiometer calibrator

Solar diffuser

Main electronics module

Space view & lunar calibration port

Blackbody reference

Radiative cooler

Double-sided scan mirror

Radiative cooler door and earth shield

Aperture cover

Thermal blanket

Mass: < 250 kg
Volume: 1 × 1 × 1.6 m
Power: < 225 W
Data rate: 10.125 mbps

Figure 4.32 Cut-away view of the MODIS sensor. Source: NASA.

But on-orbit calibrations have gone reasonably well (Wang et al., 2013). Ocean color products from VIIRS today are consistent with those from MODIS-Aqua in global deep waters, but challenges remain in turbid coastal waters.

VIIRS ocean color data processing makes use of calibrated top of the atmosphere radiances at seven spectral bands similar to those used with MODIS processing. Three additional short-wave infrared bands (M8, M10, and M11) are also used for atmospheric corrections with Case 2 waters. Table 4.12 lists the VIIRS ocean color bands and their primary use. All of these

VIIRS channels have a nadir resolution of 750 m.

The VIIRS chlorophyll concentration algorithm is similar to those used for SeaWiFS and MODIS. The algorithm, named OC3V (Ocean Chlorophyll 3-band VIIRS), uses two remote sensing reflectance ratios: 445 and 488 nm, both relative to 555 nm. The algorithm is a nonlinear equation in terms of the parameter R (Wang et al., 2017):

$$\log_{10} C_a = 0.2228 - 2.4682R + 1.5867R^2 \\ - 0.4275R^3 - 0.7768R^4 \quad (4.53)$$

Table 4.12 VIIRS channels for ocean color processing.

Primary use	VIIRS band	Central wavelength (nm)	Wavelength range (nm)
Ocean Color	M1	412	402–422
Water-Leaving Radiance	M2	445	436–454
	M3	488	478–488
	M4	555	545–565
	M5	672	662–682
Atmospheric Aerosols	M6	746	739–754
Case 1 Waters	M7	865	846–885
Atmospheric Aerosols	M8	1240	1230–1250
Case 2 Waters	M10	1610	1580–1640
	M11	2250	2230–2280

where R is defined to be the maximum value of the two remote sensing reflectance ratios:

$$R = \max \left(\log_{10} \left[\frac{R_{rs}(445)}{R_{rs}(555)} \right], \right.$$
$$\left. \log_{10} \left[\frac{R_{rs}(488)}{R_{rs}(555)} \right] \right) \qquad (4.54)$$

Additional details about the algorithms being used for other VIIRS ocean color data products, including various atmospheric correction techniques, can be found in Wang et al. (2017).

4.10 Ocean Chlorophyll Fluorescence

Fluorescence is the absorption of electromagnetic radiation at one wavelength and its reemission at another, lower energy wavelength. A trivial example is hand stamps that glow under a UV source. The reemission of light is almost instantaneous and the fluorescence stops when the illumination stops.

This is a unique phenomenon. It should not be confused with either:

- Biophosphorescence where organisms continue to emit light after the illuminating source has been taken away, or
- Bioluminescence which is the natural production of light by chemical reactions within an organism. A common example are fireflies. Bioluminescence does not involve either absorption or reemission.

All plants absorb more sunlight than they can consume through photosynthesis. Three things happens to the absorbed sunlight: 10 to 15% is used for photosynthesis, 80 to 85% is converted to heat, and 2 to 5% is reemitted as fluorescence.

The fluorescence spectrum emitted by chlorophyll in phytoplankton (single-cell plants) has a peak in the red near 683 nm. This is illustrated in Figure 4.33 showing the fluorescence of chlorophyll in a beaker. For chlorophyll concentrations greater than 0.2–0.5 mg m^{-3}, this fluorescent light can be detected by the MODIS band 14 centered at 678 nm.

Radiant energy absorbed by the phytoplankton can meet one of three fates: photosynthesis, creation of heat, or fluorescence. These three fates are in direct competition.

Fluorescence measurements:

- Can estimate the efficiency with which photosynthesis turns sunlight and nutrients into food.
- Allow remote assessment of the health and productivity of ocean phytoplankton.
- Can also be used to study how climate changes alter these processes which are at the bottom of the ocean food web.

Figure 4.34 plots the absorption due to chlorophyll as a function of wavelength. Most

Figure 4.33 Chlorophyll fluorescence in a laboratory. Source: Marie Franzen, CC BY-SA 3.0, https://commons.wikimedia.org/wiki/File:Fluorescence:of_chlorophyll_under_UV_light.jpg.

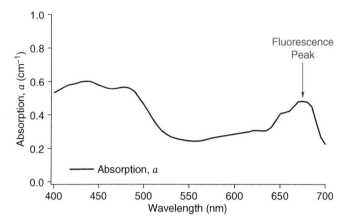

Figure 4.34 Spectral variation of chlorophyll absorption including fluorescence peak. Data from Gaigalas et al. (2009).

of the absorption occurs at blue wavelengths, but there is also a secondary peak of absorption in the red. This is about the wavelength of chlorophyll fluorescence.

This will come up again, but chlorophyll is green because it reflects in the green and absorbs in the blue and red. There is an interesting argument about whether there was some evolutionary pressure that forced plants to evolve in this way. Wouldn't it be more efficient if plants were black, especially for deeper marine plants, so they could absorb as much energy as possible?

One possible answer lies in the fact that some organisms absorb sunlight for photosynthesis using rhodopsin instead of chlorophyll.

Rhodopsin absorbs in the green, and reflects blue and red, giving it a purple color. The theory is that such organisms were the first to develop and thrived in the upper layers of the primordial ocean. Chlorophyll then developed among deeper organisms as a way to make use of the light not absorbed by the purple organisms. This is just an unproven theory at this point, but it is an interesting argument.

The left-hand panel of Figure 4.35 plots typical reflectance spectra for seawater containing only phytoplankton, while the right panel plots water-leaving radiance. The curves are plotted for different concentrations of chlorophyll based on the models contained in Hydrolight. The bump in the apparent

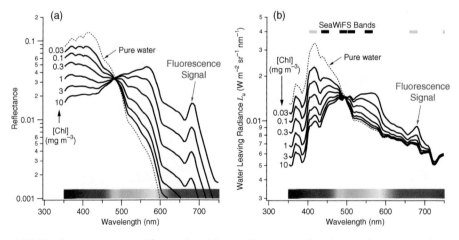

Figure 4.35 The fluorescence signal for varying chlorophyll concentrations in case 1 waters, as observed in (a) reflectance, and (b) water-leaving radiance.

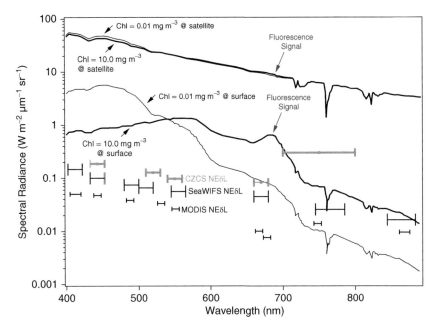

Figure 4.36 Fluorescence signal in radiance sensitivity plots. Adapted from Hooker et al. (1992).

reflectance at 680 nm that increases with increasing chlorophyll is actually not reflected energy of the same wavelength, but the fluorescence signal manifested in the reflectance and water-leaving radiances.

A fluorescence signal is also visible in Figure 4.36 showing the water-leaving radiances and sensor sensitivities. Taking into account the different units, the water-leaving radiances in this plot are substantially smaller than in the previous figure. This latter calculation was likely made assuming less illumination, possibly a lower solar angle, than for the previous figure.

Notice that CZCS was too noisy to detect anything but the strongest fluorescence and the bands were poorly placed. All of this was fixed in MODIS, and band 14 was selected to nearly match the fluorescence peak in chlorophyll. Also notice that in waters with low chlorophyll concentration, the spectrum monotonically declines from 600 to 800 nm. Even at high concentrations, there is a general decline in the spectrum that is interrupted only by the fluorescence peak. This suggests an approach to measuring the strength of the fluorescence peak.

Although there are 19 visible and near-IR bands for MODIS, only three are used to estimate fluorescence: bands 13, 14, and 15. Actually that is not quite the case. More precisely, only the water-leaving radiances at these three bands are used to estimate fluorescence. Other bands are used to estimate the atmospheric corrections needed to obtain these water-leaving radiances.

Figure 4.37 illustrates how the MODIS Fluorescence Line Height Algorithm works. The three gray regions represent the normalized transmittance of MODIS bands 13, 14, and 15.

The solid lines describe the spectral distribution of upwelling radiance above the surface of the ocean for two selected cases with chlorophyll concentrations of $0.01 \, \text{mg m}^{-3}$ and $10 \, \text{mg m}^{-3}$.

Consider the spectrum for the $10 \, \text{mg m}^{-3}$ case shown as the thick solid line. The values of the water-leaving radiances in these three bands would be as shown. The Fluorescence Line Height (FLH) algorithm assumes that there is no fluorescence signal in bands 13 or 15, and that if there was no fluorescence signal then the spectrum would be a straight line between these two measurements. Thus the

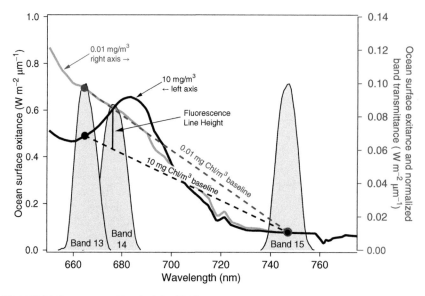

Figure 4.37 MODIS Fluorescence Line Height (FLH) algorithm. Adapted from Abbott and Letelier (n.d.)

fluorescence signal is the difference between the radiance measured in band 14 and the baseline drawn between the band 13 and 15 measurements.

Figure 4.38 illustrates the results from some example tests that have been conducted on the FLH algorithm. The left-hand plot compares drifter data obtained off the Oregon coast against MODIS data, and the level of agreement is quite good. The right-hand plot shows MODIS data compared with ship data taken along a MODIS track over the Chesapeake Bay. Again the agreement between the MODIS FLH estimates and in situ data is quite good, with a correlation coefficient of 0.87.

Figure 4.39 shows the average fluorescence for the world's oceans in January–March 2003 as estimated from MODIS. Animation of monthly-averages indicates a seasonal changes in photoplankton distribution throughout the world's oceans. The zones with no data near the

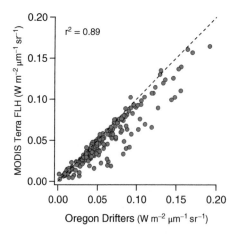
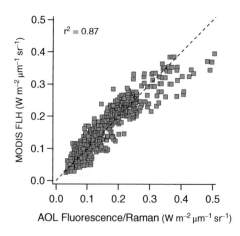

Figure 4.38 Testing MODIS Fluorescence Line Height (FLH) algorithm. From Letelier et al. (2004) and Hoge et al. (2003).

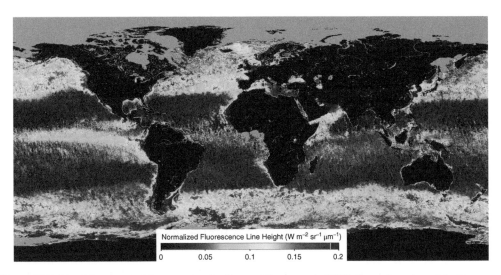

Figure 4.39 Monthly-averaged fluorescence for the world's oceans in 2004. Figure based on data from https://oceancolor.gsfc.nasa.gov/l3/.

poles also vary in latitude across the seasons, depending on solar illumination – without sufficient sunlight, the fluorescence signals are not strong enough to be observed.

Another parameter that can be derived is the fluorescence yield, which measures the efficiency which chlorophyll converts incoming photons to outgoing fluorescence. Specifically we write:

$$F = [chl](\text{PAR } a)\Phi_f \qquad (4.55)$$

where F is the fluorescence signal, chl is the chlorophyll concentration, PAR is

the photosynthetically available radiation between 400 and 700 nm, a is the chlorophyll specific absorption coefficient, and Φ_f is the fluorescence quantum yield. This equation can be inverted, so that that the fluorescence yield can be estimated from the fluorescence signal and chlorophyll concentrations obtained from MODIS, along with an estimate of the photosynthetically available radiation. The fluorescence yield is typically no more than a few percent.

Figure 4.40 shows a map of the fluorescence yields corrected for nonphotochemical

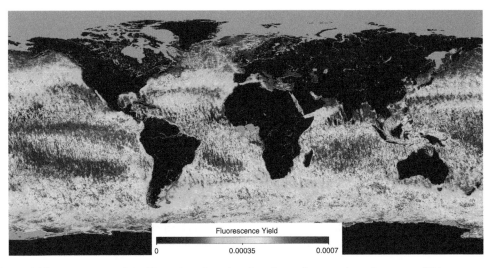

Figure 4.40 Fluorescence yield. Figure based on data from https://oceancolor.gsfc.nasa.gov/l3/.

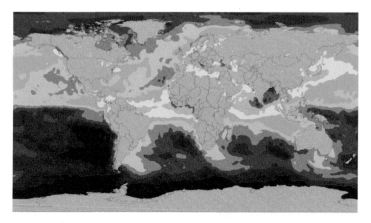

Figure 4.41 MODIS dust deposition – spring 2004. Source: NASA http://svs.gsfc.nasa.gov/3580.

quenching, which is a biological response that prevents high-light damage. This plot was based on the algorithm of Behrenfeld et al. (2009) applied to MODIS data from January–March 2004.

Fluorescence yield is a useful parameter. It turns out that high fluorescence yield occurs when plankton are stressed, often by a lack of key nutrients such as iron. The high yields in the Pacific ocean below the equator are generally indicative of the plankton being nutrient limited in these regions. Of course plankton take in CO_2 and then carry that carbon to the sea floor when they die. So data such as this has led to the suggestion that

CO_2 could be removed from the atmosphere on a grand scale by seeding the southern oceans with iron. This was tried a few years back, and a small phytoplankton bloom was created, but it turned out that the bloom was quickly limited by the absence of other nutrients.

It turns out that dust carries nutrients, especially iron needed for phytoplankton growth. So areas with low dust deposition have high fluorescence yields. This has led to attempts to estimate dust deposition using aerosol estimates and fluorescence yield data.

Figure 4.41 shows dust deposition as estimated from MODIS data.

References

Abbott, M. R., & Letelier, R. M. (n.d.). Algorithm theoretical basis document chlorophyll fluorescence (MODIS product number 20). Retrieved from https://oceancolor.gsfc.nasa.gov/docs/technical/atbd_mod22.pdf

Austin, R. W. (1974). The remote sensing of spectral radiance from below the ocean surface. In N.G. Jerlov & E. Steemann-Nielsen (Eds.), *Optical aspects of oceanography* (pp. 317–344). London, New York: Academic Press.

Barnes, W. L., Pagano, T. S., & Salomonson, V. V. (1998). Prelaunch characteristics of the moderate resolution imaging spectroradiometer (MODIS) on EOS-AM1. *IEEE Transactions on Geoscience and Remote Sensing, 36*(4), 1088–1100.

Behrenfeld, M. J., Westberry, T. K., Boss, E. S., O'Malley, R. T., Siegel, D. A., Wiggert, J. D., et al. (2009). Satellite-detected fluorescence reveals global physiology of ocean phytoplankton. *Biogeosciences, 6*(5), 779.

Gaigalas, A. K., He, H.-J., & Wang, L. (2009). Measurement of absorption and scattering with an integrating sphere detector: Application to microalgae. *Journal of Research of the National Institute of Standards and Technology, 114* (2), 69.

Gordon, H. R. (1997). Atmospheric correction of ocean color imagery in the Earth Observing System era. *Journal of Geophysical Research: Atmospheres, 102*(D14), 17081–17106.

Gordon, H. R., & Clark, D. K. (1981). Clear water radiances for atmospheric correction of coastal zone color scanner imagery. *Applied Optics, 20*(24), 4175–4180.

Gordon, H. R., & Morel, A. Y. (1983). *Remote assessment of ocean color for interpretation of satellite visible imagery: A review.* Springer Science & Business Media.

Hoge, F. E., Lyon, P. E., Swift, R. N., Yungel, J. K., Abbott, M. R., Letelier, R. M., & Esaias, W. E. (2003). Validation of Terra-MODIS phytoplankton chlorophyll fluorescence line height. I. Initial airborne lidar results. *Applied Optics, 42*(15), 2767–2771.

Hooker, S. B., Firestone, E. R., Esaias, W. E., Feldman, G. C., Gregg, W. W., & Mcclain, C. R. (1992). *SeaWiFS Technical Report Vol.1, An overview of SeaWiFS and ocean color.* National Aeronautics and Space Administration, Goddard Space Flight Center.

Hu, C., Lee, Z., & Franz, B. (2012). Chlorophyll algorithms for oligotrophic oceans: A novel approach based on three-band reflectance difference. *Journal of Geophysical Research: Oceans, 117*(C1).

Letelier, R., Abbott, M., & Nahorniak, J. (2004). Fluorescence line height (FLH). In *July 2004 MODIS science team meeting's ocean group breakout session.* Retrieved from https://modis .gsfc.nasa.gov/sci_team/meetings/200407/ presentations/oceans/Letelier.ppt

Martin, S. (2014). *An introduction to ocean remote sensing.* Cambridge University Press.

McClain, C. R. (2009). A decade of satellite ocean color observations. *Annual Review of Marine Science, 1*, 1942.

Mobley, C., Boss, E., & Roesler, C. (2021). Ocean Optics Web Book. Retrieved from http://www .oceanopticsbook.info

Mobley, C. D. (1994). *Light and water: Radiative transfer in natural waters.* Academic Press.

Mobley, C. D., Stramski, D., Bissett, W. P., & Boss, E. (2004). Optical modeling of ocean waters: Is the Case 1-Case 2 classification still useful? *Oceanography, 17*(2), 60–67.

Morel, A. et al. (1974). Optical properties of pure water and pure sea water. *Optical aspects of oceanography, 1*(1), 1–24.

Morel, A., & Prieur, L. (1977). Analysis of variations in ocean color 1. *Limnology and oceanography, 22*(4), 709–722.

National Research Council (Space Studies Board and Ocean Studies Board). (2011). *Assessing the requirements for sustained ocean color research and operations.* National Academies Press.

O'Reilly, J. E., Maritorena, S., O'Brien, M. C., Siegel, D. A., Toole, D., Menzies, D., et al. (2000). *SeaWiFS Postlaunch Technical Report Vol.11, SeaWiFS Postlaunch Calibration and Validation Analysis, Part 3.* National Aeronautics and Space Administration, Goddard Space Flight Center.

Robinson, I. S. (1983). Satellite observations of ocean colour. *Philosophical Transactions of the Royal Society of London A, 309*(1508), 415–432.

Robinson, I. S. (1985). *Satellite oceanography; An introduction for oceanographers and remote-sensing scientists.* Chichester (UK): Horwood.

Shifrin, K. S. (1998). *Physical optics of ocean water.* Springer Science & Business Media.

Smith, R. C. (1974). Structure of solar radiation in the upper layers of the sea. In N.G. Jerlov & E. Steemann-Nielsen (Eds.), *Optical aspects of oceanography* (pp. 95–119). London, New York: Academic Press.

Twardowski, M. S., Claustre, H., Freeman, S. A., Stramski, D., & Huot, Y. (2007). Optical backscattering properties of the "clearest" natural waters. *Biogeosciences, 4*(6), 1041–1058.

Wang, M., Liu, X., Jiang, L., & Son, S. (2017). *VIIRS Ocean Color Algorithm Theoretical Basis Document.* NOAA NESDIS Center for Satellite Applications and Research.

Wang, M., Liu, X., Tan, L., Jiang, L., Son, S., Shi, W., et al. (2013). Impacts of VIIRS SDR performance on ocean color products. *Journal of Geophysical Research: Atmospheres, 118*(18).

Werdell, P. J., & Bailey, S. W. (2002). *The SeaWiFS bio-optical archive and storage system (SeaBASS): Current architecture and implementation*. National Aeronautics and Space Administration, Goddard Space Flight Center.

Wilson, W. H., & Austin, R. W. (1978). Remote sensing of ocean color. In *Ocean optics V* (Vol. *160*, pp. 23–31). International Society for Optics and Photonics.

Zhang, X., & Hu, L. (2009). Estimating scattering of pure water from density fluctuation of the refractive index. *Optics express, 17*(3), 1671–1678.

Zhang, X., Hu, L., & He, M.-X. (2009). Scattering by pure seawater: Effect of salinity. *Optics Express, 17*(7), 5698–5710.

5

Optical Sensing – Land Surfaces

5.1 Introduction

Satellite imagery is also used for a variety of land remote sensing applications, including global change research, agricultural monitoring, mineral exploration, pollution monitoring, land surface change detection, and cartographic mapping. As over the ocean, infrared measurements can be used to measure land surface temperatures, although interpretation of land data is substantially more challenging than for ocean data due to the wide variety of materials that can cover the land, including rocks, soils, vegetation, snow, and ice. Optical color measurements can be used to measure vegetation coverage and health, and hyperspectral color measurements made with fine spectral resolution can be used to detect the narrow absorption and reflectance bands associated with specific minerals.

This chapter considers measurements made in both visible (400 to 700 nm) and near-infrared (700–2500 nm) wavelengths, with a focus on the measurement of land surfaces covered by vegetation.

5.2 Radiation over a Lambertian Surface

As usual we begin simple and slowly build up the complexity. In this case we start by assuming the land is well modeled as a Lambertian surface. The equation for the total radiance at a sensor looking downward through the atmosphere at the Lambertian surface can be rederived by dividing the radiance into four terms, as shown in Figure 5.1:

- L_p is the atmospheric path radiance.
- L_s is direct sunlight reflected off the surface to the sensor with no scattering in the atmosphere.
- L_{d1} is skylight reflected off the surface.
- L_{d2} is reflected light that is scattered into the sensor's field of view.

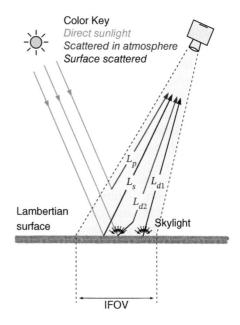

Figure 5.1 Four radiance components above a Lambertian surface. Adapted from Kaufman (1988, 1989).

Remote Sensing Physics: An Introduction to Observing Earth from Space, Advanced Textbook 3, First Edition.
Rick Chapman and Richard Gasparovic.
© 2022 American Geophysical Union. Published 2022 by John Wiley & Sons, Inc.
Companion website: www.wiley.com/go/chapman/physicsofearthremotesensing

While this is a different grouping of the radiances with slightly different notation than we used before, this approach and the notation are common to the land sensing community[1]. As before, the objective is to derive geophysical parameters, such as the spectral reflectance of the surface, from radiance measurements made at the spacecraft.

The equation for the radiance observed by the sensor can be expressed in at least three equivalent forms:

$$L(\rho, \theta, \varphi)$$

$$= L_p + L_s + L_{d1} + L_{d2} \tag{5.1a}$$

$$= L_p(\theta, \varphi) + F_d(\theta_0)\frac{\rho}{\pi(1 - s\rho)}t(\mu) \tag{5.1b}$$

$$= L_p(\theta, \varphi) + F_0(\theta_0)\mu_0$$

$$\times t(-\mu_0)\frac{\rho}{\pi(1 - s\rho)}t(\mu) \tag{5.1c}$$

where:

θ = incident angle of the sensor's line of sight.
φ = azimuth viewing angle relative to the sun.
$\mu = \cos(\theta)$ with a positive value indicating evaluation over an upward path.
θ_0 = zenith angle of the sun.
$\mu_0 = \cos(\theta_0)$ with a minus sign indicating evaluation over a downward path.
ρ = hemispherical surface reflectance (Lambertian surface).
F_d = total irradiance at surface from sun at zenith angle $\theta_0 \rightarrow F_d(\theta_0) = F_0\mu_0\, t(-\mu_0)$.
$F_0\mu_0$ = solar irradiance on a horizontal surface at the top of the atmosphere.
$t(\mu)$ = atmospheric transmittance from ground to space (direct + diffuse).
$t(-\mu_0)$ = atmospheric transmittance from space to ground (direct + diffuse).
s = reflectance of the atmosphere for isotropic light entering the base of the atmosphere (also known as the spherical albedo of the atmosphere – typical value = 0.0–0.4 depending on wavelength and atmospheric visibility).

Note that for simplicity, we have suppressed the wavelength dependence of each factor in these equations.

---

1 The land sensing community also commonly uses L_0 to designate path radiance, instead of L_p as used here.

The first term in each of these equations is the path radiance, which is a function of the sensor look angles, θ and φ, where φ is an azimuth measured relative to the sun's azimuth. Although the terms $\cos(\theta)$ and $\cos(\theta_0)$ could be written out in these equations, we use the terms $\mu = \cos(\theta)$ and $\mu_0 = \cos(\theta_0)$ because they are common in the literature.

In order for equations (5.1a) and (5.1b) to be equal, the second term in equation (5.1b) must equal the sum of the last three terms in equation (5.1a): $L_s + L_{d1} + L_{d2} = F_d(\theta_0)\frac{\rho}{\pi(1-s\rho)}t(\mu)$. This means the second term in equation (5.1b) must be the direct plus diffuse or scattered light that is reflected from the ground to the sensor. This may seem like magic, but let us walk through the individual terms.

The second term in equation (5.1b) says the direct plus diffuse radiance equals the total irradiance at the surface from the sun F_d, times the surface reflectance ρ which is divided by π to account for the assumption that the surface is Lambertian, times the atmospheric transmittance from the ground to the top of the atmosphere $t(\mu)$.

Before explaining the factor we skipped in the denominator of this equation $(1 - s\rho)$, we note that the total solar irradiance at surface F_d equals the solar irradiance on a horizontal surface at the top of the atmosphere $F_0 \cos\theta_0$ times the atmospheric transmittance from the top of atmosphere to the ground. Substituting this expression into equation (5.1b) yields equation (5.1c).

But what about the factor we skipped, $(1 - s\rho)$ – it ends up accounting for multiple scattering from the atmosphere and multiple reflections from the surface. The parameter s in this term is the effective reflectance of the atmosphere for isotropic light entering the base of the atmosphere. Consider the three photon paths illustrated in Figure 5.2. The left-hand panel shows a photon reflecting from the surface without any atmospheric scattering, and in this case the reflection coefficient is simply ρ.

The middle panel shows a photon being reflected from the surface, scattered back to the surface, and then reflecting a second

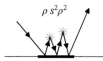

Figure 5.2 Reflection paths with 0, 1, or 2, scattering events. Adapted from Liang (2005) with permission from John Wiley & Sons.

time. Given that s is defined to be the effective reflectance of the atmosphere back towards the surface, then the effective total reflection for this single scattered photon is given by $\rho s \rho$. The photon reflects from the surface twice and the atmosphere once.

The right-hand panel shows the photon for two atmospheric scattering and three surface reflections. The effective radiance for this case is given by $\rho s^2 \rho^2$. Now even though this path is less common than the first, a fact accounted for in the reflection coefficients, we want to include all the possible multiple scattering paths.

The total reflectance from the combination of the surface and the atmosphere is given by the sum shown in the left hand side of equation (5.2). The infinite sum in the brackets is simply the Taylor series expansion of the term $1/(1 - s\rho)$, so the left hand side of equation (5.2) equals the middle term. Thus the term $1/(1 - s\rho)$ exactly accounts for the contribution of multiple scattering in the atmosphere to the net reflectance of the surface. Cool trick!

$$\rho [1 + s\rho + (s\rho)^2 + (s\rho)^3 + ...]$$
$$= \frac{\rho}{1 - s\rho} = \rho + \frac{\rho s \rho}{1 - s\rho} \qquad (5.2)$$

The second equality in equation (5.2) is simply a refactoring of the terms, which breaks the reflectance up into direct and diffuse terms.

Note that for the single-scattering approximation, $s = 0$ and the equation simplifies in a natural way:

$$L(\rho, \theta, \varphi) = L_p(\theta, \varphi) + \rho F_d(\theta_0) \, t(\mu)/\pi \qquad (5.3)$$

Figure 5.3 shows that s ranges from 0.4 at highly-scattering blue wavelengths to close to zero in the infrared where there is little scattering. Plots for two atmospheric visibilities are shown, illustrating the effect of aerosols on atmospheric reflectance. This plot was

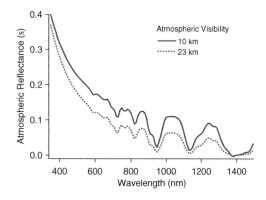

Figure 5.3 Effective reflectance of the atmosphere for isotropic light entering the base of the atmosphere as a function of wavelength for atmospheric visibilities of 10 and 23 km.

produced from MODTRAN predictions of the upward radiance just above Lambertian surfaces with reflectances of 10% and 30%. The radiances and reflectances were then combined to estimate s.

The direct and diffuse reflectance from equation (5.2) can be substituted into the reflected radiance term in the radiance equation (5.1c) to obtain:

$$L - L_p(\theta, \varphi)$$
$$= F_0(\theta_0)\mu_0 \, t(-\mu_0) \, t(\mu) \frac{\rho}{\pi(1 - s\rho)}$$
$$= F_0(\theta_0)\mu_0 \, t(-\mu_0) \, t(\mu) \frac{1}{\pi} \left[\rho + \frac{\rho s \rho}{1 - s\rho} \right] \qquad (5.4)$$

Next split the upward and downward atmospheric transmittances into direct and diffuse terms:

$$t(\mu) = t_i(\mu) + t_d(\mu)$$
$$t(-\mu_0) = t_i(-\mu_0) + t_d(-\mu_0) \qquad (5.5)$$

Here we define t_i to be the direct transmittance and t_d to be the diffuse transmittance. And we note that the downward direct transmittance is given by the usual value of $\exp(-\tau/\mu_0)$.

Again substitute terms from equation (5.5) into the right hand side of the radiance expression and regroup into the three terms: the direct reflected term L_s and the two diffuse terms: L_{d1} and L_{d2}:

$$L_s = F_0 \mu_0 \, t_i(-\mu_0) \, \rho \, t_i(\mu)/\pi \qquad (5.6)$$

$$L_{d1} = F_0 \mu_0 \, t_d(-\mu_0) \, \rho \, t_i(\mu)/\pi \qquad (5.7)$$

$$L_{d2} = F_0 \mu_0 \, \rho$$

$$\times \left[t_i(-\mu_0) \, t_d(\mu) + t_d(-\mu_0) \, t_d(\mu) \right.$$

$$\left. + \frac{t(-\mu_0) \, s \, \rho \, t(\mu)}{(1 - s\rho)} \right] /\pi \qquad (5.8)$$

Now normalize each term in equation (5.4) by the solar radiance at the top of the atmosphere $= F_0\mu_0/\pi$. Then the normalized radiance at the sensor becomes:

$$\rho^*(\rho, \theta, \varphi) = \pi L/F_0\mu_0$$

$$= \rho_0(\theta, \varphi) + \rho \, t(-\mu_0) \, t(\mu)/(1 - s\rho) \qquad (5.9)$$

where ρ_0 is the effective reflectance of the atmosphere in the absence of surface reflectance (i.e., the normalized path radiance), and the second term is the additional effective reflectance associated with the surface. Each term in this equation is nondimensional.

The value ρ^* in equation (5.9) is referred to as the "planetary albedo". The relative magnitudes of the terms contributing to the planetary albedo are shown in Figure 5.4, where $L_D = L_{d1} + L_{d2}$

These plots show the contributions to the planetary albedo from the path radiance, direct reflection and diffuse terms for a dark surface with 5% reflectance and a bright surface with 50% reflectance. These calculations were performed for a midlatitude summer atmosphere with 23 km visibility and a 45° solar zenith angle.

For a dark surface, the path radiance dominates at blue wavelengths, but is comparable to the direct surface reflection at red wavelengths. For the reflective surface, the contributions from the path radiance, direct and diffuse terms are comparable in the blue, but surface reflection dominates in the red.

We spent the time developing the theory for the Lambertian surface because it has closed-form solutions. But most real surfaces are not Lambertian. To deal with a non-Lambertian surface, the theory must be generalized to replace the surface reflectance with a bidirectional reflectance distribution function (BRDF), as will be discussed in Section 5.4. Each term in the radiance equation then involves angular integrations over the products of the BRDF and the transmittance factors.

The term L_{d2} accounts for light reflected from the surface and then scattered by the atmosphere before reaching the sensor. It turns out that this term is especially difficult to evaluate. This is because the term involves a double integration over the solar angles (θ_0, ϕ_0) and scattering angles (θ, ϕ).

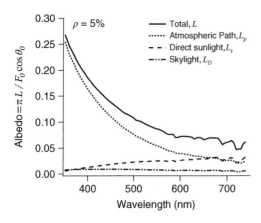 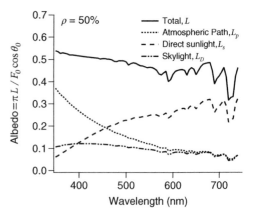

Figure 5.4 Radiance contributions above a Lambertian surface. Note different scales on the left-hand and right-hand plots.

5.3 Atmospheric Corrections

The general approach for correcting the radiance equation for atmospheric effects involves the following steps:

- First solve the radiance equation for the surface reflectance. For a Lambertian surface, the solution is:

$$\rho = \frac{L - L_p}{(L - L_p)s + F_0 \mu_0 \, t(-\mu_0) \, t(\mu)/\pi}$$
$$(5.10)$$

- Measure the sensor radiances L.
- Then use an atmospheric model such as MODTRAN to compute the path radiance, the atmospheric spherical albedo, and the transmittance functions.

Now while this works fine for Rayleigh or molecular scattering, it turns out that the variability of the aerosol scattering is too large for this approach to work at visible wavelengths. Furthermore, additional information is needed to estimate the effects of water vapor absorption at near-infrared (NIR) wavelengths. So there are two approaches that are taken:

- Look for "dark pixels" in the image at some wavelengths, where the reflected term is small. Then use a procedure similar to that used for ocean color to estimate the aerosol contributions at the dark wavelengths, and project this result to other wavelengths.
- Identify a region in the scene where other information indicates that the region should have a nearly constant surface reflectance with time. Then acquire a single image that contains the region of constant reflectance, along with information about the atmospheric conditions at the time of that image. Application of an atmospheric model to these conditions will then allow for an accurate estimation of the surface reflectance in the constant region. You can then use the atmospheric model and the known reflectance of the constant region to estimate the atmospheric correction for other images. The correction is accomplished by adjusting the atmospheric conditions until the measured reflectance of the invariant regions in the other images matches the known reflectance.

While such corrections can be effective at dealing with the average effect of the atmosphere, they are incomplete at finer scales. Consider a Lambertian surface consisting of one small dark square (say a 10×10 m square with $\rho = 0.2$), completely surrounded by a region with either the same reflectivity ($\rho = 0.2$), or a higher reflectivity ($\rho = 0.5$). In this case, the reflectance of the surrounding region will impact both the upward and downward radiance in the atmosphere above the dark square. So it is easy to see that the radiance arising from the dark square will be increased by the presence of an adjacent higher reflectivity region. This is called the adjacency effect, and is another complication in the interpretation of land reflectances. Interpretation gets even more complicated when considering real surfaces with varying topography that can produce shadows and reflections, further altering both direct and scattered light fields. For these reasons, the determination of atmospheric corrections over land is much more difficult than over water.

5.4 Scattering from Vegetation

When we were developing the theory for ocean color measurements, we treated the surface as a specular reflector governed by Fresnel's Law applied to the surface with varying slope. This approximation is based on the surface being smooth relative to the wavelength of the optical radiation. This is generally not the case for land surfaces and vegetation, where roughness occurs at scales comparable to the optical radiation. As illustrated in Figure 5.5, we need then to describe the surface in terms of its BRDF and albedo.

As we saw previously, the BRDF is defined as the ratio of the scattered radiance leaving a surface in one direction and the flux density on the surface coming from another direction.

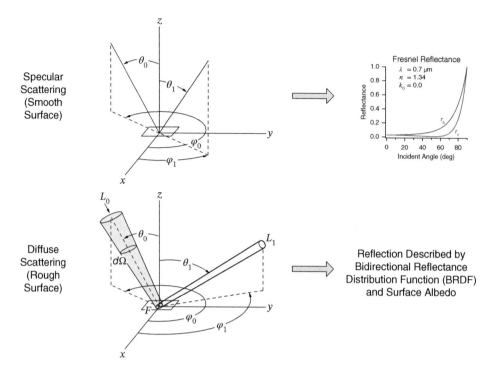

Figure 5.5 Differences between diffuse scattering and specular reflection. Adapted from Nicodemus et al. (1977).

It is designated by a capital R and has units of inverse steradians:

$$R(\theta_0, \varphi_0; \theta_1, \varphi_1)$$

$$= \frac{\text{Scattered radiance to } (\theta_1, \varphi_1)}{\text{Flux density at surface from } (\theta_0, \varphi_0)}$$

$$= \frac{L(\theta_1, \varphi_1)}{F(\theta_0, \varphi_0)\mu_0}$$

$$= \frac{L(\theta_1, \varphi_1)}{L(\theta_0, \varphi_0)\mu_0 \, d\Omega(\theta_0, \varphi_0)} \quad (\text{units} = \text{sr}^{-1})$$

$$(5.11)$$

The mean reflectance or albedo of a surface is the unit-less ratio of the radiant exitance and the incident flux density. It is designated by little r, and can be computed from the integral of the BRDF over the hemisphere where the scattered radiance is emitted. Note that the albedo is still generally dependent on the direction of the incoming radiation:

$$r(\theta_0, \varphi_0) = \frac{\text{Radiant exitance}}{\text{Incident flux density}}$$

$$= \frac{\int_0^{2\pi} \int_0^{\pi/2} L(\theta_1, \varphi_1) \cos\theta_1 \sin\theta_1 d\theta_1 d\varphi_1}{F(\theta_0, \varphi_0)\mu_0}$$

$$= \int_0^{2\pi} \int_0^{\pi/2} R(\theta_0, \varphi_0; \theta_1, \varphi_1) \cos\theta_1 \sin\theta_1 d\theta_1 d\varphi_1 \quad (5.12)$$

This observation leads to the concept of a diffuse albedo, designated by small r_d, which equals the average albedo over the hemisphere of all possible incident directions:

infrared (1.4–3 μm). The reflectance of vegetation varies but shows large common absorption bands in the infrared. Different soils generally have broadband reflectance peaking in the

r_d = Average reflectance over hemisphere of possible incident directions

$$= \frac{\int_0^{2\pi} \int_0^{\pi/2} r(\theta_0, \varphi_0) \cos \theta_0 \sin \theta_0 \, d\theta_0 \, d\varphi_0}{\int_0^{2\pi} \int_0^{\pi/2} \cos \theta_0 \, \sin \theta_0 \, d\theta_0 \, d\varphi_0}$$

$$= \frac{1}{\pi} \int_0^{2\pi} \int_0^{\pi/2} r(\theta_0, \varphi_0) \cos \theta_0 \sin \theta_0 \, d\theta_0 \, d\varphi_0 \qquad (5.13)$$

Now you will see these definitions applied to rough surfaces, but they can also be applied to volume scattering from a material that allows some penetration of the radiant energy. We had actually previously done this very thing when defining an atmospheric reflectance based on atmospheric scattering.

It turns out that volume scattering is critically important for land remote sensing. As illustrated in Figure 5.6 the optical properties of many materials such as snow and vegetation are dominated by volume scattering.

Figure 5.7 shows the wide range of albedos of some natural materials.

Figure 5.8 shows the spectral reflectances of various natural materials. Snow and clouds have the highest spectral reflectance in the visible (0.4–0.75 μm), but not necessarily in the near infrared (0.75–1.4 μm) and short-wave

infrared. Seawater and pure ice have the lowest reflectances.

The spectral characteristics of green vegetation are of special importance. Luckily the chlorophyll in vegetation has some unique spectral characteristics that are exploited to measure vegetation.

Figure 5.9 plots the general spectral reflectance of green vegetation. In the visible band from 0.4 to 0.7 μm, the reflectance of vegetation is generally between 10 and 15%, with a peak at about 0.55 μm and minima at the two chlorophyll absorption bands in the blue and red. In general the spectral reflectance in the visible is dominated by leaf pigments.

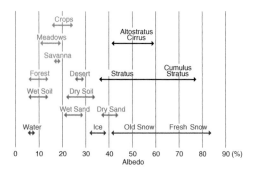

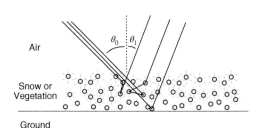

Figure 5.6 Geometry of volume scattering.

Figure 5.7 Albedos for some natural materials at visible to mid-wave wavelengths. Figure adapted from https://commons.wikimedia.org/wiki/File: Albedo-e_hg.png.

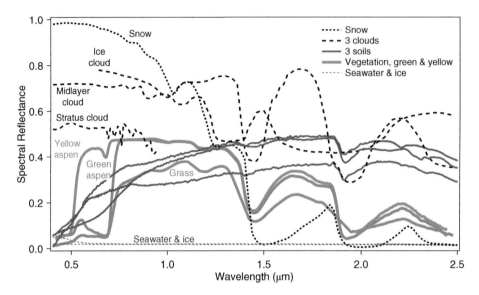

Figure 5.8 Spectral reflectances of various natural materials. Cloud data from Bowker et al. (1985), vegetation data from Kokaly et al. (2017), soil and other data from Meerdink et al. (2018).

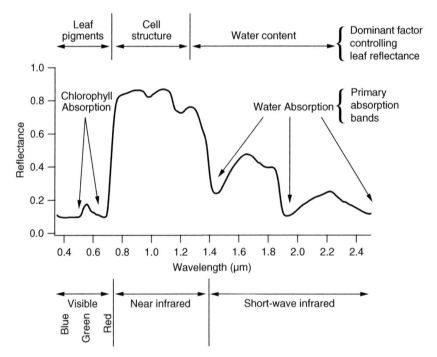

Figure 5.9 Spectral reflectance characteristics of green vegetation. Source: Liang (2005)/with permission from John Wiley & Sons.

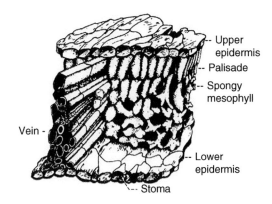

Figure 5.10 Cell structure of vegetation. From Myers et al. (1983) with permission from John Wiley & Sons.

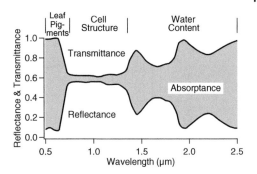

Figure 5.11 Leaf reflectance spectrum. Adapted from Gausman et al. (1970).

In the near-infrared region from 0.7 to 1.4 μm the spectral reflectance from the vegetation and the transmittance into the plant are both high, which can occur because the absorption is low. In this region the reflected radiance is predominantly due to volume scattering from the complex cell structure as illustrated in Figure 5.10.

Referring back to Figure 5.9, the character of the reflectance changes again in the short-wave infrared from 1.4 to 3.0 μm. In this region the return is dominated by the water content in the vegetation, with reduced reflectances in specific water absorption bands.

Figure 5.11 shows the spectral variation of transmittance (the top portion), absorptance (the middle portion) and reflectance (the bottom portion) for a mature orange leaf.

It turns out that it is possible to model a leaf using just four optical parameters related to water content, void percentage, effective index of refraction, and the effective absorption coefficient. We will not go through that model here, but it can be found in the literature.

Figure 5.12 shows the variation of spectral reflectance across the visible and near

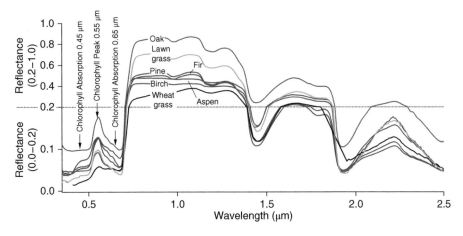

Figure 5.12 Spectral reflectance across the visible for seven types of plants. Data from Kokaly et al. (2017) and Meerdink et al. (2018).

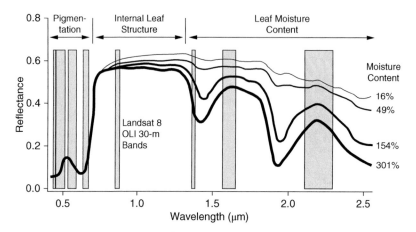

Figure 5.13 Reflectance of a sycamore leaf at different moisture levels. Based on original data from National Research Council (1976).

infrared for seven types of plants: grass as well as birch, pine, and fir trees. This suggests that visible color with sufficient resolution can be useful in determining vegetation type.

Figure 5.13 shows the spectral reflectance of a sycamore leaf with varying moisture content. The bands from Landsat 8 are overlaid on this plot, indicating that some of the short-wave infrared (SWIR) bands should be sensitive to the moisture content of vegetation, while the visible signals are not.

So while the visible signals contain information about plant chlorophyll, the SWIR contains information about plant moisture content.

Figure 5.14 shows variations in the spectral reflectance as a function of canopy thickness and green biomass, which is an integral over the height of the vegetation. Here we see that variations in the near infrared correlate with plant structure and density integrated over the sensor field of view, which in turn relates to parameters such as coverage and biomass.

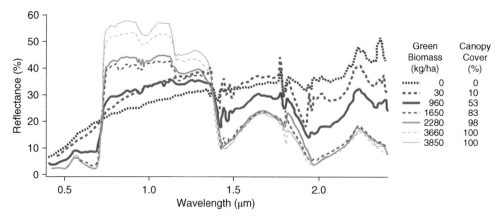

Figure 5.14 Variations in spectral reflectance as a functions of amounts of green biomass and percentage canopy cover. Adapted from Short (1982).

To summarize, vegetation has specific spectral characteristics that can be exploited to measure a variety of important parameters. Some of the algorithms that exploit these characteristics are the topic of the next two sections.

5.5 Normalized Difference Vegetation Index

The Normalized Difference Vegetation Index (NDVI) was the first metric derived to estimate the amount of growing vegetation in each pixel of multispectral land cover images. The underlying concept is that the large difference in spectral reflectance of green vegetation at visible and near-infrared wavelengths (recall Figure 5.12) provides a signature that can be exploited to identify areas of vegetation.

The original NDVI used the signals recorded from channels 1 and 2 of the AVHRR instruments on the NOAA polar-orbiting satellites. Figure 5.15 shows the wavelength response of these channels relative to a typical vegetation spectrum.

This original NDVI was computed by differencing the digital counts from the two channels, and dividing by the sum of the counts:

$$\text{NDVI} = \frac{\text{AVHRR 2} - \text{AVHRR 1}}{\text{AVHRR 2} + \text{AVHRR 1}} \quad (5.14)$$

Because one channel measures the depth of the chlorophyll absorption band, the difference is sensitive to chlorophyll concentration. Dividing by the sum tends to normalize out the largest atmospheric effects. And that is all there is to it.

The NDVI is a crude but useful vegetation measure. NDVI serves as an important parameter in assessment models for green cover, biomass, leaf area index and fraction of absorbed photosynthetically active radiation. It also provides a metric for evaluating temporal changes in vegetation growth and activity.

Operational applications of the NDVI include famine early warning systems, drought detection, assessment of land degradation and deforestation, as well as monitoring and detection of changes due to climate variations.

Generally, vegetation index products (including NDVI) are compiled for 16-day and monthly periods by compositing all cloud-free image data acquired during these periods, just like we saw for sea surface temperature estimation. The most modern algorithms go beyond NDVI to also correct for atmospheric path radiance, at least to some degree, and viewing angle effects.

NDVI varies seasonally depending on land cover type and local growing conditions. DeFries and Townshend (1994) trained a classifier to identify land cover types based on

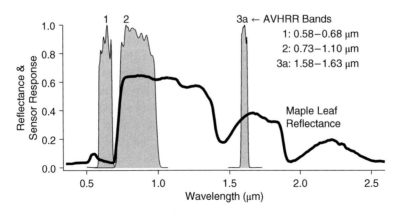

Figure 5.15 AVHRR bands and vegetation reflectance. Adapted from Peterson (1989).

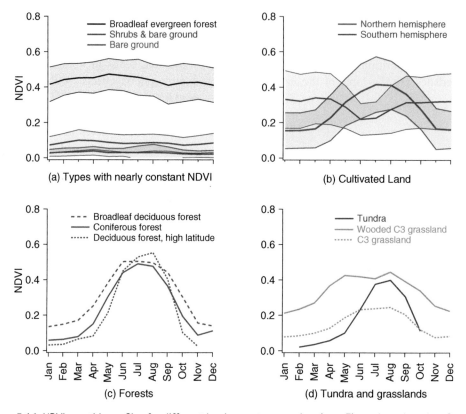

Figure 5.16 NDVI monthly profiles for different land cover types and regions. Figure based on data from NASA. Studying Earth's Environment From Space. May 2003. Accessed Oct 2018. http://www.ccpo.odu.edu/SEES/index.html.

these temporal variations. Figure 5.16 shows the NDVI monthly profiles for 1987 based on the classifier trained to identify eleven standard land cover types. Panel (a) indicates that bare ground, with or without shrubs, has low NDVI values that remain relatively constant throughout the year. Broadleaf evergreen forests are mostly tropical, and so they also have a nearly constant NDVI value, although that value is quite high.

Panel (b) shows how the NDVI for cultivated land varies seasonally, with a six-month phase difference between the northern and southern hemispheres. Panel (c) shows the similarities between three types of northern hemisphere forests. Note that the high-latitude deciduous forest has the shortest growing season and hence the narrowest NDVI peak.

Panel (d) shows the even narrower peak associated with northern tundra and the broader peaks associated with midlatitude grasslands.

Not surprisingly, the temporal variations in NDVI are not perfect discriminators of land type. Figure 5.17 contains two plots, each compares the NDVI as a function of month for two different land cover types. The plot on the left shows that broadleaf deciduous forests and wooded grasslands in South America have similar NDVI time histories. The plot on the right shows the substantial NDVI differences for broadleaf evergreen forests in South America and Africa. These data show that NDVI can provide the same measures for different vegetation and different measures for similar vegetation. Despite these issues, NDVI

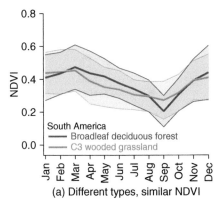

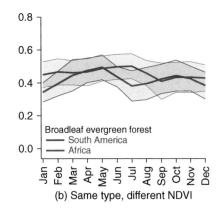

Figure 5.17 NDVI monthly profiles for two different land cover types. Figure based on data from NASA. Studying Earth's Environment From Space. May 2003. Accessed Oct 2018. http://www.ccpo.odu.edu/SEES/index.html.

remains useful as a global measure of growing vegetation.

Various approaches have been tried to improve the NDVI. Instead of using AVHRR digital counts to define NDVI, the signals in Channels 1 and 2 can be converted to radiance units using the AVHRR calibration equations:

$$L_i = \alpha_i N_i + \beta_i \qquad (5.15)$$

The radiance-based NDVI is then:

$$\mathrm{NDVI}_L = \frac{L_2 - L_1}{L_2 + L_1} \qquad (5.16)$$

This definition of NDVI removes some dependence on sun-surface-sensor geometry, and changing sensor calibration coefficients. Another refinement is to reformulate the NDVI in terms of the planetary albedo:

$$\rho_i^* = \pi \, L_i \, d^2 / F_0(\lambda_i)\mu_0 \qquad (5.17)$$

where, F_0 = solar irradiance at TOA, θ_0 = solar zenith angle, d = sun-earth distance, in astronomical units (varies $\pm 1.7\%$).

The albedo-based NDVI is then:

$$\mathrm{NDVI}_{\rho^*} = \frac{\rho_2^* - \rho_1^*}{\rho_2^* + \rho_1^*} \qquad (5.18)$$

This further reduces variations due to illumination conditions.

It can then be shown that the albedo-based NDVI is a weighted form of the radiance-based

NDVI. It is this formula that is used to compute the albedo-based NDVI. Using equations (5.17) and (5.18), we see that:

$$\mathrm{NDVI}_{\rho^*} = \frac{\dfrac{F_0(\lambda_1)}{F_0(\lambda_2)}L_2 - L_1}{\dfrac{F_0(\lambda_1)}{F_0(\lambda_2)}L_2 + L_1} \qquad (5.19)$$

The magnitude of the ratio $F_0(\lambda_1)/F_0(\lambda_2)$ can be estimated by assuming the sun to be an ideal blackbody with temperature of 5800 K and computing the spectral radiance using the Planck function for $0.58 \le \lambda_1 \le 0.68\,\mu\mathrm{m}$ and $0.72 \le \lambda_2 \le 0.94\,\mu\mathrm{m}$. The result is:

$$F_{1,2} = F_0(\lambda_1)/F_0(\lambda_2) \approx 1.48 \text{ to } 1.53 \qquad (5.20)$$

A little algebraic manipulation can relate the albedo-based NDVI to the radiance-based NDVI:

$$\mathrm{NDVI}_{\rho^*} = \frac{(F_{1,2} + 1)\mathrm{NDVI}_L + (F_{1,2} - 1)}{(F_{1,2} - 1)\mathrm{NDVI}_L + (F_{1,2} + 1)} \qquad (5.21)$$

where

$$F_{1,2} = F_0(\lambda_1)/F_0(\lambda_2) \qquad (5.22)$$

Figure 5.18 compares the original NDVI based on digital counts versus the albedo-based NDVI. In the end the albedo-based NDVI tracks the original NDVI, but with a positive offset and a larger dynamic range.

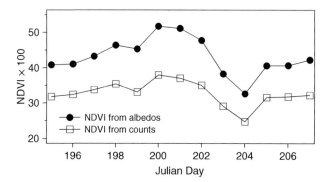

Figure 5.18 NDVI from digital counts and from albedo. Based on data from Gutman (1991).

Global NDVI data have been acquired since 1982, providing a long-term record of changes in global vegetation (Gutman et al. 1995). Figure 5.19 shows the maximum NDVI values for 1986. The green regions thus indicate the most productive locations on the planet, independent of seasonal variations.

With the launch of the MODIS instrument on the Terra satellite in 1999, vegetation researchers now had an instrument with seven wavelength channels in the visible, near IR and short-wave IR, along with higher spatial resolution (500 m vs. 1000 m with AVHRR). Figure 5.20 overlays the reflectance of corn, soybeans, bare soil, and three of the MODIS bands in the blue, red, and near infrared. Various algorithms have been developed to produce a vegetation index based on two or all three of these bands.

The original MODIS NDVI algorithm was developed to provide continuity with the AVHRR algorithms. It is albedo based using the red and near-infrared channels:

$$\text{NDVI} = \frac{\rho_{\text{NIR}} - \rho_{\text{red}}}{\rho_{\text{NIR}} + \rho_{\text{red}}} \tag{5.23}$$

Figure 5.21 shows the basic processing methodology for MODIS NDVI estimates. The same compositing scheme is used to construct two-week or monthly NDVI maps.

MODIS also introduced the Enhanced Vegetation Index (EVI) . This algorithm is based on the reflectances from all three bands:

$$\text{EVI} = \frac{2.5(\rho_{\text{NIR}} - \rho_{\text{red}})}{\rho_{\text{NIR}} + 6\,\rho_{\text{red}} - 7.5\,\rho_{\text{blue}} + 1} \tag{5.24}$$

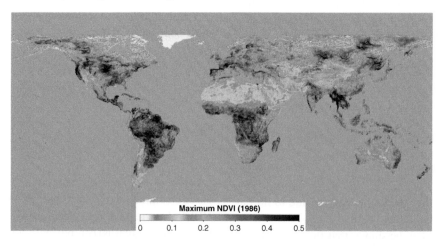

Figure 5.19 Global maximum NDVI values for 1986 from AVHRR data. Based on data from Vermote et al. (2014).

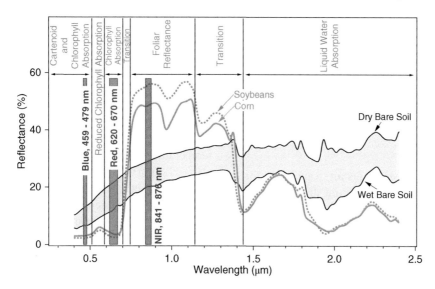

Figure 5.20 Reflectance of corn, soybeans, bare soil in three MODIS bands. Similar to Tucker and Sellers (1986) but with data from Short (1982).

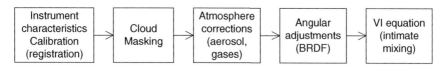

Figure 5.21 MODIS NDVI processing methodology. From Huete et al. (1999).

Note that unlike the NDVI, the reflectances used for the EVI are partially corrected for Rayleigh scattering and ozone absorption. The way to think about this algorithm is that the numerator is just like the NDVI, but the blue channel is used in the denominator to correct for the influences of aerosols.

It turns out that while the NDVI is primarily sensitive to chlorophyll, the EVI is more responsive to structural variations in the vegetation. This makes it more sensitive to such parameters as leaf area, canopy type, and canopy architecture. In the end, the two indices complement each other.

The maps in Figure 5.22 show similarities and differences between the MODIS NDVI and EVI. Although Greenland is snow covered in both metrics, notice the significant differences in northern Canada.

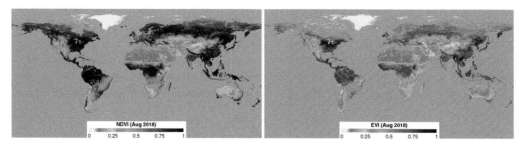

Figure 5.22 MODIS NDVI (left panel) and EVI (right panel) for August 2018. Based on data from Didan (2015).

The interested reader can find more details about the MODIS vegetation index products on the NASA MODIS Web (NASA, 2021).

5.6 Vegetation Condition and Temperature Condition Indices

Extreme weather events are responsible for a large fraction of the natural disasters experienced around the world. Drought is one of the most damaging disasters, adversely impacting the lives of large numbers of people, as well as inflicting enormous destruction annually to the economy and property of affected nations.

The productivity of vegetated areas varies from year to year due to fluctuations in temperature and rainfall. In dry and hot years, NDVI values will fall and land surface temperatures will be higher than in wet years. Intuitively, for a given area, we might expect to find maximum NDVI values in years with optimal weather conditions, and minimum values in years with drought conditions. This suggests that some combination of NDVI and remotely sensed surface temperature might be a useful tool for detecting and monitoring drought affected regions, especially for countries with limited economies and marginal weather reporting networks.

NOAA introduced the Vegetation Condition Index (VCI) and the Temperature Condition Index (TCI) in the late 1980s to make use of AVHRR data for drought monitoring. For this purpose, the inherent high frequency variations in NDVI values must first be minimized by some judicious filtering in order to enhance the low frequency variations related to weather conditions. This filtering process starts by compositing the AVHRR radiance data by saving only those values with the largest difference between counts in Channels 1 (0.58–0.68 μm) and 2 (0.73–1.10 μm) over a seven-day period for each map cell. Various filters are then applied to this weekly time series to yield a smoothed NDVI annual time series. Year-to-year differences in these annual time series are due to weather variations. In a similar fashion, the highest brightness temperature value from AVHRR Channel 4 (10.3–11.3 μm) in a seven-day period is saved for each map cell, and then filtered to produce a smoothed brightness temperature annual time series. Details about NDVI data noise sources and the filtering procedures can be found in Kogan (1995).

Drought detection and monitoring first requires calculation of smoothed NDVI and brightness temperature annual time series from several years of data that contain drought and nondrought conditions.

The Vegetation Condition Index and the Temperature Condition Index are defined in equations (5.25) and (5.26) (Kogan, 1997):

$$VCI = 100 \frac{NDVI - NDVI_{min}}{NDVI_{max} - NDVI_{min}} \quad (5.25)$$

$$TCI = 100 \frac{T_4 - T_{4\,min}}{T_{4\,max} - T_{4\,min}} \quad (5.26)$$

where $NDVI$ is the smoothed weekly NDVI for the map cell of interest, and $NDVI_{max}$ and $NDVI_{min}$ are the multiyear absolute maximum and minimum NDVI values, respectively; T_4, $T_{4\,max}$, and $T_{4\,min}$ are corresponding values from Channel 4 brightness data.

Notice that in nondrought years when $NDVI$ is large and T_4 is low, both indices will be large; in drought years with small $NDVI$ and large brightness temperature, the indices are reduced. Note also that both indices are normalized to range from 0 to 100.

Numerous validation efforts conducted in Asia, Africa, Europe, and North and South America with ground-truth data have shown these indices can successfully detect drought-induced plant stress. A detailed discussion of these validation efforts is beyond the scope of this section; the interested reader can find this information in Kogan (1997).

5.7 Vegetation Indices from Hyperspectral Data

Hyperspectral imaging collects data in hundreds of contiguous spectral bands instead of just a few (Figure 5.23). This produces a lot more data, but the detailed spectral information provides leverage for detecting various materials.

We did not discuss hyperspectral imagers for ocean color because it turns out that the spectral components of ocean color are generally broad and smooth. This means that hyperspectral imagers offer little advantage for ocean color measurements over imagers with just a few bands. Furthermore, the ocean is relatively dark and hyperspectral imagers necessarily split the radiant energy

into finer bins, reducing the flux arriving in any given band. This means that hyperspectral imagers require longer exposure times than conventional multispectral imagers.

Hyperspectral sensors are particularly well suited for vegetation measurements. They can acquire sufficient data to estimate water vapor and aerosols in the atmosphere as well as produce a wider variety of vegetation indices. Narrowband vegetation indices can:

- Indicate the overall amount and quality of photosynthetic material.
- Estimate moisture content in vegetation.
- Measure vegetation stress.

Some of the specific indices (beyond NDVI and EVI) are listed in Table 5.1.

The EO-1 spacecraft (Figure 5.24) launched in 2000, hosted NASA's first spaceborne hyperspectral imager, the Hyperion imaging spectrometer, which recorded imagery in 220 wavelength bands (Pearlman et al., 2003). EO-1 also carried the Advanced Land Imager (ALI), a nine-band imager using linear arrays of spectrometers, along with the

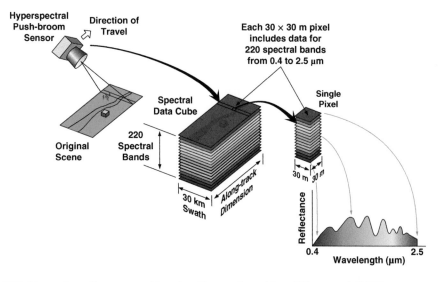

Figure 5.23 Illustration of hyperspectral imaging. Figure adapted from Wilson et al. (2000).

Table 5.1 Other vegetation indices.

– Photochemical Reflectance Index	– Plant Senescence Reflectance Index
– Structure Insensitive Pigment Index	– Carotenoid Reflectance Index
– Normalized Difference Nitrogen Index	– Anthocyanin Reflectance Index
– Normalized Difference Lignin Index	– Normalized Difference Water Index
– Cellulose Absorption Index	– Moisture Stress Index

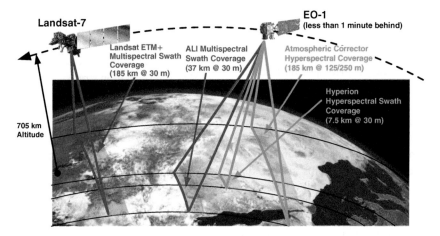

Figure 5.24 Landsat and EO-1 sensor swaths. Pearlman et al. (2000) with permission from Society of Photo-Optical Instrumentation Engineers.

Table 5.2 EO-1 instrument characteristics.

Parameters	ALI	Hyperion	LAC
Type	Multispectral	Hyperspectral	Hyperspectral
Spectral Range (μm)	0.4–2.4	0.4–2.5	0.9–1.6
Spatial Resolution (m)	30	30	250
Swath Width (km)	37	7.5	185
Spectral Resolution (nm)	Variable	10	2.6
Spectral Coverage	Discrete	Continuous	Continuous
Panchromatic Resolution	10 m	n/a	n/a
Number of Bands	10	220	256

Linear Etalon Imaging Spectrometer Array (LEISA) Atmospheric Corrector (LAC), an instrument specifically designed to acquire atmospheric data to increase the accuracy of surface reflectance a estimates.

The EO-1 spacecraft was designed to last 18 months to support a one-year mission. It finally ran out of Hydrazine in February 2011!

More details about the sensors are given in Table 5.2. The larger point is that hyperspectral sensors have been built and flown, and these sensors can provide additional information for the measurement of land surfaces and vegetation.

5.8 Landsat Satellites

The first Landsat satellite, launched in 1972, was the first satellite to use a multispectral imager. It was designed for global land observations with an 18-day repeat orbit. The Landsat program has provided continuous coverage of the Earth since that first launch (Table 5.3). Landsat 5 alone was operational for 29 years, from March 1984 to June 2013. The current operational satellites are Landsat 7 and 8, although Landsat 7 has been operating with some issues since May 2003. The major applications for Landsat are mapping and monitoring of land use and crops.

In 1985 the government awarded a ten-year contract to a private company (EOSAT) to operate the Landsat system, transferring the commercial rights to the data as part of the agreement. Data became more expensive over time, and in 1992 Congress passed the Land Remote Sensing Policy Act authorizing procurement of Landsat 7 and assuring the continued availability of Landsat digital data and images, at the lowest possible cost. As a result, the entire program transferred back to the government in 2001.

This is a history that has been repeated several times. Both Canada and Europe have tried and failed with various forms of commercialization of space remote sensing. These past attempts by governments to save money on remote sensing have failed in part because the biggest consumer of these data are governments, and the commercial uses of the data have been too small to support even the profits of the companies that have been created to run the satellites. Whether that will change with the true commercialization of space remains to be seen.

Table 5.4 shows details about the sensors used in the Landsat series. The sensors have been multispectral since day one. They have a 185 km swath with current resolutions of 15–60 m, depending on the wavelength band.

It turns out that the early sensors were not well calibrated. The first serious attempts to calibrate the sensors was made for Landsat 5. Ultimately Landsat 7, launched in 1999, was the first sensor with a good radiance calibration. The specific spectral bands for Landsat are listed in Table 5.5.

Table 5.3 History of landsat satellites.

Satellite	Launch Date	End Ops	Operating Agency
Landsat 1	July 1972	July 1978	NASA
Landsat 2	January 1975	February 1982	NASA
Landsat 3	March 1978	March 1983	NOAA
Landsat 4	July 1982	1993	Operation transferred to EOSAT Corp
Landsat 5	March 1984	December 2012	Data prices increased. Users and data coverage fell. Program transferred back to government in 2001
Landsat 6	October 1993	Launch failed	–
Landsat 7	April 1999	Still operating with anomalies	USGS operates and distributes data
Landsat 8	February 2013	Operational	USGS
Landsat 9	September 2021[1]		USGS

[1] Planned launch date.

Table 5.4 Landsat sensors.[1]

	Multispectral Scanners, MSS (Landsat 1-5)	Thematic Mapper, TM (Landsat 4,5)	Enhanced Thematic Mapper Plus, ETM+ (Landsat 7)	Operational Land Imager + Thermal InraRed Sensor (Landsat 8, 9)
Spectral Bands	3 Vis; 1 NIR; 1 LWIR(Landsat 3)	3 Vis; 1 NIR; 2 SWIR; 1 LWIR	3 Vis; 1 NIR; 2 SWIR; 1 LWIR; 1 panchromatic	4 Vis; 1 NIR; 3 SWIR; 2 LWIR; 1 panchromatic
Spatial Resolution (IFOV)	79 m (LS 1–3) 83 m (LS 4,5) 240 m LWIR	30 m Vis-SWIR 120 m LWIR	15 m pan 30 m Vis-SWIR 60 m LWIR	15 m pan 30 m Vis-SWIR 100 m LWIR
Sampling	1.4 samples/IFOV along-track	1 sample/IFOV along-track	1 sample/IFOV along-track	7000 samples/IFOV along-track
Cross-track coverage (km)	185	185	185	185
Radiometric Resolution (bits)	6	8	8 (2 gain states)	12
Radiometric Calibration	Internal lamp and shutter; partial aperture solar (Landsat 1–3)	Internal lamp, shutter and blackbody	Internal lamp, shutter and blackbody, partial aperture solar, full aperture solar diffuser	Internal lamp, shutter and blackbody, solar diffuser, lunar collects
Scanning Mechanism	Unidirectional scanning	Bidirectional scanning with scan line corrector	Bidirectional scanning with scan line corrector	nonscanning push-broom

[1]Landsat 9 is a copy of LandSat 8 except the radiometric resolution will be increased to 14 bits for improved viewing of dark targets. In addition the Thermal InfraRed Sensor will be modified to eliminate a stray light issue that limits the accuracy of LandSat 8 thermal measurements.

Table 5.5 Landsat bands (all Wavelengths in μm).

	MSS (LS-1-5)	TM (LS-4/5)	ETM (LS-6)	ETM+ (LS-7)	LS-8
Pan			P) 0.52–0.90	P) 0.52–0.90	8) 0.500–0.680
Vis	1) 0.5–0.6	1) 0.45–0.52	1) 0.45–0.52	1) 0.45–0.52	1) 0.433–0.453
	2) 0.6–0.7	2) 0.52–0.60	2) 0.52–0.60	2) 0.53–0.61	2) 0.450–0.515
	3) 0.7–0.8	3) 0.63–0.69	3) 0.63–0.69	3) 0.63–0.69	3) 0.525–0.600
					4) 0.630–0.680
NIR	4) 0.8–1.1	4) 0.76–0.90	4) 0.76–0.90	4) 0.78–0.90	5) 0.845–0.885
					9) 1.360–1.390
SWIR		5) 1.55–1.75	5) 1.55–1.75	5) 1.55–1.75	6) 1.560–1.660
		7) 2.08–2.35	7) 2.08–2.35	7) 2.08–2.35	7) 2.100–2.300
LWIR		6) 10.4–12.5	6) 10.4–12.5	6) 10.4–12.5	10) 10.30–11.30
					11) 11.50–12.50

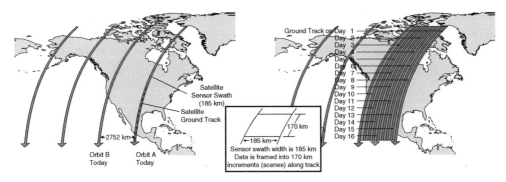

Figure 5.25 Landsat Ground Track. From USGS (2018).

All of the Landsat satellites follow sun-synchronous orbits that are inclined approximately 8° to the poles (Figure 5.25).

Landsats-1, 2, and 3 orbited 14 times per day at a 920-km altitude. Over a period of 252 orbits, or every 18 days, a satellite covered every portion of the Earth, except for polar regions above 82° latitude. Landsats-4, 5, 7, and 8 were designed to orbit at 705-km altitude with a 16-day repeat interval (232 orbits) for global coverage.

Figure 5.26 shows the Landsat-7 satellite, with its one big solar cell array. The imager worked by scanning back and forth in a cross-track direction while the satellite flew down track, a design known as a whisk-broom scanner. In order to form images without gaps at the end of each line, a clever optical component called the scan line corrector (or SLC) was designed into the system. In May 2003 the SLC failed.

From May 2003 to December 2012 Landsat-7 was operated with the broken SLC, but Landsat-5 was still operational. Ultimately a series of mechanical issues in Landsat-5 finally forced the USGS to decommission it and move it into a lower disposal orbit. Landsat-8 was then launched in February 2013.

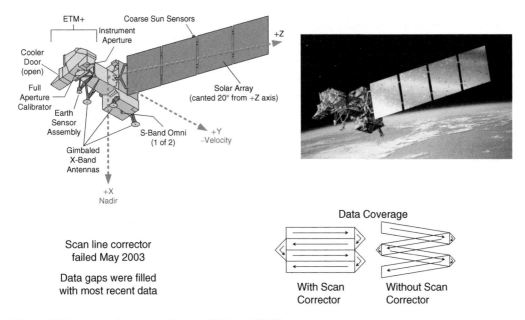

Figure 5.26 Landsat-7 overview. Source: NASA and USGS.

Landsat-8 is also in a 705 km, 16-day repeat orbit. But it has new imagers, the OLI and the TIRS (Irons et al., 2012). The Operational Land Imager is a push-broom design with no scan mirror. The design maintains the 185 km swath to keep with the legacy requirements. The sensor supports the nine bands. The linear detector array in the OLI has 2 rows with seven detector modules each. The eight color bands have almost 7000 detector elements in each band while the panchromatic band (Band 8) has over 13,000 elements. The sensor also has on-board calibration with both an internal source and a solar diffuser. TIRS is a thermal infrared imager with two channels in the 10.3–12.5 µm band for split-window operation.

5.9   High-resolution EO sensors

5.9.1   Introduction

There has been an explosion of commercial high-resolution EO imagers over the past two decades. While these instruments were designed to provide imagery for a broad range of commercial and government sponsors, their data can also be useful for scientific purposes.

Nearly all of the imagers are push-broom scanners with CCD linear arrays (Figure 5.27). Multiple CCD arrays are used in each focal plane to provide both panchromatic and multispectral capabilities. They all use large focal length telescopes to get high resolution, and this means that the spacecraft attitude must be adjusted to image left or right of track. On-board data compression is used to reduce storage requirements and downlink data rates. All of these satellites are in sun-synchronous orbits.

5.9.2   First-Generation Systems

There are three systems that can be considered to be part of the first generation of high-resolution EO sensors: IKONOS, Quick-Bird and OrbView-3. The key sensor parameters for these satellites are given in Table 5.6. Notice the similarity in specifications: panchromatic with 0.6 to 1-m resolution and four spectral bands with 2.4 to 4-m resolution. The orbital parameters for these satellites are given in Table 5.7.

IKONOS

Figure 5.28 is a diagram of the optical design of the IKONOS sensor, a design that is typical of all of the systems we are going to discuss. This telescope had a 10-m focal length, but the path is folded multiple times to get it to fit into a reasonable volume. This particular system had almost 17,000 elements in the linear CCD arrays. The sensor's mass was 170 kg and required a few hundred watts to operate.

Table 5.8 provides the basic specifications for the IKONOS sensor.

Figure 5.29 is a diagram of the IKONOS satellite. Note that the entire satellite is specifically designed to support the one EO instrument.

Figure 5.30 is an image from IKONOS. The detail is remarkable.

QuickBird

QuickBird had better resolution than IKONOS (0.6 m vs 1.0 m), but only had ±25° field of regard. This illustrates just one of the trade-offs made in the design of a remote sensing satellite.

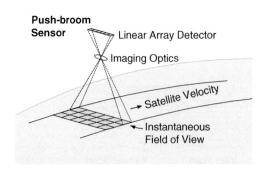

Figure 5.27 Push-broom configuration used in most high-resolution EO sensors. Adapted from Elachi and van Zyl (2006).

Table 5.6 Sensor specifications for first generation commercial EO sensors.

	IKONOS	QuickBird	OrbView-3
Panchromatic Nadir Resolution (m)	1.0	0.6	1.0
Panchromatic Band (nm)	526–929	450–900	450–900
Multispectral Nadir Resolution (m)	4.0	2.4	4.0
Blue Band (nm)	445–516	450–520	450–520
Green Band (nm)	506–595	520–600	520–600
Red Band (nm)	632–698	630–690	625–695
NIR Band (nm)	757–853	760–900	760–900
Aperture Size (m)	0.7	0.6	0.45
Focal Length (m)	10	8.8	2.78
FOV (deg)	0.95	2.12	0.98
Swath Width @ nadir (km)	11.3	16.5	8
Viewing Angle (deg)	±60	±25	±50
Launched	Sep 1999	Oct 2001	Jun 2003
Retired	Mar 2015	Jan 2015	Mar 2007

Table 5.7 Orbit specifications for first generation commercial EO sensors.

	IKONOS	QuickBird	OrbView-3
Altitude (km)	681	450	470
Inclination (deg)	98.1	97.2	97
Orbit Period (min)	98	93.5	94
Revisit Period (days)	≈3	1 to 3.5	<3
Type		Sun-synchronous	
Nodal Crossing		10:30 AM	

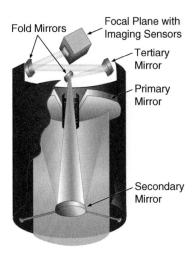

Figure 5.28 IKONOS optical subsystem. Figure from DigitalGlobe/ScapeWare3d/Getty Images.

Figure 5.31 shows the configuration of the QuickBird Detector Array. It consists of a number of linear CCD arrays carefully arranged in the focal plane. The image moves vertically past this focal plane as the satellite moves down track, as in Figure 5.32. Notice the arrays are offset from each other so that they can provide 100% coverage across the entire swath.

Figure 5.33 shows the same scene in four different bands from QuickBird. Obviously such multicolor imagery has value for detection, classification, and interpretation of the imagery.

Figure 5.34 is a QuickBird image showing two working boats tending to a nearly-capsized ship.

Table 5.8 IKONOS system specifications.

Telescope Assembly	Focal length - 10 m, f/14.3; Primary mirror − 0.7 m diameter; Size − 61 in long × 31 in wide; Weight − 240 lbs
Imaging Sensor	Linear CCD Array; Panchromatic - 13,500 detectors; Multispectral − 3,375 detectors
Digital Processor	Compression rate − 11 bits per pixel compressed to 2.6 bits per pixel
Total System	Weight − 376 lbs; Power − 350 watts

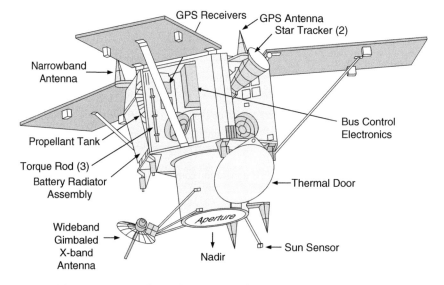

Figure 5.29 IKONOS satellite. DigitalGlobe/ScapeWare3d/Getty Images.

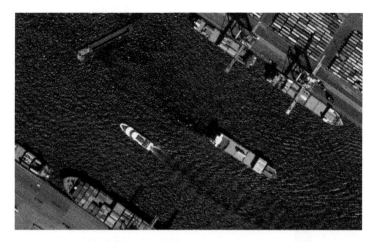

Figure 5.30 Example image from IKONOS. Source: DigitalGlobe/ScapeWare3d/Getty Images.

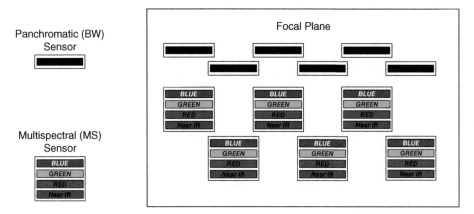

Figure 5.31 QuickBird focal plane array configuration. Figure from DigitalGlobe/ScapeWare3d/Getty Images.

Figure 5.32 Illustration of Quickbird spectral imaging. DigitalGlobe/ScapeWare3d/Getty Images.

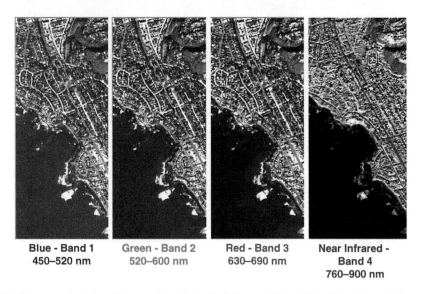

Figure 5.33 Example single band images from QuickBird. Source: DigitalGlobe/ScapeWare3d/Getty Images.

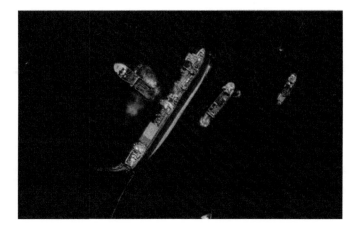

Figure 5.34 Example color image from QuickBird. Source: DigitalGlobe/ScapeWare3d/Getty Images.

OrbView-3

OrbView-3 offered 1.0 m resolution with a shorter telescope than IKONOS, which means the detector pixels were smaller.

Figure 5.35, taken from the Orbview-3 promotional material, illustrates Orbview-3's field of regard. It also illustrates that the sensor could look forward and backwards to obtain stereo image pairs – images of the same scene from different directions. These are useful for high-resolution measurement of the height of objects. The table included in this figure lists the per kilometer price of images at the time of the satellite's operation in the early 2000s.

5.9.3 Second-Generation Systems

Three systems are considered second generation: WorldView-1, WorldView-2 and

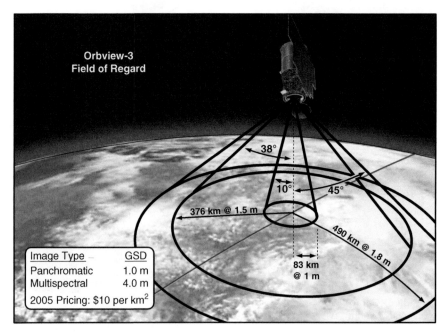

Figure 5.35 OrbView-3 field of regard. Figure adapted from https://directory.eoportal.org/web/eoportal/satellite-missions/o/orbview-3.

Table 5.9 Sensor specifications for second generation commercial EO sensors.

	WorldView-1	WorldView-2	GeoEye-1
Panchromatic Nadir Resolution (m)	0.5	0.46	0.41
Multispectral Nadir Resolution (m)	–	1.84	1.65
Multispectral Bands	0	8	4
Aperture Size (m)	0.6	1.10	0.60
Focal Length (m)	8.8	13.3	8.8
FOV (deg)	2.12	1.28	2.12
Swath Width @ nadir (km)	17.6	16.4	15.2
Max Single Frame Collect (km)	–	–	15.2×15.2
Max Image Area (strip collect) (km)	17.6×330	16.4×250	–
Max Large Area Collect (km)	60×110	65.6×110	50×1300
Viewing Angle (deg)	±45	±45	±60
Launched	Sep 2007	Oct 2009	Sep 2008

Table 5.10 Orbit specifications for second generation commercial EO sensors.

	WorldView-1	WorldView-2	GeoEye-1
Altitude (km)	496	770	682
Inclination (deg)	97.2	97.8	98.1
Orbit Period (min)	94.6	100	98
Revisit Period (days)	1.7–4.6	1.1–3.7	<3
Type		Sun-synchronous	
Nodal Crossing		10:30 AM	

GeoEye-1. Key specifications for these systems are listed in Table 5.9. Notice (Table 5.9) that the panchromatic resolution is now about 0.5 m for all these sensors, with a little better than 2-m multispectral resolution. The second generation systems also had long focal lengths to support improved resolution, and large apertures to support reasonable exposure times. All of the second generation sensors could look fore and aft to support stereo viewing. Orbit specifications are shown in Table 5.10.

WorldView-1
WorldView-1 was unusual in that it only had a panchromatic imager with a passband

of 400–900 nm. The spectral passband of WorldView-1 is shown in Figure 5.36.

WorldView-2
WorldView-2 had eight bands with the pass bands illustrated in Figure 5.37.

Figure 5.38 shows a WorldView-2 image that looks like it was taken from an aircraft.

GeoEye-1
The GeoEye-1 Sensor had a more standard five bands, as shown in Figure 5.39.

Figure 5.40 shows a ship image from GeoEye-1.

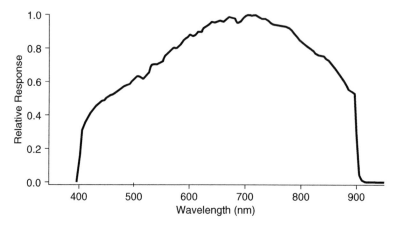

Figure 5.36 WorldView-1 spectral band. Figure from DigitalGlobe/ScapeWare3d/Getty Images.

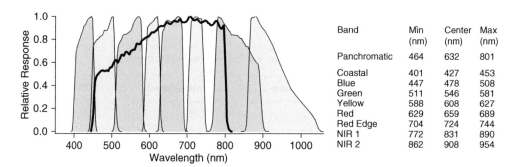

Band	Min (nm)	Center (nm)	Max (nm)
Panchromatic	464	632	801
Coastal	401	427	453
Blue	447	478	508
Green	511	546	581
Yellow	588	608	627
Red	629	659	689
Red Edge	704	724	744
NIR 1	772	831	890
NIR 2	862	908	954

Figure 5.37 WorldView-2 bands. DigitalGlobe/ScapeWare3d/Getty Images.

Figure 5.38 Example image from WorldView-2. Source: DigitalGlobe/ScapeWare3d/Getty Images.

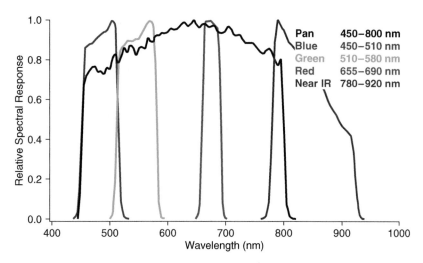

Figure 5.39 GeoEye-1 bands. DigitalGlobe/ScapeWare3d/Getty Images.

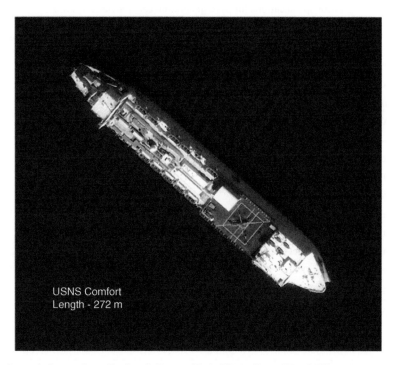

Figure 5.40 Example image from GeoEye-1. Source: DigitalGlobe/ScapeWare3d/Getty Images.

5.9.4 Third-Generation Systems

A third generation of systems has been recently launched, with resolutions down to 30 cm in panchromatic and 1.2 m in multispectral. These include WorldView-3 and WorldView-4, which started life as GeoEye-2 but was renamed when GeoEye merged with Digital Globe in 2013. WorldView-4 ceased operation in January 2019 but archived data are still available from that system. Specifications are shown in Tables 5.11 and 5.12.

WorldView-3

WorldView-3 supports a large number of bands extending into the short-wave IR and includes a 30-m resolution imager with wider swath. Table 5.13 lists the specifications for the WorldView-3 sensor bands. The panchromatic and visible multispectral bands are digitized to 11 bits/pixel while the SWIR bands are digitized to 14 bits/pixel.

The spectral bands supported by WorldView-3 are illustrated in Figure 5.41.

In 2017, Digital Globe began development of WorldView Legion, a constellation of satellites that will replace the WorldView-1, GeoEye-1, and WorldView-2 satellites. It has been reported that the constellation will consist of six satellites, each with imaging resolution comparable to that of WorldView-4. The satellites are planned to fly in a mix of polar and midlatitude orbits, providing an ability to

Table 5.11 Sensor specifications for third generation commercial EO sensors.

	WorldView-3	WorldView-4
Panchromatic Nadir Resolution (m)	0.31	0.30
Visible Multispectral Bands/Resolution	8 @ 1.24 m	3 @ 1.2 m
Infrared Bands/Resolution	8 @ 3.7 m	1 @ 1.2 m
CAVIS Bands/Resolution[1]	12 @ 30 m	–
Aperture Size (m)	1.1	1.1
Focal Length (m)	13.3	–
FOV (deg)	1.2	1.2
Swath Width @ nadir (km)	13.1	13.2
Max Single Frame Collect (km)	13.1 × 13.1	13.1 × 13.1
Max Image Area (strip collect) (km)	13.1 × 360	13.1 × 360
Max Large Area Collect (km)	65.5 × 112	66.5 × 112
Viewing Angle (deg)	±20	–
Launched	Aug 2014	Nov 2016

[1]CAVIS – Clouds, Aerosols, Water Vapor, Ice and Snow

Table 5.12 Orbit specifications for third generation commercial EO sensors.

	WorldView-3	WorldView-4
Altitude (km)	617	611
Inclination (deg)	98	98
Orbit Period (min)	97	97
Type	Sun-synchronous	
Nodal Crossing	10:30 AM	

Table 5.13 WorldView-3 sensor bands.

Band name	Spectral band (nm)	Nadir GSD (m)
Panchromatic	450–800	0.31
Coastal Blue	400–450	1.24
Blue	400–510	1.24
Green	450–580	1.24
Yellow	510–625	1.24
Red	585–690	1.24
Red Edge	705–745	1.24
Near-IR1	770–895	1.24
Near-IR2	860–1040	1.24
SWIR-1	1195–1225	3.70
SWIR-2	1550–1590	3.70
SWIR-3	1640–1680	3.70
SWIR-4	1710–1750	3.70
SWIR-5	2145–2185	3.70
SWIR-6	2185–2225	3.70
SWIR-7	2235–2285	3.70
SWIR-8	2295–2365	3.70
CAVIS Desert Clouds	405–420	30
CAVIS Aerosols-1	459–509	30
CAVIS Green	525–585	30
CAVIS Aerosols-2	620–670	30
CAVIS Water-1	845–885	30
CAVIS Water-2	897–927	30
CAVIS Water-3	930–965	30
CAVIS NDVI SWIR	1220–1410	30
CAVIS Cirrus	1350–1410	30
CAVIS Snow	1620–1680	30
CAVIS Aerosols-3	2105–2245	30

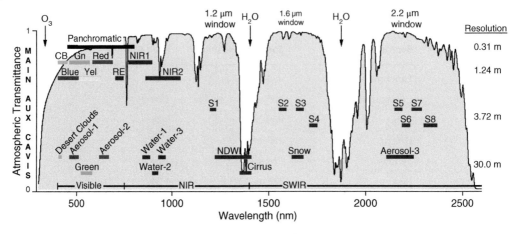

Figure 5.41 WorldView-3 bands. DigitalGlobe/ScapeWare3d/Getty Images.

observe rapidly changing areas as frequently as every 20–30 minutes.

5.9.5 Commercial Smallsat Systems

The satellites discussed so far have had a launch mass of 800–2800 kg. These are large satellites, which makes them expensive to build and expensive to launch. The commercial space industry has also been developing alternative EO remote sensors based on larger constellations of smaller satellites. Examples of these include the five RapidEye satellites, each with a mass of 150 kg on launch, the 13 Skysat-C satellites, each with a mass of 110 kg on launch, and the dozens of Planet Dove (Flock-1) satellites, each with a mass of 5 kg on launch. Figure 5.42 illustrates the general trend.

Table 5.14 lists key sensor parameters and Table 5.15 lists key orbital parameters for the RapidEye, Skysat-C, and the Planet Dove satellites.

The five RapidEye satellites were launched in 2008 and deployed to be uniformly spaced around a common orbital plane. These are conventional push-broom sensors, like the larger satellites discussed in previous sections. The unique aspect of the RapidEye constellation is that it was designed to produce large coverage rates at moderate resolution in a

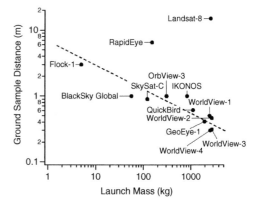

Figure 5.42 GSD vs mass for selected remote sensing satellites. Adapted from Murthy et al. (2014).

smaller satellite. The RapidEye constellation was retired in March 2020.

Two prototype SkySat satellites were launched in 2013/2014, but the 13 current SkySat-C satellites were deployed in launches in 2016 and 2018. These satellites use a frame sensor that can obtain still panchromatic frames or video[2] over a 2.5 × 1 km area. In addition, the system can be operated in a novel push-frame mode. As illustrated in Figure 5.43, the SkySat focal plane consists of three imaging

---

2 SkySat is capable of delivering 30 fps panchromatic video for durations of up to 90 s with a 1.1 m resolution.

Table 5.14 Sensor specifications for commercial smallsat EO sensors.

	RapidEye	SkySat	Dove
Panchromatic Nadir Resolution (m)	–	0.9	–
Visible Multispectral Bands/Resolution	4 @ 6.5 m	3 @ 1 m	3 @ 3.7 m
Near-Infrared Bands/Resolution	1 @ 6.5 m	1 @ 1 m	1 @ 3.7 m
Aperture Size (m)	0.145	0.35	0.091
Focal Length (m)	0.633	3.6	1.14
Swath Width @ nadir (km)	77	8	
Max Single Frame Collect (km)	–	–	24.6 × 16.4
Max Image Area (strip collect) (km)	77 × 300	6.6 × 20	–
Video Area Collect (km)	–	2 × 1.1	–
Viewing Angle (deg)	±22	±1	±1
Launched	Aug 2008	2016–2018	2013–2021
Size (m)	0.8 × 0.9 × 1.2	0.6 × 0.6 × 0.8	0.1 × 0.1 × 0.3
Launch Mass (kg)	156	110	5

Table 5.15 Orbit specifications for commercial smallsat EO sensors.

	RapidEye	SkySat	Dove
Number in Constellation	5	13	dozens
Altitude (km)	630	575	475
Inclination (deg)	97.8	98	98
Orbit Period (min)	96.7	96.1	94.0
Type		Sun-synchronous	
Nodal Crossing	11:00 AM	10:30 & 13:30 AM	09:30-11:30 AM

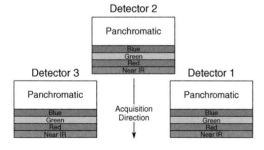

Figure 5.43 Focal plane for SkySat-C. From d'Angelo et al. (2016) with permission from Taylor & Francis.

detectors, each having 2560 pixels across the direction of motion and 2160 pixels in the direction of motion. The top half of each detector is used used to acquire panchromatic data, while color filters on the bottom half are used to acquire multispectral data. In the push-frame mode, image sequences are taken at high rates with typical integration times of 0.4 ms. Acquiring the images at high rates minimizes motion blurring, and assures that there is considerable overlap between each image, but the SNR in each individual image is relatively low. All of the images are transmitted to the ground, correlations in the panchromatic sequence are used to precisely align the images, and then the panchromatic and multispectral data can be stacked and processed to improve both SNR and resolution (Murthy et al., 2014).

The Planet Dove satellites are likely to be the extreme example of this trend. The Dove satellite is a 3U CubeSat, meaning that it fits into a $10 \times 10 \times 30$ cm volume (almost) and is designed to a common standard, allowing the CubeSats to easily utilize any spare launch capabilities. The telescope for this design is oriented along the length of the satellite, so the 9.1 cm aperture is as large as possible in this form factor. The Dove satellites are mass produced using commercial parts, which is part of the reason they can be inexpensive to manufacture in large numbers. The earliest flock (the term Planet uses to describe a group of Dove launched at the same time) was launched from the International Space Station at an altitude of 400 km. The newer flocks are being launched into 475 km sun-synchronous orbits. Planet's stated goal was to place 130 of these in orbit to provide imaging of everywhere on the Earth (outside the polar regions) at least once per day.

As of December 2018, Planet had lost 36 Doves due to two launch failures, lost one to a failed deployment, and had 100 Doves reenter the atmosphere. There are 32 Doves operational in 400-km altitudes that were launched in 2015–2016, and 123 Doves operational in 475-km altitudes (Flocks 2p, 3p, 3p', 3r, and 3s).

CubeSats suffer from several significant shortcomings. Their size limits the size of the telescope and optics, so they are most effective when placed in lower Earth orbits. They are also too small to generally carry fuel for orbit adjustment or corrections. And their ability to scan or change pointing is quite limited. The PlanetLab Dove satellites, for example, have no propellant and no reaction wheels, relying on slow and low power magnetic torquers to maintain a nadir-looking orientation. The 100 kg-class satellites are large enough to include propellant to maintain an orbit, reaction wheels for quickly reorienting the spacecraft, and steerable optics to be able to sense over a wider field of regard.

It is easy to anticipate that smallsats of various classes are likely to play an increasingly important role in Earth remote sensing.

References

Bowker, D. E., Davis, R. E., Myrick, D. L., Stacy, K., & Jones, W. T. (1985). *Spectral reflectances of natural targets for use in remote sensing studies*. NASA.

d'Angelo, P., Máttyus, G., & Reinartz, P. (2016). Skybox image and video product evaluation. *International Journal of Image and Data Fusion, 7* (1), 3–18.

DeFries, R. S., & Townshend, J. R. G. (1994). NDVI-derived land cover classifications at a global scale. *International Journal of Remote Sensing, 15* (17), 3567–3586.

Didan, K. (2015). MOD13C2 MODIS/Terra vegetation indices monthly L3 global 0.05deg CMG v006 [data set]. *NASA EOSDIS LP DAAC. doi: 10.5067/MODIS/MOD13C2.006.*

Elachi, C., & van Zyl, J. J. (2006). *Introduction to the physics and techniques of remote sensing.* John Wiley & Sons.

Gausman, H., Cardenas, R., & Hart, W. G. (1970). Aerial photography for sensing plant anomalies. *NASA Manned Spacecraft Center 3rd Annual Earth Resources Program Review, 2.*

Gutman, G. G. (1991). Vegetation indices from AVHRR: An update and future prospects. *Remote Sensing of Environment, 35*(2-3), 121–136.

Gutman, G., Tarpley, D., Ignatov, A., & Olson, S. (1995). The enhanced NOAA global land dataset from the Advanced Very High Resolution Radiometer. *Bulletin of the American Meteorological Society, 76*(7), 1141–1156.

Huete, A., Justice, C., & Van Leeuwen, W. (1999). MODIS vegetation index (MOD13) algorithm theoretical basis document. Greenbelt: National Aeronautics and Space Administration, Goddard Space Flight Center.

Irons, J. R., Dwyer, J. L., & Barsi, J. A. (2012). The next Landsat satellite: The Landsat data continuity mission. *Remote Sensing of Environment, 122,* 11–21.

Kaufman, Y. J. (1988). Atmospheric effect on spectral signature-measurements and corrections. *IEEE Transactions on Geoscience and Remote Sensing, 26*(4), 441450.

Kaufman, Y. J. (1989). The atmospheric effect on remote sensing and its correction. *Theory and application of optical remote sensing*, 336–428.

Kogan, F. N. (1995). Application of vegetation index and brightness temperature for drought detection. *Advances in space research, 15* (11), 91–100.

Kogan, F. N. (1997). Global drought watch from space. *Bulletin of the American Meteorological Society, 78*(4), 621–636.

Kokaly, R. F., Clark, R. N., Swayze, G. A., Livo, K. E., Hoefen, T. M., Pearson, N. C., et al. (2017). USGS Spectral Library Version 7: U.S. Geological Survey Data Series 1035. Retrieved from https://speclab.cr.usgs.gov/spectral-lib .html

Liang, S. (2005). *Quantitative remote sensing of land surfaces*. John Wiley & Sons.

Meerdink, S. K., Hook, S. J., Abbott, E. A., & Roberts, D. A. (2018). The ECOSTRESS spectral library 1.0. Retrieved from https:// speclib.jpl.nasa.gov

Murthy, K., Shearn, M., Smiley, B. D., Chau, A. H., Levine, J., & Robinson, M. D. (2014). Skysat-1: Very high-resolution imagery from a small satellite. In Meynart, R., Neeck, S. P., & Shimoda, H. (Eds.), *Sensors, systems, and next-generation satellites xviii, Proceedings of SPIE* (Vol. 9241, 92411E). International Society for Optics and Photonics.

Myers, V. I., Bauer, M. E., Gausman, H. W., Hart, W. G., & Heilman, J. L. (1983). Remote sensing applications in agriculture. *Manual of remote sensing, 2*, 2111–2228.

NASA. (2021). NASA MODIS vegetation index products (NDVI and EVI). Retrieved from https://modis.gsfc.nasa.gov/data/dataprod/ mod13.php

National Research Council. (1976). Resource and environmental surveys from space with the Thematic Mapper in the 1980's. NEC/CORSPERS-76/1. Washington, DC: National Academy of Sciences.

Nicodemus, F. E., Richmond, J. C., Hsia, J. J., Ginsberg, I. W., & Limperis, T. (1977). *Geometrical considerations and nomenclature for reflectance*. US Department of Commerce, National Bureau of Standards.

Pearlman, J., Segal, C., Liao, L., Carman, S., Folkman, M., Browne, B., et al. (2000). Development and operations of the EO-1 Hyperion imaging spectrometer. In Barnes, W. L. (Ed.), *Earth observing systems V, Proceedings of SPIE* (Vol. 4135, pp. 243–253). Bellingham, WA: The International Society for Optics and Photonics.

Pearlman, J. S., Barry, P. S., Segal, C. C., Shepanski, J., Beiso, D., & Carman, S. L. (2003). Hyperion, a space-based imaging spectrometer. *IEEE Transactions on Geoscience and Remote Sensing, 41*(6), 1160–1173.

Peterson, D. L. (1989). Applications in forest science and management. *Theory and applications of optical remote sensing*, 429–473.

Short, N. M. (1982). *The Landsat tutorial handbook: Basics of satellite remote sensing*. National Aeronautics and Space Administration, Scientific and Technical Information Branch.

Tucker, C. J., & Sellers, P. J. (1986). Satellite remote sensing of primary production. *International journal of remote sensing, 7*(11), 1395–1416.

USGS. (2018). *Landsat 7 (L7) data user's handbook*. USGS.

Vermote, E., Justice, C., Csiszar, I., Eidenshink, J., Myneni, R., Baret, F., et al. (2014). NOAA climate data record of normalized difference vegetation index (NDVI), version 4. *NOAA National Centers for Environmental Information. doi:10.7289/ V5PZ56R6*.

Wilson, T., Felt, R., & Baugh, R. (2000). *Naval EarthMap Observer (NEMO) Hyperspectral Remote Sensing Program*. Paper presented at the RTO Sensors and Electronics Technology Panel (SET) Symposium, Space-Based Observation Technology. Samos, Greece, 16–18 October. NATO.

6

Microwave Radiometry

6.1   Introduction to Microwave Radiometry

Microwave radiometers consist of an antenna, a microwave receiver, and a processor. They are designed to measure the thermal radiation emitted in the microwave spectrum by the scene within the antenna's field of view.

Satellite radiometers operate at multiple frequencies and polarizations to measure the oceans, land and atmosphere:

- Ocean measurements include surface temperature and salinity, wind speed and direction, sea ice coverage, ice type and concentration.
- Land measurements include surface temperature, snow cover, and soil moisture.
- Atmospheric measurements include water vapor content, liquid water content, precipitation rate, temperature and moisture profiles.

It should not be surprising that the surface observations are made in atmospheric windows. Figure 6.1 shows the transmission through the Earth's atmosphere along a vertical path under clear sky conditions. Surface measurements are generally made in

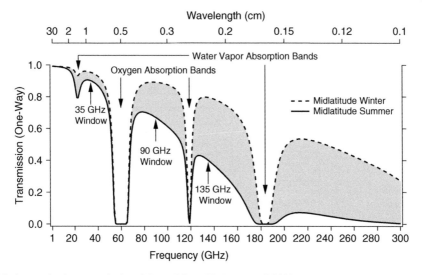

Figure 6.1 Atmospheric transmission. Adapted from Ulaby et al. (1981).

Remote Sensing Physics: An Introduction to Observing Earth from Space, Advanced Textbook 3, First Edition.
Rick Chapman and Richard Gasparovic.
© 2022 American Geophysical Union. Published 2022 by John Wiley & Sons, Inc.
Companion website: www.wiley.com/go/chapman/physicsofearthremotesensing

the 90-GHz window (W-band), the 35-GHz window (Ka-band), and the bands from 1–15 GHz, where the atmosphere is mostly transparent.

Atmospheric water and temperature measurements are made along the edges of the water vapor and oxygen absorption bands from 22 to 200 GHz.

6.2 Microwave Radiometers

The technology for Earth observations with microwave radiometers has developed over decades based on advances made in radar and radio astronomy. During World War II, major advances in radar technology led to the development of highly-sensitive microwave receivers. Astronomers combined these receivers with large dish antennas to detect microwave emissions from stars and planets. Aircraft systems were then flown in the 1950s and 1960s to study emissions from the land and sea. These became the prototypes for the

early satellite instruments, the first of which was launched in 1968.

Figure 6.2 is a block diagram of a total-power radiometer. The general design is an example of a superheterodyne receiver. We will consider the receiver design one stage at a time, describing the signal voltage and corresponding spectrum at each stage in the receiver.

Any electromagnetic flux impinging on the antenna is converted to an electrical voltage. The antenna is designed to be maximally efficient in a specified band, but generally the signal coming out of the antenna is weak (typically 0.1–1 μV) and has a broad frequency spectrum.

The first stage of the receiver is an RF amplifier that amplifies the signal with gain G_{RF} and applies a spectral filter to limit the signal to the specific band the instrument is designed to measure, specified by center frequency f_{RF} and bandwidth B.

The signal is then passed to a mixer, which is a two-input device with an output equal to a weighted sum of the two inputs and the

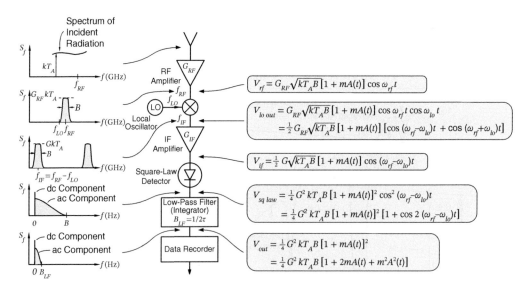

Figure 6.2 Block diagram of a microwave radiometer. Adapted from Ulaby et al. (1981).

product of the two inputs[1]. The second input to the mixer is driven by the local oscillator (LO), with a frequency f_{LO} set to just below the band of interest, as shown in the second spectral diagram from the top in Figure 6.2.

The effect of the mixer is easiest to see by considering the output resulting from an input of two sine waves of differing amplitudes with frequencies f_1 and f_2. In general, the output will contain four components: sinusoids at the two original frequencies, f_1 and f_2, a sine wave at the sum frequency $f_1 + f_2$, and a sine wave at the difference frequency $f_1 - f_2$. A low-pass filter is then used to eliminate the f_1, f_2 and $f_1 + f_2$ components, and all that is left is the spectral component that has been shifted down in frequency. The same holds true for each frequency in a band of frequencies. So the mixer, LO, and subsequent low-pass filter effectively shift the incoming RF energy, which might be at 15 or 95 GHz, down to an intermediate frequency (IF) such as 200 MHz.

The low-pass filter is usually built into the next stage, the IF amplifier. This whole process of mixing the signal down to an intermediate frequency is done because it is easier to amplify, filter, and measure signals at lower frequencies than directly at microwave frequencies. Mixing to a lower frequency is called heterodyning, and following this stage with an IF amplifier is the defining characteristic of a superheterodyne receiver.

The signal is then passed into a square law detector, which simply squares the signal. A very low pass filter is then applied (with a time constant on the order of milliseconds) to average the squared signal. Any DC offset in the electronics is subtracted off. The resulting output is proportional to the total in-band power in the original signal.

---

1 An ideal mixer would be a perfect multiplier, but the output of real mixers contains remnants of the original signals, which is known as leakage.

6.3 Microwave Radiometry

Microwave radiometers measure thermal noise in the microwave band of interest. This thermal noise can be approximately modeled as blackbody radiation. Recall that the spectrum of blackbody radiation is the Planck function:

$$L_{BB}(\lambda, T) = \frac{2hc^2}{\lambda^5} \left[exp\left(\frac{hc}{\lambda k_B T} \right) - 1 \right]^{-1}$$

$$(6.1)$$

In this expression, h is Planck's constant, c is the speed of light, λ is the wavelength, k_B is Boltzmann's constant and T is the source temperature, measured in Kelvin.

We also previously showed that the ratio in the exponential is much less than one at microwave frequencies. Thus the exponential simplifies and the Planck function becomes:

For $\dfrac{hc}{\lambda k_B T} \ll 1$,

$$exp\left(\frac{hc}{\lambda k_B T} \right) \cong 1 + \frac{hc}{\lambda k_B T}$$

so $\quad L_{BB}(\lambda, T) = \dfrac{2c}{\lambda^4} k_B T \qquad (6.2)$

Application of the Jacobian between wavelength and frequency allows the Planck function to be expressed as a function of frequency:

$$L_{BB}(\nu, T) = \left| \frac{d\lambda}{d\nu} \right| L_{BB}(\lambda, T)$$

where $\quad \nu \lambda = c$

$$= \frac{2\nu^2}{c^2} k_B T \qquad (6.3)$$

While blackbody radiation is a good model for an ideal emitter, real materials are modeled as a having a radiance equal to the emissivity of the material ε_s times the blackbody radiation arising from a source of the same physical temperature T_s:

$$L_s(\nu, \theta, \varphi, T_s) = \varepsilon_s(\nu, \theta, \varphi, T_s) L_{BB}(\nu, T_s)$$

$$(6.4)$$

In general this emissivity is less than one.

The brightness temperature, T_B, is defined to be the emissivity times the temperature:

$$T_B(\nu, \theta, \varphi, T_s) = \varepsilon_s(\nu, \theta, \varphi, T_s)T_s \quad (6.5)$$

The radiance is then given in terms of the brightness temperature by:

$$L_s(\nu, \theta, \varphi, T_s) = \frac{2\nu^2}{c^2}k_B T_B(\nu, \theta, \varphi, T_s)$$

$$(6.6)$$

Note the brightness temperature depends on frequency, polarization[2], viewing direction, source temperature, and other source properties, such as surface roughness or other factors that affect the dielectric constant and hence emissivity.

If the scene being viewed consists of multiple sources, the apparent, or radiometric, temperature of the scene is the sum of the brightness temperatures weighted by the fraction of the instrument's field of view each source covers.

$$T_{app} = \sum_i^n f_i T_{Bi} \quad (6.7)$$

The scene radiance is then given by this expression of the apparent radiometric temperature.

$$L_{scene}(\nu, T_{app}) = \frac{2\nu^2}{c^2}k_B T_{app}(\nu, \theta, \varphi)$$

$$(6.8)$$

Equation (3.16) derived in Chapter 3.3.1 is perfectly general, so the radiant flux at the entrance of an aperture can be written:

$$\Phi_0 = A_0 \, \Omega_{FOV} \int L_{scene}(\nu, T_{app}) \, d\nu$$

$$(6.9)$$

---

2 Polarization refers to the orientation of the plane of the electric field with respect to the radiometer. Normally the radiometer will be oriented such that it can receive radiation with an electric field plane that is parallel or perpendicular to the Earth's surface. These are referred to as horizontal and vertical polarization, respectively. A more complete description of polarization is provided in Section 6.4.

where A_0 is the entrance aperture area and Ω_{FOV} is the solid angle of the sensor field of view.

6.3.1 Antenna Pattern

For a microwave antenna, the sensor field of view is determined by an integral of the normalized antenna pattern, G_n, over 4π steradians:

$$\Omega_{FOV} = \int_{4\pi} G_n(\theta, \varphi) \, d\Omega \quad (6.10)$$

The normalized antenna pattern, G_n is, in turn:

$$G_n(\theta, \varphi) = \frac{G(\theta, \varphi)}{G_{max}} \quad (6.11)$$

where G_{max} is the maximum value of the antenna pattern, normally found at the center of the main beam. By definition, the peak value of G_n is 1.

Equation (6.9) for the flux at the entrance aperture now becomes:

$$P_{rec} = \frac{1}{2} \int_{\nu}^{\nu+\Delta\nu} \int_{4\pi} A_e \, L(\nu, \theta, \varphi, T)$$

$$\times G_n(\theta, \varphi) \, d\Omega \, d\nu \quad (6.12)$$

where A_e is the effective antenna area.

Note that the factor of 1/2 in this expression accounts for the antenna being sensitive to radiation of only a single polarization, horizontal or vertical. Thermal radiation is generally unpolarized, hence the antenna collects only half of the emitted radiation.

Three forms for displaying the normalized antenna pattern are shown in Figure 6.3. The left-hand panel is a rectilinear plot with dB response plotted as a function of angle. This form is used for circularly symmetric antennas or to display cuts taken through primary axes. The middle panel shows a polar diagram of a single cut through the pattern, usually taken at zero elevation angle. And the right-hand panel shows a two-dimensional representation that

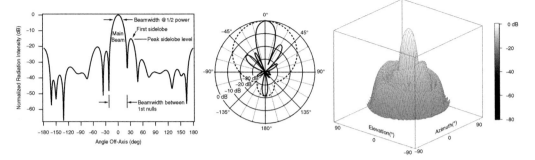

Figure 6.3 Example antenna patterns.

obviously contains more information on the structure of the antenna pattern away from the main beam, but is generally more difficult to use quantitatively.

The effective beamwidth of the antenna is usually designated by the full angular width at the point where the response is a full 3 dB down from the peak, as shown. Rarely you will see someone designate the beamwidth by the width between the first nulls in the response.

The secondary peaks common to virtually all real antennae are known as sidelobes. The peak of the first sidelobe and the integral under all of the sidelobes are important measures of how much energy may get into the signal from directions other than that encompassed by the main beam.

Consider a rectangular antenna with sides a and b. If the antenna is diffraction limited, then the angular resolution $\delta\theta$ from each side (measured in radians, not degrees!) is approximately:

$$\delta\theta_a = \lambda/a \quad \text{and} \quad \delta\theta_b = \lambda/b \qquad (6.13)$$

A more accurate derivation of equation (6.13) for a fully-filled aperture produces an additional factor of 1.2, but this factor is ignored

here, reflecting the common practice in the literature.

For narrow-beam antennae, a good approximation for the antenna field of view is given by:

$$\Omega_{FOV} \cong \delta\theta_a \, \delta\theta_b = \frac{\lambda^2}{ab} = \frac{\lambda^2}{A} \qquad (6.14)$$

where A is the physical area of the antenna aperture.

The effective area of the antenna is given by the physical area of the antenna times the aperture efficiency, $A_e = A\epsilon_{ap}$, where the aperture efficiency accounts for the antenna's imperfect conversion of the radiation impinging on its aperture to power at the antenna terminals. Manipulation of equation (6.14) produces an expression for the effective area of an antenna:

$$A_e = \frac{\epsilon_{ap}\lambda^2}{\Omega_{FOV}} = \left(\frac{c^2}{\nu^2}\right) \frac{\epsilon_{ap}}{\int\limits_{4\pi} G_n(\theta,\varphi)d\Omega} \qquad (6.15)$$

The power received by the antenna now becomes:

$$P_{rec} = \frac{\epsilon_{ap}\, c^2 \int_{\nu}^{\nu+\Delta\nu} \int\limits_{4\pi} (1/\nu^2)L(\nu,\theta,\varphi,T)G_n(\theta,\varphi)d\Omega d\nu}{2\int\limits_{4\pi} G_n(\theta,\varphi)d\Omega} \qquad (6.16)$$

The outer integration is over frequency. The inner integral is over a solid angle, and is nothing more than the average of the radiance from all directions weighted by the normalized antenna pattern.

6.3.2 Antenna Temperature

Consider the example of an antenna arranged in such a way that it sees nothing but a blackbody cavity. The radiance of the cavity is:

$$L_{BB}(v, T_{BB}) = \frac{2v^2}{c^2} k_B T_{BB} \qquad (6.17)$$

Then consider a receiver with a bandwidth narrow enough that the spectral radiance is constant over the bandwidth. In this case the integrals simplify, and the received power is:

We then define the antenna temperature as the weighted integral:

$$T_A(v) = \frac{\int\limits_{4\pi} T_{app}(v, \theta, \varphi) G_n(\theta, \varphi) d\Omega}{\int\limits_{4\pi} G_n(\theta, \varphi) d\Omega} \qquad (6.21)$$

The received power is then:

$$P_{rec} = k_B T_A(v) \Delta v \qquad (6.22)$$

where the temperature is the antenna temperature.

It is important to recognize that the antenna temperature T_A is not the physical temperature of the antenna, but is instead related to the apparent temperature of the scene.

$$P_{rec} = \left(\frac{c^2}{2}\right) \frac{\int_v^{v+\Delta v} \int\limits_{4\pi} (1/v^2) L(v, T_{BB}) G_n(\theta, \varphi) d\Omega dv}{\int\limits_{4\pi} G_n(\theta, \varphi) d\Omega} = k_B T_{BB} \Delta v \qquad (6.18)$$

This is a very important result. In the case of a microwave radiometer, the received signal power is proportional to the blackbody temperature times the radiometer bandwidth. In the case of a radar, this is the thermal noise floor, usually stated as $k_B TB$, where B is the bandwidth of the noise that the signal must compete against.

Next consider a real world case, with the radiometer looking at a material with an emissivity of less than one. This scene is characterized by its apparent temperature T_{app}, and the scene radiance becomes:

$$L_{scene}(v, T_{app}) = \frac{2k_B}{\lambda^2} T_{app}(v, \theta, \varphi) \qquad (6.19)$$

Replacing L_{BB} with L_{scene}, it follows that the received power is now a weighted integral over the apparent temperature of the scene:

Conceptually this seems clear, but it is unfortunate that the technical definition of antenna temperature differs from its plain language meaning.

In the end, the scene information is convolved with the antenna pattern. For a homogeneous scene (e.g., sea surface) at a constant temperature T_s, the antenna temperature is an integral over the product of emissivity, physical temperature, and antenna pattern, normalized by the integral over the antenna pattern:

$$T_A(v) = \frac{\int\limits_{4\pi} \varepsilon_s(v, \theta, \varphi, T_s) T_s G_n(\theta, \varphi) d\Omega}{\int\limits_{4\pi} G_n(\theta, \varphi) d\Omega} \qquad (6.23)$$

$$P_{rec} = \frac{k_B \int_v^{v+\Delta v} \int\limits_{4\pi} T_{app}(v, \theta, \varphi) G_n(\theta, \varphi) d\Omega dv}{\int\limits_{4\pi} G_n(\theta, \varphi) d\Omega} = \frac{k_B \Delta v \int\limits_{4\pi} T_{app}(v, \theta, \varphi) G_n(\theta, \varphi) d\Omega}{\int\limits_{4\pi} G_n(\theta, \varphi) d\Omega} \qquad (6.20)$$

So a radiometer measures received power. Dividing by Boltzmann's constant and bandwidth yields the measured antenna temperature. Correcting this for the antenna pattern can provide an estimate of the brightness temperature of the scene. Correcting the brightness temperature for scene emissivity then provides an estimate for the physical temperature of the surface.

> **Antenna Temperature** corrected for antenna pattern = **Brightness Temperature**
> **Brightness Temperature** corrected for emissivity = **Surface Temperature**

6.3.3 Examples

As an example of a typical radiometer, we use information from Hollinger et al. (1990) where brightness temperatures measured by the SSM/I radiometer at 19 and 37 GHz are tabulated for (1) a calm ocean, (2) the Amazon rain forest, and (3) the Arabian desert. The results for both V-pol and H-pol radiation are shown in Table 6.1. The received power in this table was calculated based on frequency, bandwidth and temperature.

As the table shows, the total received power over these bands is of the order of a picowatt. Making precise measurements at these power levels requires state-of-the-art engineering.

6.4 Polarization

Up to this point, our discussion of polarization has been limited to considering vertical and horizontal polarization. Some remote sensors exploit additional information in the polarization of the received signals to extract geophysical parameters. To understand these sensors we need a more complete description of the polarization of an electromagnetic signal.

6.4.1 Basic Polarization

We saw earlier that a plane wave solution to Maxwell's equation is:

$$\vec{E} = \vec{E}_0\, e^{i(kz - \omega t + \delta)} \tag{6.24}$$

Table 6.1 Example SSM/I brightness temperatures at 19 and 37 GHz. Adapted from Hollinger et al. (1990)

	Freq./Pol. (GHz)	Bandwidth (MHz)	T_B (K)	P_{rec} (10^{-12} W)
Calm Ocean	19 V	240	179	0.59
	19 H	240	101	0.34
	37 V	900	202	2.51
	37 H	900	130	1.62
Amazon Rain Forest	19 V	240	282	0.93
	19 H	240	282	0.93
	37 V	900	282	3.50
	37 H	900	282	3.50
Arabian Desert	19 V	240	300	0.99
	19 H	240	257	0.85
	37 V	900	293	3.64
	37 H	900	257	3.20

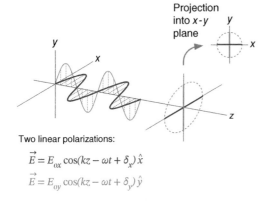

Two linear polarizations:

$$\vec{E} = E_{ox} \cos(kz - \omega t + \delta_x)\,\hat{x}$$
$$\vec{E} = E_{oy} \cos(kz - \omega t + \delta_y)\,\hat{y}$$

Figure 6.4 Horizontal and vertical polarization.

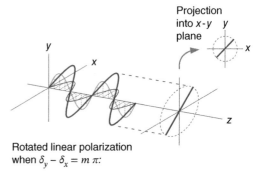

Rotated linear polarization
when $\delta_y - \delta_x = m\,\pi$:

$$\vec{E} = E_{ox} \cos(kz - \omega t + \delta)\,\hat{x} + E_{oy} \cos(kz - \omega t + \delta)\,\hat{y}$$

Figure 6.5 Rotated linear polarization.

where δ is an arbitrary phase offset and we have ignored attenuation along the direction of propagation, as if the wave were propagating in free space. Figure 6.4 shows the two solutions: horizontal polarization (drawn in blue), where the E-field is parallel to the ground defined by the x-z plane, and vertical polarization (drawn in green), where the E-field is perpendicular to the ground. While each of these solutions has arbitrary amplitude and phase, the figure illustrates the case where the amplitudes are equal and both phases equal $-\pi/2$.

Maxwell's equations are linear, so a sum of these two solutions is also a solution. In the case where $\delta_y - \delta_x$ equals a multiple of π, the sum is also a linearly polarized wave, but with the E-field aligned at an angle to the ground. Figure 6.5 illustrates the case of 45° polarization that occurs when $\delta_x = \delta_y = -\pi/2$ and $E_{ox} = E_{oy}$.

Circular polarization, illustrated in Figure 6.6 occurs when $\delta_y - \delta_x = \pm\pi/2$ and $E_{ox} = E_{oy}$. The two signs for the phase difference correspond to the E-field rotating in either the clockwise or counterclockwise direction.

Finally, any other combination of arbitrary phase differences and equal or unequal amplitudes produces elliptical orientation, as shown in Figure 6.7.

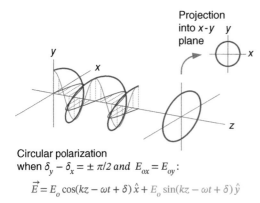

Circular polarization
when $\delta_y - \delta_x = \pm\,\pi/2$ *and* $E_{ox} = E_{oy}$:

$$\vec{E} = E_o \cos(kz - \omega t + \delta)\,\hat{x} + E_o \sin(kz - \omega t + \delta)\,\hat{y}$$

Figure 6.6 Circular polarization.

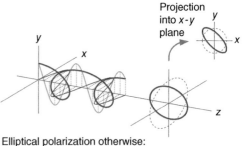

Elliptical polarization otherwise:

$$\vec{E} = E_{ox} \cos(kz - \omega t + \delta_x)\,\hat{x} + E_{oy} \sin(kz - \omega t + \delta_y)\,\hat{y}$$

Figure 6.7 Elliptical polarization.

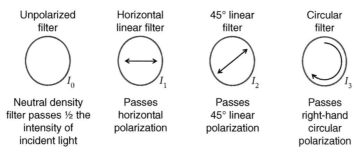

Figure 6.8 Four polarized filters used for defining the Stokes parameters.

6.4.2 Jones Vector

The most compact general representation of the polarization is given by the Jones vector $\underline{E}$:

$$\underline{E} = \begin{bmatrix} E_{ox}\, e^{i\delta_x} \\ E_{oy}\, e^{i\delta_y} \end{bmatrix} \tag{6.25}$$

A general geometrical description of polarization is then given by the ellipse amplitude A, orientation ϕ, and ellipticity e:

$$A = \sqrt{E_{oy}^2 + E_{ox}^2}$$

$$\tan 2\phi = 2\frac{E_{oy}E_{ox}}{E_{ox}^2 - E_{oy}^2}\cos(\delta_y - \delta_x) \tag{6.26}$$

$$\sin 2e = 2\frac{E_{oy}E_{ox}}{E_{ox}^2 + E_{oy}^2}|\sin(\delta_y - \delta_x)|$$

The complex polarization ratio ρ is then defined by:

$$\rho = \frac{E_{oy}}{E_{ox}}e^{i(\delta_y - \delta_x)} = \frac{\sin\phi\cos e + i\cos\phi\sin e}{\cos\phi\cos e - i\sin\phi\sin e} \tag{6.27}$$

and the degree of linear polarization P is defined such that P varies from 0 to 1, with 0 corresponding to unpolarized or circularly polarized radiation and 1 to linearly polarized radiation:

$$P = \frac{|E_{oy}^2 - E_{ox}^2|}{E_{ox}^2 + E_{oy}^2} \tag{6.28}$$

6.4.3 Stokes Parameters

While it is possible to fully characterize the polarization of a received signal by measuring the amplitudes of the received signal (E_{ox} and E_{oy}) at two polarizations and the phase difference between them ($\delta_y - \delta_x$), it is also possible

to characterize the polarization by making a set of power measurements. Stokes was the first to consider the outputs of four distinct polarizing filters, each that would pass exactly half of the intensity of an unpolarized beam (Figure 6.8).

Stokes then defined four parameters based on the measured intensities, as shown in the third column vector:

$$\begin{bmatrix} S_0 \\ S_1 \\ S_2 \\ S_3 \end{bmatrix} = \begin{bmatrix} I \\ Q \\ U \\ V \end{bmatrix} = \begin{bmatrix} 2I_0 \\ 2I_1 - 2I_0 \\ 2I_2 - 2I_0 \\ 2I_3 - 2I_0 \end{bmatrix}$$

$$= \begin{bmatrix} E_{ox}^2 + E_{oy}^2 \\ E_{ox}^2 - E_{oy}^2 \\ 2E_{ox}E_{oy}\cos(\delta_y - \delta_x) \\ 2E_{ox}E_{oy}\sin(\delta_y - \delta_x) \end{bmatrix}$$

$$= \begin{bmatrix} \langle E_x E_x^* \rangle + \langle E_y E_y^* \rangle \\ \langle E_x E_x^* \rangle - \langle E_y E_y^* \rangle \\ 2\,\mathrm{Re}\,\langle E_y E_x^* \rangle \\ 2\,\mathrm{Im}\,\langle E_y E_x^* \rangle \end{bmatrix} \tag{6.29}$$

The fourth column vector lists the equivalent relations to the complex polarimetric parameters, and the fifth column vector lists the equivalent relations to the complex electric field components E_x and E_y. If the y-axis of the measurements are aligned with the vertical, then E_y is the vertical field E_v and E_x is the horizontal field E_h.

The exploitation of the Stokes parameters to estimate wind direction is discussed in the following section.

6.5 Passive Microwave Sensing of the Ocean

Microwave radiometers are commonly used for remote sensing of the ocean surface. As in infrared and visible wavelength radiometers, the signal from a microwave radiometer viewing the ocean surface consists of contributions from downwelling radiation reflected from the sea surface, radiation emitted from the surface, and atmospheric emissions (Figure 6.9).

There are a number of sources of downwelling microwave radiation, including galactic noise, cosmic radiation, solar radiation and downwelling atmospheric emissions:

- Galactic noise is not strong at microwave frequencies, being less than 1 K for frequencies greater than 3 GHz. It becomes more important at lower RF frequencies and it is somewhat directional, peaking towards the galactic center.
- Cosmic radiation is the remnant 2.7 K radiation from the big bang. It is reasonably isotropic.

Table 6.2 Solar radiometric temperature.

Frequency (GHz)	Temperature (K)
1	200,000
6	22,000
10	15,000
37	7,000

- Solar radiation is quite bright and highly frequency dependent, with the radiometric temperatures given in Table 6.2. In order to measure surface emissions, we want to avoid looking at the reflection of the sun.
- Atmospheric emissions from water vapor, liquid water, and oxygen vary significantly in a highly frequency-dependent manner.

The microwave brightness temperature observed by a satellite radiometer is given by:

$$T_B(v, \theta, p) = \tau_A T_{Bs} + \tau_A r_s [T_{sky} + T_{cos} \tau_A] + T_u \qquad (6.30)$$

for frequency v, sensor look angle θ and polarization p, where:

T_{Bs} = surface brightness temperature
τ_A = atmospheric transmittance = $\exp(-\tau_a)$
τ_a = atmospheric opacity (optical depth)
T_{sky} = sky brightness temperature
T_{cos} = cosmic radiation (2.7 K)
r_s = surface reflectivity (Fresnel equations)
T_u = upwelling atmospheric temperature.

The first term is the surface brightness temperature times the atmospheric transmittance. This is the one term that contains the signal of interest for ocean measurement.

The second term is the brightness temperature reflected from the surface. The downwelling brightness temperature is the sum of the sky brightness temperature plus cosmic radiation attenuated by the downward trip through the atmosphere. This downwelling brightness temperature is then multiplied by

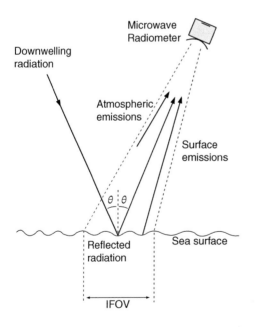

Figure 6.9 Energy sources observed by a downward-looking microwave radiometer.

the surface reflectivity and the attenuation from the trip back up through the atmosphere.

The last term is the brightness temperature of upwelling from the atmosphere.

6.5.1 Atmospheric Transmission

At 6–18 GHz, where atmospheric absorption is relatively low, the upwelling from the atmosphere is about equal to the downwelling sky temperature, which is given by an effective atmospheric temperature T_{Ae} times one minus the atmospheric transmittance:

$$T_u \cong T_{sky} \cong T_{Ae}[1 - \exp(-\tau_a)] \quad (6.31)$$

The effective atmospheric temperature is a weighted-mean temperature of the microwave-absorbing region of the atmosphere. It is frequency and polarization dependent, and depends on the vertical distribution of temperature, humidity, and liquid water.

The optical depth of the atmosphere is given by a sum of the nadir looking opacities due to oxygen, water vapor, and liquid water, times $\sec\theta$ to account for the length of the slant path:

$$\tau_a = (\tau_o + a_v q_v + a_l q_l)\sec\theta \quad (6.32)$$

where:

τ_o = nadir oxygen opacity
q_v = vertical-column water vapor
a_v = water vapor nadir opacity coefficient
q_l = vertical-column cloud liquid water
a_l = cloud liquid water opacity coefficient.

Referring back to Figure 6.1, the transmissivity of a nadir path through the atmosphere under clear sky conditions includes water vapor absorption bands at 22 and 180 GHz and oxygen absorption bands at 60 and 120 GHz. In between these absorption bands the atmosphere is relatively transparent.

Clouds and rain have a significant impact on transmissivity, especially at higher frequencies. Figure 6.10 shows the transmission through clouds and rain as a function of frequency and wavelength. The left-hand panel is for clouds with varying liquid water content. The right-hand panel is for light to heavy rain.

Direct to home television transmissions from a broadcast satellite are made at 18.5 GHz, with a wavelength of 1.7 cm. This frequency was chosen so the dishes could be small, but satellite transmissions at this frequency can be affected by rain. For this reason, the direct broadcast beams are shaped over the United States to provide the most power to regions that have the highest occurrence of heavy rains, such as Florida.

6.5.2 Seawater Emissivity

The right-hand panel in Figure 6.11 shows the real and imaginary parts (ε_r and ε_i) of the complex dielectric constant for fresh water and salt water, as a function of frequency and wavelength. Salinity does not have a significant impact on the dielectric constant above

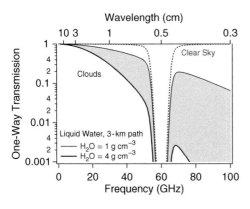
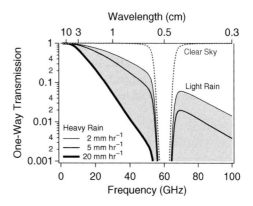

Figure 6.10 Transmission through clouds and rain.

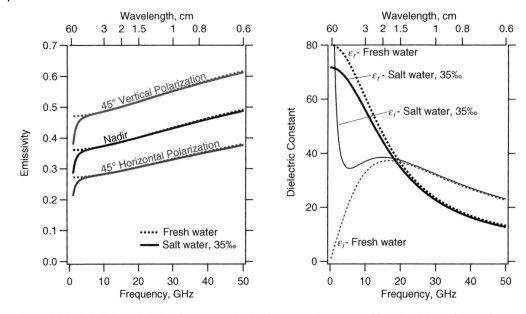

Figure 6.11 Emissivity and dielectric constant for fresh water and seawater. Note that 35 ppt (also written 35‰) is a typical surface salinity for the world's oceans. Adapted from Wilheit (1972).

20 GHz, but it has a significant impact on the imaginary part of the dielectric constant at frequencies around 1 GHz.

The left-hand panel in Figure 6.11 shows the emissivity for fresh and seawater for measurements at nadir and 45° with horizontal and vertical polarization. In general the emissivity rises with frequency. V-pol at 45° has higher emissivity than at nadir, which is higher than H-pol at 45°. This suggests polarization effects are measurable and may provide leverage for distinguishing between competing effects.

The left-hand panel also shows that salinity has a significant impact on emissivity at frequencies around 1 GHz, but not at higher frequencies. This suggest salinity may be measurable at frequencies of 1–2 GHz.

6.5.3 Fresnel Reflection Coefficients, Emissivity, and Skin Depth

The Fresnel reflection coefficients (horizontal r_h, vertical r_v, and unpolarized r) can be computed from the dielectric constant

$\tilde{\varepsilon} = \varepsilon_r + i\varepsilon_i$ which is frequency dependent:

$$r_h(\theta) = \left| \frac{\cos\theta - \sqrt{\tilde{\varepsilon} - \sin^2\theta}}{\cos\theta + \sqrt{\tilde{\varepsilon} - \sin^2\theta}} \right|^2$$

$$r_v(\theta) = \left| \frac{\tilde{\varepsilon}\cos\theta - \sqrt{\tilde{\varepsilon} - \sin^2\theta}}{\tilde{\varepsilon}\cos\theta + \sqrt{\tilde{\varepsilon} - \sin^2\theta}} \right|^2 \quad (6.33)$$

$$r(\theta) = (r_h(\theta) + r_v(\theta))/2$$

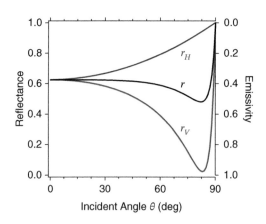

Figure 6.12 Fresnel reflection coefficients for sea water (20°C, 35‰) at X-band (10 GHz).

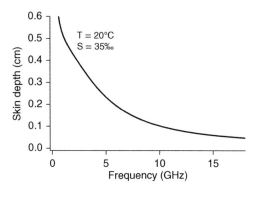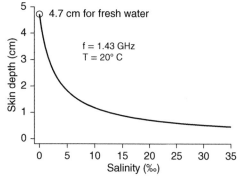

Figure 6.13 Skin depth of microwave radiation into seawater. Adapted from Swift (1980).

The reflection coefficients can also be expressed in terms of the complex index of refraction $\tilde{n} = n + ik_0 = \sqrt{\tilde{\varepsilon}}$ using:

$$n = \frac{\varepsilon_i}{2k_0} \tag{6.34}$$

$$k_0^2 = \frac{\varepsilon_r}{2}\left[\left(1 + \frac{\varepsilon_i^2}{\varepsilon_r^2}\right)^{1/2} - 1\right] \tag{6.35}$$

and application of equations (3.101) to (3.104).

Figure 6.12 shows the reflection coefficients for seawater at 10 GHz (X-band); here $\varepsilon_r = 55.9$, $\varepsilon_i = 37.7$, and the Brewster's angle is about 84°. The V-pol reflectance at the Brewster's angle is close to zero, so the emissivity, which is 1 minus the reflectance, is nearly 1. Notice the emissivity is about 0.37 at nadir, and for H-pol decreases monotonically to 0 at the horizon.

The left-hand panel in Figure 6.13 shows the skin depth[3] of microwave radiation in seawater is about 0.7 cm at 1 GHz and declines at higher frequencies at a rate inverse to frequency. The panel on the right shows that salinity, which controls conductivity in the water, has a major impact on skin depth at 1.43 GHz. This particular frequency is of interest because it corresponds to a portion of the electromagnetic spectrum where transmissions are prohibited to allow for radio astronomy measurements.

---

3 This is the skin depth of power, also known as the flux absorption depth, which was derived in equation (3.96) to be $\delta = \lambda/4\pi k_0$.

6.5.4 Sky Radiometric Temperature

Both plots in Figure 6.14 show the sky radiometric temperature that would be observed by a radiometer looking upwards from the ground as a function of frequency.

The left-hand panel shows the sky radiometric temperature for frequencies from 1 to 350 GHz for a dry atmosphere (bottom curve), and atmospheres with 2 and 7.5 gm cm^{-2} of liquid water along the path (top curves). Absorption and reemission by liquid water has a significant impact on the sky brightness at all frequencies other than those dominated by oxygen absorption. For a dry atmosphere, the sky is generally cold at frequencies where the atmosphere is reasonably transparent, reaching a minimum driven by the 3K cosmic background. The maximum radiometric temperature at microwave frequencies is about the physical temperature of the atmosphere (300 K) in the center of the oxygen absorption bands. Radiometric temperatures are generally higher in a humid atmosphere.

The right-hand panel shows the sky radiometric temperature for frequencies from 0.1 to 100 Ghz. The five solid curves are for zenith angles of 0°, 60°, 75° , 85°, and 89.5°, where the 89.5° case is looking at the horizon. Galactic noise dominates at frequencies below 1 GHz. The five solid curves have been computed for the sensor pointing towards the galactic pole, representing the minimum

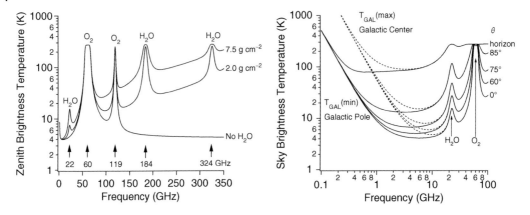

Figure 6.14 Sky radiometric temperature. Based on the model presented in Ulaby et al. (1981).

noise level. The dashed curves represent the noise maximum when the sensor points at the galactic center.

From 1–10 GHz, there is a general $\sec\theta$ dependence on zenith angle that disappears in the absorption bands at higher frequencies. Notice that the other effect is that lower paths are not just longer, but the paths spend more time in the warmer lower atmosphere, so the radiometric temperature is higher.

Figure 6.15 compares measured and computed sky temperature as a function of $\sec\theta$, where θ is the zenith angle. Notice

the agreement is quite good from nadir to 80–85°, depending on frequency. The higher temperatures in the lower atmosphere lead to the discrepancy at angles closer to the horizon.

Figure 6.16 contains plots of sky temperature as a function of zenith angle for clear sky, moderate cloud cover, and moderate rain plus cloud cover. Again we see that sky temperature increases towards the horizon. The sky temperature peaks in the oxygen absorption line at 0.43 cm wavelength, corresponding to 70 GHz. It is also clear that water vapor makes a large

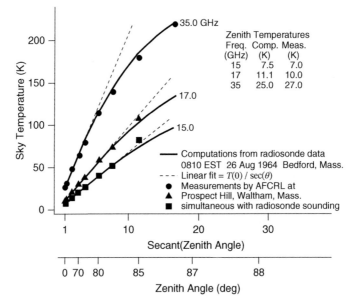

Figure 6.15 Measured and computed sky temperature as a function of $\sec\theta$. Crane (1971) with permission from IEEE.

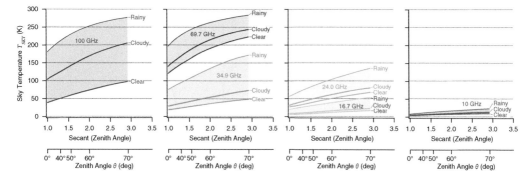

Figure 6.16 Sky temperature for clear sky (7.5 g m^{-3} water vapor at sea level), moderate cloud cover (0.3 g water m^{-3}), and cloud plus moderate rain (4 mm hr^{-1}). Adapted from Weger (1960).

difference at 100 GHz (0.3 cm) and 35 GHz (0.86 cm).

6.5.5 Sea Surface Brightness Temperature

The left-hand panel in Figure 6.17 shows the brightness temperature of a specular sea surface as a function of incident angle for H- and V-pol at four microwave frequencies. The increase in the brightness temperature at V-pol near 80° incident angle is a direct result of the increased emissivity due to the Brewster's angle. Also notice that the brightness temperature falls to zero at the horizon.

The right-hand panel plots the brightness temperature vs. sea surface temperature for a smooth sea. The plot was constructed for normal incidence at a frequency of 1.43 GHz, with multiple curves for various values of salinity. Note that the sensitivity to salinity varies in a nonlinear way with surface temperature. The first-order sensitivity of the brightness temperature to SST goes through zero at 1.43 GHz for typical values of ocean salinity. This makes this the ideal frequency to make salinity measurements.

The plots in Figure 6.18 show the specular brightness temperature sensitivity to a 7° change in surface temperature, as a function of frequency for two polarizations (vertical and horizontal) and three incident angles. The plots were made for a salinity of 33 ppt.

The plots indicate that the most sensitive measurement of surface temperature can be made at 5–6 GHz with vertical polarization and an incident angle of around 60°, while rain acts to reduce the sensitivity to surface temperature.

The plots in Figure 6.19 show the dependence of the brightness temperature of the surface as a function of incident angle for four wind speeds. The left-hand two plots are for

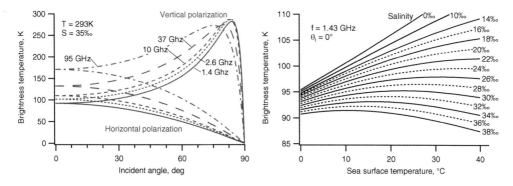

Figure 6.17 Brightness temperature of a specular sea surface as a function of incident angle (left) and salinity (right). Adapted from Ulaby et al. (1982) and Swift (1980) but based on the model in Wentz (2004).

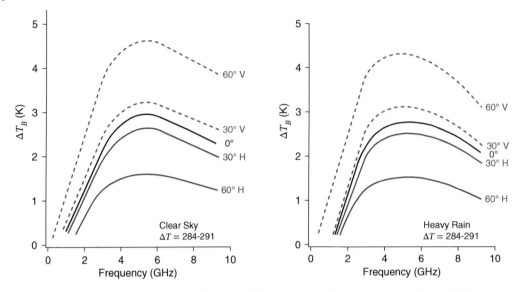

Figure 6.18 Brightness temperature sensitivity to a 7° change in surface temperature. Source: Alishouse (1975).

19.35 GHz and the right-hand two plots are for 37 GHz. The top two plots are V-pol and the bottom two plots are H-pol.

The open circles indicate an incident angle of 53°, the angle at which the SSM/I sensor operates.

There is a cross-over point in the V-pol data around 53°. At this angle the V-pol measurements are independent of wind speed, while the H-pol measurements continue to exhibit a monotonic dependence on wind speed. This cross-over or pivot point allows the

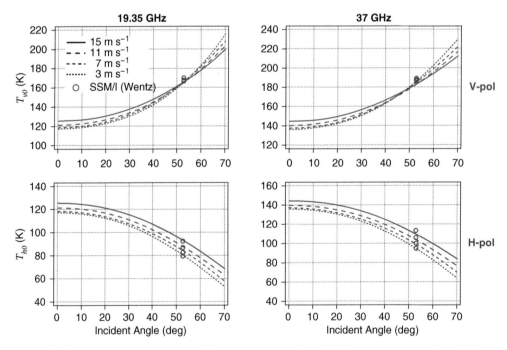

Figure 6.19 Brightness temperature sensitivity to wind speed. Yueh (1997) with permission from IEEE.

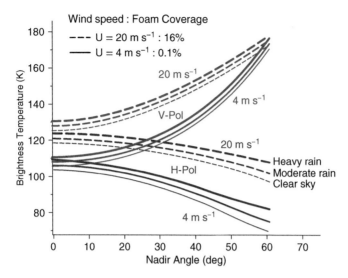

Figure 6.20 Brightness temperature sensitivity to foam at 4 GHz. Source: Alishouse (1975).

isolation of all contributions to the received signal other than wind speed. These contributions can then be removed from the H-pol signal, isolating out just the wind speed dependence. For this reason, most microwave radiometers are designed to work at an incident angle of about 53°.

Foam on the surface caused by breaking waves has a high emissivity. Figure 6.20 shows the brightness temperatures of a rough foam-covered ocean surface. The top-most curve is heavy rain, with moderate rain in the middle, and clear sky on the bottom.

The lower set of curves is for a 4 m s⁻¹ wind speed, which is the about the highest wind speed before the onset of whitecaps. The upper set of curves is for 20 m s⁻¹ wind speed, which corresponds to foam covering about 16% of the surface[4]. Thus increasing foam coverage with increasing wind speed explains some of the dependence of the surface brightness temperature on wind speed.

Figure 6.21 summarizes the relative sensitivity of the brightness temperature T_B to various

phenomena, as a function of frequency. P_i in this figure is a generic measurement of the individual phenomena, for example salinity, wind speed, sea surface temperature, and so forth.

To summarize, brightness temperature is sensitive to:

- Salinity but not sea surface temperature at 1.4 GHz; this zero sensitivity to SST makes this the preferred frequency to make salinity measurements.
- Sea surface temperature and some wind speed at 4–5 GHz.
- Predominantly wind speed at 18 GHz, H-pol, with some small contributions from SST and atmospheric water content.
- Water vapor at 22 GHz.
- Liquid clouds and wind speed at 35 GHz.

Brightness temperature is not sensitive to wind speed for 19 and 37 GHz, V-pol measurements made at an incident angle of 53°.

The algorithms used to derive geophysical quantities from passive microwave measurements work by combining measurements at various frequencies and polarizations to isolate out the contributions from individual phenomena.

---

4 The fractional foam coverage of the surface is generally modeled as a power law in wind speed with an exponent of between 3 and 3.5.

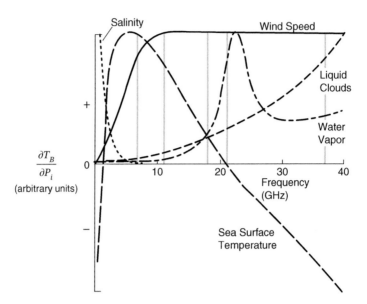

Figure 6.21 Relative sensitivity of brightness temperature to various phenomena. The vertical gray bands indicate the frequencies observed by the Scanning Multichannel Microwave Radiometer (SMMR), the first space-based microwave radiometers. Source: Wilheit (1979).

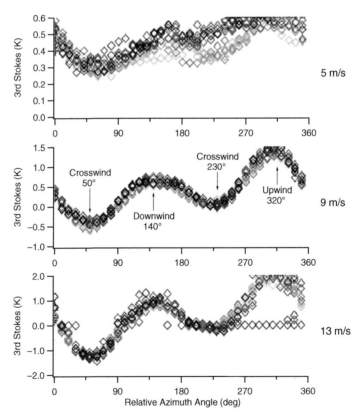

Figure 6.22 Variation in third Stokes parameter with wind direction. From Gaiser et al. (2005) with permission from IEEE.

6.5.6 Wind Direction from Polarization

Up to this point we have discussed utilizing measurements at multiple frequencies, incident angles, and two linear polarizations to estimate various geophysical parameters. It turns out that estimates of wind direction also can be made from passive microwave data that include additional polarimetric measurements. The wind direction dependence in the surface emissions arises from the anisotropic distribution of slopes associated with wind-driven waves.

Section 6.4 described how the full polarization of an electromagnetic signal can be characterized by the Stokes parameters. This is important, because the directional anisotropy of wind wave slopes causes the third and fourth Stokes parameters to depend on the look angle relative to the wind. These variations are small, but they are measurable.

The variation of the third Stokes parameter with look azimuth relative to the wind at

10.7 GHz is shown in Figure 6.22. From top to bottom the plots are for wind speeds of 5, 9, and 13 m s^{-1}. The data have a noise level of 0.37 K, but it is clear that the brightness temperature varies with look direction with an amplitude that increases with wind speed. At 5 m s^{-1} the signal is at most 0.3 K, while at 13 m s^{-1} the signal is over 3 K.

The emissions have a peak when looking into the wind, with a secondary peak looking downwind, and a minimum looking crosswind.

This signal is well modeled by a sum of three terms:

$$U = U_0 + U_1 \cos \varphi + U_2 \cos 2\varphi \quad (6.36)$$

where φ is the azimuth angle between the wind direction and the look direction. Fitting this model to the data then provides an estimate of the wind direction.

Figure 6.23 shows the same plots for the fourth Stokes parameter, which is also dependent on look direction relative to the wind.

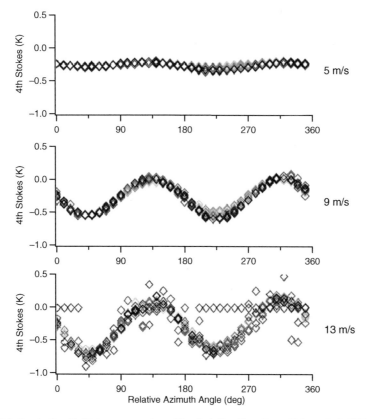

Figure 6.23 Variation in fourth Stokes parameter with wind direction. From Gaiser et al. (2005) with permission from IEEE.

WindSat is a fully polarimetric, passive microwave radiometer designed to exploit these signatures to measure wind vectors. Design features of the radiometer are discussed in the next section.

6.6 Satellite Microwave Radiometers

The United States has fielded four types of microwave radiometers on satellites.

The Scanning Multichannel Microwave Radiometer (SMMR) was the first instrument of its type, deployed on Seasat and Nimbus-7, which were both launched in 1978. Table 6.3 shows that it had dual-pol capability at 6.6, 10.7, 18, 21, and 37 GHz. The SMMR sensors proved the utility of passive microwave sensing from space.

The Special Sensor Microwave/Imager (SSM/I) was then deployed on a series of six Defense Meteorological Satellite Program (DMSP) satellites with launch dates from 1987 to 1999. The capabilities of SSM/I were then subsumed into the SSMI/S sensor, which has even more channels and has flown on four satellites with launch dates from 2003 to 2014. The S in SSMI/S stands for sounder, and we will discuss this sensor in detail when we get to atmospheric sounders.

The Tropical Rainfall Measuring Mission (TRMM), a NASA satellite carrying the TRMM Microwave Imager (TMI), was launched in 1997. It was in an orbit inclined to the equator by just 35°, because it was designed

to concentrate its measurements over the tropical regions.

Finally, the Advanced Microwave Scanning Radiometer-EOS (AMSR-E) was a Japanese sensor deployed on NASA's Aqua satellite launched in 2002. The sensor scanner unit for AMSR-E failed in 2011.

Key characteristics of these instruments are summarized in Table 6.3.

6.6.1 SMMR

The Scanning Multichannel Microwave Radiometer (SMMR) was the first scanning radiometer deployed in space. Figure 6.24 shows the instrument package, the bottom of which points towards nadir, and the large offset reflector mounted on top of the package. The reflector was spun, completing a circular scan every 4.1 seconds as the satellite moved forward at a ground speed of $6.5 \, \text{km s}^{-1}$.

There was a single feed horn common to all five bands. Because the effective size of the antenna was fixed to the reflector diameter, the beamwidth and footprint on the ground scaled inversely with frequency, as shown in Table 6.4. Note these are not high-resolution instruments. The reflector boresight was set to an angle of 42°, resulting in an incident angle on the ground of 50.3°.

6.6.2 SSM/I and SSMI/S

The satellites deployed for the Defense Meteorological Satellite Program (DMSP) are weather satellites used by the U.S. Department

Table 6.3 U.S. satellite microwave radiometers.

Instrument	Frequencies (GHz) and polarizations						Incident angle (deg)
SMMR	6.6 V,H	10.7 V,H	18.0 V,H	21.0 V,H	37.0 V,H		51°
SSM/I			19.3 V,H	22.2 V	37.0 V,H	85.5 V,H	53°
TMI		10.7 V,H	19.3 V,H	21.3 V	37.0 V,H	85.5 V,H	53°
AMSR-E	6.9 V,H	10.7 V,H	18.7 V,H	23.8 V,H	36.5 V,H	89.0 V,H	55°

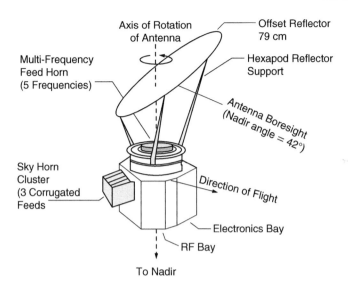

Figure 6.24 Scanning Multichannel Microwave Radiometer (SMMR). Adapted from NASA drawing.

Table 6.4 SMMR antenna footprints.

Channel frequency (GHz)	Polarization	Along-track footprint (km)	Cross-track footprint (km)
6.6	H, V	148	95
10.7	H, V	90	60
18.0	H, V	45	30
21.0	H, V	40	25
37.0	H, V	20	15

of Defense. They carry either the SSM/I or the more advanced SSMI/Sounder radiometers. Both instruments are designed around a large rotating offset reflector to scan the beam in a circular arc about the satellite ground track. The SSM/I sensor package including feed horn and offset dish are illustrated in Figure 6.25.

Figure 6.26 illustrates the scan geometry for the SSM/I. The reflector is offset from nadir to produce a beam at 45°, which results in an incident angle of 53.1°. This angle corresponds to the point where there is no wind speed dependence in V-pol at 19 and 37 GHz, as previously discussed.

The antenna footprints are inversely related to frequency. Table 6.5 lists the sensor footprints for SSM/I.

Figure 6.25 SSMI/S sensor. Source: ESA (2001).

The reflector is rotated at a rate of once every 1.9 seconds. The spacecraft moves forward a distance of 12.5 km in the time of a single scan. In order to prevent any spatial aliasing, the 85 GHz channel, which has a footprint of 15 × 13 km, is sampled on every scan. But the 19-, 22-, and 37-GHz channels, which have large footprints, are sampled on every other

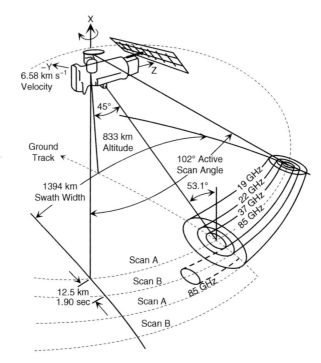

Figure 6.26 SSM/I circular scan geometry. Adapted from Hollinger and Lo (1983) and Hollinger et al. (1990).

Table 6.5 SSM/I footprints for each channel.

Channel frequency (GHz)	Along-track footprint (km)	Cross-track footprint (km)
19.35	69	43
22.235	60	40
37.0	37	28
85.5	15	13

Table 6.6 SSM/I sampling for each channel.

Channel	Samples/ Scan	Sample Interval (ms)	Cross-track Sample spacing (km)
85.5 GHz	128	4.2	12.5
Others	64	8.4	25.0

scan, and hence have down-track samples every 25 km. The SSM/I sampling rates are given in Table 6.6.

Because the DMSP are large satellites, it actually obstructs the view of the SSM/I during a portion of the scan. The design limits the scan to acquiring data over a total extent of 102°, aft of the satellite direction. In the end, the SSM/I samples a swath width of almost 1400 km on every pass. The SSMI/S scans a larger range of angles, resulting in a 1700-km swath.

Figure 6.27 illustrates the SSM/I scan pattern for the 37-GHz channel.

The DMSP satellites are generally placed in sun-synchronous orbits. Figure 6.28 shows the SSMI/S coverage for one of the satellites for a 24-hour period. The dashed lines are the sub-satellite track, and the solid lines are the edges of the 1700-km wide swath. One specific pass is highlighted in light blue. While the daily coverage is complete near the poles, there are small daily coverage gaps illustrated in black about the equator.

6.6.3 SSM/I Wind Algorithm

The algorithm used to extract wind speed from SSM/I data estimated wind speed from a linear combination of the brightness temperatures

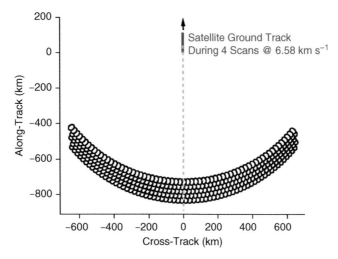

Figure 6.27 SSM/I scan pattern.

Figure 6.28 Daily coverage for a single SSMI/S sensor.

with vertical polarization at 19 and 22 GHz, and vertical and horizontal polarizations at 37 GHz:

$$WS\text{(m/s)} = a_0 + a_1 T_B(19V)$$
$$+ a_2 T_B(22V)$$
$$+ a_3 T_B(37V)$$
$$+ a_4 T_B(37H) \qquad (6.37)$$

where, $a_0 = 147.90$, $a_1 = 1.0969$, $a_2 = -0.4555$, $a_3 = -1.7600$, $a_4 = 0.7860$, and WS = wind speed at 19.5 m height (m s^{-1}).

The coefficients as shown were derived by best fits of SSM/I data and wind speeds measured by buoys. It turns out that winds vary with height in a predictable way, so this algorithm was developed for a standard wind speed measurement height of 19.5 m. This odd height arose historically from a mean height of ship anemometers of about 64 ft. More modern algorithms are referenced to a wind speed measurement height of 10 m. In general, the differences at these two heights amounts to 10–15% of the wind speed.

As we previously saw, the brightness temperatures for V-pol at this incident angle are insensitive to wind speed. So in this algorithm, the sensitivity to wind speed mostly comes from a single term, the brightness temperature with horizontal polarization at 37 GHz. The 19V and 37V terms compensate for sea surface temperature and total water content, while the 22V channel compensates for water vapor.

Figure 6.29 shows the mean bias and standard deviation of the wind speed difference between the SSM/I wind speed algorithm and buoy measurements. The bias in this algorithm is generally less than 1 m s^{-1} with a standard deviation of about 2 m s^{-1}. We need to note that these errors are known to increase with increasing rain rate, and can be as large as 5 m s^{-1}. The errors also increase at very high wind speeds, such as in hurricanes. Algorithms that work better in rain and high winds have been an emphasis of research programs over the past two decades.

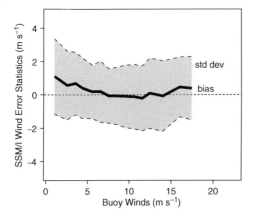

Figure 6.29 Mean bias and standard deviation of wind speed estimates from SSM/I. Adapted from Goodberlet et al. (1990).

Figure 6.30 plots sample SSM/I wind fields for the ascending passes from one satellite on a single day.

SSM/I data are also used to estimate the total precipitable water content (water vapor plus liquid water in clouds) and liquid water (clouds only) in the atmosphere (Alishouse et al., 1990) (Figure 6.31). The algorithm for total precipitable water content is:

$$WV(\text{kg/m}^2) = a_0 + a_1 T_B(19\text{V})$$
$$+ a_2 T_B(22\text{V})$$
$$+ a_3 T_B(37\text{V})$$
$$+ a_4 T_B^2(22\text{V}) \qquad (6.38)$$

where, $a_0 = 232.89393$, $a_1 = -0.148596$, $a_2 = -1.829125$, $a_3 = -0.36954$, $a_4 = -0.006193$, and WV = atmospheric column water vapor (kg m^{-2}).

This algorithm is nonlinear and only dependent on V-pol measurements at 19, 22, and 37 GHz. The coefficients were determined from comparison with data obtained from radiosonde measurements made from islands small enough to not overly disturb the SSM/I measurement. The water vapor sensitivity comes predominantly from the 22 GHz measurements. This algorithm is nearly unbiased with an estimated standard deviation of 3 kg m^{-2}.

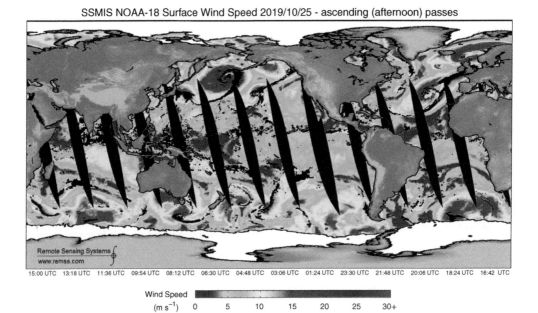

Figure 6.30 SSM/I wind fields for the ascending passes on a single day. Source: RemSS, Surface Wind Speed. http://images.remss.com/ssmi/ssmi_data_daily.html with permission from Remote Sensing Systems.

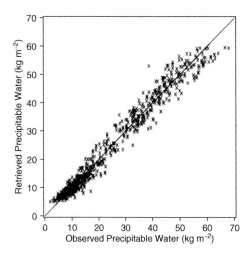

Figure 6.31 SSM/I total precipitable water measurements. Source: Alishouse et al. (1990) with permission from IEEE.

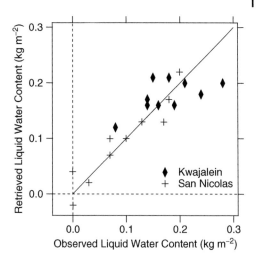

Figure 6.32 SSM/I cloud liquid water content measurements. Source: Alishouse et al. (1990) with permission from IEEE.

A companion algorithm was developed to estimate liquid water content in the atmosphere (Alishouse et al., 1990). The sensitivity to liquid water in this algorithm arises from the measurements made at 85 GHz. The coefficients for this algorithm were determined by comparison of the SSM/I data with radiometric sounders that were deployed in special tests conducted from San Nicolas Island off the California coast and Kwajalein island in the Pacific (Figure 6.32). For this reason there are far fewer data points available for comparison.

$$CLW(\text{kg m}^{-2}) = a_0 + a_1 T_B(19H)$$
$$+ a_2 T_B(22V)$$
$$+ a_3 T_B(37V)$$
$$+ a_4 T_B(85H) \qquad (6.39)$$

where, $a_0 = -3.14559$, $a_1 = 6.0257x10^{-3}$, $a_2 = -4.8803x10^{-3}$, $a_3 = 1.9595x10^{-2}$, $a_4 = -3.0107x10^{-3}$, and CLW = cloud liquid water (kg m^{-2}).

It turns out that these same ground-based radiometers were regularly deployed at four locations in Colorado, but no satisfactory SSM/I algorithm could be developed to estimate liquid water content over land, with or without snow cover.

6.6.4 AMSR-E

Figure 6.33 shows the AMSR-E imager that was flown on the Aqua satellite (Kawanishi et al., 2003). This instrument used multiple feed horns to illuminate the large rotating reflector. The feed horns were arranged along the scan direction as shown, so the centers of the beams were aligned.

The general characteristics of the antennas are:

- Reflector Diameter – 1.6 m.
- Rotation rate – 40 rpm.
- LOS nadir angle – 47.4°
- LOS incident angle – 55°.
- 6 horns – 89 GHz (2), 18 and 23 GHz (1), one each for other frequencies.
- Swath – ±61° about ground track = 1445 km.
- Spacecraft advances 10 km per scan.

This sensor acquired data from 2002 until the scanner failed in December 2011. Table 6.7 lists the AMSR-E bands and footprints.

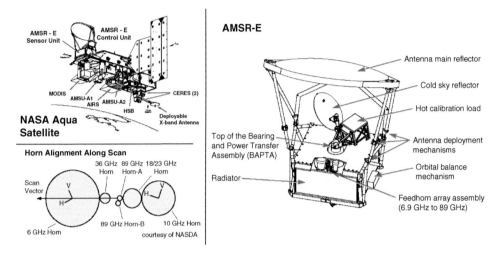

Figure 6.33 AMSR-E microwave scanning radiometer.

Table 6.7 AMSR-E bands and footprints.

Frequency (GHz)	Polarization	Along-track footprint (km)	Cross-track footprint (km)
6.9	H, V	75	43
10.7	H, V	51	29
18.7	H, V	27	16
23.8	H, V	32	18
36.5	H, V	14	8
89.0	H, V	6	4

- Weight = 675 lbs
- Power = 295 W
- Spin rate = 31.6 rpm
- RF = 22 Channels, 5 Frequencies.

6.6.5 WindSat

WindSat is a sensor developed by the Naval Research Laboratory for the U.S. Navy and the NPOESS Integrated Program Office. It was designed to exploit polarimetric signatures to demonstrate the ability of a fully polarimetric radiometer to measure wind vectors.

Much of the design is similar to earlier systems, with the antenna being a large offset reflector that physically rotated to provide a conical scan (Figure 6.34).

The general characteristics of the WindSat payload were:

- Height = 10.5 ft
- Width = 8.25 ft

WindSat was deployed on the DOD Coriolis satellite that was launched in January 2003 (Figure 6.35). It was designed for a three-year mission, but ceased operation in 2020.

WindSat supports the channels listed in Table 6.8. A full six polarimetric measurements are made in the 10.7, 18.7, and 37 GHz bands. These are the bands that are used for wind vector estimation.

Considerable effort was put into the design of the receivers, which were built as carefully matched pairs. Significant effort was also put into the on-orbit calibration of the system. This level of effort was required because of the low levels of the signals that were being measured.

Unlike the previous instruments we discussed, the incident angle for Windsat varies a bit from band to band because of the physical layout of the feed horns, as illustrated in Figure 6.36.

These offset horns create a complex footprint on the ground as the satellite orbits and the reflector spins. The beam footprints for a few samples of a few scans are shown

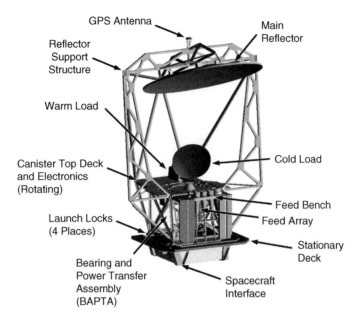

Figure 6.34 WindSat payload. Source: Naval Research Laboratory.

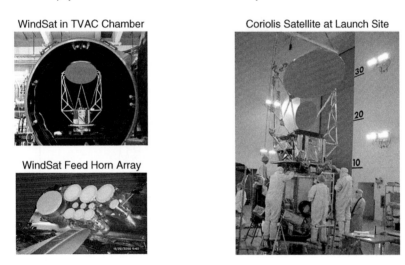

Figure 6.35 Coriolis satellite. Source: Naval Research Laboratory.

Table 6.8 WindSat channel specifications.

Freq. (GHz)	Channels	BW (MHz)	T (msec)	NEDT (K)	Inc Ang (deg)	IFOV (km)
6.8	v, h	125	5.00	0.48	53.5	40×60
10.7	v, h, ±45,*	300	3.50	0.37	49.9	25×38
18.7	v, h, ±45,*	750	2.00	0.39	55.3	16×27
23.8	v, h	500	1.48	0.55	53.0	12×20
37.0	v, h, ±45*	2000	1.00	0.45	53.0	8×13

*lc, rc = left and right circular polarization.

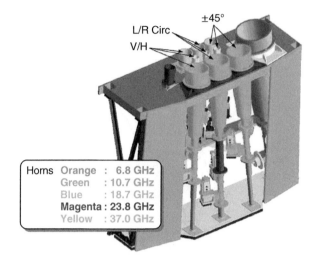

Figure 6.36 WindSat horn assembly. The coloring in the CAD drawing indicates the horns for the various bands. From Gaiser et al. (2004) with permission from IEEE.

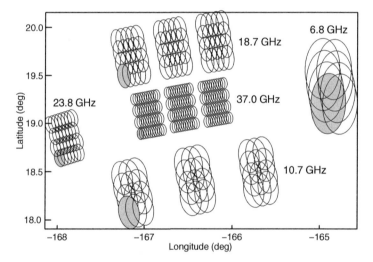

Figure 6.37 Ground footprint of WindSat beams from four consecutive scans. Source: Gaiser et al. (2004) with permission from IEEE.

in Figure 6.37. These measurements need to be aligned to a common grid for processing, which places stringent requirements on pointing accuracy.

Figure 6.38 illustrates the circular scanning pattern of the WindSat[5]. This satellite is unusual in that it takes observations during both the forward-looking and aft-looking scans. This makes the WindSat

swath geometry quite different from, and significantly more complicated to work with, than other passive microwave sensors. This geometry can actually provide two looks at many locations in the swath from different directions. This occurs when the aft measurements overlay the forward-looking measurements.

Figure 6.39 is an example WindSat data product for all of the ascending passes during one day in 2015. The data products are available for download over the internet.

5 The scan pattern in Figure 6.38 is not exact. The scan pattern actually varies with individual feeds, and hence frequency and polarization.

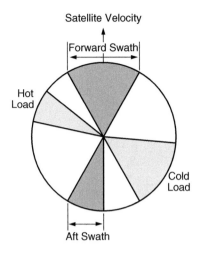

Figure 6.38 Circular scanning pattern of the WindSat. From Jones et al. (2006) with permission from IEEE.

Figure 6.40 is just a reminder that circulation in the atmosphere is driven by differential heating of the Earth, which causes pressure variations that induce flows. These flows organize into large cells in order to balance the pressure forces with the Coriolis force. Thus the flows organize into large zonal cells.

WindSat is capable of monitoring these global circulations. We will later see that there are active scatterometers that are designed to do the same, but WindSat has some advantages as a lower power, passive sensor.

Figure 6.41 ends this section with a full-resolution image of a wind field produced by WindSat. In this plot the color indicates wind speed, but the vectors indicate wind direction, which is the unique capability that WindSat brings to passive microwave sensing.

6.7 Microwave Radiometry of Sea Ice

Sea ice is a significant component of the Earth's thermal system. This section reviews the characteristics of sea ice that are observable by a microwave radiometer.

Although sea ice coverage is regularly monitored using infrared, optical, and active radar, microwave radiometers offer several advantages for sea ice measurement in that they can make observations both day and night and also through clouds, an important capability for monitoring polar regions where cloud cover can be persistent. Moreover, satellite-based microwave

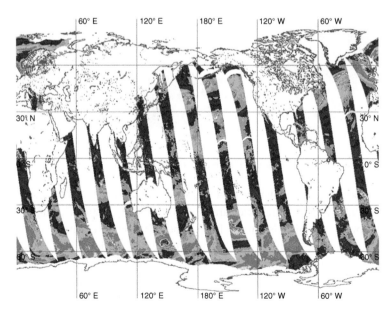

Figure 6.39 Example WindSat data from ascending passes during one day. Figure courtesy of the U.S. Naval Research Laboratory.

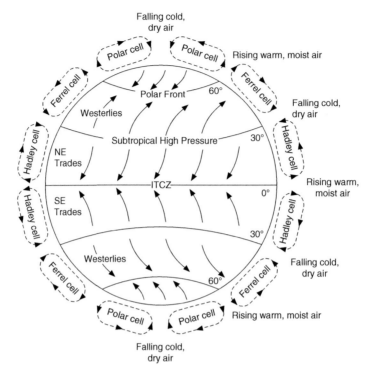

Figure 6.40 Atmospheric circulation. Figure courtesy of the U.S. Naval Research Laboratory.

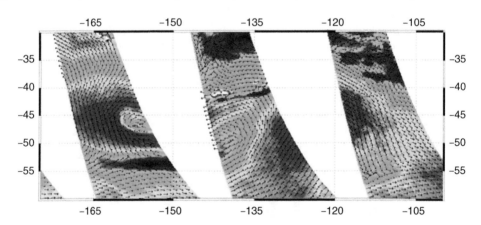

Figure 6.41 Vector wind field from WindSat. Figure courtesy of the U.S. Naval Research Laboratory.

radiometers have been used for polar ice measurements since 1979, so long time series are available, an example of which is shown in Figure 6.42.

Nevertheless, microwave radiometers also suffer from some limitations. They have coarser spatial resolution than other sensors. While they can measure through clouds, the observations are affected by weather. And as

we will see, microwave radiometers also can have difficulty distinguishing between certain types of sea ice and water.

Generally sea ice is categorized as first-year ice or multiyear ice, which is ice that has persisted for more than one year (Figure 6.43). These categories are important for ice research because the characteristics of ice changes significantly as it ages.

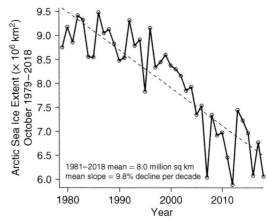

Figure 6.42 Sea ice map and coverage time series. Source: National Snow and Ice Data Center http://nsidc .org/arcticseaicenews/category/analysis/.

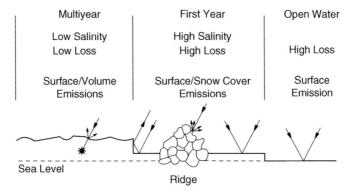

Figure 6.43 Overview of sea ice types. Source: Carsey (1992) with permission from American Geophysical Union (AGU).

First-year ice is generally thin, lies low in the water and has relatively high salinity. Multiyear ice is generally thicker, rides higher in water and has lower salinity.

Microwave radiometric measurements of ice rely on the fact that emissivity and brightness temperature are different between water and different types of ice. The highest emissivity, and hence brightness temperature, usually comes from first-year ice, which is brighter than multiyear or sea ice, which in turn is brighter than open water. There are also differences in the penetration depth of microwave radiation into sea ice due to the differences in salinity.

It is usual to think of sea ice in terms of a layered model, as shown in Figure 6.44. Often

the sea ice will be covered by a layer of snow, which consists of low-salinity ice crystals with a significant void fraction.

The sea ice itself lies partly above and partly below the mean sea surface. The physical characteristics of the sea ice (structure, air content, and salinity) can vary significantly in the vertical dimension. The air–snow and snow–ice interfaces are often quite rough.

First-year ice contains many brine pockets[6] at all depths, but few relatively small air bubbles. In older ice, the upper layers contain many relatively large air bubbles but the ice has relatively low salinity. The lower layers in old ice often contain brine pockets. All of this structure affects penetration depth and emissivity.

---

6 Brine pockets are local regions of higher salinity.

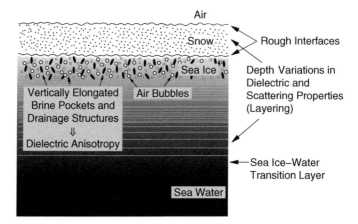

Figure 6.44 Vertical structure of sea ice. From Carsey (1992) with permission from American Geophysical Union (AGU).

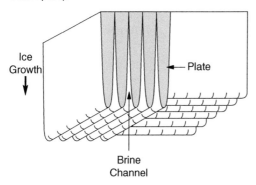

Figure 6.45 Sea ice growth structure. From Carsey (1992) with permission from American Geophysical Union (AGU).

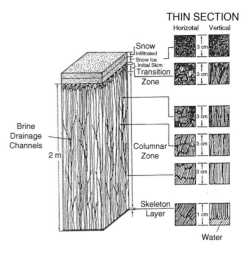

Figure 6.46 Characteristics of sea ice brine channels. From Schwarz and Weeks (1977) with permission from International Glaciological Society.

The general characteristics of sea ice are driven by how ice crystals tend to form and grow in sea water. Ice crystals initially form at the sea surface in thin layers (Figure 6.45). The ice crystals then grow downward forming vertical plates. The gaps in between the plates act as channels for brine to drain from the ice as it grows.

Crystals and drainage channels grow rapidly when small, but then the growth slows as the ice plates grow larger. The drainage channels become increasingly elongated, and the ice–water interface becomes deeply corrugated. The thin sections on the right of Figure 6.46 show how the structure becomes more elongated with increasing depth in the ice.

Ultimately, pressure ridges form in multiyear ice making the surface rough and irregular.

The diagram in Figure 6.47 is a conceptual view of the changes in ice depth and salinity with ice age. First-year ice starts out as thin, saline layers (curve a), but its salinity falls as it thickens from a few centimeters to maybe 100 cm. In contrast, multiyear ice can be several meters thick and has lower salinity.

If a snow layer is present, then any microwave radiation emitted by the ice must pass through this snow layer before it can be

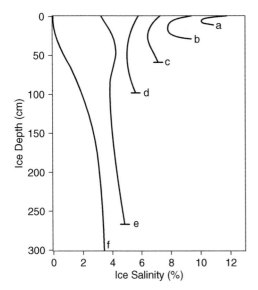

Figure 6.47 Thickness and salinity of sea ice of varying age. From Carsey (1992) with permission from American Geophysical Union (AGU).

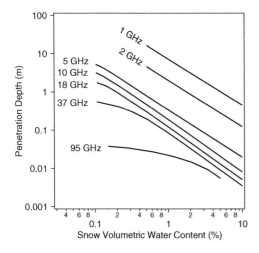

Figure 6.48 Microwave penetration depth into snow. Source: Carsey (1992) with permission from American Geophysical Union (AGU).

detected. Figure 6.48 shows the penetration depth in snow of microwave radiation as a function of the snow's volumetric water content.

Lower frequencies generally penetrate deeper than higher frequencies. Luckily the atmosphere is so dry in the polar regions that the snow is not very thick on most sea ice.

Figure 6.49 compares models and observations of the emissivity of open water and first-year ice for an incident angle of 50°. The dashed curves are for horizontal polarization and the solid curves are for vertical polarization.

In general, the first-year ice has higher emissivity than open water. The polarization differences for open water are driven by the Fresnel reflection curves.

Figure 6.50 compares the brightness temperature for open water, first-year ice, and multiyear ice at an incident angle of 53°. At 20 GHz we see that the first-year ice is brighter than the multiyear ice, which is brighter than the open water. At 37 GHz the story changes and the brightness temperature of the multiyear ice falls to between the V- and H-pol signals from open water.

There is significantly less difference in polarization in first-year ice because multiple scattering from within the ice structure depolarizes the radiation. The brightness temperature measurements in this plot were made from an aircraft. Airborne ice measurements can be more accurate than satellite measurements for a variety of reasons: the reduced resolution of most satellites increases the amount of inhomogeneities within each field of view, satellite data often include higher wind conditions, and satellite data include extra atmospheric path radiance.

The range of emissivity observed for multiyear sea ice is plotted in Figure 6.51 as a function of frequency. The measurements shown here were obtained from multiple years and locations. These data indicate that significant variability can occur in the emissivity of multiyear ice, especially at higher frequencies. This result is an indication of the variability in internal structure, salinity, and other physical characteristics of the multiyear ice.

Despite the variability, these curves indicate that it is generally possible to distinguish

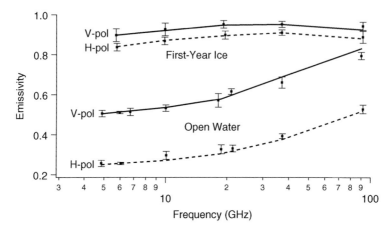

Figure 6.49 Emissivity for open water and first-year ice as a function of frequency and polarization. Source: Carsey (1992) with permission from American Geophysical Union (AGU).

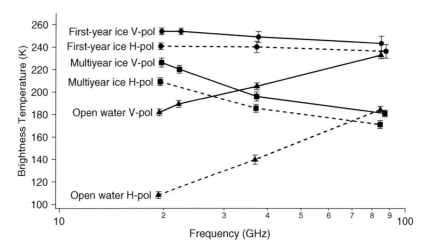

Figure 6.50 Brightness temperature for open water, first-year ice, and multiyear ice. Source: Carsey (1992) with permission from American Geophysical Union (AGU).

between open water, first-year ice and multiyear ice using brightness temperatures measured at multiple frequencies and polarizations.

The plots in Figure 6.52 show the basis for some early algorithms used to classify open water, first-year ice, and multiyear ice. The data were acquired from the DMSP SSM/I sensor during passes over the northern hemisphere ocean areas poleward of 50° latitude.

The left-hand panel shows the 18-GHz H-pol data on the vertical axis versus 18-GHz V-pol on the bottom axis. At this frequency, a simple threshold on temperature can be used to separate open water from ice, but it is difficult to separate first-year from multiyear ice.

The right-hand panel shows the 18-GHz V-pol data on the vertical axis versus 37-GHz V-pol on the bottom axis. At this frequency, the first-year ice is easily separable from the multiyear ice, with a separate cluster for the open water.

In general, the algorithms use combinations of these parameters to classify the surface.

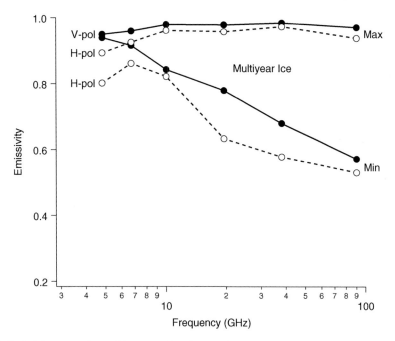

Figure 6.51 Emissivity range for multiyear sea ice. Source: Carsey (1992) with permission from American Geophysical Union (AGU).

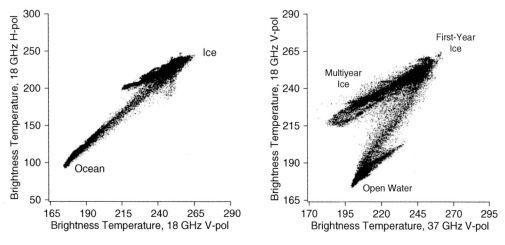

Figure 6.52 Relative brightness temperatures from multiple channels. Source: Carsey (1992) with permission from American Geophysical Union (AGU).

6.8   Sea Ice Measurements

While sea ice can completely cover a portion of the surface, it generally provides only partial coverage of the sea. As illustrated in Figure 6.53, there are various measures that are applied to describe sea ice coverage.

The diagram on the left provides a schematic representation of a region of the sea that is partly covered by sea ice. The region is described by a 4 × 4 grid of 1 km square cells, and the sea ice is indicated as gray blocks. The ice area in this diagram is the actual area covered by ice. In this example there are five

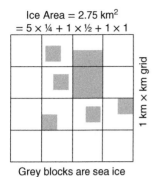

Ice Area = 2.75 km²
= 5 × ¼ + 1 × ½ + 1 × 1

1 km × km grid

Grey blocks are sea ice

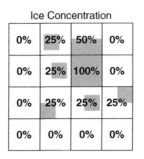

Ice Concentration

0%	25%	50%	0%
0%	25%	100%	0%
0%	25%	25%	25%
0%	0%	0%	0%

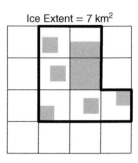

Ice Extent = 7 km²

Figure 6.53 Sea ice measures: area, concentation, and extent.

cells that are quarter covered, one cell that is half covered, and one cell that is fully covered, yielding an ice area of 2.75 km².

The middle diagram illustrates ice concentration, which is defined to be the percentage of each grid cell that is covered by ice. The numbers indicate the ice concentration that would be estimated from the ice shown in this figure. Ice concentration varies from 0 to 100%.

The diagram on the right illustrates the last measure of ice coverage: sea ice extent. The ice extent, illustrated by a thick boundary encompassing seven grid cells, is defined to be the cumulative area of grid cells having an ice concentration greater than or equal to 15%. In this example, the ice extent is 7 km².

Figure 6.54 shows examples of ice concentration at northern latitudes for March and September 2017. The color indicates ice concentration, which is the percentage of each grid cell covered by ice. These data were acquired from 1979–1996 from the Nimbus 7 SMMR sensor and the SSM/I instruments on the DMSP F8, F11, and F13 satellites. These figures indicate the obvious – there are large seasonal variations in polar ice concentration.

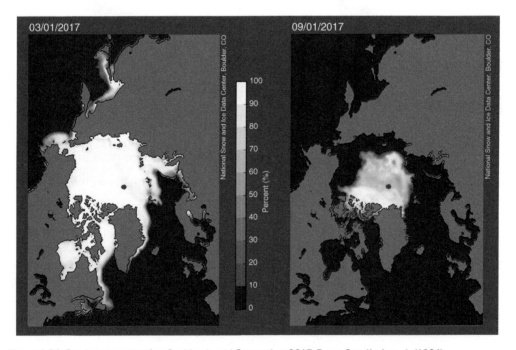

Figure 6.54 Sea ice concentration for March and September 2017. From Cavalieri et al. (1996).

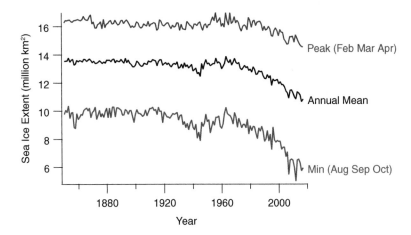

Figure 6.55 Northern hemisphere sea ice extent versus time. Based on data from Walsh et al, 2019 at https://nsidc.org/data/g10010/.

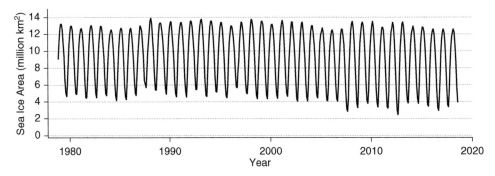

Figure 6.56 Northern hemisphere sea ice area. Based on data from NSDIC, Artic Sea news. https://nsidc .org/arcticseaicenews/sea-ice-tools/.

March and September typically represent the maximum and minimum coverage at northern latitudes.

Figure 6.55 plots the northern hemisphere sea ice extent, the cumulative area of grid cells having ice concentration ≥ 15%, from 1850 to 2017. The black curve is the annual mean, while the color curves are the means for the minimum and maximum seasons. The ice extent was computed based on 25 km × 25 km grid cells, a common grid size used with microwave image data. The early data on this plot were estimated from ground observations, the middle data from airborne observations, and the later data from satellites.

The time series show there have been large decreases in coverage over the last few decades,

especially in the minimum extent that occurs during the summer. The ice extent in winter is still quite large, but there are indications that the winter ice has been getting younger and thinner[7].

Figure 6.56 plots the northern hemisphere ice area, the cumulative area of ice coverage calculated as a sum over all grid cells with ice concentration ≥ 15% of the product of grid cell area and ice concentration. These data are based entirely on satellite data from 1979 to 2018.

The two things evident in this plot are the large annual cycle in sea ice area and the general decline of the minimum area of the sea ice.

───────

7 A fate the authors wish would happen to them.

Figure 6.57 is a time series of the northern hemisphere sea ice extent anomaly, which is the monthly sea ice extent minus the mean of all the monthly data. Removal of the mean monthly signal helps reduce most of the annual variation, making the variations from the mean conditions more easily seen.

These data show a steep decline in ice extent since the 1980s.

Images of the maximum and minimum sea ice extent in 2018 are shown in Figure 6.58. The line in these plots indicates the median

sea ice extent for the selected day of year during the 20-year period from 1991 to 2010. The maximum sea ice extent in 2018 was the second smallest observed during the 39-year satellite record, while the minimum summer sea ice extent was the sixth lowest on record.

This trend is continuing. Figure 6.59 is a plot of the Arctic sea ice extent as a function of time for the years 1979–2018. Years 1979–2008 are plotted in grey, while years 2009–2018 are plotted in black. Notice that the sea ice extents

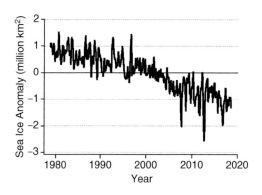

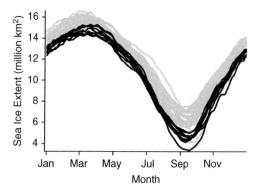

Figure 6.57 Northern hemisphere sea ice extent anomaly. Based on data from NSDIC, Artic Sea news. https://nsidc.org/arcticseaicenews/sea-ice-tools/.

Figure 6.59 Arctic sea ice extent as a function of time of year. Years 1979–2008 are plotted in grey, years 2009–2018 are plotted in black. Based on data from NSDIC, Artic Sea news. https://nsidc.org/arcticseaicenews/sea-ice-tools/.

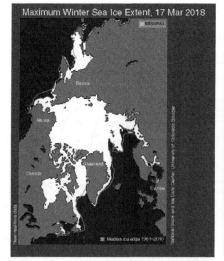

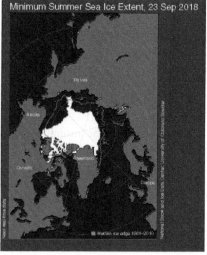

Figure 6.58 Maximum and minimum ice extent in 2018. Source: National Snow and Ice Data Center http://nsidc.org/arcticseaicenews/sea-ice-tools/.

in the last few years are the lowest on record. The absolute summertime sea ice minimum was in 2012. The wintertime sea ice maximum was near an absolute minimum in 2017.

The Arctic ice extent observed in the month of February for the years 1979–2018 is plotted in Figure 6.60. These data indicate the annual maximum extent is declining about 45,800 km^2 per year, which amounts to 3% of the ice extent per decade.

Figure 6.61 shows the ice extent for the month of September from 1979 to 2018 when the ice extent is a minimum. These data indicate a decline of 83,200 km^2 per year, equivalent to about 14% per decade.

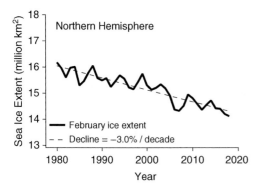

Figure 6.60 Arctic sea ice extent in February for the years 1979–2018. Based on data from NSDIC, Artic Sea news. https://nsidc.org/arcticseaicenews/sea-ice-tools/.

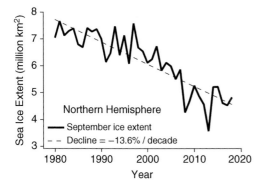

Figure 6.61 Arctic sea ice extent in September for the years 1979–2018. Based on data from NSDIC, Artic Sea news. https://nsidc.org/arcticseaicenews/sea-ice-tools/.

The motion of the ice can be tracked by measuring horizontal displacements of features in the ice from image to image. Continuous, year-to-year tracking of ice motion and coverage has allowed researchers to measure the age of each parcel of ice (Maslanik et al. 2011). Figure 6.62 show the trends of the extent of first-year and multiyear ice for the Arctic from 1985 to 2018. The ice concentration threshold for the data in this figure was 40% of the grid cell, rather than the more typical 15%. This was done to minimize uncertainties in estimating sea ice age, and accounts for the difference in the ice extent values shown in this figure and those in Figure 6.60.

Notice that while the mean ice extent has been declining, the mix of ice ages, and hence ice thickness, has been shifting dramatically. The extent of older ice has been declining at a higher rate than total ice extent. From 1989 to 2019 the percentage of the ice extent older than four years declined from 30% to 2%. And while the total ice extent declined at a rate of about 3% per decade over this time period, the multiyear ice declined at a rate of nearly 15% per decade. This implies the average thickness of the ice cover, and hence the ice volume, has been in sharp decline.

The plots we have seen in this chapter are consistent with the shrinkage of most of the glaciers around the world, the loss of landed ice from Antarctica and Greenland, the rising sea level and increasing atmospheric temperatures. All are signs of climate change and specifically global warming.

The correlation of these unprecedented signals of global warming with the rise in greenhouse gases also leaves no scientific doubt that global warming is being driven by human activity. These are scientific findings that are not open to political debate.

To be complete, we need to point out that the trends in sea ice coverage are very different in the Antarctic. Figure 6.63 shows the mean Antarctic sea ice extent has varied above and below historical values in recent years.

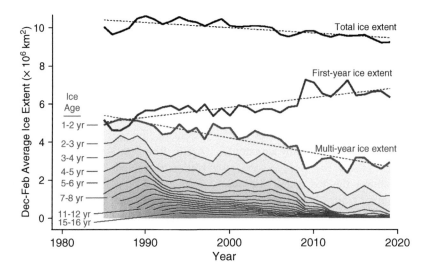

Figure 6.62 Total, first-year and multiyear ice extent for the Arctic from 1985 to 2018, based on a 40% ice concentration threshold. Data from Tschudi et al. (2019).

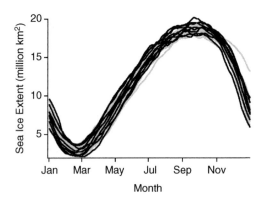

Figure 6.63 Antarctic sea ice extent as a function of time of year. Years 1979–2008 are plotted in grey, years 2009–2018 are plotted in black. Based on data from NSDIC, Artic Sea news. https://nsidc.org/arcticseaicenews/sea-ice-tools/.

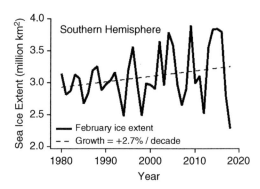

Figure 6.64 Antarctic sea ice extent in February for years 1979–2018. Based on data from NSDIC, Artic Sea news. https://nsidc.org/arcticseaicenews/sea-ice-tools/.

The difference is that a lot of Antarctic sea ice originates from landed ice extending over the sea and then breaking off. So Antarctic sea ice extent is at least partly a manifestation of the increased flow of landed ice into the sea.

Figure 6.64 shows that both the mean and variability of the Antarctic sea ice extent are increasing. The physical basis for these trends is not well understood.

6.9 Microwave Radiometry of Land Surfaces

Three parameters can be retrieved from microwave radiometer observations of vegetation-covered surfaces:

- Surface temperature – T_s.
- Surface soil moisture – w_s (gm cm^{-3}).
- Vegetation water content – w_c (kg m^{-2}).

The estimation of these parameters from measurements of brightness temperature is more challenging than ocean or ice measurements, but these are important parameters for monitoring vegetation.

As before, the microwave brightness temperature observed by a satellite radiometer is the sum of multiple terms:

$$T_B(v, \theta, p) = \tau_A T_{Bs} + \tau_A r_s [T_{sky} + T_{cos} \tau_A]$$
$$+ T_u \qquad (6.40)$$

where:

T_{Bs} = surface brightness temperature
τ_A = atmospheric transmittance = $\exp(-\tau_a)$
T_{sky} = sky brightness temperature
τ_a = atmospheric opacity (optical depth)
T_{cos} = cosmic radiation (2.7 K)
r_s = surface reflectivity (Fresnel equations)
T_u = upwelling atmospheric brightness temperature.

The first term in equation 6.40 is the surface brightness temperature T_{Bs} times the atmospheric transmittance. The second and third terms account for sky radiation T_s and cosmic radiation T_{cos} that are reflected from the surface with reflection coefficient r_s and then attenuated by the atmosphere. The fourth term is the upwelling brightness temperature of the atmosphere T_u, another term for the path radiance.

Note the land surface parameters are incorporated into this equation via the terms for surface brightness temperature, T_{Bs}, and the surface reflectivity, r_s.

We will now consider the individual factors in this equation, beginning with the reflectivity and emissivity of soil. Figure 6.65 plots laboratory measurements of the real and imaginary parts of the dielectric constant for three soils as a function of moisture content at a wavelength of 21 cm. The smooth curves are the predictions of an empirical model.

Data of this type indicate that the best frequencies to use for measuring moisture content are L-band (1.4 GHz as shown in the plot) and C-band (5 GHz). The sensitivity to moisture content is small above 5 GHz.

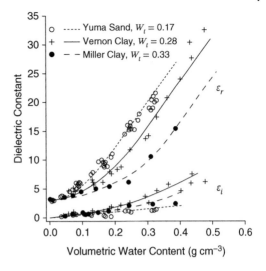

Figure 6.65 Real and imaginary parts of the dielectric constant for three soils as a function of moisture content. From Schmugge (1983) with permission from IEEE.

The Fresnel reflectivities for a *flat surface* are given by the usual expressions:

$$r_v(\theta) = \left| \frac{\tilde{\varepsilon} \cos \theta - \sqrt{\tilde{\varepsilon} - \sin^2 \theta}}{\tilde{\varepsilon} \cos \theta + \sqrt{\tilde{\varepsilon} - \sin^2 \theta}} \right|^2$$

$$r_h(\theta) = \left| \frac{\cos \theta - \sqrt{\tilde{\varepsilon} - \sin^2 \theta}}{\cos \theta + \sqrt{\tilde{\varepsilon} - \sin^2 \theta}} \right|^2 \qquad (6.41)$$

But soil is not flat, so adjustments need to be made.

The dependence of the complex dielectric constant on soil moisture w_s implies that the surface reflectivity and emissivity are also functions of soil moisture, along with being dependent on frequency and polarization. Figure 6.66 shows the measured emissivity of wet and dry soil and grass as measured at L-band (1.4 GHz) with horizontal polarization. Note the wet surfaces are more reflective and less emissive than the dry surfaces.

Figure 6.67 shows the skin depth as a function of soil volumetric water content and frequency for three different soil types. These data indicate that while the skin depth may be large for low-frequencies looking at very dry soil, most microwave radiation will be

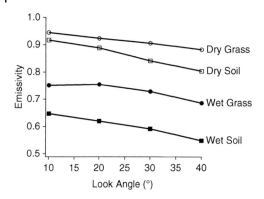

Figure 6.66 Emissivity of wet and dry soil and grass. From Jackson and Schmugge (1991).

emitted from the upper 5 to 50 centimeters of soil.

Now consider the surface brightness temperature, which originates from three sources: soil emissions, upward vegetation emissions, and downward emissions reflected from the surface. We will model each source in turn.

The brightness temperature of the soil emissions can be modeled as the effective emissivity of the rough surface ε_s times the soil temperature times the transmittance through the vegetation. Of course the effective emissivity is just one minus the effective reflectivity.

$$
\begin{aligned}
T_{Be}(T_s, w_s, w_c, \theta) &= \varepsilon_s(\theta) \, T_s \exp(-\tau_c(\theta)) \\
&= [1 - r_s(\theta)] \, T_s \\
&\quad \times \exp(-\tau_c(\theta)) \quad (6.42)
\end{aligned}
$$

where:

T_s = soil temperature
r_s = rough surface reflectivity of soil
τ_c = vegetation optical depth.

A simple empirical model can then be derived for effective reflectivity based on the reflectivity of a smooth surface modified by a parameter $\exp(-h)$ that accounts for surface roughness:

$$
r_s = r_p(\theta) \exp(-h) \quad (6.43)
$$

$$
h = \left(\frac{4\pi\sigma \cos\theta}{\lambda} \right)^2 \quad (6.44)
$$

where:

σ = surface height standard deviation
λ = wavelength
r_p = Fresnel flat surface reflectivity for polarization p.

Typical values of h range from 0 to 0.3. This model for the effect of roughness on reflectivity is derived from an assumption that the surface slopes are Gaussian distributed. The exact form then accounts for the probability density of slopes that can specularly reflect energy towards the downward looking sensor.

For the next terms in the surface brightness formula we need to model surface vegetation. Figure 6.68 shows the optical depth τ_c as a function of vegetation water content for corn, grass, and soybean covered fields. Based on data like these, the optical depth is modeled as

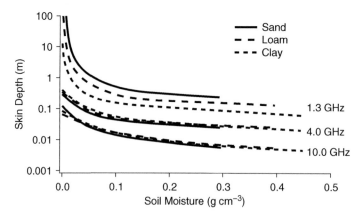

Figure 6.67 Soil skin depth as a function of soil volumetric water content and frequency. From Cihlar and Ulaby (1974).

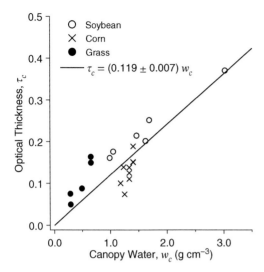

Figure 6.68 Vegetation optical depth measured at L-band. From Schmugge (1983) with permission from IEEE.

a simple linear relationship involving the water content and a secant theta factor accounting for the difference between a vertical and slant path though the vegetation:

$$\tau_c(\theta) = b \, w_c \sec \theta \qquad (6.45)$$

where w_c is the vegetation columnar water content, measured in $kg \, m^{-2}$. The linear coefficient b is then determined for the specific frequency, and in this case has a value of about 0.1.

Figure 6.69 illustrates the three sources of soil brightness temperature: soil emissions, upward vegetation emissions, and downwelling vegetation emissions reflected from the soil.

The simplest model for the upward vegetation emissions is the product of three terms: the vegetation brightness temperature, the canopy emissivity, and the canopy transmissivity:

$$T_{Bu} = (1 - \omega_p)(1 - e^{-\tau_c}) \, T_v \qquad (6.46)$$

where:

ω_p = vegetation single-scattering albedo (typically between 0 and 0.3)
T_v = vegetation canopy temperature (typically close to T_s)
$1 - \exp(-\tau_c)$ = canopy emissivity.

The canopy temperature is usually about the same as the soil temperature. The canopy emissivity is one minus the canopy absorptance, which is given as an exponential of the vegetation optical depth. And the canopy transmissivity is given by one minus the single-scattering albedo. This albedo typically has a value of between 0 and 0.3.

A model for the downward vegetation emissions reflected from surface contains the same three terms because the emissions from the vegetation are assumed to be isotropic:

$$T_{Bd} = (1 - \omega_p)(1 - e^{-\tau_c}) r_s(\theta) \, e^{-\tau_c} \, T_v \quad (6.47)$$

but also includes the soil reflectivity r_s and a factor accounting for the transmissivity of the canopy.

The surface brightness temperature then becomes a simple sum of these three terms:

$$T_{Bs} = T_{Be} + T_{Bu} + T_{Bd} \qquad (6.48)$$

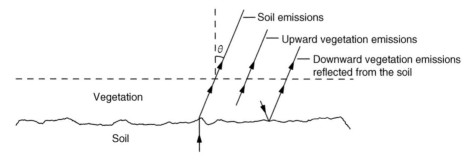

Figure 6.69 Components of the emissions from soil and vegetation. From Schmugge (1983) with permission from IEEE.

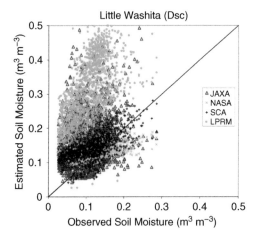
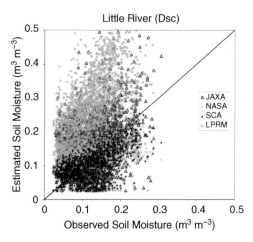

Figure 6.70 Model/data comparisons for AMSR-E soil moisture retrieval algorithms. From Jackson et al. (2010) with permission from IEEE.

Of course all of these terms are functions of soil moisture, vegetation water content, and surface temperature.

The total brightness temperature then becomes:

$$T_B(v, \theta, p) = \tau_A T_{Bs} + \tau_A r'_s [T_{sky} + T_{cos} \tau_A] + T_u \qquad (6.49)$$

where the surface reflectivity is the rough surface reflectivity r'_s, which is then modified by two passes through the vegetation canopy.

After atmospheric correction, the expression for the total brightness temperature we just developed depends on three parameters: soil temperature T_s, soil moisture w_s, and water content in the vegetation w_c. In general, the retrieval algorithms make use of total brightness temperature at two frequencies (6.9 and 10.7 GHz) and two polarizations (H and V). Modeling the measurements, provides four equations and three unknowns.

One approach is to compute the brightness temperatures in equation (6.49) at each frequency and polarization, using best guesses as a starting point for all parameters (Njoku and Li, 1999). The weighted error between the measurements and model predictions are then computed:

$$\chi^2 = \sum_i \left(\frac{T_{Bi}^{obs} - T_{Bi}^{calc}}{\sigma_i} \right)^2 \qquad (6.50)$$

where σ_i is the standard deviation of the measurement noise in the i-th channel. The process is then iterated to minimize the weighted errors.

The expected accuracies for the AMSR-E retrieval algorithms based on prelaunch simulations were about 20% for soil moisture, $0.15 \, \mathrm{kg \, m^{-2}}$ for vegetation water content, and 2°C for soil temperature. As illustrated in Figure 6.70, the actual accuracy has not been quite so good, but improved algorithms are still being developed.

6.10 Atmospheric Sounding

For measurements of water and land, atmospheric path radiance is a nuisance that needs to be removed. We have also discussed how path radiance can be used to estimate integrated atmospheric quantities, such as total

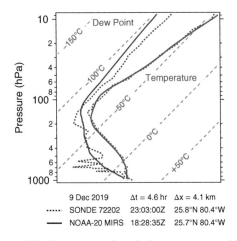
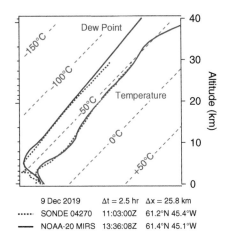

Figure 6.71 Comparison of vertical temperature and humidity profiles derived from atmospheric sounding and radiosondes. The left-hand was taken from Miami, FL, and the right-hand panel from Narsarsuaq, Greenland. Data from NOAA (2019) and Reale et al. (2012).

water vapor. It turns out there is another use of passive microwave radiometry – frequency variations in path radiance can be used to estimate vertical profiles of atmospheric parameters. This is referred to as atmospheric sounding. Some example vertical temperature profiles derived from atmospheric sounding are shown in Figure 6.71.

Figure 6.71 is called a skew-T log-P diagram. The vertical axis is logarithmic in pressure, and the temperatures are plotted on a diagonal axis. Such figures often include other axes indicating dry and moist adiabats or lines of constant entropy. The temperature axis is skewed on this plot so that the dry adiabats are nearly orthogonal to the temperature axis. The adiabat lines are not included on these plots in order to simplify the presentation.

Recall our discussion of the theory of radiative transfer. Figure 6.72 illustrates the spectral radiance $L_\lambda(h)$ seen by a sensor at height h viewing down at a surface with spectral radiance of $L_\lambda(0)$. The tube in this figure indicates the absorbing path between the surface and

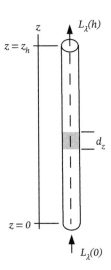

Figure 6.72 Geometry of radiant flux tube. Source: Stewart (1985) with permission from University of California Press.

the sensor. Absorption along this path is characterized by a total absorption coefficient $\kappa(z)$, which consists of contributions from water vapor, CO_2, and ozone.

Consider a small absorbing volume in the path at height z along the path with infinitesimal length dz. The radiance L leaving this volume will be attenuated by an amount dL equal to the absorption coefficient times the radiance entering the volume times the length of the volume dz:

$$dL_\lambda^a(z) = -\kappa(z)L_\lambda(z)dz \qquad (6.51)$$

By Kirchoff's law, the emissivity of the material in the infinitesimal volume element is simply $\kappa(z)dz$. Hence the emitted radiance from the volume element is given by:

$$dL_\lambda^e(z) = \kappa(z)L_{BB}(T(z))dz \qquad (6.52)$$

where $T(z)$ is the temperature of the gas at height z above the surface.

Then the net change in radiance from the volume element is simply the emissions minus the attenuation:

$$dL_\lambda = \kappa(z)L_{BB}(T(z))dz - \kappa(z)L_\lambda(z)dz \qquad (6.53)$$

This is a differential equation with the solution:

$$L_\lambda(h) = L_\lambda(0)\exp\left[\tau(0,h)\right]$$
$$+ \int_0^h \kappa(z)L_{BB}(T(z))$$
$$\times \exp\left[\tau(z,h)\right]dz \qquad (6.54)$$

where $\tau(z_1,z_2) = \int_{z_1}^{z_2}\kappa(z')dz'$ is the optical thickness or optical depth of the layer between z_1 and z_2. The first term is the surface radiance attenuated by the total path and the second term is the integral of the attenuated emissions over the path.

We can convert the previous equation into brightness temperatures:

$$T(h) = T(0)\exp[-\tau(0,h)]$$
$$+ \int_0^h \kappa(z)\,T(z)$$
$$\times \exp[-\tau(z,h)]dz \qquad (6.55)$$

This expression determines what part of the atmosphere contributes to the measured brightness temperature.

Next define a weighting function $W(z)$ corresponding to two of the three terms in the integrand for the path radiance:

$$W(z) = \kappa(z)\exp[-\tau(z,h)]$$
$$= \kappa(z)\exp\left[-\int_0^h \kappa(z')dz'\right]$$
$$\qquad (6.56)$$

According to its definition, the optical depth can be expressed in terms of an integral over altitude of the absorption coefficient κ.

With this definition of the weighting function, the atmospheric brightness temperature then becomes a simple integral over altitude of the weighting function times the temperature:

$$T_a(h) = \int_0^h W(z)\,T(z)\,dz \qquad (6.57)$$

Thus the shape of the weighting function given in equation 6.56 tells us the relative contribution of various altitudes in the atmosphere to the measured brightness temperature.

In the simplest model of the atmosphere, κ is an exponential function of altitude $\kappa(z) = a_0 e^{-z/H}$, where a_0 is a fixed scale factor, and H is a scale height, which is approximately 8 km for oxygen. Substituting this formula into the definition of the weighting function yields the following expressions for the weighting function:

$$W(z) = a_0 \exp\left[-\frac{z}{H} - a_0 H e^{-z/H}\right]$$
$$= a_0 \exp\left[-\frac{z}{H} - \tau_m e^{-z/H}\right] \qquad (6.58)$$

where $\tau_m = a_0 H$. Note that a_0 is a frequency dependent scale factor, so the temperature weighting function is dependent on frequency and depth in the atmosphere.

The four plots in Figure 6.73 show the relative behavior of the temperature weighting function $W(v,z)$ for an exponentially-decaying atmosphere and different values of τ for a downward looking sensor. The left two curves indicate that for values of τ less than one, the brightness temperature is predominantly sensitive to temperature near the surface. The right two curves show that temperatures higher up in the atmosphere can be sensed for values of τ greater than one.

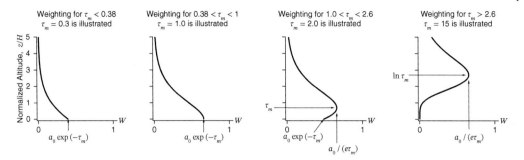

Figure 6.73 Example temperature weighting functions $W(\nu, z)$ for an exponentially decaying atmosphere. Adapted from Elachi and van Zyl (2006).

So the temperature weighting function limits the range of altitudes that contribute to the observed brightness temperature and the four plots in Figure 6.73 indicate that the altitude range to be sampled varies with frequency via $\tau_m(f)$.

Thus to measure a variety of altitudes we need to make measurements at a variety of values of τ. There are three reasons why brightness temperature measurements are made at various frequencies on the edge of an oxygen absorption band: the absorption characteristics are well known, the large gradient in absorption at the edge of the band provides a wide range of sampling depths in the atmosphere, and the ground radiance is negligible where the absorption is high.

Figure 6.74 shows how atmospheric sounding works for the AMSU-A and Microwave Humidity Sounder (MHS) sensors. The left-hand plot shows the specific zenith opacity (τ) for oxygen and water vapor. AMSU-A channels 3 to 14 are closely spaced along the lower edge of the oxygen absorption band centered at 60 GHz. The middle plot shows the temperature weighting functions determined by application of radiative transfer theory to a standard atmosphere. It is clear that each specific channel has a different opacity, and hence samples temperature at different altitudes in the atmosphere. This plot illustrates just how closely spaced the bands need to be in order to make measurements sensitive to different altitudes in the atmosphere. The right-hand

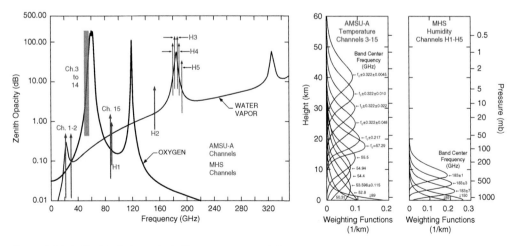

Figure 6.74 AMSU/MHS channels (left), temperature weighting functions (middle), and humidity weighting functions (right). Left-hand panel from Robel (2009). Middle and right-hand panels. Source: Karbou et al. (2005) with permission from American Geophysical Union (AGU).

plot shows the humidity weighting functions for the 180-GHz channels.

Data from the AMSU-A channels can be inverted in a tomographic algorithm to estimate the most likely atmospheric temperature profiles consistent with the measurements. Likewise, the similar MHS data acquired along the edges of the water vapor absorption band can be used to estimate vertical profiles of atmospheric water vapor.

References

Alishouse, J. C. (1975). *A summary of the radiometric technology model of the ocean surface in the microwave region.* NOAA.

Alishouse, J. C., Snider, J. B., Westwater, E. R., Swift, C. T., Ruf, C. S., Snyder, S. A., et al. (1990). Determination of cloud liquid water content using the SSM/I. *IEEE Transactions on Geoscience and Remote Sensing, 28*(5), 817–822.

Alishouse, J. C., Snyder, S. A., Vongsathorn, J., & Ferraro, R. R. (1990). Determination of oceanic total precipitable water from the SSM/I. *IEEE Transactions on Geoscience and Remote Sensing, 28*(5), 811–816.

Carsey, F. D. (ed.) (1992). *Microwave remote sensing of sea ice.* AGU Monograph 68. Washington, DC: American Geophysical Union.

Cavalieri, D. J., Parkinson, C. L., Gloersen, P., & Zwally, H. J. (1996). Sea ice concentrations from Nimbus-7 SMMR and DMSP SSM/I-SSMIS passive microwave data, version 1. NASA National Snow and Ice Data Center Distributed Active Archive Center, Boulder, CO, https://nsidc.org/data/NSIDC-0051/versions/1.

Cihlar, J., & Ulaby, F. T. (1974). *Dielectric properties of soils as a function of moisture content.* NASA.

Crane, R. K. (1971). Propagation phenomena affecting satellite communication systems operating in the centimeter and millimeter wavelength bands. *Proceedings of the IEEE, 59*(2), 173–188.

Elachi, C., & Van Zyl, J. J. (2006). *Introduction to the physics and techniques of remote sensing.* John Wiley & Sons.

Gaiser, P. W., Bettenhausen, M., Jelenak, Z., Twarog, E. M., & Chang, P. (2005).

WindSat – Space Borne Remote Sensing of Ocean Surface Winds. NOAA/NASA Workshop on Ocean Vector Winds, 8–10 February, Miami, FL.

Gaiser, P. W., St Germain, K. M., Twarog, E. M., Poe, G. A., Purdy, W., Richardson, D., et al. (2004). The WindSat spaceborne polarimetric microwave radiometer: Sensor description and early orbit performance. *IEEE Transactions on Geoscience and Remote Sensing, 42*(11), 2347–2361.

Goodberlet, M. A., Swift, C. T., & Wilkerson, J. C. (1990). Ocean surface wind speed measurements of the Special Sensor Microwave/Imager (SSM/I). *IEEE Transactions on Geoscience and Remote Sensing, 28*(5), 823–828.

Hollinger, J. P., & Lo, R. C. (1983). *SSM/I (Special Sensor Microwave/Imager) project summary report.* Naval Research Laboratory, Washington.

Hollinger, J. P., Peirce, J. L., & Poe, G. A. (1990). SSM/I instrument evaluation. *IEEE Transactions on Geoscience and Remote Sensing, 28*(5), 781–790.

Jackson, T. J., & Schmugge, T. J. (1991). Vegetation effects on the microwave emission of soils. *Remote Sensing of Environment, 36*(3), 203–212.

Jackson, T. J., Cosh, M. H., Bindlish, R., Starks, P. J., Bosch, D. D., Seyfried, M., et al. (2010). Validation of advanced microwave scanning radiometer soil moisture products. *IEEE Transactions on Geoscience and Remote Sensing, 48*(12), 4256–4272.

Jones, W. L., Park, J. D., Soisuvarn, S., Hong, L., Gaiser, P. W., & St. Germain, K. M. (2006). Deep-space calibration of the WindSat radiometer. *IEEE Transactions on*

Geoscience and Remote Sensing, 44(3), 476–495.

Karbou, F., Aires, F., Prigent, C., & Eymard, L. (2005). Potential of Advanced Microwave Sounding Unit-A (AMSU-A) and AMSU-B measurements for atmospheric temperature and humidity profiling over land. *Journal of Geophysical Research: Atmospheres, 110*(D7).

Kawanishi, T., Sezai, T., Ito, Y., Imaoka, K., Takeshima, T., Ishido, Y., et al. (2003). The Advanced Microwave Scanning Radiometer for the Earth Observing System (AMSR-E), NASDA's contribution to the EOS for global energy and water cycle studies. *IEEE Transactions on Geoscience and Remote Sensing, 41*(2), 184–194.

Maslanik, J., Stroeve, J., Fowler, C., & Emery,W. (2011). Distribution and trends in Arctic sea ice age through spring 2011. *Geophysical Research Letters, 38*(13).

Meissner, T., & Wentz, F. J. (2004). The complex dielectric constant of pure and sea water from microwave satellite observations. *IEEE Transactions on Geoscience and Remote Sensing, 42*(9), 1836–1849.

Njoku, E. G., & Li, L. (1999). Retrieval of land surface parameters using passive microwave measurements at 6-18 GHz. *IEEE Transactions on Geoscience and Remote Sensing, 37*(1), 79–93.

NOAA. (2019). *NOAA Products Validation System*. Retrieved from https://www.star .nesdis.noaa.gov/smcd/opdb/nprovs/ (accessed December 20, 2019).

Reale, T., Sun, B., Tilley, F. H., & Pettey, M. (2012). The NOAA Products Validation System (NPROVS). *Journal of Atmospheric and Oceanic Technology, 29*(5), 629–645.

Robel, J. (2009). NOAA KLM users guide. National Environmental Satellite, Data, and Information Service.

Schmugge, T. J. (1983). Remote sensing of soil moisture: Recent advances. *IEEE Transactions on Geoscience and Remote Sensing, GE-21*(3), 336–344.

Schwarz, J., & Weeks, W. F. (1977). Engineering properties of sea ice. *Journal of Glaciology, 19*(81), 499–531.

Stewart, R. H. (1985). *Methods of satellite oceanography*. University of California Press.

Swift, C. T. (1980). Passive microwave remote sensing of the ocean: A review. *Boundary-Layer Meteorology, 18*(1), 25–54.

Tschudi, M., Meier, W. N., Stewart, J. S., Fowler, C., & Maslanik, J. (2019). EASE-Grid Sea Ice Age, Version 4. doi: https://doi.org/10.5067/UTAV7490FEPB (accessed December 2019).

Ulaby, F. T., Moore, R. K., & Fung, A. K. (1981). *Microwave remote sensing: Active and passive. Vol. I. Microwave remote sensing fundamentals and radiometry*. Reading, MA: Addison-Wesley.

Ulaby, F. T., Moore, R. K., & Fung, A. K. (1982). *Microwave remote sensing: Active and passive. Vol. II. Radar remote sensing and surface scattering and emission theory*. Reading, MA: Addison-Wesley.

NASA. (2001). Retrieved from https://directory .eoportal.org/web/eoportal/satellite-missions/ d/dmsp-block-5d

Walsh, J. E., Chapman, W. L., Fetterer, F., & Stewart, J. S. (2019). Gridded Monthly Sea Ice Extent and Concentration, 1850 Onward, Version 2. National Snow and Ice Data Center, Boulder, CO, doi: https://doi.org/10.7265/jj4s-tq79.

Weger, E. (1960). Apparent sky temperatures in the microwave region. *Journal of Meteorology, 17*(2), 159–165.

Wilheit, T. (1972). The Electrically Scanning Microwave Radiometer (ESMR) experiment. *The Nimbus-5 users guide.*

Wilheit, T. T. (1979). A model for the microwave emissivity of the ocean's surface as a function of wind speed. *NASA Technical Memorandum*, 80278.

Yueh, S. H. (1997). Modeling of wind direction signals in polarimetric sea surface brightness temperatures. *IEEE Transactions on Geoscience and Remote Sensing, 35*(6), 1400–1418.

7

Radar

7.1 Radar Range Equation

While the first half of this book is devoted to passive remote sensing, the second half is devoted to active remote sensing starting with various forms of radar. At their core, radars are simple devices. Referring to Figure 7.1, radars transmit pulses out of an antenna pointing at the object to be sensed, which is in this case the surface. The area illuminated on the surface is determined by the antenna beam pattern.

The energy in each pulse scatters off the surface in all directions. If it is the sea surface being illuminated, then most of the energy scatters forward according to Snell's Law, and only some of the energy is scattered back towards the radar. This latter energy is referred to as backscatter.

The backscattered power is received at the antenna, amplified and detected within the radar receiver, then processed and transmitted to the ground. As in all remote sensing systems, the extraction of geophysical information from the radar data then depends on our understanding of how the backscattered signal is affected by the surface properties.

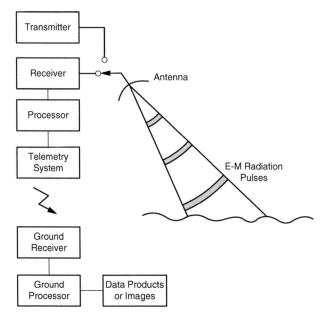

Figure 7.1 Block diagram for radar remote sensing.

Remote Sensing Physics: An Introduction to Observing Earth from Space, Advanced Textbook 3, First Edition.
Rick Chapman and Richard Gasparovic.
© 2022 American Geophysical Union. Published 2022 by John Wiley & Sons, Inc.
Companion website: www.wiley.com/go/chapman/physicsofearthremotesensing

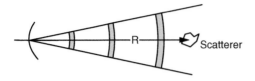

Figure 7.2 Radar scatterer in the main beam.

Let us begin to derive the radar equation. As illustrated in Figure 7.2, consider an isolated scatterer illuminated by a radar beam from an antenna at distance R. The illuminated area at a distance R from the antenna is:

$$A_i = \Omega_{FOV}^{MB} R^2 \qquad (7.1)$$

where Ω_{FOV}^{MB} is the solid angle subtended by the main beam of the antenna.

We define the antenna directivity D to be the ratio of 4π steradians to the solid angle subtended by the main beam:

$$D = \frac{4\pi}{\Omega_{FOV}^{MB}} \qquad (7.2)$$

As follows from the previous discussion of antenna gain in Section 6.3.1, the antenna gain G is the product of the aperture efficiency ϵ_{ap} and the directivity: $G = \epsilon_{ap} D$. This means that the gain is large for an antenna with a narrow beam, while the gain would be one for a nonrealizable omnidirectional antenna with perfect efficiency. Aperture efficiency for a typical horn antenna varies from 0.4 to 0.8, but can be much lower for other types of antennas.

The power density measured at the scatterer is then given by the transmitted power divided by the illuminated area:

$$PD_{scatt} = \frac{P_t}{\Omega_{FOV}^{MB} R^2} = \frac{P_t G}{4\pi R^2} \qquad (7.3)$$

which works out to be the transmit power times the antenna gain divided by $4\pi R^2$. In this formula, P_t is the transmitted power. This can be the peak transmit power to compute the peak power density at the scatterer, or the average transmit power to compute the average power density.

We previously discussed the plots in Figure 7.3 showing the different forms of antenna beam pattern. Just a reminder that

while the gain of an antenna may be high, the beam patterns are often shown in a normalized form. Remember that only a nonrealizable omnidirectional antenna would have a peak gain of one.

The power reflected from the scatterer back toward the radar is the product of a constant, σ, that has units of area, and the power density at the scatterer:

$$P_{ref} = \sigma PD_{scatt} \qquad (7.4)$$

We call σ the backscatter or radar cross-section, often abbreviated RCS.

Just to get you calibrated, it turns out that the RCS of a sphere of radius r is simply the projected area of the sphere πr^2, at least as long as the wavelength is much less than the radius[1]. Yet for other simply shaped objects (e.g. flat plates, cubes, trihedrals) the RCS may be much larger or smaller than the physical cross-section depending on the shape and orientation of the object.

The scattered power density back at the radar's antenna is then the transmit power times the antenna gain divided by one factor of $4\pi R^2$ to account for the one-way spreading loss to the target, times the RCS of the target, divided by one more factor of $4\pi R^2$ to account for the one-way spreading loss back to the radar:

$$PD_{ant} = \frac{P_{ref}}{4\pi R^2} = \frac{\sigma PD_{scatt}}{4\pi R^2} = \frac{P_t G \sigma}{(4\pi R^2)^2} \qquad (7.5)$$

The power received by the antenna is then given by equation (7.5) times the effective receiving area of the radar's antenna, A_e:

$$P_{rec} = \frac{P_t G \sigma A_e}{(4\pi R^2)^2} \qquad (7.6)$$

---

1 All forms of the radar equation are simple products of a number of terms. To simplify calculations, radar engineers often work in logarithmic units, converting the radar equation into a sum. In the logarithmic form, power is expressed as decibels relative to 1 watt, which is written dBW = $10\log_{10}(P_t)$. Gain is expressed as decibels relative to an isotropic antenna, which is written dBi = $10\log_{10}(G)$. And RCS is expressed as decibels relative to one square meter, which is written: dBsm = $10\log_{10}(\sigma)$.

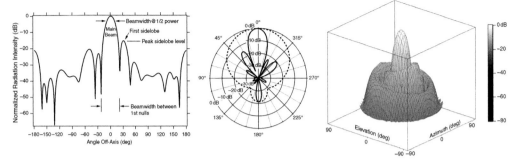

Figure 7.3 Example antenna beam patterns.

Equation (7.6) is the simplest form of the radar equation. It describes how much power is received by the radar based on parameters determined by the radar design (transmit power P_t, antenna gain G, and antenna effective area A_e), the scatterer properties (RCS, σ), and the distance from the radar to the scatterer (R).

Radars often use the same antenna for transmit and receive. Such a radar is called a monostatic radar. Now let us relate the antenna gain and the antenna effective area.

For a narrow beamwidth circular dish with diameter d, the illuminated solid angle is approximately the angular FOV squared:

$$\Omega_{FOV}^{MB} \cong \frac{\pi}{4}(\delta\theta)^2 \cong \frac{\pi}{4}\left(\frac{\lambda}{d}\right)^2 \cong \frac{\pi^2}{16}\frac{\lambda^2}{A} \tag{7.7}$$

where A is the physical area of the antenna aperture. For a narrow beamwidth rectangular antenna the illuminated solid angle is approximately the product of the angular FOV's in the azimuth and elevation directions:

$$\Omega_{FOV}^{MB} \cong \delta\theta_{az}\,\delta\theta_{el} \cong \left(\frac{\lambda}{d_{az}}\right)\left(\frac{\lambda}{d_{el}}\right) \cong \frac{\lambda^2}{A} \tag{7.8}$$

Strictly speaking, equations (7.7) and (7.8) are only valid if the antenna aperture is uniformly illuminated. Many antennas are designed with aperture illumination that is reduced towards the edges of the aperture. This tapered illumination has the impact of reducing antenna sidelobes at the cost of increasing the solid

angle subtended by the main beam. For simplicity, only uniformly-illuminated apertures are consider here.

The effective area of the antenna is related to the physical area of the antenna aperture and the aperture efficiency by: $A_e = A\epsilon_{ap}$. Recalling the definition of antenna gain ($G = \epsilon_{ap}D = 4\pi\epsilon_{ap}/\Omega_{FOV}^{MB}$) we obtain an expression for the effective area of an antenna in terms of its gain and the radar's wavelength:

$$A_e = \frac{\lambda^2}{4\pi}G \tag{7.9}$$

Using this expression for the effective antenna area, the radar equation for a monostatic radar becomes:

$$P_{rec} = \frac{P_t G^2 \lambda^2 \sigma}{(4\pi)^3 R^4} \tag{7.10}$$

which is a bit more common form than the first version we derived.

When the antenna illuminates multiple isolated scatterers, the radar equation simply becomes the sum of the received powers from each of the scatters:

$$P_{rec} = \frac{P_t \lambda^2}{(4\pi)^3} \sum_i \frac{G_i^2 \sigma_i}{R_i^4} \tag{7.11}$$

So far we have treated the scatterer as some isolated object. Things change a bit when illuminating an extended target such as land or the sea surface. In these cases, the scatterer cross-section is replaced by the scattering cross-section per unit surface area. This is a dimensionless quantity known as the normalized radar cross-section or NRCS and is designated σ_0. To get back to radar cross-section, we simply multiply σ_0 by the illuminated area. In this case, the received power becomes:

$$P_{rec} = \frac{\lambda^2}{(4\pi)^3} \int \frac{P_t G^2 \sigma_0}{R^4} dA \qquad (7.12)$$

NRCS is often expressed as $10\log_{10}(\sigma_0)$. To be precise we could refer to σ_0 values as "dB relative to $1\,m^2$ of RCS per $1\,m^2$ of area", but they are most commonly referred to as just "dB".

Notice that σ_0 is the only parameter in the radar equation that contains information about the scene – it plays a role similar to that of scene reflectivity in passive remote sensing. Geophysical information extracted from the mean backscatter power received by an antenna must be encoded somehow in the normalized radar cross-section or other properties of the radar signal, such as the return pulse waveform, Doppler frequency of the return pulse, or return pulse time delay.

It is worth looking at some quick examples of antenna gains. Consider the antenna gain for a 1.0-m diameter dish at 10 GHz, which equals a 3-cm wavelength. Antenna directivity was defined to be: $D = 4\pi/\Omega_{FOV}^{MB}$ and the solid angle of the antenna main beam is: $\Omega_{FOV}^{MB} = \frac{\pi}{4}\delta\theta_{az}\,\delta\theta_{el} = \frac{\pi}{4}\left(\frac{\lambda}{d}\right)^2$, where d is the antenna diameter. Substitution then yields:

$$D = \frac{16d^2}{\lambda^2} \qquad (7.13)$$

Assuming an aperture efficiency of 0.5, the gain for the 1-m dish at 10 GHz is thus:

$$G(dB) = 10\log_{10}\epsilon_{ap} + 10\log_{10} D$$
$$= 10\log_{10}\epsilon_{ap} + 10\log_{10}(16)$$
$$+ 20\log_{10}(d) - 20\log_{10}(\lambda)$$
$$= 39.5\,dB \quad \rightarrow \quad G = 10^{3.95} = 8912 \qquad (7.14)$$

which is the gain relative to an isotropic antenna.

Be careful not to mix dB units and regular units in your calculations. Radar engineers will often do the entire radar equation calculation in dB so all of the terms will all add or subtract. But if you are working in linear units then the gain of this 39.5 dB antenna would be about 8900.

It is interesting to apply the same formula to compute the gain of a large optical telescope. After all, telescopes measure electromagnetic waves that obey the same properties as microwaves, so there is nothing stopping us from doing such a calculation.

The Mount Palomar telescope has a diameter of 5 m, and we will assume a wavelength of 500 nm, a value in the middle of the visible band. Here we have used a factor of 1.22 that arises from a more detailed analysis of the gain arising from a filled circular aperture:

$$\Omega_{FOV}^{MB} = \frac{\pi}{4}(\delta\theta)^2 = \frac{\pi}{4}\left(\frac{1.22\lambda}{d}\right)^2 \qquad (7.15)$$

$$G = \frac{16d^2}{1.22^2\,\lambda^2} = \frac{16 \times 5^2}{1.22^2\,(500 \times 10^{-9})^2}$$
$$= 1.1 \times 10^{15} \qquad (7.16)$$

In the end the gain of the Mount Palomar telescope is an amazing 150.3 dB!

7.2 Radar Cross-Section

We previously mentioned that the RCS of a sphere of radius r is πr^2, as long as the wavelength is much less than the radius. Figure 7.4 shows the actual radar cross-section of a perfectly conducting sphere as a function of the size of the sphere.

The units for this plot have been nondimensionalized by dividing the RCS on the vertical axis by πr^2, and by dividing the sphere radius on the horizontal axis by λ.

The RCS of a sphere breaks down into three regions. For very small spheres, there is the Rayleigh limit where the RCS is proportional to (r^6/λ^4). The nature of the scattering changes when the circumference of the sphere becomes comparable to the wavelength of the radiation.

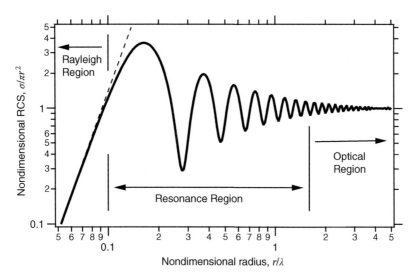

Figure 7.4 RCS of a sphere of radius r in nondimensional units.

Resonances can then occur with surface waves propagating around the circumference of the sphere. The RCS is oscillatory in this regime where (r/λ) ranges from 0.1 to about 2. This is the resonant or Mie scattering regime. The oscillations in the RCS dampen at values of (r/λ) greater than 2, approaching the value of πr^2. This is referred to as the optical regime.

Figure 7.5 shows the radar cross-section of a large naval vessel, which is not a sphere. The measurements were made at a frequency of 9.2 GHz with horizontal polarization. Constructive and destructive interference of bright targets on the ship cause the RCS to vary with minor changes in look direction. So three curves capture the median RCS and the RCS of the twentieth and eightieth percentile.

The possibly surprising thing here is that the peak RCS, which occurs when the ship is viewed broadside, is about 64 dBsm, which is about 2.5 km^2. It turns out that ships are constructed of many flat plates and corners that are far better radar reflectors than a sphere. So much so that it would take a 1600-m diameter sphere to equal the RCS of this ship.

We will end this section by examining the normalized RCS of a few natural surfaces as a function of frequency, incident angle, and polarization. We have seen multiple examples of the reflection coefficient and emissivity of a natural surface depending on polarization, where polarization refers to the orientation of the plane of the electric field with respect to the sensor. Passive radiometers can be oriented so that the polarization is defined as the orientation of the received electric field relative to the Earth's surface. In this case, the electric field is parallel to the surface for horizontal polarization and perpendicular to the surface for vertical polarization.

Polarization is more complicated for an active sensor like radar. The radar can transmit a signal of one polarization and receive the returned signal on one or more channels using either the same or different polarization. The signal arising from a radar that transmits a horizontally-polarized signal and receives a horizontally-polarized signal is referred to as HH. A vertically-polarized transmit

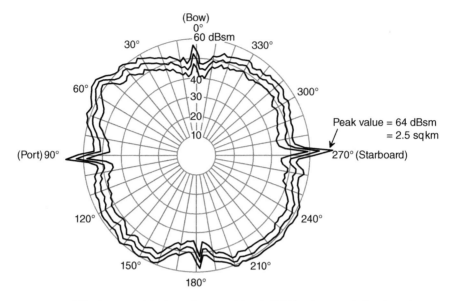

Figure 7.5 Example RCS of a large ship. From Eaves and Reedy (2012) with permission from Springer Nature.

and vertically-polarized received channel is referred to as VV, while a vertically-polarized transmit and horizontally-polarized received channel is referred to as VH. HH and VV polarization channels are referred to as copolarization, because the transmit and receive channels utilize the same polarization, while VH and HV channels are referred to as cross-polarization.

Polarization can be a powerful tool in discriminating different types of scatterers, so the RCS in the following plots are described in terms of their polarization. A more complete description of polarization, including the use of phase measurements between polarizations, is provided in Section 6.4.

Figure 7.6 contains plots of various statistical measures of σ_0 (horizontal polarization transmit and receive) over vegetated surfaces as a function of frequency for three different incident angles. These plots show that the variability of σ_0 is large, especially for nadir viewing, and that σ_0 decreases significantly with increasing incident angle. Here $\bar{\sigma}$ is the mean RCS, σ_{50} the median, and σ_5 and σ_{95} are the 5th and 95th percentiles, respectively.

Figure 7.7 contains other plots of the normalized RCS of vegetated surfaces, but this time as a function of incident angle for four different frequencies. The sharp falloff in NRCS with incident angle is clearly shown. Note that σ_0 increases with increasing frequency.

A similar pattern can be seen in the normalized RCS of snow, which increases at higher frequencies and decreases at larger incident angles (Figure 7.8).

Finally we come to the normalized RCS of the sea surface. Setting aside the L-band (1.2 GHz) data, Figure 7.9 shows that over the middle range of possible incident angles the normalized RCS remains relatively constant over a wide range of frequencies. Note that when the radar polarization is referred to as simply V or H-polarization, as in this figure,

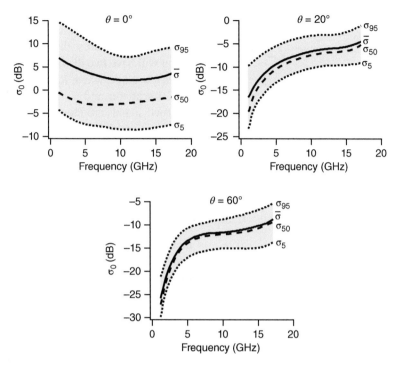

Figure 7.6 Normalized HH-polarized RCS of vegetated surfaces as a function of frequency. Adapted from Ulaby et al. (1982).

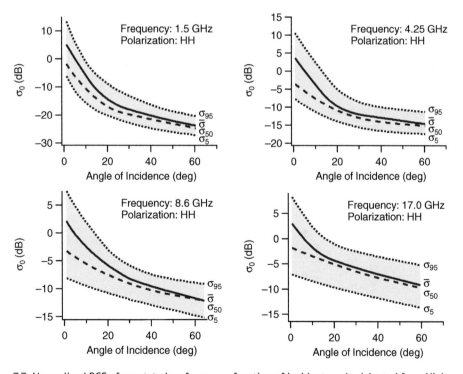

Figure 7.7 Normalized RCS of vegetated surfaces as a function of incident angle. Adapted from Ulaby et al. (1982).

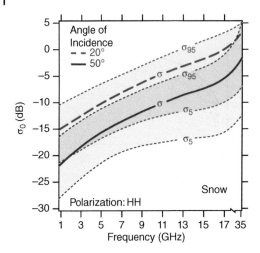

Figure 7.8 Normalized RCS of snow. Adapted from Ulaby et al. (1982).

you should just assume the measurement was made using copolarization.

Figure 7.10 shows the normalized RCS at 4.5 GHz decreases at larger incident angles. In the case of the sea surface the backscatter can be broken into at least three regimes based on the physical characteristics of the surface that are driving the backscatter. We will discuss the details of these regimes in Sections 8.2 and 9.2.

7.3 Radar Resolution

Consider a radar that transmits a single pulse at time t_0 when there are two point scatterers in the radar beam, as shown in Figure 7.11.

We will designate the duration or pulse width of the transmitted pulse as τ.

If the scatterers are well separated, the return signal will have two distinct pulses arriving at times $t_i = 2R_i/c$, where R_i is the slant range to the i-th scatterer, and c is the speed of light. The returned signals are illustrated in Figure 7.12. The duration of the signal returned from each scatterer will be the pulse width τ.

Now allow scatterer #2 to be moved closer to the antenna until the return signal from scatterer #2 begins just as the return from scatterer #1 ends. In other words, move the scatterers closer until the two return pulses start to merge. At this range R_2, we have $t_2 = t_1 + \tau$, and R_2 ends up being $R_1 + c\tau/2$. In this case, the return signal would appear as a single pulse of duration 2τ, and the radar operator would not be able to distinguish between the presence of two point scatterers or the return from a single target with a range extent of $c\tau$.

Slant range resolution δ_{sr} is defined as the minimum scatterer separation in the line of sight of the radar needed for the return from one scatterer to end before the return from a second scatterer begins. From the equation above, it follows that: $\delta_{sr} = R_2 - R_1 = c\tau/2$. Thus the slant range resolution is determined only by the duration of the transmitted pulse. And we need short pulses to get good resolution. As an example, the speed of light is $0.3\,\text{m ns}^{-1}$. So a 2-nanosecond pulse has a slant range resolution of 30 cm.

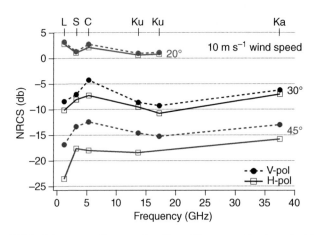

Figure 7.9 Normalized RCS of the sea surface as a function of frequency for three incident angles. Based on data from Unal et al. (1991) and the 35-GHz model of Yurovsky et al. (2017).

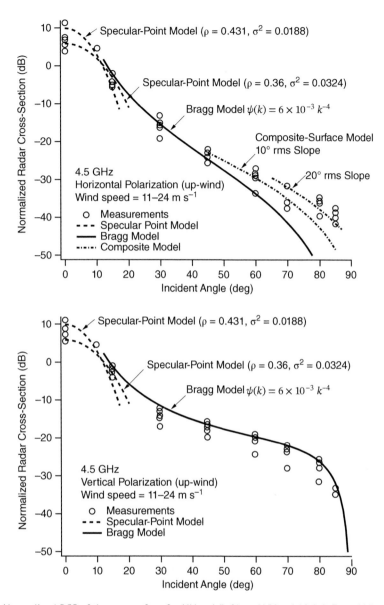

Figure 7.10 Normalized RCS of the sea surface for HH-pol (left) and VV-pol (right). From Valenzuela (1978) with permission from Springer Nature.

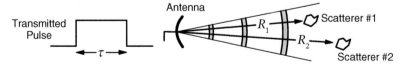

Figure 7.11 Two point scatterers in a radar beam.

Slant range resolution is measured along the radar line of sight, but we often want to know the effective resolution of a radar along the ground. Consider the two scatterers located on the ground shown in Figure 7.13. They are separated in slant range by a distance equal to the slant range resolution δ_{sr}. It can be seen from the geometry that the ground resolution δ_g equals the slant range resolution divided by $\sin\theta$:

$$\sin\theta = \frac{\delta_{sr}}{\delta_g} \quad \rightarrow \quad \delta_g = \frac{\delta_{sr}}{\sin\theta} = \frac{c\tau}{2\sin\theta}$$

(7.17)

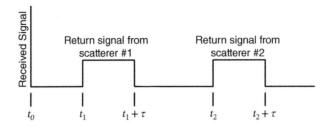

Figure 7.12 Radar returns from two scatterers.

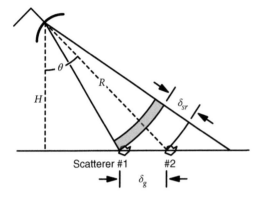

Figure 7.13 Relationship between slant range and ground range resolution.

Notice that the ground resolution depends on the pulse width and the incident angle θ of the antenna line of sight.

As the angle of incidence approaches 90°, the ground resolution becomes approximately the same as the slant range resolution. Conversely, as the incident angle becomes small, the ground resolution becomes large until at nadir, it becomes undefined as both scatterers are at the virtually the same range.

As an example, at an incident angle of 30°, $\delta_g = 2\delta_{sr}$, so a 1-m slant range resolution equals a 2-m ground resolution.

Now let us consider the resolution in the cross-range or azimuth direction. The diagram in Figure 7.14 shows the antenna beam in the azimuth or cross-range direction as viewed from above. The return signals from all scatterers located along an arc at constant range R will arrive at the same time and will be unresolved.

The extent of the resolution cell in the azimuth direction δ_{az} is determined by the extent of the antenna beam in the azimuth direction, which is the slant range times the azimuth beamwidth θ_{az}, where the slant range is given by the antenna height divided by cosine of the incident angle θ:

$$\delta_{az} = R\theta_{az} = \frac{H\theta_{az}}{\cos\theta} \qquad (7.18)$$

where, as before, H is the height of the antenna above the ground. Note this calculation has been made assuming a flat Earth. In general the ranges will be larger for off-nadir viewing on a spherical Earth as can be computed from equations (2.24)–(2.27).

In previous discussions we showed that the antenna beamwidth is related to the radar wavelength λ and the length of the antenna in the direction of interest, namely $\theta_{az} = \lambda/D_{az}$, where D_{az} is the antenna dimension in the azimuth direction.

Combining the above yields this expression for the azimuth extent of the resolution cell:

$$\delta_{az} = \frac{H\lambda}{D_{az}\cos\theta} \qquad (7.19)$$

Thus the azimuth extent of the resolution cell, referred to as the azimuth resolution, is determined by antenna height, radar wavelength, the azimuth dimension of the antenna, and the incident angle of the line of sight.

Let us look at two examples. In each case we consider an X-band radar with a wavelength of 3 cm transmitting 3.33-ns pulses from a circular dish antenna with a diameter of 1.0 m. The antenna line of sight is 30° from nadir. We can

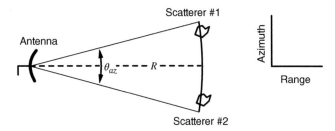

Figure 7.14 Azimuth resolution.

then apply the previous equations:

$$\delta_{sr} = c\tau/2 \tag{7.20}$$

$$\delta_g = \frac{c\tau}{2\sin\theta} \tag{7.21}$$

$$\delta_{az} = \frac{H\lambda}{D_{az}\cos\theta} \tag{7.22}$$

to compute values of the slant, ground and azimuth resolutions.

In the first example, for an aircraft platform at an altitude of 5 km, we get a slant range resolution of 0.5 m, a ground resolution of 1.0 m, and an azimuth resolution of 173 m.

For the second example, consider the same radar mounted on a satellite orbiting at an altitude of 800 km. While the altitude makes no difference for the slant range or ground resolutions, the azimuth resolution for the satellite example is 27.7 km![2] Obtaining azimuthal resolution from space seems challenging.

Next consider how large an antenna would be required to have an azimuth resolution of 1.0 m at X-band from the satellite at an altitude of 800 km looking down at 30° from nadir. Rearranging equation (7.22) we obtain:

$$D_{az} = \frac{H\lambda}{\delta_{az}\cos\theta} = 27.7\text{ km} \tag{7.23}$$

Thus achieving 1.0-m resolution from space would require an antenna that was 27.7 km long! It seems unlikely that an antenna this long would physically fit onto a satellite.

------

2 Equations 2.24–2.27 can be used to compute the actual parameters for this example over a spherical Earth. For example, the range for a spherical Earth is 948 km, instead of 923 km; the incidence angle is 35° for a 30° sensor look angle, and the azimuth resolution is 28.4 km.

Almost as bad is the fact that the satellite is much farther away from the Earth's surface than the aircraft. In fact, the received power falls off as the cube of the antenna height:

$$P_{rec} \propto \frac{Area_{resolution\ cell}}{Height^4} \propto \frac{1}{Height^3} \tag{7.24}$$

That is because while the spreading loss goes as range to the fourth power, the resolution cell size grows proportionally with range since the cell is beam limited in one direction. So in the end, a satellite at an altitude of 800 km would receive about 0.024% of the power received by an aircraft at an altitude of 5 km:

$$\frac{P_{rec}^{sat}}{P_{rec}^{a/c}} = 2.4 \times 10^{-4} = -36.2\text{ dB} \tag{7.25}$$

So it is not unreasonable to ask, how do we get enough power and azimuth resolution to acquire useful data from a satellite radar?

7.4 Pulse Compression

We talked about a radar sending out a pulse, but what are the spectral characteristics of this pulse?

Consider a single radar pulse, as illustrated in Figure 7.15[3] , with frequency f_0, amplitude $A(t)$, and duration τ. Mathematically the signal can be expressed:

$$s(t) = \begin{cases} A\cos\omega_0 t & -\tau/2 \le t \le \tau/2 \\ 0 & \text{otherwise} \end{cases} \tag{7.26}$$

------

3 Note that a real pulse will have many more cycles than shown in the sketch in Figure 7.15.

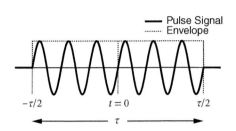

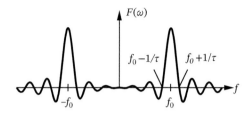

Figure 7.15 Short pulse and its Fourier transform.

The frequency spectrum of this pulse is given by the Fourier transform of the pulse, which can be shown to be the sum of two sinc functions, one centered at $+\omega_0$ and one at $-\omega_0$:

$$F(\omega) = \int_{-\infty}^{\infty} A(t)e^{-i\omega t}\,dt$$

$$= A\int_{-\tau/2}^{\tau/2} \cos(\omega_0 t)e^{-i\omega t}\,dt$$

$$= \frac{A\tau}{2}\left[\frac{\sin[(\omega_0 - \omega)\tau/2]}{(\omega_0 - \omega)\tau/2}\right.$$

$$\left. + \frac{\sin[(\omega_0 + \omega)\tau/2]}{(\omega_0 + \omega)\tau/2}\right] \qquad (7.27)$$

Notice that the null-to-null bandwidth of this signal is $2/\tau$. It can be shown that the half-amplitude bandwidth $B_{1/2}$ is: $B_{1/2} = 1.2/\tau$. But by convention, the pulse bandwidth B is universally defined as: $B = 1/\tau$.

It follows that the slant range and ground resolutions can be rewritten in terms of the pulse bandwidth:

$$\delta_{sr} = \frac{c\tau}{2} = \frac{c}{2B} \qquad (7.28)$$

$$\delta_{g} = \frac{c\tau}{2\sin\theta} = \frac{c}{2B\sin\theta} \qquad (7.29)$$

Although it may not seem so, the pulse bandwidth is actually fundamental to the resolution, a fact that is justified later in this section. Thus the pulse length is determined by its bandwidth, and the slant range and ground resolutions are expressed in terms of the bandwidth of the transmitted pulses.

Let us work an example. Assume an X-band radar with a wavelength of 3 cm transmitting

3.33-ns pulses. The antenna line of sight is 30° from nadir. The pulse bandwidth is:

$$B = 1/\tau = 300\ \text{MHz} \qquad (7.30)$$

The slant and ground range resolution are then:

$$\delta_{sr} = \frac{c}{2B} = 0.5\ \text{m} \qquad (7.31)$$

$$\delta_{g} = \frac{c}{2B\sin\theta} = 1.0\ \text{m} \qquad (7.32)$$

Unfortunately, it can be difficult to build a transmitter capable of putting enough power into a short time interval (3.3 ns) to get a received signal large enough for many applications. The solution is pulse compression.

The energy E in a pulse is $E = P_t\tau$ where P_t is the peak transmitted power. To maximize the energy incident on a target, there are two choices:

- Increase the transmitted power, and/or.
- Increase the pulse length.

Hardware limitations frequently constrain the transmitted power to be less than desired. But if we just lengthen a simple pulse, then our resolution degrades. The solution to this problem is to transmit a modulated pulse and then receive that pulse with a modified receiver. This solution can provide a longer pulse length without sacrificing range resolution.

While many types of modulated signals can be used, one of the simplest is a linear-frequency-modulated, or LFM signal. In

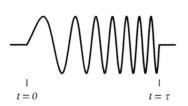 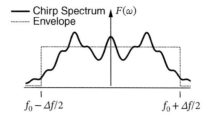

Figure 7.16 Chirped pulse and its spectrum.

an LFM pulse, the transmitted frequency ramps linearly with time from $f_0 - \Delta f/2$ to $f_0 + \Delta f/2$. Specifically, the dependence of the instantaneous frequency on time is given by:

$$f(t) = f_0 - \Delta f/2 + \frac{\Delta f}{\tau} t \qquad 0 \le t \le \tau$$

(7.33)

The pulse waveform and its frequency spectrum are illustrated in Figure 7.16. Such pulses are typically microseconds long and may chirp, or change frequency, over hundreds of MHz.

Modifications must be made to the receiver in order to obtain the best resolution from this long, modulated pulse. There are various forms that such pulse-compression receivers can take, and while these forms may appear to be quite different, they are mathematically equivalent.

Here we will first describe the pulse compression-receiver that uses a dispersive filter to compress an LFM pulse. In this type of receiver, the received pulse is passed through a dispersive filter that delays each frequency in the pulse by a decreasing amount such that the output signal is sharply peaked in time. The left diagram in Figure 7.17 shows the frequency of the transmitted signal varies as a function of time. The middle plot shows that the dispersive or pulse compression filter inserts an inverse delay into the signal, with a propagation time for the highest frequencies that is τ less than the propagation time for the lowest frequencies.

The effective temporal width of the pulse as measured at the output of the dispersive filter can be shown to be $1/\Delta f$. And as the pulse is compressed, the peak amplitude of the output pulse increases.

The compressed pulse has the same spectrum as a pulse with length $= 1/\Delta f$, hence the same range resolution. The radar illustrated in Figure 7.18 uses a chirped waveform and a pulse compression receiver to create a long, low-power chirp pulse with large bandwidth. Such a long, low-power chirp pulse can be equivalent to a short, high-power pulse – it can have the same energy on the target and the same resolution.

So how much gain do we get from pulse compression? The pulse compression filter is a frequency dispersive device – propagation velocity through it depends on frequency. As the received frequency increases, the propagation speed increases, causing all the frequency components in the pulse to arrive at the same time.

The ideal pulse compression filter has no losses, so the energy of the compressed pulse must equal the energy of the input pulse:

$$P_c \tau_c = P_i \tau \quad \rightarrow \quad P_c = P_i \frac{\tau}{\tau_c} = P_i \tau \Delta f$$

(7.34)

where P_i is power of the input pulse and P_c is the power of the compressed pulse. Note the compressed pulse length has been reduced by a factor of $\tau \Delta f$. This is referred to as the time-bandwidth product, or the pulse compression factor.

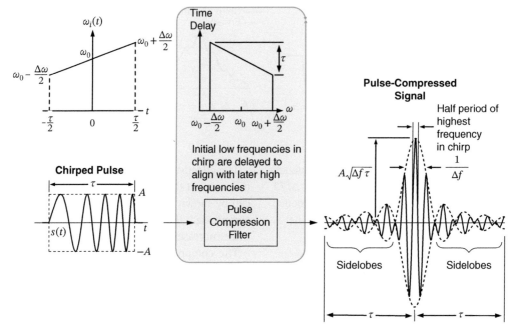

Figure 7.17 Pulse compression using a dispersive filter. Source: From P. Z. Peebles, Radar Principles with permission from John Wiley & Sons.

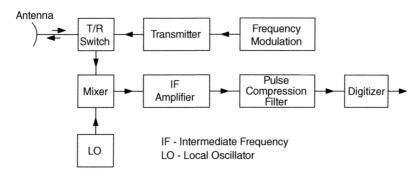

Figure 7.18 Pulse compression radar. Adapted from Skolnik (2001).

For example, a 10-μs pulse with chirp bandwidth of 150 MHz has a pulse compression factor of 1500 and provides a slant range resolution of 1 m. Furthermore, the output pulse power equals the pulse compression factor ($\tau\Delta f$) times the input pulse power and the output pulse amplitude equals the square root of the pulse compression factor ($\sqrt{\tau\Delta f}$) times the input pulse amplitude.

For this example, a 1-W radar with a 10-μs pulse with a chirp bandwidth of 150 MHz would have exactly the same pulse compressed output power as a 1500-W radar with a 6.6-ns pulse, and yet both would provide exactly the same resolution.

There are at least two other ways to implement a pulse compression receiver:

- A matched-filter convolution of the received signal with a replica of the transmitted signal, known as a correlation receiver, can be implemented in either the receiver hardware or postprocessing software. This approach is discussed below in conjunction

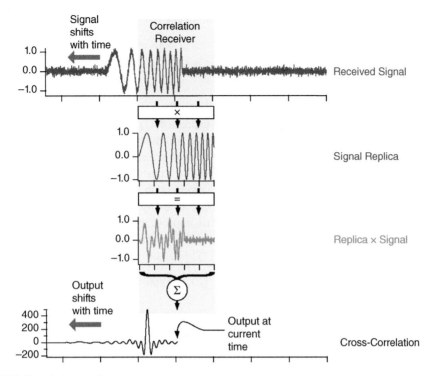

Figure 7.19 Correlation receiver.

with Figure 7.19, but details can be found in any radar text.

- The receiver can be designed to mix the received pulse with a signal identical to the transmitted pulse with a time delay equal to the expected range to the target. This mixing with a signal replicate shifts all of the incoming chirp down to a single frequency, where the energy in the pulse can all be summed together. This is the approach generally used in precision radar altimeters.

All of these approaches are mathematically equivalent, although there are detailed engineering reasons why one of these approaches might be preferred in a given design. All three approaches are used in practice.

Figure 7.19 illustrates the operation of a correlation receiver. The time series of the received signal, which consists of a copy of the transmitted signal along with some additive noise, is shown in red. These data scroll to the left as new data arrive at the receiver.

The signal replica, which is perfectly known, is shown in blue. At every step, the receiver forms the product of the two signals as shown in green, all the elements of which are then summed to produce the output.

When the received signal starts to overlap with the signal replica, the oscillations are positive and negative, and these oscillations nearly cancel out. But when the signal aligns with the replica, the product of the received signal and the replica form a noisy version of the square of the replica, so the green signal is nearly everywhere positive and it sums to a large value. Thus it is easy to see that the peak occurs when the signal is aligned with the replica.

It turns out that this is equivalent to a matched filter, and hence an optimal solution for detection of signals in additive white noise.

Linear frequency modulation has been used here to illustrate pulse compression, but any form of modulation can be used to obtain improved resolution. While the details of the modulation determine the characteristics of range sidelobes, the bandwidth of the modulation determines the slant range resolution.

It should not be surprising that a radar receiver is like any other microwave receiver. In other words, it receives thermal noise based on where the antenna is pointed with a power given by $N_p = kTB$ where:

- k = Boltzmann's constant.
- T = Brightness temperature of the scene ≈ 300 K if the radar is looking at the Earth or lower atmosphere.
- B = Receiver bandwidth in Hz.

In order to maximize the received signal and minimize the received noise, the bandwidth of the receiver should be set to the bandwidth of the transmitted signal: $B = \Delta f$.

Real receivers are not perfect, so they add some noise to the measurement. This degradation is accounted for by inclusion of a multiplicative noise figure, designated *NF*:

$$N_{rec} = k\,TB(\text{NF}) = k\,T\Delta f(\text{NF}) \qquad (7.35)$$

Typical noise figures might add 5–7 dB to the thermal noise measured by a system.

The signal-to-noise (SNR) ratio for a monostatic radar with a pulse compression receiver then becomes:

$$\text{SNR} = \frac{P_{rec}}{N_{rec}} = \frac{P_t\,G^2\lambda^2\sigma\Delta f\tau}{(4\pi)^3R^4\,k\,TB(\text{NF})}$$

$$= \frac{P_t\,G^2\lambda^2\sigma\tau}{(4\pi)^3R^4\,k\,T(\text{NF})} \qquad (7.36)$$

Note that signal-to-noise ratios are always expressed as power ratios.

In summary, the SNR of a pulse compression receiver scales with pulse length, but not bandwidth. So using modulated pulses with a pulse compression receiver effectively separates the effects of pulse length, which can be used to improve received power, and bandwidth, which sets resolution.

7.5 Types of Radar

There are at least four types of radars that are regularly used in remote sensing of the Earth: altimeters, scatterometers, synthetic aperture radars (SAR), and high-frequency radars. Table 7.1 lists the viewing geometry and some of the primary characteristics of these types of radars. It is worth noting that while altimeters are designed to work looking straight down, scatterometers and SARs are designed to work

Table 7.1 Applications of remote sensing radars.

Radar type	Incident angles	Ocean measurements	Land measurements
Altimeter	Nadir (0°)	Significant wave height, wind speed, surface elevation, sea ice extent and currents	Land and ice sheet topography
Scatterometer	30–50°	Wind speed and direction	Soil moisture, snow accumulation, rainforest and desert monitoring
SAR	20–65°	Wave spectrum, bathymetry, surface currents, wind speed, sea ice, oil slicks, ship wakes	Vegetation type and cover, rainforest monitoring, snow and ice sheet topography and texture, flood detection, ground movement
HF Surface Wave	Near 90°	Surface current vector	

at moderate incident angles, and so rely on a different scattering mechanism. Altimeters, scatterometers, and SARs have been deployed from satellites, and are the topics to be covered in the following three chapters.

7.6 Example Terrestrial Radars

While the primary focus of this text is on satellite remote sensing, the use of radar for terrestrial remote sensing is significant. The brief descriptions below of weather radars and HF surface wave radars serve to illustrate the variety of terrestrial radar remote sensors. Satellite-based radars are discussed in the following three chapters.

7.6.1 Weather Radars

Modern weather radars are land-based systems designed to measure the location, type and rate of precipitation. They basically work by measuring and analyzing the varying reflectivity of precipitation. They are usually designed as coherent, pulse-Doppler radars, allowing them to also measure wind velocity and turbulence. The most modern versions also measure returns at multiple polarizations, providing improved discrimination between types of precipitation.

Weather Radar Range Equation
The radar equation for a monostatic radar is:

$$P_{rec} = \frac{P_t G^2 \lambda^2 \sigma}{(4\pi)^3 R^4} \tag{7.37}$$

The wavelengths of weather radars are an order of magnitude larger than the dimensions of the largest raindrops, so scattering from a single raindrop is in the Rayleigh regime with RCS:

$$\sigma_1 = \frac{\pi^5}{\lambda^4} |K|^2 D^6 \tag{7.38}$$

where K is the dielectric constant of the raindrops and D is the drop diameter.

The radar cross-section for a weather radar is determined by the sum of all the cross-sections of the drops within the sampled volume defined by the radar beam area A_{beam} and range resolution Δr:

$$\sigma = A_{beam} \, \Delta r \int_0^{D_{max}} \sigma_1(D) N(D) \, dD \tag{7.39}$$

where $N(D)dD$ is the number of drops in each size increment in a cubic meter of rain and D_{max} is the maximum rain drop size. Substitution then yields:

$$\begin{aligned} \sigma &= \frac{\pi R^2 (\delta\theta)^2}{4} \frac{c\tau}{2} \frac{\pi^5}{\lambda^4} \\ &\quad \times \int_0^{D_{max}} |K|^2 D^6 N(D) \, dD \\ &= \frac{\pi^6 R^2 (\delta\theta)^2 c\tau}{8\lambda^4} Z \end{aligned} \tag{7.40}$$

where we define the radar reflectivity per unit volume to be:

$$Z = \int_0^{D_{max}} |K|^2 D^6 N(D) \, dD \tag{7.41}$$

Furthermore, based on equation (7.7) and the definition of antenna gain:

$$G \approx \frac{16}{(\delta\theta)^2} \tag{7.42}$$

Combining equations (7.37), (7.40), and (7.42) yields:

$$P_{rec} = \frac{\pi^3 P_t G^2 (\delta\theta)^2 c\tau}{512 R^2 \lambda^2} Z \approx \frac{\pi^3 P_t G c\tau}{32 R^2 \lambda^2} Z \tag{7.43}$$

The radar is detecting volume scattering, so the received power is proportional to antenna gain G instead of G^2, with a range dependence of R^2 instead of R^4. These dependencies make sense. Increased antenna gain implies a smaller beam, a smaller sampling volume and, hence, lower radar cross-section. Similarly, longer ranges imply larger sampling volumes, and higher radar cross-section.

Note the received power is proportional to the parameter Z, which encodes all of the information about the rain drop size distribution. The parameter Z is so fundamental to radar meteorology that the radar measurements are often expressed in dBZ, which are power units relative to the return of a droplet of rain with a diameter of 1 mm ($1\,\text{mm}^6$ per m^3).

Marshall and Palmer (1948) reported the drop size distribution in a cubic meter of stratiform rain is given by: $N(D) = N_0 e^{-\Lambda D}$, where $N_0 = 8000\ \text{m}^{-3}\text{mm}^{-1}$, $\Lambda = 4.1R^{-0.21}\text{mm}^{-1}$, R is the rain rate in millimeters per hour, and D is the raindrop diameter in mm. Thus the reflectivity is given by:

$$Z_{\text{rain}} = |K|^2 \int_0^{D_{max}} N_0 e^{-\Lambda D} D^6\, dD \quad (7.44)$$

Using the same drop size distribution, the rain rate can be expressed:

$$R = \int_0^{D_{max}} N_0 e^{-\Lambda D} \frac{\pi D^3}{6} v(D)\, dD \quad (7.45)$$

where $v(D)$ is the fall rate of a drop of diameter D. The fall rate of a drop depends on the balance of the acceleration due to gravity and drag, and is often expressed as a power law in drop diameter, $v = \alpha D^\gamma$. In this case, Z and R can be related as power law:

$$Z_{\text{rain}} = aR^b \quad (7.46)$$

Different values of a and b are used for different types of precipitation, but this is the basis for rain rate measurements from weather radars.

Example Weather Radars

Two of the most common weather radars in the United States are NOAA's WSR-88D (NEXRAD) radars and the FAA's Terminal Doppler Weather Radar (TDWR) that are installed around many larger airports. Table 7.2 lists some key parameters for these two radars.

One major difference is that the WSR-88D operates at S-band with a wavelength of 10.5 cm, while the TDWR operates at C-band

Table 7.2 Example weather radars.

Parameter	WSR-88D	TDWR
Band (Wavelength, cm)	S (10.5)	C (5.3)
Frequency (GHz)	2.7–3.0	5.60–5.65
Peak power (kW)	750	250
Max Doppler Range (km)	230	90
Pulse repetition frequency (Hz)	320–1300	1066–1930
Pulse width (µs)	1.57 or 4.57	1.1
Bandwidth (MHz)	0.63	1.0
Resolution (m)	250 – 1000	150 – 300
Antenna	8.2-m circular dish	8.2-m circular dish
Antenna beamwidth (deg)	0.95	0.55
Max antenna rotation rate (rpm)	6	4.4
Volume scan time (min)	4 – 10	1 – 5.4
Antenna elevation (deg)	0.5–19.5	0–60
Polarization	Dual	Horizontal
Max Unambiguous Speed (m s^{-1})	32.8	14 to 25

with a wavelength of 5.3 cm. The RCS for a unit volume is about 16 times higher for the TDWR than for the WSR-88D. This allows the TDWR to operate with finer resolution in range, a smaller beamwidth, and lower peak power. But the TDWR can suffer from signal loss in heavy rain due to too much scattering at near ranges, reducing the illuminating power at the farthest ranges.

Weather Radar Pulse-Doppler Measurements

Virtually all weather radars are coherent, allowing them to measure the relative phase of each returned pulse. The phase differences between adjacent pulses varies because of the Doppler effect. For pulse-pair measurements, each phase measurement reflects the two-way path length between the radar and the scatterer at the time of that pulse. The second phase measurement then differs from the first by the change in the two-way path during the time between the two measurements. Note that if the scatterers move a distance Δx toward the radar between the two measurements, the two-path length decreases by a factor of $2\Delta x$. The phase changes by an amount $\Delta\varphi = 2k\Delta x = \frac{4\pi}{\lambda}\Delta x$, with radar wavenumber k, or radar wavelength λ. If the time between the two-pulses is Δt_p, then the relative velocity of the scatterer and the radar is:

$$v = \frac{\Delta x}{\Delta t_p} = \frac{\lambda\Delta\varphi}{4\pi\Delta t_p} \qquad (7.47)$$

Because phase measurements are unique over only a range between $\pm\pi$ radians, the measurement of scatterer velocity are unambiguous only over a limited range. For example, consider a target that moves 1 cm towards the radar in the 1 ms between between pulses from a 10-cm wavelength radar. The target is moving at a relative speed of 10 m s^{-1}, and the phase difference between the two measurements will be $\Delta\varphi = 0.4\pi$. The magnitude of this phase difference is less than π radians, so substitution into equation (7.47) yields the correct answer.

Next suppose the target was moving at 40 m s^{-1}, so it moves 4 cm towards the radar in between pulses. In this case, the phase difference between the two measurements will be 1.6π, which is outside the principal interval of $\pm\pi$, so the value will be wrapped to $\Delta\varphi = -0.4\pi$. Substitution then yields a speed of -10 m s^{-1}, with the minus sign indicating motion away from the radar. The maximum unambiguous speed is in general given by $\frac{\lambda}{4t_p}$, which evaluates to 25 m s^{-1} for this example.

Note the TDWR has a maximum unambiguous speed as low as 14 m s^{-1} for the lowest pulse-repetition frequency (PRF), which should seem low for a radar that is designed to detect and measure storm winds. The radar can be operated at higher PRF to increase the maximum unambiguous speed, but this comes at a cost. The time between pulse transmissions is 800 µs for a PRF of 1250 Hz. The pulses are 1.1 µs wide, so the time in between pulses when the transmitter is off and the receiver can be operated is 799 µs. At the speed of light, this period of time corresponds to a two-way path length of 240 km. Thus a signal from a target more than 120 km from the radar will return after the second pulse is transmitted. If the signal is strong enough it could be confused with a nearby return. The unambiguous range for a PRF of 1250 Hz is thus 120 km.

The radar operator can increase the maximum unambiguous range for the radar by decreasing the PRF, but this will decrease the maximum unambiguous velocity. To beat this problem, the current WSR-88D utilizes multiple PRFs and sophisticated processing to remove some of the range ambiguities. Velocity ambiguities are dealt with using the correlation of phase in adjacent cells in conjunction with phase unwrapping algorithms. These are some of the constraints a radar designer has to consider.

The examples above have described Doppler-phase measurements as if the radar were measuring the velocity of a hard target. It turns out, the same approach also works for a distributed target of raindrops that are all moving with a distribution of speeds. In this case, the pulse-pair measurement can be shown to estimate the mean Doppler velocity of the observed scatterers. In fact three pulses can be used to estimate the difference in two velocities from which an estimate of the velocity spread can be obtained. The mathematics of these spectral moment estimators was first described by Miller and Rochwarger (1972).

Weather Radar Polarization Measurements

The original WSR-88 radars used a single horizontally-polarized signal, but the WSR-88D radars were all upgraded to dual polarization by 2013. Rain drops flatten as they fall, so polarization can be used to better estimate the size, shape, and composition of precipitation. Whereas single channel radars measure reflectivity Z, Doppler velocity, and spectrum width, the dual-polarization radars can also measure differential reflectivity (Z_{DR}), copolar cross-correlation coefficient (ρ_{hv}), differential phase shift (φ_{DP}), and specific differential phase shift (K_{DP}).

Parameter Z_{DR} is the logarithmic ratio of the received power at horizontal and vertical polarization, and measures the oblateness of the scatterers. Parameter ρ_{hv} is near unity for homogenous scatterers, but varies from unity for mixed scatterers, such as mixed rain and snow. Parameters φ_{DP} and K_{DP} measure the phase difference between the horizontally and vertically polarized radiation, which also depends on the oblateness of the precipitation. They can be used to better localize regions of heavy rain.

Carlin (2015) provides a good, short introduction to weather radar phase measurements. More detailed information can be found in Kumijian (2013a, 2013b, 2013c).

7.6.2 HF Surface Wave Radar

HF Surface Wave Radars (HFSWR) are widely used to remotely sense ocean currents. Such radars generally operate within the HF band from 10 to 30 MHz. The antennas for these radars are mounted very close to the shore so the radar energy can be efficiently coupled into a surface mode that can propagate over the horizon, following the curved surface of the ocean. (While electromagnetic waves in a vacuum travel in straight lines, vertically-polarized waves can also efficiently diffract around a curved conducting surface, such as the ocean surface. This is referred to as surface wave propagation. This mode is very lossy over a nonconducting surface such as land, so HFSWR antennas are installed right long the shore line.)

HFSWR radars are imaging radars. They utilize timing and pulse bandwidth to obtain range resolution. One common variety of HFSWR utilizes a transmit antenna with a broad beam and phased-array receive antenna to achieve azimuth resolution. Because of the low frequency of operation, these are not high-resolution systems. A resolution cell size of 1 km on a side is typical.

HFSWRs utilize the fact that HF backscatter from the ocean at near-grazing angles is predominantly due to Bragg scattering (discussed in Section 9.2). Thus the Doppler spectrum of the HF radar return contains sharp peaks associated with those surface gravity waves with a wavelength of half of the radar wavelength that are either approaching toward, or receding from, the radar. The frequency of these Doppler peaks reflect the phase velocity of the scattering waves (see Chapter 9.1). By subtracting the known phase velocity of the scattering wave in the absence of current, $c_p = \sqrt{gk}$, from the measured phase velocity, an estimate of the component of the surface current in the look direction of the radar can be obtained. Two or more radars, viewing a particular measurement region from different directions, can then be used to estimate the two-dimensional surface current vector.

The shift in the Doppler spectrum due to a current of a few centimeters per second is exceedingly small, which means that long observations are required to make this measurement. A typical HFSWR might utilize a full 20 minutes of data to make a single vector current map.

The currents associated with ocean waves decay exponentially with the product of depth and wavenumber. So the propagation of ocean waves are affected mostly by near-surface currents, with an effective mean depth of about 1/6 of the wave's wavelength. Thus the frequency of the HF radar's operation sets the ocean wave wavelength being sensed, which impacts the effective depth of the current measurement.

A more detailed overview of HF radar is provided in Barrick et al. (1977).

References

Barrick, D. E., Evans, M. W., & Weber, B. L. (1977). Ocean surface currents mapped by radar. *Science, 198*(4313), 138–144.

Carlin, J. (2015). Weather radar polarimetry. *Physics Today*. Retrieved from https://physicstoday.scitation.org/do/10.1063/PT.5.4011/full/

Eaves, J., & Reedy, E. (2012). *Principles of modern radar*. Springer Science & Business Media.

Kumjian, M. R. (2013a). Principles and applications of dual-polarization weather radar. Part I: Description of the polarimetric radar variables. *Journal of Operational Meteorology, 1*.

Kumjian, M. R. (2013b). Principles and applications of dual-polarization weather radar. Part II: Warm-and cold-season applications. *Journal of Operational Meteorology, 1*.

Kumjian, M. R. (2013c). Principles and applications of dual-polarization weather radar. Part III: Artifacts. *Journal of Operational Meteorology, 1*.

Marshall, J. S., & Palmer, W. M. K. (1948).The distribution of raindrops with size. *Journal of Meteorology, 5*(4), 165–166.

Miller, K., & Rochwarger, M. (1972). A covariance approach to spectral moment estimation. *IEEE Transactions on Information Theory, 18*(5), 588–596.

Skolnik, M. (2001). *Introduction to radar systems*. McGraw-Hill.

Ulaby, F. T., Moore, R. K., & Fung, A. K. (1982). *Microwave remote sensing: Active and passive. Vol. II. Radar remote sensing and surface scattering and emission theory*. Reading, MA: Addison-Wesley.

Unal, C. M., Snoeij, P., & Swart, P. J. F. (1991). The polarization-dependent relation between radar backscatter from the ocean surface and surface wind vector at frequencies between 1 and 18 GHz. *IEEE Transactions on Geoscience and Remote Sensing, 29*(4), 621–626.

Valenzuela, G. R. (1978). Theories for the interaction of electromagnetic and oceanic waves: A review. *Boundary-Layer Meteorology, 13*(1), 61–85.

Yurovsky, Y. Y., Kudryavtsev, V. N., Grodsky, S. A., & Chapron, B. (2017). Ka-band dual copolarized empirical model for the sea surface radar cross section. *IEEE Transactions on Geoscience and Remote Sensing, 55*(3), 1629–1647.

8

Altimeters

8.1 Introduction to Altimeters

Radar altimeters used in remote sensing illuminate the sea surface in the nadir direction in order to make three fundamental measurements:

- They measure the travel time for a pulse to reach and return from surface. This distance to the surface is then combined with a precision orbit determination to precisely measure the surface elevation along the ground track of the altimeter. It turns out that surface elevation over the ocean is a surprisingly useful measurement.
- They measure mean backscattered power which provides an estimate of surface wind speeds along the ground track of the altimeter.
- They measure the shape of the return pulse to estimate the significant wave height along the ground track of the altimeter.

As shown in Table 8.1, there is a wide range of applications for altimetry measurements:

In this and the following section we discuss how altimeters work, and how they make these specific geophysical measurements. For more detail, Fu and Cazenave (2001) provide an excellent overview of the breadth of satellite altimetry application and techniques.

Most radar altimeters share a common set of design parameters:

- Radar frequency is about 14 GHz (2.2 cm radiation).
- Transmitted pulse length is about 100 μs.
- Pulse bandwidth ≈300 MHz (chirped pulse is used).
- Range resolution is 0.5 meter.
- Pulse rate is about 1,000 pulses per second.
- Antenna is a dish with a diameter about 0.7 to 1.0 meter.

While the radar frequency is almost always about 14 GHz (2.2 cm wavelength), many of the more modern radars also have a second, lower frequency. The transmitted pulse length is often about 100 microseconds, but FM chirps are used with a pulse bandwidth

Table 8.1 Applications for altimetry.

– Ocean currents, eddies, and tides	– Sea surface wave height
– Sea level change	– Large-scale ocean circulation
– Marine geoid	– Sea floor bathymetry
– Continental ice sheet dynamics and mass balance	– Surface wind speed

Remote Sensing Physics: An Introduction to Observing Earth from Space, Advanced Textbook 3, First Edition.
Rick Chapman and Richard Gasparovic.
© 2022 American Geophysical Union. Published 2022 by John Wiley & Sons, Inc.
Companion website: www.wiley.com/go/chapman/physicsofearthremotesensing

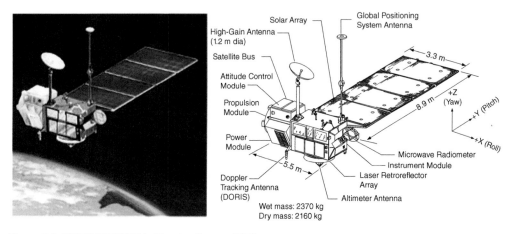

Figure 8.1 TOPEX/POSEIDON altimeter. Source: NASA.

Table 8.2 Satellite altimeters that are no longer active.

Satellite altimeter/ Agency	Mission period	Frequency (GHz)	Altitude (km)	Height precision (cm)	Orbit repeat (days)	Orbit accuracy (cm)
S-193 NASA	5/73–2/74	13.9	438	100	–	–
GEOS-3 NASA/APL	4/75–12/78	13.9	838	25	–	500
Seasat NASA/APL	7/78–10/78	13.5	799	5	24	20
Geosat Navy/APL	3/85–12/89	13.5	794	4	17	10–20
ERS-1 & 2 ESA/Italy	7/91–5/96 8/95–2007	13.8	783	3	35	5
Topex/Poseidon[1] NASA/APL/France	10/92–10/05	13.6/5.3	1340	2	9.916	2–3
Geosat Follow-On Navy/E-Systems	2/98–12/08	13.5	794	3.5	17	5
Envisat[2] ESA/Italy	3/02–5/12	13.6/3.2	773	3	35	3
Jason-1 NASA/France	12/01–6/13	13.6	1332	2	9.916	2

[1]Zieger et al. (1991)
[2]Resti et al. (1999)

of about 300 MHz. This produces a slant range resolution of 0.5 meter, although these instruments end up achieving a measurement precision of a few centimeters by clever processing.

Most altimeter satellites have been placed in orbits at an altitude of about 800 km, but a few have orbited at 1300 km. The time for a pulse to travel to the Earth's surface and return to an altimeter at 800-km altitude is about 5.3 ms. So a typical altimeter will have multiple pulses in the air at one time

Figure 8.1 shows the fully deployed configuration of the TOPEX/POSEIDON satellite. The altimeter antenna at the bottom of the figure points straight down and the positive *x*-axis points towards the flight direction.

Table 8.2 lists some of the satellite altimeters that are no longer operational. There are several things to note on this list. For example, the first four U.S. altimeters were designed

at The Johns Hopkins University Applied Physics Laboratory. The TOPEX/POSEIDON altimeter, launched in 1992, was the first dual frequency altimeter, the second frequency being used to estimate and remove certain ionospheric effects (Fu et al., 1994). Also note that the height precision and orbit accuracy improved to a few centimeters by the 1990s.

Topex/Poseidon was also notable for another reason. For that mission, NASA made the decision that one instrument would be designed and built at JHU/APL, and the other built by the French National Centre for Space Studies (CNES). Since then, ESA has produced the highest quality altimeters in the world. The precision of the measurements and the orbit determination have continued to improve, so height precision is now better than 2 cm. Table 8.3 lists all of the currently operating satellite radar altimeters.

Table 8.3 Active satellite altimeters.

Satellite altimeter/ Agency	Mission period	Frequency (GHz)	Altitude (km)	Height precision (cm)	Orbit repeat (days)	Orbit accuracy (cm)
Jason-2	6/08–Present	13.575/5.3	1332	2	9.916	2
Jason-3	1/16–Present					
NASA/France						
CryoSat-2	4/10–Present	13.575	725	1–2	369 with 30-	2
ESA/Italy					day subcycle	
Sentinel-6	11/20–Present	13.575/5.3	1336	1–2	10	2
ESA						
SARAL/ALtiKa	2/13–Present	35.75	790	0.5–1.5	35	3
India/France						
HY-2A	8/11–Present	13.58/5.25	971	<4	14/168[1]	1–3
HY-2B	10/18–Present				14	
HY-2C	09/20–Present				drifting	
China						
SRAL/Sentinel-3A	2/16–Present	13.575/5.41	805	3.5	27	2
SRAL/Sentinel-3B	4/18–Present					
ESA						

[1] HY-2A orbit repeat was 14 days for the first 4.5 years; since then the repeat is 168 days.

8.2 Specular Scattering

Different backscatter mechanisms contribute to the radar cross-section of the sea surface depending on the incident angle of the radar illumination. Figure 8.2 illustrates the three primary regimes:

- In the regime of near-vertical illumination, from nadir looking out to about 15°, the dominant mechanism is specular scattering. By specular we mean the returns are dominated by returns from those small patches of the ocean that are nearly horizontal at wave crests and within wave troughs.
- In the mid incident angle regime, with incident angles ranging from 20° to 75°, the returns are dominated by diffuse scattering, driven by a Bragg resonant condition.
- Finally, at large incident angles, say angles of greater than 80°, wave shadowing and diffraction effects become important. This is known as the low grazing angle regime, where grazing angle is defined relative to the horizon, which means that grazing angle equals 90° minus incident angle.

At radar frequencies, electromagnetic fields do not penetrate through the sea surface to any appreciable depth. Recall that the incident flux $\Phi(0)$ is attenuated with depth z as: $\Phi(z) = \Phi(0)\exp(-z/\delta)$ where "skin depth" δ is given by:

$$\delta = \frac{\lambda}{4\pi k_0} \tag{8.1}$$

and k_0 = imaginary part of the index of refraction. The complex index of refraction $\tilde{n}$ is in turn related to the complex dielectric constant by: $\tilde{n}^2 = (n + ik_0)^2 = \tilde{\varepsilon} = \varepsilon_r + i\varepsilon_i$.

Recall that we can convert between the complex index of refraction and dielectric constant by:

$$\varepsilon_r = n^2 - k_0^2 \quad \text{and} \quad \varepsilon_i = 2nk_0 \tag{8.2}$$

$$n^2 = \frac{\varepsilon_r}{2}\left[\left(1 + \frac{\varepsilon_i^2}{\varepsilon_r^2}\right)^{1/2} + 1\right]$$

$$k_0^2 = \frac{\varepsilon_r}{2}\left[\left(1 + \frac{\varepsilon_i^2}{\varepsilon_r^2}\right)^{1/2} - 1\right] \tag{8.3}$$

The complex dielectric constant for seawater can be computed from the Debye theory for

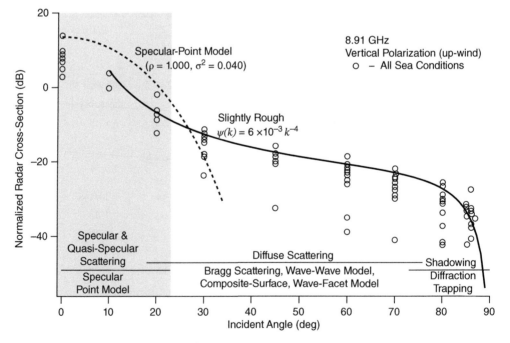

Figure 8.2 Normalized radar (backscatter) cross-section of the sea surface. Source: Valenzuela, G. R. (1978) with permission from Springer Nature.

Table 8.4 Microwave skin depths in seawater.

Frequency (GHz)	δ (cm)	δ/λ_{RF}
1.25	0.49	0.020
3.0	0.35	0.035
10	0.10	0.033

the dielectric constant of an ionic solution. Figure 8.3 plots the complex dielectric constant for seawater as a function of frequency. Table 8.4 indicates that the penetration depths are quite small, approximately 2–3% of the wavelength at microwave frequencies.

The small flux penetration depth means that volume scattering can be ignored. Thus returns from the ocean surface are dominated by surface scattering, which depends on surface shape. In contrast, volume scattering turns out to be important for scattering from relatively fresh sea ice and continental ice sheets at lower frequencies.

Estimates of the sea surface radar cross-sections are computed from rough-surface electromagnetic scattering theory applied to surfaces with length scales and slopes appropriate to surface waves. In general, the surface is assumed to be a dielectric with dielectric constant determined by radar frequency, seawater conductivity, and temperature.

It turns out that exact solutions of the wave equation for the electric and magnetic fields scattered from a realistic sea surface

do not exist, so approximate solutions are used to account for many features of measured data. The two most popular approximate solutions are geometric optics and physical optics:

- *Geometric optics* models give reasonable results for specular scattering cases where the electromagnetic wavelengths are small compared to the length scales of surface roughness. Thus this type of model is often used to predict backscatter for a nadir-looking altimeter.
- *Physical optics* models are applicable to a wider range of cases, but require finding approximate solutions to complex integral equations. These integral equations are often solved in an iterative fashion, but the calculations are still challenging.

Developing more accurate approaches to numerical solutions is an active area of current research.

Because we are exploring altimeters, let us discuss the concept of specular scattering in more detail (Figure 8.4). Assume the plane of incidence is the x-z plane, and that the local radius of curvature of the surface r satisfies the inequality: $2\pi r \cos^3 \theta \gg \lambda$. This means the surface is smooth enough that the total returns from the surface can be treated as the sum of all the rays coming from each separate element of the surface. For backscatter this means the returns only come from each location where the surface is orthogonal to the incoming ray;

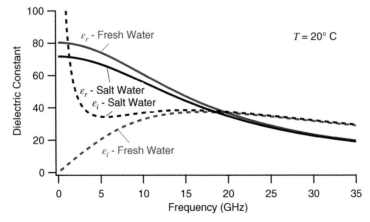

Figure 8.3 Complex dielectric constant for seawater and fresh water.

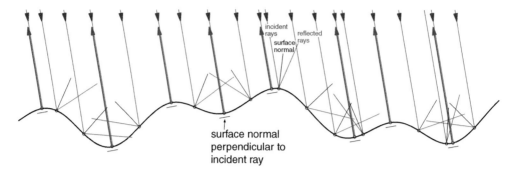

incident rays
reflected rays
surface normal

surface normal
perpendicular to
incident ray

Figure 8.4 Specular scattering occurs from the portion of the surface normal to the radar. The incoming rays are indicated by arrows. The surface normals are shown as the shortest lines, and the reflected rays are longer lines with thick lines indicating those rays that are specularly reflected back towards the radar.

at all other locations, the tilted surface reflects the incoming ray off into a direction away from the radar.

In this case, the normalized radar cross-section can be shown to be:

$$\sigma_0 = \pi r(0) \sec^4 \theta \, p(\tan \theta, 0) \qquad (8.4)$$

where $r(0)$ is the Fresnel power reflection coefficient[1], which has a value of 0.61 for $\lambda = 3$ cm, and $p(\eta_x, \eta_y)$ is the joint probability distribution function for slopes η_x and η_y. Thus the expression $p(\tan \theta, 0)$ is just the probability of having a slope perpendicular to the radar.

The sun glitter measurements made by Cox and Munk in 1954 show that the probability distribution function for surface slope is approximately Gaussian with a mean-square slope varying linearly with wind speed (Cox & Munk, 1954, 1956). Their results are illustrated in Figure 8.5. The top curve is the PDF of the crosswind slopes, which is well approximated by a Gaussian distribution. The bottom curve shows that the upwind/downwind slopes are nearly Gaussian, but with some nonzero skewness. It turns out the skewness does not affect near-nadir backscatter, so the deviations from a Gaussian distribution can be ignored for this analysis.

Thus Cox and Munk found that the probability distribution function for surface slope is approximately given by:

$$p(\tan \theta, 0) = \frac{1}{\pi \sigma^2} \exp\left(\frac{-\tan^2 \theta}{\sigma^2}\right) \qquad (8.5)$$

where σ^2 is the mean-square slope of sea surface, which is related to wind speed U_w by:

$$\sigma^2 = 0.003 + 0.0051 \, U_w(\text{m/s}) \qquad (8.6)$$

This relationship was developed based on sun glint images acquired over deep water, in the absence of strong currents and for a naturally-clean surface[2].

Applying the Cox–Munk slope statistics, the NRCS of the rough surface is given by:

$$\sigma_0 = \frac{r(0)}{\sigma^2} \sec^4 \theta \, \exp\left(\frac{-\tan^2 \theta}{\sigma^2}\right) \qquad (8.7)$$

Cox and Munk used optical measurements to estimate the slope distribution, which means that their measurements apply to very small spatial scales. It turns out that this result needs to be modified a bit for use at microwave wavelengths. Specifically the mean-square slope has to be reduced a bit because only slopes from waves that are much longer than the microwave wavelength are to be included in the mean-square slope in the NRCS equation.

---

1 The Fresnel power reflection coefficient sometimes appears in the literature as $|r(0)|^2$, where $r(0)$ is the Fresnel amplitude reflection coefficient.

2 Cox and Munk also made measurements within oil-slicks, reporting a significant reduction in the mean-square slope of sea surface within the slick.

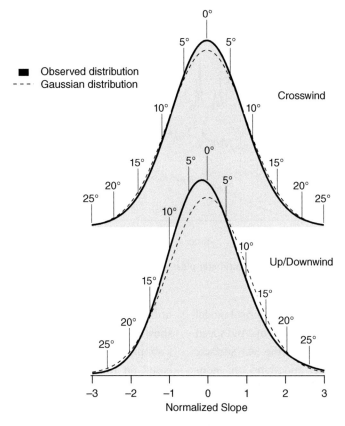

Figure 8.5 Gaussian distribution of sea surface slopes as observed by Cox and Munk. Adapted from Cox and Munk (1954).

8.3 Altimeter Wind Speed

Radar altimeters view the sea surface in the nadir direction, so the primary backscatter mechanism is specular scattering from horizontal wave facets. For this specific geometry the NRCS is given by:

$$\sigma_0 = \frac{r(0)}{\sigma^2} = \frac{r(0)}{0.003 + 0.0051 \; U_w(\text{m/s})} \tag{8.8}$$

where $r(0)$ is the Fresnel power reflection coefficient at 0° incident angle and σ^2 is the mean square wave slope which is proportional to wind speed.

This result indicates that NRCS depends inversely on wind speed – calm seas reflect more power back to the radar than rough seas.

It turns out that the actual dependence of the NRCS on wind speed is a bit more complicated than shown here because only wave facets that are large compared to the radar wavelength contribute to the mean-square slope in the NRCS. Nevertheless, the altimeter measures the *total backscatter power*, which is used with the radar equation to estimate NRCS, which in turn is used to estimate the surface wind speed.

As illustrated in Figure 8.6, the returned signal from a single altimeter pulse is generally quite noisy, but many pulses can be averaged together to obtain good estimates of the mean backscattered power. The NRCS of the surface is then computed from the measured power, the characteristics of the radar and the radar equation.

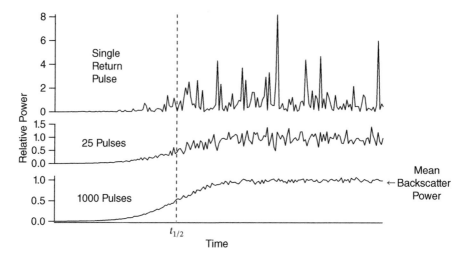

Figure 8.6 Returned signal from a single altimeter pulse and multipulse averages. Modified from Chelton et al. (1989).

Figure 8.7 compares several empirical model functions for altimeter estimates of wind speed at a height of 10 m above the sea surface. In general, the model functions are non-linear because RCS only depends on waves longer than the radar wavelength. Brown et al. (1981) proposed one of the earliest empirical geophysical model functions that utilized a three-branch logarithmic model (not shown). Chelton and McCabe (1985) proposed a simpler form ($\sigma_0(\text{dB}) = 10[A + B\log_{10}(U_w)]$) based on the scatterometry model. Chelton and Wentz (1986) proposed a table-driven model based on much larger data sets than had previously been used. Dobson et al. (1987) proposed a smoothed version of the Brown

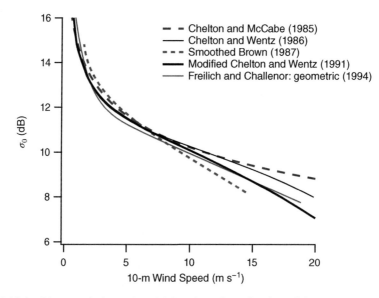

Figure 8.7 Multiple altimeter wind speed model functions. A good review of these models is provided in based on Chelton et al. (2001).

model that fixed some issues in the earlier model but was only appropriate for the σ_0 range from 7 to 15 dB. Witter and Chelton (1991) proposed the modified Chelton–Wentz model based on a reanalysis using both Geosat and Seasat data. Freilich and Challenor (1994) proposed a model of the simple form:

$$\sigma_0(dB) = a - b\, U_w + c \exp(-d\, U_w) \tag{8.9}$$

with constants a, b, c, and d determined by fitting the observed distributions of wind speed and σ_0. Some of the curves in this plot have been adjusted to correspond to the modern standard measurement height of 10 m.

Although the equation is nonlinear at lower winds speeds, note that $\sigma_0(dB)$ varies linearly with wind speed for winds above 5 m s^{-1}. It is the exponential term in the formula that makes the model deviate from linearity at low wind speeds.

Figure 8.8 shows comparisons of buoy-measured wind speed with Topex altimeter-derived wind speed using two different wind retrieval algorithms: the Gourrion et al. (2002) Ku-band model (GO2) on the left and the dual-frequency linear composite model (LCM) proposed by Chen et al. (2002) on the right.

Table 8.5 Altimeter wind speed algorithm errors. Data from Freilich and Challenor (1994).

Model function	Mean error $\langle e \rangle$ (m s^{-1})	Root-mean square error $\sqrt{\langle e^2 \rangle}$ (m s^{-1})	Standard deviation $\sqrt{\langle e^2 \rangle - \langle e \rangle^2}$ (m s^{-1})
B81	0.23	1.70	1.68
SB	−0.016	1.68	1.68
MCW	0.48	1.75	1.68
NWP3	1.25	2.50	2.16
GG	0.03	1.64	1.61
Geometric	−0.12	1.72	1.72
Rayleigh-based	−0.16	1.69	1.68
Buoy-based	−0.26	1.60	1.58

Table 8.5 shows the mean bias, the RMS error and the standard deviation of eight model functions, some based on theory, others based on empirical fits. The best algorithms have near zero bias and an RMS error of 1.6–1.7 m s^{-1}. This is quite good considering that the buoy data are not perfect – they have errors of their own. Furthermore buoys measure wind at a single location, while the altimeter measures an area-averaged wind. In the end, all of the

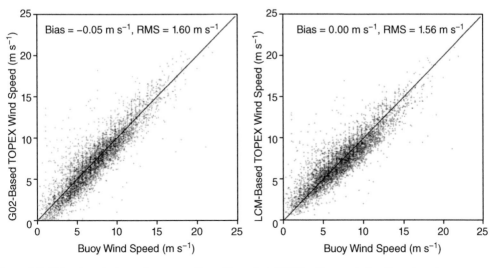

Figure 8.8 Altimeter wind speed accuracy. From Chen et al. (2002) with permission from American Geophysical Union.

errors shown in the plots cannot be attributed to the altimeter, but altimeter measurements are judged by comparison with buoys.

8.4 Altimeter Significant Wave Height

Altimeters can also measure significant wave height, which is defined as the average height of the $1/3$ highest waves in a wave spectrum ($SWH = H_{1/3}$). A common rule of thumb is that the significant wave height equals about four times the rms wave height.

The altimeter measures the significant wave height from the slope of the leading edge of the average return pulse. Approximately 1000 pulses (one second of data) must be averaged to define the slope of the leading edge.

The shape of the return pulse can be explained by examining the schematic representation in Figure 8.9 illustrating how an altimeter pulse interacts with the surface.

The upper row of Figure 8.9 shows a sequence of five sketches of a wide-beamwidth, short pulse propagating from the satellite to the sea surface. In the left-hand sketch, the pulse has not yet hit the surface. Then each sketch to the right shows the pulse at later times as the pulse hits the surface. The shading of the pulse indicates the relative power in the beam.

The second row shows the footprint on the sea surface contributing to the radar return at each illustrated time. After the leading edge of the pulse intersects the surface, the illuminated region is a circle. The circle grows until the trailing edge of the pulse intersects the surface. After that point the illuminated footprint becomes a growing annulus.

The third row indicates the beam pattern, with shading to represent the portion of the beam that is intersecting the surface.

The bottom row plots the area of the footprint contributing to the radar return as a function of time in red and the received power in blue. The illuminated area grows linearly with time until the trailing edge of the pulse intersects the surface. From then on, the area of the illuminated annulus remains about constant as a function of time. The RCS of the surface is initially proportional to the illuminated area, so the returned power mimics the shape of

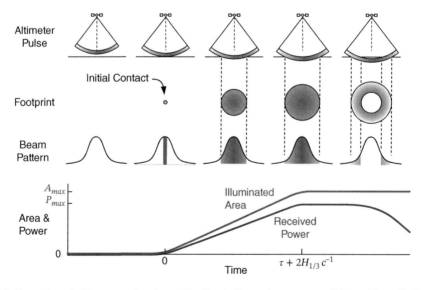

Figure 8.9 Formation of altimeter pulse shape. Shading indicates beam power. Adapted from Chelton et al. (1989).

the illuminated area curve. Eventually the power in the returned pulse falls off due to the antenna beam pattern as well as reduced RCS at increasing angles.

The growth of the illuminated area can be easily computed for a flat surface. For height R_0, the transmitted pulse begins to intersect the surface at time $t_c = c/R_0$. For time t' after the intersection, the radius of the illuminated circle is $r = \sqrt{(t' + t_c)^2 - R_0^2} = \sqrt{2R_0 ct'}\sqrt{1 + ct'/2R_0}$. The term $ct'/2R_0 \ll 1$ for the time of the pulse duration, so $r \approx \sqrt{2R_0 ct'}$, and the illuminated area $A = 2\pi R_0 ct'$. Thus to lowest order the illuminated area initially grows linearly with time. After the trailing edge of the pulse intersects the surface, the illuminated area becomes: $A = 2\pi R_0 c(t' + \tau) - 2\pi R_0 ct' = 2\pi R_0 c\tau$. Thus after the trailing edge of the pulse intersects the surface, the illuminated area is constant. Since the returned power is proportional to illuminated area, the normalized power for a smooth surface is given by:

$$P(t) = \begin{cases} 0 & \text{if } \frac{1}{\tau}\left[t - \frac{2R_0}{c}\right] \leq 0 \\ \frac{t}{\tau} & \text{if } 0 < \frac{1}{\tau}\left[t - \frac{2R_0}{c}\right] \leq 1 \\ 1 & \text{if } 1 < \frac{1}{\tau}\left[t - \frac{2R_0}{c}\right] \end{cases}$$

$$(8.10)$$

where the time t is measured with respect to the start of the transmitted pulse.

This analysis shows that for a calm sea surface, the area rise time is equal to the compressed effective pulse duration τ. For a rough sea surface with significant wave height $H_{1/3}$, this rise time increases by an amount $2c^{-1}H_{1/3}$. Figure 8.10 shows the time evolution of the average illuminated area for a pulse of duration $\tau = 3.125$ ns and Gaussian sea surface height distributions with $H_{1/3} = 0$, 1, 5, and 10 m. Time is displayed in nanoseconds relative to the two-way arrival time of the midpoint of the pulse reflected from the mean sea level at nadir. The reduction in the slope of the leading edge with increasing sea surface height is quite evident.

The rise time increase with increasing sea surface height is due to the scattered power being distributed across the crests, slopes, and troughs of the waves. Since to lowest order the surface height is a Gaussian-distributed random variable, the returned signal is nothing more than the convolution of the smooth surface response and the Gaussian distribution of surface heights. A more detailed analysis for the returned signal for a rough surface is provided in Brown (1977).

We have seen that by averaging a number of pulses the altimeter can measure both wind

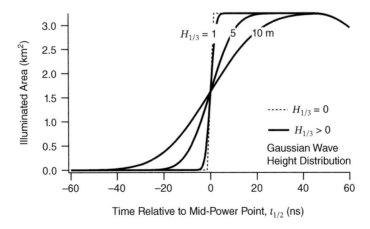

Figure 8.10 Effect of surface roughness on leading edge of returned pulse. Adapted from Chelton et al. (2000).

speed and significant wave height. Yet the footprint of the altimeter varies quickly with time. So what is the effective footprint for these measurements?

The wind speed measurement is made at the maximum of the returned power. In a calm sea, the instantaneous footprint at maximum power is the circle that occurs just as the trailing edge of the pulse intersects the surface. In general, the maximum illuminated area for a single pulse is:

$$A_{max} = \frac{\pi R_0(c\tau + 2H_{1/3})}{1 + R_0/R_e} \qquad (8.11)$$

where R_e is the radius of the Earth. Each measurement is made by averaging 1000 pulses. So the measurement footprint becomes the oval-like shapes as shown in Figure 8.11, reflecting the forward motion of the altimeter during the time it takes to receive 1000 pulses.

But as we saw, wave roughness also acts to delay the time of maximum power, which effectively broadens the footprint. The four sets of ovals in Figure 8.11 are the footprints at maximum power computed for significant wave heights of 0, 1, 5, and 10 m. The solid curves are for an 800-km orbit and the dashed curves are for a 1340-km orbit. In both cases, one second of averaging was assumed. Table 8.6 shows

Table 8.6 Altimeter footprints for various significant wave heights. From Chelton et al. (1989).

$H_{1/3}$ (m)	Effective footprint diameter (800 km altitude) (km)	Effective footprint diameter (1340 km altitude) (km)
0	1.7	2.1
1	2.9	3.6
3	4.4	5.6
5	5.6	7.0
10	7.7	9.7
15	9.4	11.7
20	10.8	13.5

the effective footprint diameters as a function of significant wave height for an 800 km or 1340 km orbit.

The significant wave height is computed on board the satellite using a real-time processor to compute the slope of the leading edge of a one-second average pulse. A look-up table is then used to derive the significant wave height and the values are sent to the ground as part of the data stream. In addition, the satellite averages the returns from 100 pulses and then sends the averaged waveforms to the ground every tenth of a second. These data are the basis for the development and application of more sophisticated analysis algorithms.

Ultimately the significant wave height is compiled into a global wave climatology. Near real-time wave data are also provided as inputs to wave forecast models. It is important to realize that the observations are limited to the ground track of the satellite, but geophysical models can be used to effectively interpolate between the measurements.

Altimeter-derived surface wind speed and significant wave height measurements can be used to monitor ocean swell, which can be combined with models to predict areas of dangerous seas for ships to avoid.

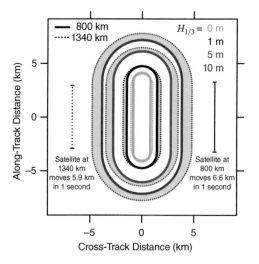

Figure 8.11 Radar altimeter footprint. Modified from Chelton et al. (2000).

8.5 Altimeter Sea Surface Height

8.5.1 Introduction

Extracting oceanographic information other than wind speed and SWH from altimeter measurements requires accurate determination of two things:

- The altimeter range to the mean sea surface.
- The distance of the radar to the Earth's center determined from a precise orbit determination.

The details are surprisingly complex.

Altimeters use timing to measure height. Specifically the altimeter measures the round-trip travel time for pulses transmitted by the radar and reflected from the sea surface. It does this by measuring the time delay between the pulse transmission and the returned signal crossing a threshold set to half of a running average of the maximum returned power.

The pulse propagation speed through the atmosphere is then needed to convert the propagation time measured by the altimeter to an estimate of height above the surface. This is complicated at the level of precision achieved by altimeters because the pulse is being transmitted through the atmosphere and ionosphere. Specifically, the propagation velocity is determined by the atmospheric index of refraction, which in turn depends on:

- Atmospheric gases
- Water vapor
- Cloud liquid water
- Electron density in the ionosphere.

The other issue is that the altimeter measures the range to the median scattering surface, but we want to know the range to the mean sea surface. The difference between the two is known as the sea state bias. This is yet another significant correction that needs to be made.

8.5.2 Pulse-limited vs Beam-limited Altimeter

Altimeters make their measurements looking straight down for a good reason. Consider the left-hand panel of Figure 8.12, which represents an altimeter with a very narrow main beam. It should be easy to see that this design is very sensitive to any off-vertical tilting of the antenna. Even small tilt errors in a narrow beam that is nominally pointed straight down lead to large altitude errors. For example, at the Topex altitude of 1336 km, a pointing error of 0.05 degree corresponds to a range measurement error that differs from the true altitude by about 50 cm.

Moreover, to get an illuminated spot diameter of 5 km with a 0.3 degree beamwidth, the Topex antenna would require a diameter of about 7.5 meters.

A pulse-limited altimeter avoids this antenna pointing angle sensitivity because the instrument only uses data from the leading edge

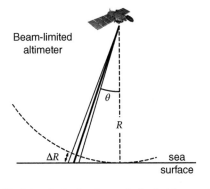
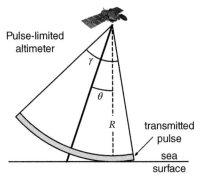

Figure 8.12 Pulse-limited vs beam-limited altimeter. Modified from Chelton et al. (1989).

of the returned pulse, which is guaranteed to come from the surface at the nearest range. Pulse-limited altimeters also provide an illuminated spot diameter of a few kilometers with a practical antenna size.

For example, the Topex antenna diameter is 1.5 meters, resulting in an antenna beamwidth of 1.1 degrees. When combined with a pulse effectively shortened by pulse compression, this gives a flat-ocean illuminated spot diameter of about 2 km with 0 SWH. The spot becomes larger as the SWH increases.

8.5.3 Altimeter Pulse Timing Precision

The primary drivers of the shape of the returned pulse were discussed in Section 8.4. Figure 8.13 provides a little more detail about the characteristics of that pulse. The pulse does not ramp from zero, but it starts at the background noise power, P_{noise}, before ramping up to the maximum received power, P_{max}, which in fact is signal plus noise. The two-way travel time for the leading edge of the pulse to propagate from the satellite to the wave crests and back is t_c. For a calm sea, the power rise time is the compressed effective pulse duration τ. For a rough sea surface with significant wave height $H_{1/3}$, this rise time increases by an amount equal to $2H_{1/3}/c$. This is because the troughs of the waves are effectively $H_{1/3}$

further away from the satellite. The timing is based on the round trip time, which means the path increases by twice this amount, and the speed of light converts the extra range to extra time.

The travel time of the pulse is measured from the start of the transmit pulse to the half-power point of the returned pulse. The height precision of the altimeter is determined by the standard deviation in determining the half-power point of the return waveform including the effects of averaging for one second. This is not an accuracy as there are still many corrections that need to be made, but it does set a lower bound on the noise of the measurement.

8.5.4 Altimeter Range Corrections

As we mentioned before, the altimeter measures the time of flight from the radar to the surface and back. As illustrated in Figure 8.14, we perform a series of corrections to convert this measurement to an estimate of the range of the altimeter antenna to the mean sea surface. We call this range R.

The altimeter's orbit also has to be determined to great accuracy. This orbit determination allows one to estimate the range from the altimeter antenna to the center of the Earth at every time in the satellite's orbit. This is then

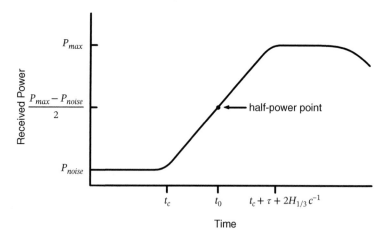

Figure 8.13 Altimeter pulse shape. From Chelton et al. (1989) with permission from American Meteorological Society (AMS).

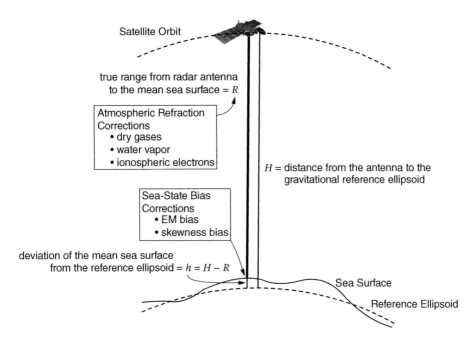

Figure 8.14 Altimeter measurement geometry. Adapted from Chelton, Walsh, and MacArthur (2000).

combined with a model of the gravitational reference ellipsoid to estimate the height of the altimeter over that ellipsoid. We call this height H. The difference between the mean sea surface and the reference ellipsoid, designated h, then provides a wealth of geophysical information about the lumpiness of the Earth's gravity field, atmospheric pressure variations and ocean currents, all of which affect sea surface height.

The altimeter measures the round-trip travel time t_0. Multiplying this by c and dividing by 2 yields what is called the apparent range, which would be the range to a flat surface in a vacuum:

$$R_{ap} = \frac{ct_0}{2} \tag{8.12}$$

Of course the true range R could be determined if we knew the effective propagation velocity v over the path:

$$R = \frac{vt_0}{2} \tag{8.13}$$

Instead, we estimate the true range by making three corrections to the apparent range:

$$R = R_{ap} - \Delta R_{atm} - \Delta R_{ion} - \Delta R_{ssb} \tag{8.14}$$

where:

ΔR_{atm} is the correction to the apparent range due to atmospheric refraction.

ΔR_{ion} is the correction due to propagation of the pulse through the ionosphere.

ΔR_{ssb} is the sea state bias correction.

The first correction accounts for the delay in the pulse arrival due to atmospheric refraction, which has the effect of lengthening the path. The second correction is for delay in the propagation of the pulse as it passes through the ionosphere. And the final correction is for what is called sea state bias.

Once all these corrections have been made, the range can be subtracted from the range to the reference ellipsoid to obtain the deviation of the mean sea surface from the gravitational reference ellipsoid:

$$H - R = h \tag{8.15}$$

where H is the distance from the reference ellipsoid to the antenna (determined from precision orbit data) and h is the deviation of the mean sea surface from the gravitational reference ellipsoid.

The atmospheric refraction correction – also known as the path delay – is the path length to be subtracted from the range to the sea surface that is estimated from the round-trip travel time and the free space value of the speed of light. The correction has two components: a dry tropospheric correction and a wet tropospheric correction.

The dry tropospheric correction is due to the refractive effects of the dry gas constituents in atmosphere. This is the single largest correction made to the apparent range. It is on the order of 225 cm depending on surface pressure and latitude. The dry tropospheric correction is computed from:

$$\Delta R_{dry}(\text{cm}) = 222.74 P_0/g_0(\varphi)$$
$$= 0.2271 P_0(1 + 0.0026 \cos 2\varphi)$$
$$(8.16)$$

where P_0 is the surface pressure in mbar and $g_0(\varphi)$ is the surface gravitational acceleration (cm sec^{-2}) at latitude φ. The surface pressure values are obtained from observations and numerical weather prediction models, while the gravitational acceleration is obtained from a simple model.

The wet tropospheric correction accounts for the atmospheric refraction due to water vapor and cloud liquid water droplets. This path correction is of order 10 to 30 cm, with the cloud liquid water typically contributing 1 cm or less. On Topex/Poseidon the columnar water vapor estimates were obtained from a three-frequency microwave radiometer operating at 18, 21, and 37 GHz. For other systems, the data are obtained from weather models.

The ionospheric correction accounts for path delays due to the dielectric properties of free electrons in the ionosphere. This correction is of order 5 to 15 cm. Single-frequency altimeters make this correction based on this integral over range R of the electron density in the ionosphere, with the electron content obtained from ionospheric models:

$$\Delta R_{ion}(\text{cm}) = \frac{40.3 \times 10^6}{f^2} \int_0^R n_e(z)dz$$
$$(8.17)$$

where $n_e(z)$ is the electron density (electrons/cm^3).

Topex/Poseidon was the first two-frequency altimeter, with the second frequency used to provide a direct measurement of the ionospheric correction. Note the equation for the ionospheric correction has a simple dependence on frequency, so two measurements at different frequencies were used to estimate the correction.

Sea state bias is the subtlest of the corrections. It arises because of differences between the distributions of scatterers and the sea surface height. The dark black curve in Figure 8.15 is an example surface elevation profile that includes a cnoidal component that makes the profile sharply peaked with rounded troughs. This profile has been constructed so the mean height is 0. The mean slope of the surface at each location was then computed numerically (not shown). The relative NRCS (bottom red curve) was then computed at each location based on the mean slopes and an assumed wind speed of 10 m/s using equation 8.7. Notice that the NRCS is only large for segments of the surface that are horizontal – the definition of specular for a nadir-looking instrument. The vertical location of the high RCS flashes is shown in the red plot on the right. This makes it is easy to see that there are more glints in the troughs of the waves than near the crests. The mean and median scattering heights can then be computed from this distribution (blue horizontal lines).

The range to the sea surface is estimated from the time interval from when a pulse is transmitted to the time when the mid-point of the leading edge of the returned waveform is received. This half-power point is determined by the median height of the specular scatterers, that height where half the scatterers are above the height and half are below (blue dashed line in Figure 8.15). Unfortunately we want to estimate the level of the mean sea, not the median of the scatterers. The sea state bias is the term that makes this correction. One common empirical form of the correction takes the simple form of just 2% of the significant wave height. Thus the correction would be 5 cm for a typical significant wave height of 2.5 m.

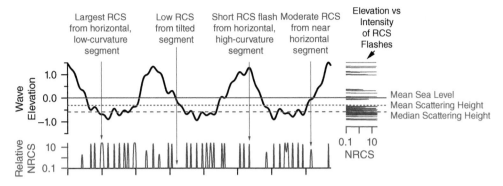

Figure 8.15 Sea state bias. Modified from Chelton, Walsh, and MacArthur (1989).

A more detailed analysis of this correction shows that two distinct physical processes contribute to the correction:

- The electromagnetic or EM bias is due to the difference between the mean sea surface and the mean scattering surface. This arises because waves are nonlinear so that wave troughs tend to be smoother than wave crests. This means the returns from the troughs are stronger than the returns from the crests, biasing the altimeter measurements.
- The skewness correction comes from the difference between the mean scattering surface and the median scattering surface. This comes about because the scatterers are attached to the wave and wave elevations are non-Gaussian which shifts the median from the mean sea level down toward the wave troughs.

Considerable research has been conducted on the estimation of sea state biases. This is certainly not the largest correction, but it turns out that uncertainty in the sea state bias correction is the single largest contributor to the uncertainty in range estimation.

Figure 8.16 compares some theoretical models, the lines, and some empirical models, the circles, for the sea-state bias coefficient. The dashed curve was an old theoretical model by Glazman, Fabrikant, and Srokosz (1996). The solid and dotted lines are one of the current best theories by Elfouhaily, Thompson, Chapron, and Vandemark (2000).

Table 8.7 lists typical values for each of the altimeter corrections, along with the

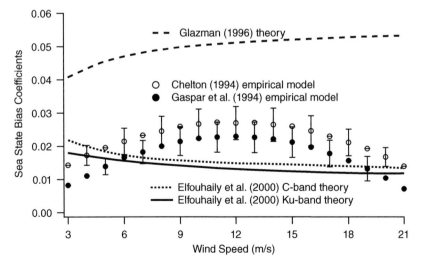

Figure 8.16 Sea state bias coefficient. From Elfouhaily, Thompson, Chapron, and Vandemark (2000) with permission from American Geophysical Union.

Table 8.7 Typical values of altimeter corrections.

Range Correction	Total (cm)	Variable (cm)	Uncertainty (cm)
Dry tropospheric refraction			~1
Latitudes 30°N to 30°S	226	0.5	
Latitudes 30° to 60°	226	2	
Wet tropospheric refraction			~1
Latitudes 30°N to 30°S	24	6	
Latitudes 30° to 60°	10	5	
Ionospheric refraction			<1
Latitudes 30°N to 30°S	12	5	
Latitudes 30° to 60°	6	2	
Sea state bias			~2
Latitudes 30°N to 30°S	4	1	
Latitudes 30° to 60°	6	3	

typical variations in those corrections and the uncertainties in those corrections. The dry tropospheric refraction is the largest correction by far, but has low variability and almost no uncertainty. The sea-state bias is the smallest correction, but actually has the largest uncertainty. In any case, you can begin to appreciate how hard it is to accurately estimate the height of an altimeter operating at a mean altitude of 1300 km to a precision of a few centimeters.

8.6 Sea Surface Topography

As we discussed before, the altimeter measures R, we get H from the satellite orbit and the reference ellipse, then the mean sea surface height above the reference ellipsoid $h = H - R$.

Now we mentioned the reference ellipsoid before, but didn't really describe it. The Earth is approximately an oblate spheroid. So instead of modeling the Earth's surface or gravitational field as a sphere, it is modeled as an idealized ellipsoid. All topography measurements are then made relative to this ellipsoid. For example, heights reported by GPS are all measured relative to the WGS 84 reference ellipsoid.

As illustrated in Figure 8.17 the height of the mean sea surface above the reference ellipsoid can in turn be expressed as the sum of 4 components:

$$H - R = h = h_g + h_t + h_a + h_d \quad (8.18)$$

The first component h_g is the geoid height relative to the reference ellipsoid. It turns out that the Earth's gravity field is lumpy. There might be an undersea mountain of a particularly dense material where the gravity field is a bit stronger and another region with less dense materials in the crust where the gravity is weaker. The geoid is defined to be the surface of equal gravitational potential that undulates above and below the reference ellipsoid. The geoid would define the height of the sea surface if there were no tides, ocean currents or atmospheric pressure variations.

Because the fluctuations in the geoid make satellites speed up and slow down, our current knowledge of the geoid comes mostly from precise orbital information that has been encapsulated into geoid models. The current

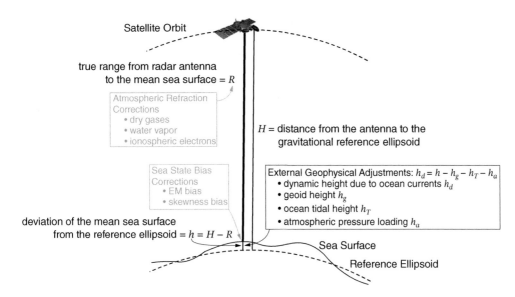

Figure 8.17 Components of altimeter measurements. Adapted from Chelton et al. (2000).

standard geoid model is EGM2020, with the 2020 indicating the year of revision. Typical geoid undulations are on the order of tens of meters, with a maximum of about 100 m. The geoid model does not change on time scales that we are concerned about here.

The second component, h_t, is the height variations due to the tides, which can be predicted by a tidal model. Many people think of tides as a coastal phenomenon but tides exist throughout the world's oceans. The typical standard deviations of the tides are 10–60 cm, and of course the actual tidal height varies with location and time. While not strictly applicable, it is interesting to note that there is also an Earth tide, which induces the ground you are standing on to go up and down by as much as a meter during a single day.

The third component, h_a, is the sea surface height variations due to atmospheric pressure variations. This is an inverse barometer effect, with a sensitivity of about −1 cm per mbar. This parameter is based on atmospheric pressure obtained from weather models. Typical values are 2–16 cm depending on season and latitude.

The final component to the sea surface height, h_d, is the dynamic surface elevation or sea surface topography. This is caused

by ocean currents and the Coriolis force, as will be explained. Typical values are tens of centimeters to 150 cm.

Ocean current systems are major contributors to the sea surface topography. To first order, these current systems are in geostrophic equilibrium, which implies that the horizontal pressure gradients in the water column are balanced by Coriolis forces. Of course the Coriolis force, a rather alliterative phrase, is due to the rotation of the Earth.

With some approximation, the equation of motion for a parcel of water moving at constant velocity on the surface of a rotating Earth is given by three terms: the pressure gradient, the force of gravity and the Coriolis force:

$$\frac{1}{\rho}\nabla p = \vec{g} - 2\vec{\omega}_E \times \vec{v} \qquad (8.19)$$

where:

ρ = water density
p = pressure
g = gravitational acceleration
ω_E = Earth angular rotation rate
  $= 7.27 \times 10^{-5}$ rad s^{-1}
v = water velocity at sea surface, with x, y components (u, v).

Equation (8.19) is a simplified version of the Navier–Stokes equations with approximations appropriate for large-scale flows in the deep ocean.

The standard coordinate system for these equations, as illustrated in Figure 8.18, is:

x-axis = positive to east (into the page in this illustration)
y-axis = positive to north
z-axis = positive to local zenith
θ = latitude.

The gravitational force is purely in the z-direction, so it drops out of the x and y components of the equation of motion yielding these two expressions:

$$\frac{1}{\rho}\frac{\partial p}{\partial x} = -2(\vec{\omega}_E \times \vec{v})_x = +(2\omega_E \sin\theta)v$$

$$= +fv \qquad (8.20)$$

$$\frac{1}{\rho}\frac{\partial p}{\partial y} = -2(\vec{\omega}_E \times \vec{v})_y$$

$$= -(2\omega_E \sin\theta)u = -fu \qquad (8.21)$$

where $f = 2w_E \sin\theta$ = Coriolis parameter ($= 1 \times 10^{-4}$ rad s^{-1} at 45° N latitude).

These equations say that the pressure gradient in the x-direction (east) is balanced by the Coriolis parameter times the north component of the current, and the pressure gradient in the y-direction (north) is balanced by the Coriolis parameter times the west component of the current.

Now any change in surface elevation η, is linearly related to a change in pressure at any

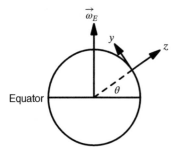

Figure 8.18 Coordinate system for Navier–Stokes equations.

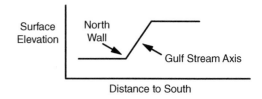

Figure 8.19 Gulf Stream topography.

depth: $p = \rho g \eta$. Hence:

$$\frac{\partial p}{\partial x} = \rho g \frac{\partial \eta}{\partial x} = +\rho fv \quad \rightarrow \quad \frac{\partial \eta}{\partial x} = +\frac{f}{g}v$$

$$(8.22)$$

$$\frac{\partial p}{\partial y} = \rho g \frac{\partial \eta}{\partial y} = -\rho fu \quad \rightarrow \quad \frac{\partial \eta}{\partial y} = -\frac{f}{g}u$$

$$(8.23)$$

The gradients of both pressure and surface elevation are proportional to the currents at a 90° angle to the gradients. In other words, a north-flowing current v produces an east–west surface slope, and an east-flowing current u produces a north–south slope. Note that the surface elevation is higher on the right-hand side of the current.

Consider the Gulf Stream where it flows toward the east. Then $v = 0$ and u is positive:

$$\frac{\partial \eta}{\partial y} = -\frac{f}{g}u \quad \rightarrow \quad \text{slope } \frac{\partial \eta}{\partial y} < 0 \quad (8.24)$$

Hence, as you go north, the surface elevation decreases (Figure 8.19).

For example, consider the Gulf Stream flowing east at a latitude of 38° at about 2 m s^{-1}. The width of the Gulf Stream is about 55 km (0.5° latitude). At that latitude, the Coriolis parameter f is about 9×10^{-5} rad s^{-1}. Substitution then yields the probably surprising estimate that the surface elevation change across the Gulf Stream is:

$$\Delta\eta = -\frac{fu}{g}\Delta y = 1.0 \text{ m} \qquad (8.25)$$

with the south side being higher than the north wall. Now a 1-m surface elevation change over a 55-km distance is not enough of a slope to surf down, but it is measurable.

To generalize this result, we note that altimeter measurements of the along-track slope of the surface elevation can be used to estimate the cross-track current. And while the altimeter is not sensitive to the along-track current, the current vector can be measured by combining cross-track measurements at the intersection points of ascending and descending passes. Altimeters are often placed into repeating orbits so these intersection points occur at fixed locations on the Earth allowing for the construction of time series of currents.

Figure 8.20 is an image of sea surface temperature that show Gulf Stream eddies, essentially large spinning blobs of fluid that can be spun off either north or south of the Gulf Stream. Eddies spun off to the north are called warm core rings because they consist of warmer Gulf Stream water surrounded by colder shelf water. Eddies spun off to the south are cold core rings, and are less evident in thermal imagery because of their vertical structure.

In either case the rings are on the order of 100 km across and can last for weeks to months. Some dissipate, some unwind when they interact with the shelf, and others get reabsorbed into the Gulf Stream.

The development of rings is highly nonlinear, but easy to conceptualize. While the Gulf Stream is roughly in geostrophic balance, the flow turns out to be unstable to some types of meanders. Figure 8.21 illustrates that rings start as a meander of the Gulf Stream. Nonlinearities in the flow then cause the amplitude of the meander to increase.

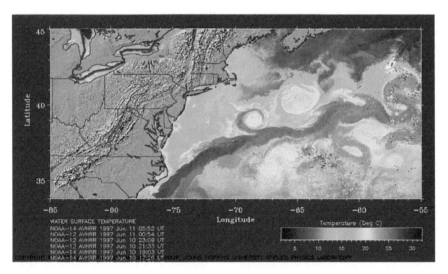

Figure 8.20 Gulf Stream rings. Source: JHU/APL http://fermi.jhuapl.edu.

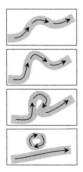

1. Gulf Stream meanders develop

2. Amplitude of the meander increases

3. Meander forms a loop and begins to pinch off

4. Ring separates from the Gulf Stream

Figure 8.21 Formation of Gulf Stream rings. Figure from Costal Carolina university, Ring Formation in Gulf Stream http://marine.coastal.edu/gulfstream/ with permission from Coastal Carolina University.

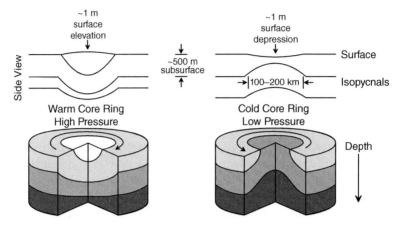

Figure 8.22 Cross-sectional views of the subsurface structure of a warm and cold ring. Source: SFSU, MEANDERS AND EDDIES IN THE OCEAN http://tornado.sfsu.edu/geosciences/classes/m415_715/ Monteverdi/Satellite/Oceanography/eddy.htm with permission from San Francisco State University.

Eventually the meander forms a loop and begins to pinch off. Finally, the ring separates from the Gulf Stream.

Figure 8.21 shows the development of a warm core ring, but meanders to the south can create cold core rings. Notice that the circulation in the warm core ring is always clockwise, which we refer to as anticyclonic. The circulation for cold core rings is cyclonic.

Figure 8.22 provides cross-sectional views of the subsurface structure of a warm and cold ring. The top plots show isopyncnals, or lines of constant density in the water column, as viewed from the side. The curves at the top of the plot labeled surface are not drawn to the same vertical scale, but are meant to represent the sea surface expression of the ring.

The warm core ring is circulating clockwise, meaning that a pressure gradient exists with the center of the ring about 1 m higher than the edge. The underwater expression of this high-pressure cell takes just the opposite shape with a magnitude of a few hundred times that of the surface expression. So it is easy to see that the warm core ring takes the form of a lens of trapped fluid. The cold core ring is very different with a subsurface maximum current accompanied by a dip in the surface above a low-pressure cell.

Figure 8.23 is a plot of the sea surface height measured by the Seasat altimeter on a single pass in 1978. From left to right we can see a part of a warm core ring, the edge of the Gulf Stream, and two cold core rings.

Figure 8.24 contains maps of sea surface height on the left and sea surface temperature anomaly on the right during a period of El Niño. El Niño, and its counterpart La Niña, are

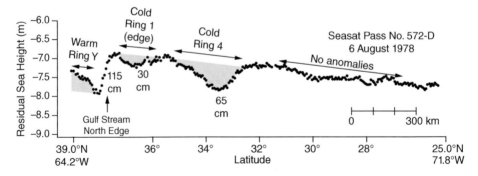

Figure 8.23 Seasat altimeter measurements of warm and cold core rings. Source: Cheney and Marsh (1981) with permission from John Wiley & Sons.

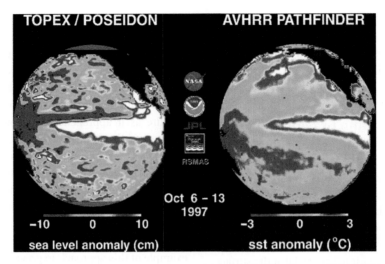

Figure 8.24 Sea surface height and temperature anomaly during El Niño. Source: NASA/NOAA.

weather patterns that periodically occur across the equatorial Pacific. These phenomena are the results of a complex set of interactions between the ocean and atmosphere.

Under normal conditions, a steady westward wind piles water up on the western side of the Equatorial Pacific. The cycle begins when these winds relax, releasing this wave of water to propagate east across the Pacific. The wave is equatorially trapped by dynamic balances driven by the Coriolis acceleration. Much of the wave is subsurface, so it takes the form of an equatorially-trapped Kelvin wave. When the wave reaches the coast of South America, it causes a major upwelling event in the coastal

waters off of South America. The oceanic upwelling brings nutrients to the surface, making this one of the most biologically productive regions of the oceans. The extra heat at the ocean surface also has a huge impact on evaporation, which then produces unusually heavy rains over the continent. There is a complex interaction with the planetary oscillations in the atmosphere that starts the cycle over again.

Figure 8.25 contains maps of sea surface height and temperature anomaly during the La Niña phase of this oscillation, showing how the ocean rebounds from forcing caused by the El Niño event.

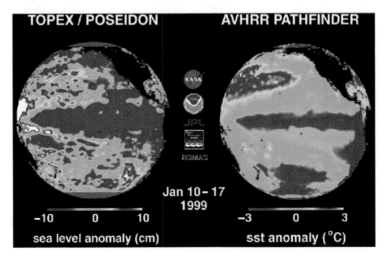

Figure 8.25 Sea surface height and temperature anomaly during La Niña. Source: NASA/NOAA.

In the end we see that satellite altimeters have an ability to measure sea surface topography, providing a unique view into large-scale ocean dynamics.

8.7 Measuring Gravity and Bathymetry

In the absence of ocean currents, tides, and storms, the height of the sea surface follows the geoid. So if the effects of currents, tides, and storms can be removed, altimeter measurements can be used to estimate the geoid and hence the gravitational potential at the surface. Storms are transient and their effects can be removed by modeling and filtering. Ground repeat orbits with orbital periods chosen to avoid aliasing of tidal energy, are used to obtain time series that can be averaged to remove the effect of tides. Permanent ocean currents, such as western boundary currents present a more fundamental challenge to estimating the geoid because it is difficult to separate the dynamic topography associated with steady currents from the geoid signal.

For this reason, modern geoid models are mostly based on a different approach to gravity-sensing from satellites. In this approach, a pair of satellites are placed in the same orbit, one following the other. The lead satellite speeds up relative to the trailing satellite as it approaches a region of higher gravity, and then slows on the other side of the anomaly. The gravity field is then estimated by analysis of precision measurements of the variations in range between the two satellites. Such measurements are immune to the effects of currents, storms, tides, and EM-bias. GRACE, more fully described in Chapter 12.1, is an example of this approach to gravity sensing.

Gravity-sensing satellites, although highly accurate, only provide limited spatial resolution. For GRACE, the best resolution was about 250 km (Vishwakarma et al., 2018). So while modern geoid models (EGM2008, (Pavlis et al., 2012)) are derived from gravity-sensing satellites, altimeters still provide global gravity measurements at the finest spatial scales. Figure 8.26 illustrates the gravity anomaly map (the differences between the gravity potential and the gravity associated with the reference

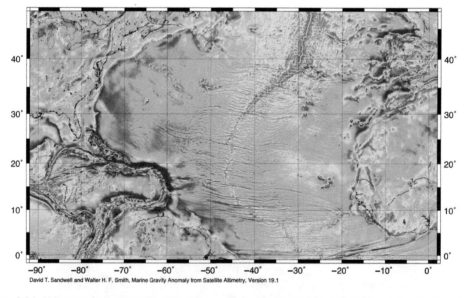

David T. Sandwell and Walter H. F. Smith, Marine Gravity Anomaly from Satellite Altimetry, Version 19.1

Figure 8.26 Altimeter-derived gravity anomaly map of the North Atlantic region. Source: Sandwell and Smith. See Sandwell et al. (2014)/with permission of Sandwell and Smith.

ellipsoid) based on data from the Jason-1 and CryoSat altimeters (Sandwell et al., 2014). Such measurements of the geoid now extend to spatial scales of a few kilometers.

It has long been known that there is a strong correlation between bathymetry and the geoid. A paper by Dixon et. al. illustrated this correlation using data from the Seasat altimeter (Dixon et al., 1983). Figure 8.27 illustrates how the extra mass of a seamount causes a local increase in gravity. This increased gravity draws water toward the anomaly, causing the sea surface to bulge upwards above the seamount. A seamount with a height of 2 km can induce a sea level bulge of a 20 cm, with smaller features producing smaller effects (Sandwell et al., 2002).

At the finest resolution, the satellite altimeter is sensitive to the along-track change in gravity. The along-track tilt in the local gravity field caused by a seamount leads to an along-track sea surface slope. One second of averaging provides an along-track sample every 6–12 km, depending on altitude and sea state. A 20-cm height difference made across a separation of 20 km corresponds to a mean surface slope of 10 μrad. As a rule of thumb, the most recent reports suggests an accuracy of about 2 mGal can be achieved in gravity anomaly mapping, corresponding to surface slope measurements of about 2 μrad. Thus the accuracy of the gravity field recovery is controlled by the accuracy of the sea surface slope measurement.

The correlation between bathymetry and the geoid is strongest over spatial wavelengths between 15 and 200 km (Smith & Sandwell, 1997), with the resolution ultimately limited by the local depth of the ocean. However variations in sediment thickness in some regions of the oceans complicate the estimation of topography from gravity. In these areas, combining the altimeter-based gravity measurements with estimates of bottom topography obtained from other measurements can lead to estimates of additional parameters of interest to geologists.

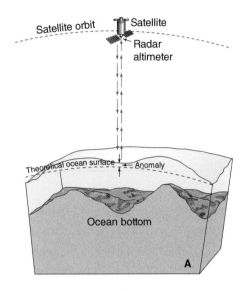

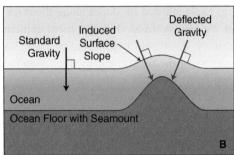

Figure 8.27 Basis for measuring bathymetry from an altimeter. From Sandwell et al. (2002) with permission from ELSEVIER.

8.8 Delay-Doppler Altimeter

As we have discussed, all pulse-limited altimeters measure the height to the nearest points on the surface. When flying over any reasonably level surface, like the ocean, this point is located on the ground track of the satellite just below the satellite. But if you try to use an altimeter over a sloped surface, say land or landed ice, the closest point to the altimeter will come from a location some distance from the subsatellite point and the measurement will not represent the height of the surface at the location of the altimeter (Figure 8.28).

Over a flat surface (e.g. water), an altimeter measures height exactly below the antenna

Over a sloped surface (e.g. landed ice), an altimeter measures closest point, so measurement location depends on slope

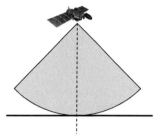 

Figure 8.28 Altimeter location uncertainty on sloped surfaces.

A few years ago Raney was thinking about how to use a radar altimeter to measure the thickness of landed ice, such as over Greenland. Raney realized that the measurements were badly affected by slopes along the track. But he also realized that the measurement accuracy could be improved, at least for slopes along the flight track, by improving the along-track resolution of the instrument.

So in 1998 Keith Raney invented a new class of altimeter. The Delay-Doppler altimeter uses Doppler processing to sharpen the temporal response function, improve SNR, and reduce measurement noise (Raney, 1998).

If we imagine applying a Doppler filter to the returns, then the return from the zero-Doppler bin will ramp up as the zero-Doppler area grows, but then decays quickly as the area shrinks. To see this, first consider the radial extent of the illuminated area for a conventional altimeter illustrated in the left-hand panel of Figure 8.29. The radius of the initially illuminated circle grows as $\sqrt{2hct}$, and the illuminated ring has constant area with width proportional to $\sqrt{2hc}(\sqrt{t} - \sqrt{t - \tau})$. As shown in the right-hand panel of Figure 8.29, the Doppler bins are constant width, so the power in the zero-Doppler bin is proportional to:

$$P(t) = \begin{cases} 0 & \text{if } \frac{1}{\tau}\left[t - \frac{2R_0}{c}\right] \le 0 \\ \sqrt{\frac{t}{\tau}} & \text{if } 0 < \frac{1}{\tau}\left[t - \frac{2R_0}{c}\right] \le 1 \\ \sqrt{\frac{t}{\tau}} - \sqrt{\frac{t}{\tau} - 1} & \text{if } 1 < \frac{1}{\tau}\left[t - \frac{2R_0}{c}\right] \end{cases} \quad (8.26)$$

where the time t is measured with respect to the start of the transmitted pulse.

The illustration in Figure 8.29 shows the surface response of a conventional pulse-limited altimeter on the left and a Delay-Doppler altimeter on the right. The response at the half-power point and a time after the peak are highlighted in gray, with the highlighting applied only to the zero Doppler bin for the Delay-Doppler altimeter. Bins forward of this zero Doppler bin will have some small motion toward the radar and hence exhibit a positive Doppler shift, and those aft will have a negative Doppler shift.

The analysis is straightforward to show that the other Doppler bins have similar behavior, just advanced or delayed in time. So Raney's insight was to Doppler process the altimeter data, then sum together all of the Doppler returns with proper delays to increase the SNR. The result is a return with a faster rise time, greater amplitude, and significantly improved along-track resolution.

It is interesting to note that Raney was not the first one to consider the application of

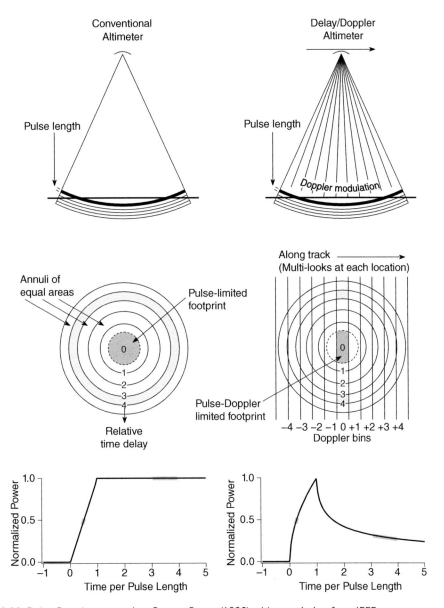

Figure 8.29 Delay Doppler processing. Source: Raney (1998) with permission from IEEE.

Doppler processing to altimeters, but he was the first person to succeed. This is because all of the previous attempts had jumped to the idea of summing the bins together coherently. When you do this, you find that a fully coherent measurement is Chi-squared distributed with only two-degrees of freedom. This speckle noise associated with the random placement of scatterers in the field of view absolutely dominates the noise for the instrument, so that coherent altimetry does not work. The Delay-Doppler altimeter succeeds because it performs an incoherent power sum, which effectively reduces the speckle noise to manageable levels.

Ultimately a Delay-Doppler/Phase Mono-pulse (D2P) altimeter was built at JHU/APL as a proof-of-concept instrument and flown multiple times to measure ice thickness, over both sea ice and landed ice. The phase monopulse part of the design was due to Robert Jensen, also of JHU/APL. It was an old radar trick that uses two receive antennas separated in the cross-track direction. The phase between the two measurements provided information about where the returns were coming from in the cross-track direction, allowing the instrument to determine the location of the altitude measurement even in the presence of a large cross-track surface slope.

The European Space Agency recognized the value of Raney's invention. The SAR mode in current ESA altimeters on the CryoSat-2 and Sentinel-3 satellites may be misnamed, but it is based on Raney's Delay-Doppler design (Raynal et al., 2018).

References

Brown, G. (1977). The average impulse response of a rough surface and its applications. *IEEE Transactions on Antennas and Propagation, 25*(1), 67–74.

Brown, G., Stanley, H., & Roy, N. (1981). The wind-speed measurement capability of spaceborne radar altimeters. *IEEE Journal of Oceanic Engineering, 6*(2), 59–63.

Chelton, D. B., & McCabe, P. J. (1985). A review of satellite altimeter measurement of sea surface wind speed: With a proposed new algorithm. *Journal of Geophysical Research: Oceans, 90*(C3), 4707–4720.

Chelton, D. B., Ries, J. C., Haines, B. J., Fu, L.-L., & Callahan, P. S. (2001). Satellite altimetry. In *International Geophysics* (Vol. 69, pp. 1–ii). Elsevier.

Chelton, D. B., Walsh, E. J., & MacArthur, J. L. (1989). Pulse compression and sea level tracking in satellite altimetry. *Journal of Atmospheric and Oceanic Technology, 6*(3), 407–438.

Chelton, D. B., Walsh, E. J., & MacArthur, J. L. (2000). Satellite altimetry. *Satellite Altimetry and Earth Sciences.*

Chelton, D. B., & Wentz, F. J. (1986). Further development of an improved altimeter wind speed algorithm. *Journal of Geophysical Research: Oceans, 91* (C12), 14250–14260.

Chen, G., Chapron, B., Ezraty, R., & Vandemark, D. (2002). A dual-frequency approach for retrieving sea surface wind speed from TOPEX altimetry. *Journal of Geophysical Research: Oceans, 107* (C12).

Cheney, R. E., & Marsh, J. G. (1981). Seasat altimeter observations of dynamic topography in the Gulf Stream region. *Journal of Geophysical Research: Oceans, 86* (C1), 473–483.

Cox, C., & Munk, W. (1954). Measurement of the roughness of the sea surface from photographs of the sun's glitter. *Journal of the Optical Society of America, 44*(11), 838–850.

Cox, C., & Munk, W. (1956). Slopes of the sea surface deduced from photographs of sun glitter. *Bulletin of the Scripps Institution of Oceanography, 6*(9), 401–488.

Dixon, T. H., Naraghi, M., McNutt, M. K., & Smith, S. M. (1983). Bathymetric prediction from Seasat altimeter data. *Journal of Geophysical Research: Oceans, 88*(C3), 1563–1571.

Dobson, E., Monaldo, F., Goldhirsh, J., & Wilkerson, J. (1987). Validation of Geosat altimeter-derived wind speeds and significant wave heights using buoy data. *Journal of Geophysical Research: Oceans, 92*(C10), 10719–10731.

Elfouhaily, T., Thompson, D. R., Chapron, B., & Vandemark, D. (2000). Improved electromagnetic bias theory. *Journal of Geophysical Research: Oceans, 105*(C1), 1299–1310.

Freilich, M. H., & Challenor, P. G. (1994). A new approach for determining fully empirical altimeter wind speed model functions. *Journal of Geophysical Research: Oceans, 99*(C12), 25051–25062.

Fu, L.-L., & Cazenave, A. (2001). *Satellite altimetry and earth sciences: A handbook of techniques and applications.* Elsevier.

Fu, L.-L., Christensen, E. J., Yamarone, C. A., Lefebvre, M., Menard, Y., Dorrer, M., & Escudier, P. (1994). TOPEX/POSEIDON mission overview. *Journal of Geophysical Research: Oceans, 99*(C12), 24369–24381.

Glazman, R., Fabrikant, A., & Srokosz, M. (1996). Numerical analysis of the sea state bias for satellite altimetry. *Journal of Geophysical Research: Oceans, 101*(C2), 3789–3799.

Gourrion, J., Vandemark, D., Bailey, S., Chapron, B., Gommenginger, G., Challenor, P., & Srokosz, M. (2002). A two-parameter wind speed algorithm for Ku-band altimeters. *Journal of Atmospheric and Oceanic Technology, 19*(12), 2030–2048.

Pavlis, N. K., Holmes, S. A., Kenyon, S. C., & Factor, J. K. (2012). The development and evaluation of the Earth Gravitational Model 2008 (EGM2008). *Journal of Geophysical Research: Solid Earth, 117*(B4).

Raney, R. K. (1998). The delay/Doppler radar altimeter. *IEEE Transactions on Geoscience and Remote Sensing, 36*(5), 1578–1588.

Raynal, M., Labroue, S., Moreau, T., Boy, F., & Picot, N. (2018). From conventional to delay Doppler altimetry: A demonstration of continuity and improvements with the Cryosat-2 mission. *Advances in Space Research, 62*(6), 1564–1575.

Resti, A., Benveniste, J., Roca, M., Levrini, G., & Johannessen, J. (1999). The Envisat radar altimeter system (RA-2). *ESA bulletin, 98*(8).

Sandwell, D. T., Gille, S. T., & Smith, W. H. F. (2002). Bathymetry from space: Oceanography, geophysics, and climate. *Geoscience Professional Services, Bethesda, MD.* Retrieved from https://www.geo-prose.com/pdfs/bathy_from_space.pdf.

Sandwell, D. T., Müller, R. D., Smith, W. H. F., Garcia, E., & Francis, R. (2014). New global marine gravity model from CryoSat-2 and Jason-1 reveals buried tectonic structure. *Science, 346*(6205), 65–67.

Smith, W. H. F., & Sandwell, D. T. (1997). Global sea floor topography from satellite altimetry and ship depth soundings. *Science, 277*(5334), 1956–1962.

Valenzuela, G. R. (1978). Theories for the interaction of electromagnetic and oceanic waves: A review. *Boundary-Layer Meteorology, 13*(1), 61–85.

Vishwakarma, B. D., Devaraju, B., & Sneeuw, N. (2018). What is the spatial resolution of GRACE satellite products for hydrology? *Remote Sensing, 10*(6), 852.

Witter, D. L., & Chelton, D. B. (1991). A Geosat altimeter wind speed algorithm and a method for altimeter wind speed algorithm development. *Journal of Geophysical Research: Oceans, 96*(C5), 8853–8860.

Zieger, A. R., Hancock, D. W., Hayne, G. S., & Purdy, C. L. (1991). NASA radar altimeter for the TOPEX/POSEIDON project. *Proceedings of the IEEE, 79*(6), 810–826.

9

Scatterometers

9.1 Ocean Waves

Wind stress on the air–water interface creates ocean waves (Figure 9.1). The kinematics of ocean waves, in other words how they propagate, is determined by gravity, pressure, and the properties of water. The dynamics of ocean waves, in other words their amplitude, is determined by a balance between energy input from the wind, dissipation from breaking, and nonlinear wave–wave interactions.

Ocean waves drive many forms of remote sensing, including scatterometry.

The development of a mathematical theory of surface gravity waves from the Navier–Stokes equations with appropriate boundary conditions can be found in any oceanography or most fluid mechanics texts. We will not review the details here, but we note that the base solution usually begins by

assuming the surface elevation η measured relative to mean sea level takes the form of a propagating sinusoid:

$$\eta = A \sin(kx - \omega t + \phi) \quad \text{where}$$
$$k \equiv \frac{2\pi}{\lambda} \quad \text{and} \quad \omega \equiv 2\pi f \quad (9.1)$$

with amplitude A, spatial wavenumber k, radian frequency ω, and initial phase ϕ.

We designate the vertical velocity of the fluid to be w. The vertical velocity at the surface equals the time derivative of the surface elevation: $w(0) = \frac{\partial \eta}{\partial t}$. A more general expression for the vertical velocity in the fluid at any depth z is given by:

$$w(z) = -A\omega \cos(kx - \omega t + \phi) \exp(-kz) \quad (9.2)$$

In this solution the vertical velocity decays exponentially with depth at a rate proportional to the wavenumber.

Figure 9.1 Three images of large breaking waves. The left-hand panel shows *The Great Wave off Kanagawa* by Katsushika Hokusai. The middle panel is a near re-creation of Hokusai's great wave in a laboratory wave tank (McAllister et al., 2019). The right-hand panel is a photograph of a similar wave in Lake Erie taken by Dave Sandford (https://www.davesandfordphotos.com).

Remote Sensing Physics: An Introduction to Observing Earth from Space, Advanced Textbook 3, First Edition. Rick Chapman and Richard Gasparovic.
© 2022 American Geophysical Union. Published 2022 by John Wiley & Sons, Inc.
Companion website: www.wiley.com/go/chapman/physicsofearthremotesensing

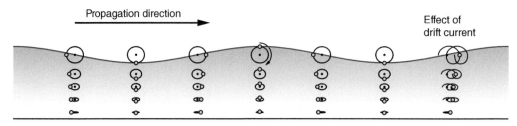

Propagation direction

Effect of drift current

Figure 9.2 Illustration of wave motions. Figure adapted from original by Kraaiennest https://commons .wikimedia.org/wiki/File:Orbital_wave_motion.svg.

In two dimensions the continuity equation reduces to $u_x + w_z = 0$, with the subscripts indicating derivatives. So the horizontal velocity u is given by:

$$u = -A\omega \sin(kx - \omega t + \phi)\exp(-kz) \tag{9.3}$$

Note that the vertical and horizontal velocities have the same amplitude at any given depth but are 90° out of phase. This means that to lowest order the flows move in closed circles.

Actual flows are bit more complex because of two factors. First, horizontal momentum at the surface creates vertical shear that acts on the fluid. Second, the sinusoidal solution is only strictly correct in the limit of small amplitudes. In general, wave nonlinearities will also cause an apparent drift, called the Stokes drift, in the orbits of particles.

Figure 9.2 shows the orbital currents associated with a finite amplitude surface gravity wave. Ignoring drift, the orbital currents are closed circles, as shown on the left-side of the figure. Yet there is some small drift in the direction of the wave propagation caused by the effects we just described, which opens the circles as shown in the rightmost column of the figure.

Ocean gravity waves obey a dispersion relationship that relates frequency and wavenumber. Equation (9.4) is the general dispersion relation formulated for gravity waves propagating on the surface of an ocean of depth d:

$$\omega = \sqrt{gk}\tanh(kd) \tag{9.4}$$

The two interesting limits to this solution are shown in equation (9.5). In the deep water limit, $kd > \pi \rightarrow \tanh(kd) \approx 1$. This approximation is good for most gravity waves in the middle of the ocean. In the shallow water limit, when $kd < 1 \rightarrow \tanh(kd) \approx \sqrt{kd}$. This is somewhat applicable to waves in shallow coastal waters, although such shoaling waves are usually highly nonlinear. It is far more applicable in the case of extremely long waves in the open ocean, such as tsunamis.

$$\omega \approx \begin{cases} \sqrt{gk} & \text{for deep water } (kd > \pi) \\ k\sqrt{gd} & \text{for shallow water } (kd < 1) \end{cases} \tag{9.5}$$

We will limit ourselves to the deep water dispersion relation from here on.

The phase speed for surface gravity waves, c_p, is the speed at which phase fronts move:

$$c_p \equiv \frac{\omega}{k} = \sqrt{\frac{g}{k}} = \frac{g}{\omega} \tag{9.6}$$

The energy in gravity waves does not propagate at the phase speed, but instead at the group speed. The group speed of gravity waves c_g is given by the derivative of frequency with respect to wavenumber:

$$c_g \equiv \frac{d\omega}{dk} = \frac{1}{2}\sqrt{\frac{g}{k}} = \frac{1}{2}c_p \tag{9.7}$$

The group speed of deep water gravity waves is thus 1/2 of the phase speed.

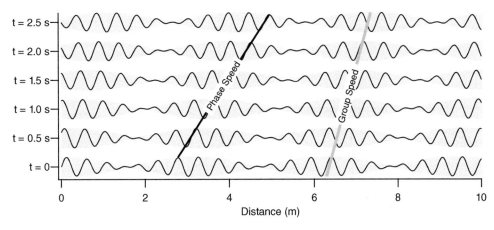

Figure 9.3 Illustration of wave group and phase speeds.

Figure 9.3 illustrates phase and group speeds for two waves. The black line indicates the phase speed of the waves and the gray line, which is aligned to the peaks in the envelope of the two waves, indicates the group speed.

Note that everything we have discussed pertains to the kinematics of gravity waves – surface waves where the primary restoring force is gravity. Surface waves with wavelengths smaller than 1.7 cm are so highly curved that surface tension becomes the primary restoring force. These are called capillary waves and their kinematics are quite distinct from gravity waves. For example, shorter capillary waves propagate faster than longer capillary waves, so the minimum phase speed of surface waves occurs at a wavelength of 1.7 cm. While most remote sensors primarily respond to surface gravity waves, systems operating above 35 GHz can be sensitive to capillary waves, which can have implications for Doppler sensing systems. The reason for this frequency-dependent sensitivity is explained in Section 9.2.

The generation and growth of ocean waves is very complex, but a few basics are worth noting. Small ripples are initially created by flow instabilities as the wind passes over a wave-less surface. These ripples are typically created near the minimum phase speed that occurs in the transition between gravity and capillary waves. This effect can be observed on a calm day with the waves created by a weak off-shore breeze at a beach.

These small waves then undergo some strong wave–wave interactions that create slightly longer waves heading at angles to the left and right of the wind. This creates a diamond pattern on the surface.

More regular surface gravity waves then grow from wave–wave interactions which tend to pump energy to longer wavelengths, and atmospheric pressure fluctuations which begin to arise because of the coupling between the wind and the growing waves (Figure 9.4).

We usually model ocean waves as a sum of sinusoids, encompassing all possible directions and wavenumbers:

$$\eta_{model}(t) = \sum (a_n \sin \omega_n t + b_n \cos \omega_n t)$$

$$(9.8)$$

Note that here, the wavenumbers k_x are implied by the $\omega_n t$ term, which is not unusual for this type of analysis. Sometimes we will assume that each wavenumber has a unique phase, but often we will model each wavenumber as the sum of a sine and a cosine term, each with a unique amplitude. It can be shown that the two formulations are equivalent.

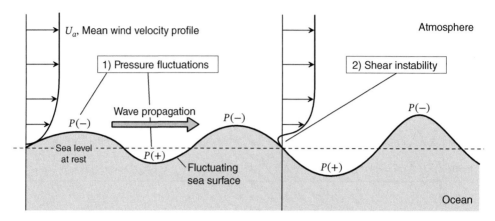

Figure 9.4 Schematic of near-surface boundary layer fluctuations that lead to wave growth. Figure by Seung Joon Yang, https://commons.wikimedia.org/wiki/File:Sjyang_waveGeneration.png.

One reason for expressing the surface as a sum of sine and cosines is that it makes it clear that the amplitude coefficients a_n and b_n can be computed directly from a Fourier decomposition of the surface:

$$a_n = \frac{2}{T} \int_{-T/2}^{T/2} \eta(t) \cos n\omega t \, dt$$

$$(n = 0, 1, 2, ...)$$

$$b_n = \frac{2}{T} \int_{-T/2}^{T/2} \eta(t) \sin n\omega t \, dt$$

$$(n = 0, 1, 2, ...) \qquad (9.9)$$

where T is the duration of the analysis window.

The frequency spectrum of the surface is then given at each frequency by the sum of the squares of the amplitude coefficients:

$$\Psi(\omega_n) = a_n^2 + b_n^2 \qquad (9.10)$$

The spectrum can be generalized to a directional spectrum that describes a time-varying surface with the magnitude of k and ω related by the dispersion relation: $\Psi(k_x, k_y, \omega)$.

Ocean wave data are then used to construct wave spectral models that describe the energy in each ocean wave spectral component (Figure 9.5). The simplest of these models, those that depend only on wind speed, apply to equilibrium situations where the wind has been blowing steady for a long time and over a long distance. Many modern wave spectral models also include the effects of limited fetch or duration. There are many such models, and if you're seeking fame in oceanographic circles, just create a new wave spectral model.

One simple model spectrum was proposed by Pierson and Moskowitz (1964). The full expressions for the Pierson–Moskowitz model are:

$$\Psi(\omega) = \frac{\alpha g^2}{\omega^5} \exp\left[-\beta(\frac{\omega_0}{\omega})^4\right] \qquad (9.11)$$

$$\alpha = 8.1 \times 10^{-3} \qquad \beta = 0.74$$

$$\omega_0 = \frac{g}{U_{19.5}} \qquad (9.12)$$

$U_{19.5}$ = wind speed at 19.5 m height

$$= 1.03 \, U_{10} \qquad (9.13)$$

$$H_{1/3} \approx 0.22 \frac{(U_{10})^2}{g} \qquad (9.14)$$

Despite its age, the Pierson–Moskowitz model is still used for some studies because it is so simple. But most modeling today utilizes more sophisticated models.

It is interesting to note that none of these models accurately describes the shape of the fully-developed wave spectrum from first principles. The development of such a model is still an unsolved problem. All of the practical models use some combination of approximations and parameters adjusted to data.

NOAA's National Data Buoy Center maintains over a thousand ocean buoys. These are

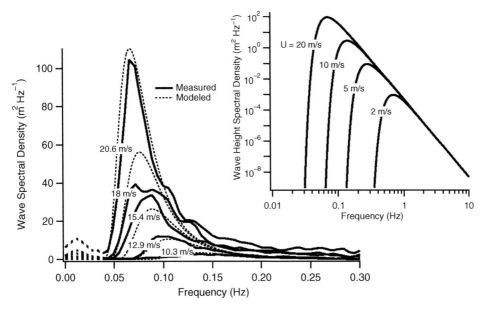

Figure 9.5 Measured and modeled ocean spectra for varying wind speeds. Source: Moskowitz (1964) with permission from John Wiley & Sons.

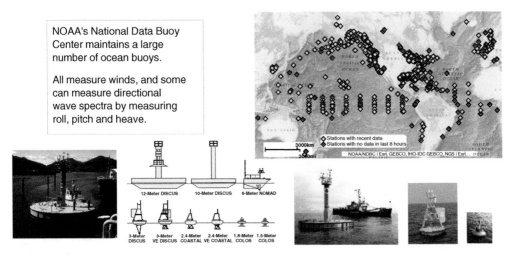

NOAA's National Data Buoy Center maintains a large number of ocean buoys.

All measure winds, and some can measure directional wave spectra by measuring roll, pitch and heave.

Figure 9.6 NOAA NDBC buoys. http://www.ndbc.noaa.gov.

concentrated around U.S. coasts, but many are also deployed in deeper water (Figure 9.6).

All of these buoys measure winds along with air and sea temperature. Some of them can also measure the directional wave spectra by measuring roll, pitch, and heave. The data are reported back via satellite and are available for download from the web. The buoys range in size from 3 to 12 m. The 10 and 12 m buoys are far more rugged than the smaller buoys, but are expensive to maintain and must be towed into position.

Measurements made near NDBC buoys are a primary means for validating many remote sensing data.

Ocean waves are also regularly predicted. The NOAA Wave Watch III model is an open-source wave spectrum prediction model that is run operationally with global predictions every three hours. Example outputs are shown in Figure 9.7. The model inputs include global winds, ice coverage, and sea surface temperature. The model takes into account wave growth from wind forcing, nonlinear wave–wave interactions, wave dissipation, and wave propagation.

Figure 9.8 shows some of the Wave Watch III predictions for the Pacific.

The left-hand panel shows the predicted vector winds that are inputs to the model. Several storms at various stages of development can be seen in the Northern Pacific and the Southern Ocean.

The middle panel shows the significant wave heights predicted by Wave Watch III. Comparing the two panels allows you to see that the largest waves originate in the storms with the highest wind speeds.

And the right-hand panel shows the predicted peak wave period, measured in seconds. This is an interesting plot that clearly illustrates the propagation of long swell that are created by distant storms. Storms create high amplitude waves over a wide range of wavelengths

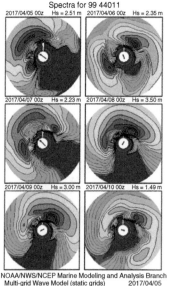

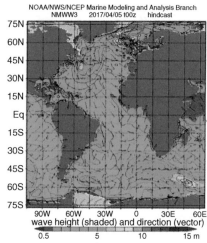

Figure 9.7 Example Wave Watch III predictions. Source: NOAA https://polar.ncep.noaa.gov/waves.

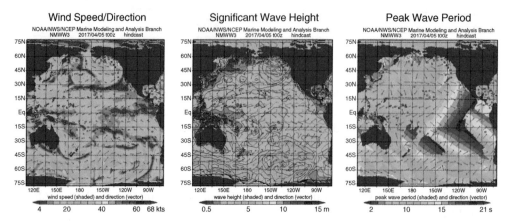

Figure 9.8 Example Wave Watch III predictions for the Pacific. (Source: NOAA https://polar.ncep.noaa.gov/waves/).

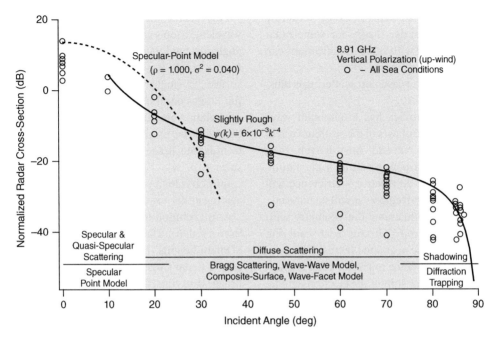

Figure 9.9 Bragg scattering dominates NRCS at moderate incident angles. Source: Valenzuela (1978) with permission from Springer Nature.

and periods, but the longest waves are only created at the highest wind speeds. Wave energy propagates at the group speed, which is proportional to wave period. Thus longer period waves propagate faster than shorter period waves. So a storm will create a spectrum of waves, but the longer periods will propagate faster than the shorter period waves. This dispersion of gravity waves causes the color banded structure in the predictions of peak wave period. Measurements of the change in the period of swell waves at a point can be used to estimate the distance to the storm that created the waves.

9.2 Bragg Scattering

We have previously seen in Section 8.2 that radar scattering breaks down into three regimes: the near-nadir specular regime that is used by radar altimeters; the low-grazing-angle regime near the horizon, where shadowing and wave breaking dominate; and the resonant or Bragg-scattering regime illustrated in Figure 9.9 that occurs at moderate

incident angles. This section discusses the physics of Bragg scattering and how it is modeled[1] .

For viewing geometries with incident angles between 25° and 75°, specular scattering is negligible and diffuse scattering theory must be used. The lowest-order approximation is to assume that the surface is "slightly rough", that is, we assume that:

- The wave height is much less than the RF wavelength.
- The slopes of the waves contributing to the backscatter are small compared to $2\sin\theta$ where θ is the incident angle.

The second assumption is always satisfied for ocean waves, but the first assumption is valid only for low-frequency (tens of MHz) radars. Scattering that satisfies these conditions is known as resonant or Bragg scattering. Despite the fact that microwave radars with centimeter wavelengths do not strictly satisfy the first

---

1 Bragg scattering is named after William Lawrence Bragg and his father William Henry Bragg who explained the scattering of X-rays from crystalline structures (Bragg & Bragg, 1913).

assumption, the concept of Bragg scattering is a useful first-order theory for some radar applications, and with some modifications can yield accurate predictions.

The geometry for resonant scattering is illustrated in Figure 9.10.

Assume the surface has a dominant wave component. If the phases of the scattered waves from each wave crest add constructively, then the backscattered power will be a maximum. Otherwise, some destructive interference will occur and the received power will be reduced.

The diagram illustrates the geometry with rays originating from a distant radar scattering from points on the near and far crest. The extra two-way path distance to the far crest is given by $2d$. So the phase shift of the electromagnetic wave scattered from the second crest, relative to the phase of the electromagnetic wave from the first crest, is $\Delta\varphi = 2k_{RF}\,d$.

Constructive interference occurs when the extra phase in the second path is an integer multiple of 2π:

$$\Delta\varphi = 2k_{RF}\,d = \pm2\pi m \qquad (9.15)$$

where m = 1,2,3…

If the wavelength of the surface wave is λ_{SW}, then the distance d equals $\lambda_{SW}\sin\theta$, so the lowest order condition for Bragg scattering becomes:

$$\lambda_{sw}\sin\theta = \pm\frac{1}{2}\lambda_{RF} \quad \text{or}$$

$$k_{sw} = \pm2k_{RF}\sin\theta = k_B \qquad (9.16)$$

where k_B is the Bragg wavenumber.

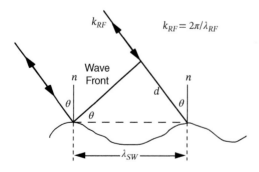

Figure 9.10 The geometry for Bragg or resonant scattering.

This means that the returns for a given radar wavelength operated at a particular incident angle are predominantly due to the scattering from a single wave component on the ocean surface, a single wavenumber that matches the Bragg condition. Surface waves that satisfy this relation are known as "Bragg waves". Note that while we used a one-dimensional example to illustrate Bragg scattering, in actuality the surface is two dimensional and the Bragg condition is for a particular wavenumber propagating either towards or away from the radar.

A full solution to electromagnetic scattering from a rough surface can be derived in the limit of small heights. This limit leads to a solution for the Radar Cross-Section (RCS) arising from Bragg scattering (Wright, 1966)[2], with the normalized RCS given by:

$$\sigma_{0,p}^{Bragg} = 8\pi k_{RF}^4\cos^4\theta|g_p(\theta)|^2$$
$$\times\,[\Psi(k_B,0) + \Psi(-k_B,0)] \quad (9.17)$$

where:

the subscript p refers to polarization – either horizontal or vertical.

$g_p(\theta)$ is a function of the complex dielectric constant and the incident angle.

$\Psi(k_x,k_y)$ is the two-dimensional wavenumber spectrum for a surface wave with components k_x and k_y.

k_B is the wavenumber of the Bragg wave, defined as $k_B = \pm2k_{RF}\sin\theta$.

In this model, the RCS arises from the sum of two wave spectral amplitudes: one for the Bragg wave component propagating towards the radar and one for the Bragg wave component propagating away from the radar. Also note that these Bragg resonant waves have wavelengths that are of the same order as the wavelength of the radiation transmitted by the radar.

Equations (9.18) and (9.19) provide the details on the polarization dependent terms

---

2 We have referenced Wright's paper on the subject, although the original small perturbation formulation was actually derived by Rice (1951).

g_{HH} and g_{VV}. That is it, a complete description of Bragg scattering:

$$g_{HH}(\theta) = \frac{\varepsilon_r - 1}{\left[\cos\theta + (\varepsilon_r - \sin^2\theta)^{\frac{1}{2}}\right]^2} \quad (9.18)$$

$$g_{VV}(\theta) = \frac{(\varepsilon_r - 1)\left[\varepsilon_r(1 + \sin^2\theta) - \sin^2\theta\right]}{\left[\varepsilon_r \cos\theta + (\varepsilon_r - \sin^2\theta)^{\frac{1}{2}}\right]^2}$$

$$(9.19)$$

Figure 9.11 is a plot of Bragg wavelengths for L-band (23 cm), C-band (5 cm), and X-band (3 cm) as a function of incident angle. Note that at moderate incident angles, the Bragg wavelength is within a factor of two of the radar wavelength. In fact, the Bragg wavelength exactly equals the radar wavelength at an incident angle of 30°.

The next level of model sophistication involves combining features of Bragg scattering from short-wavelength surface waves with features of facet scattering from the longer surface waves (Figure 9.12).

In this approach the surface is divided into long waves and short waves. The long waves are then modeled as a set of connected tilted facets, each locally tangent to the long waves. These facets have tilt angles given by the distribution of the slopes of the long waves.

Bragg scattering then occurs locally from the short waves that are sitting on each facet. The facets tilt these short waves towards or away from the radar, thus each facet has its own local incident angle. The Bragg returns from all the facets are then summed together. This approximation is known as the composite, two-scale or "tilted-Bragg" scattering model.

A few details are worth noting. For example, the division between long and short surface waves is typically taken to be between 3 and 10 times the RF wavelength. The division is arbitrary, although it can be roughly justified by analysis using physical optics. The short waves that do not meet the Bragg condition do not contribute to this solution.

It should also be noted that the long waves hydrodynamically modulate the amplitude of the short waves. This hydrodynamic modulation increases the short wave amplitude on the front face of the surface waves, as illustrated

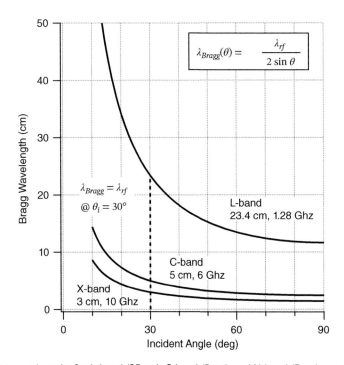

Figure 9.11 Bragg wavelengths for L-band (23 cm), C-band (5 cm), and X-band (3 cm) as a function of incident angle.

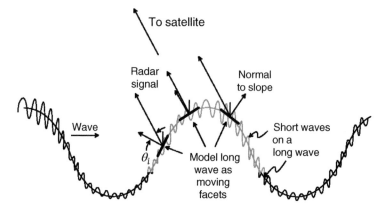

Figure 9.12 Two-scale model applies Bragg scattering to varying slopes of longer surface waves. Adapted from Stewart (1985).

in Figure 9.12. This causes a difference in the RCS depending on whether the radar is looking upwind or downwind[3].

The full two-scale RCS calculation involves (1) an expression for the local Bragg RCS from a facet allowed to tilt in two directions and (2) an integration over the facet tilt angles, weighted by the joint probability distribution function for slopes of the long waves. Bragg scattering from a tilted facet selects a component from the surface wave spectrum with particular k_x, k_y values that meet the Bragg resonance condition, as before.

The long wave slope distribution function is generally assumed to be Gaussian as in specular scattering. Note that the wind speed dependence of the RCS comes from the wind dependence of the surface wave spectrum. Hence surface roughness enters via the effect of the wind on the surface wave spectrum.

An expression for the RCS of a single facet at HH polarization is given below. The expression for the RCS at VV polarization is of similar complexity.

Let the normal to the surface area illuminated by the radar tilt by an angle ψ in the plane of incidence, and by an angle δ in the plane perpendicular to the plane of incidence (Figure 9.13). Then the local incident angle θ_i is given by: $\cos\theta_i = \cos(\theta + \psi)\cos\delta$ where $\theta =$ angle of radar line of sight to the mean surface.

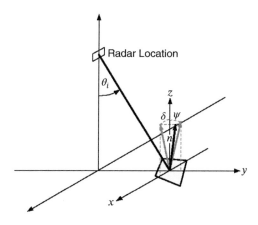

Figure 9.13 Three-dimensional geometry of the local incident angle.

The composite model RCS for a single facet observed in HH polarization is then:

$$\sigma_0^H(\theta_i) = 8\pi k_{rf}^4 \cos^4\theta$$
$$\times |A\, g_{HH}(\theta_i) + B\, g_{VV}(\theta_i)|^2$$
$$\times \Psi(2k_{rf}\,\alpha, 2k_{rf}\,\gamma\sin\delta) \quad (9.20)$$

where $\Psi(k_x, k_y)$ is the surface wave height spectrum and:

$$\alpha = \sin(\theta + \psi) \quad (9.21)$$
$$\gamma = \cos(\theta + \psi) \quad (9.22)$$
$$A = \left[\frac{\sin(\theta + \psi)\cos\delta}{\sin\theta_i}\right]^2 \quad (9.23)$$
$$B = \left[\frac{\sin\delta}{\sin\theta_i}\right]^2 \quad (9.24)$$

In the RCS expression, we took care to properly define the Bragg resonant condition on the

---

3 Hydrodynamic modulation is discussed further in Chapter 10.9.1.

tilted surface. The components (k_x, k_y) in the wave height spectrum were derived from these expressions:

$$k_x = 2k_{rf}\sin(\theta + \psi) = \text{local } k_{Bragg}$$
(9.25)

$$k_y = 2k_{rf}\cos(\theta + \psi)\sin\delta \qquad (9.26)$$

Now the previous expression was for the RCS from a single tilted facet. The total RCS is then computed by weighting the local RCS by the long wave joint probability distribution function for wave slopes, and then integrating over all wave slopes:

$$\sigma_0^H(\theta) = \int_{-\infty}^{\infty}\int_{-\infty}^{\infty} \sigma_0^H(\theta_i)\, p(\tan\psi, \tan\delta)$$
$$\times d\tan\psi\, d\tan\delta \qquad (9.27)$$

where $p(s_x, s_y)$ is the joint slope probability distribution function for the long waves.

Composite Bragg scattering theory formed the basis for nearly all remote sensing work from the 1970s to mid 2000s. Many attempts were made in the 1990s to incorporate improved scattering models based on more accurate approximations to the fully non-linear EM scattering theory. For example, see Voronovich's work on small-slope approximate (SSA) methods (Voronovich, 1994). These methods proved to be complex and offered improved accuracy only for limited applications such as altimeter EM bias calculations.

Present day models of scattering at moderate incident angles incorporate two important additional factors: ocean surface nonlinearities and microscale breaking. It turns out that breaking waves occur at all scales, and small-scale breaking is constantly occurring to redistribute and balance energy input from the wind. It also appears that microscale breaking makes a significant difference in electromagnetic scattering. For example, see recent work by Johannessen et al. (2005) and Kudryavtsev et al. (2003a, 2003b, 2005), which probably represent close to the state of the art.

9.3 RCS Dependence on Wind

Radar cross-section depends on wind speed and direction, providing a means to estimate winds from RCS measurements. Figure 9.14

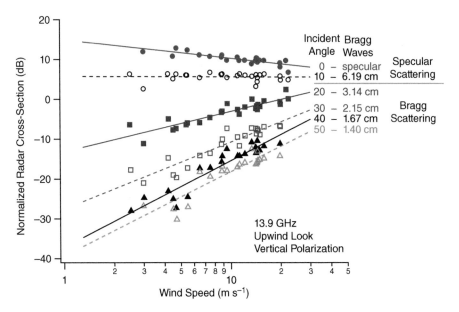

Figure 9.14 Measured NRCS at 13.9 GHz as a function of incident angle and wind speed. Data from Schroeder et al. (1984).

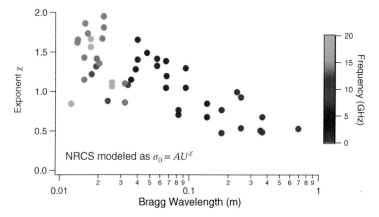

Figure 9.15 The exponent χ as a function of the Bragg wavelength. Adapted from Jones and Schroeder (1978) with additional data from Thompson et al. (1983) and Unal et al. (1991).

shows the scattering cross-section per unit area of the sea at 13.9 GHz as a function of incident angle and wind speed. These data are for V-pol and the radar was looking upwind.

Note the NRCS is in dB, so both axes are logarithmic. The data are clearly organized along straight lines depending on incident angle, so it is natural to fit the data to a power law of the form: $\sigma_0 = AU^\chi$ where U is the wind speed with constants A and χ dependent on incident angle.

Also note that the Bragg wavelengths at 13.9 GHz for the illustrated incident angles have been included on this plot. It is actually more useful to relate χ to the Bragg wavelength than to the incident angle.

Figure 9.15 is a plot of the parameter χ as a function of the Bragg wavelength, made from data acquired at a variety of frequencies and incident angles. Fundamentally this plot shows that χ varies with the ocean wavelength that scatters the signal. It also shows that scattering is more sensitive to the wind at

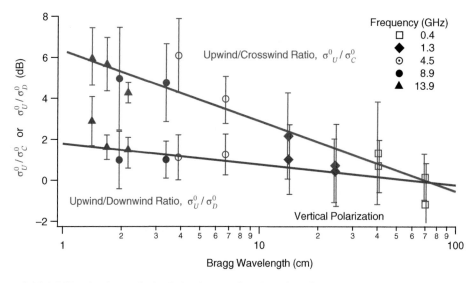

Figure 9.16 NRCS ratios for vertical polarization as a function of the Bragg wavelength. Modified from Jones and Schroeder (1978).

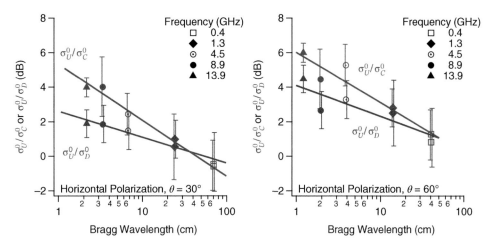

Figure 9.17 NRCS ratios as a function of Bragg wavelength for horizontal polarization at incident angles of 30° and 60°. Modified from Jones and Schroeder (1978).

shorter wavelengths (higher frequencies) than at longer wavelengths. This is the reason that most scatterometers work at C-band (5 GHz) or Ku-band (13 GHz).

Let us next consider the effect of wind direction on NRCS. Figure 9.16 shows ratios of upwind[4]/crosswind NRCS and upwind/downwind NRCS for vertical polarization as a function of the Bragg wavelength. This plot says that at higher frequencies, which are on the left-hand side of the plot, the upwind NRCS is much higher than the crosswind, but only slightly higher than downwind. This suggests that wind direction may be determined by measurements made at various frequencies and look directions.

Figure 9.17 contains similar plots showing NRCS ratios as a function of Bragg wavelength for horizontal polarization at incident angles of 30° and 60°. The story is the same for these plots with upwind NRCS being a little higher than the downwind and much higher than the crosswind.

Figure 9.18 delivers on what the last two figures hinted at. Here scattering cross-section per unit area of the sea is plotted as a function of angle relative to the mean wind. The data were obtained from a rotating airborne

scatterometer operating at 13.9 GHz with a 40° incident angle. In this case upwind is defined as 0° azimuth relative to the wind. The open circles are V-pol and the closed circles are H-pol.

In Figure 9.18 we can see that the upwind NRCS is a little higher for downwind and much higher than for crosswind. The dashed lines are simply aids to seeing the relative levels of the peaks. We also see that the directional dependence is almost sinusoidal with peaks upwind and downwind and minima crosswind.

9.4 Scatterometer Algorithms

As we have seen, the normalized radar cross-section of the ocean surface is a function of at least four parameters: wind speed, radar look direction relative to the wind direction, incident angle, and radar frequency.

We saw before that the scatterometer NRCS seemed to scale as the wind speed raised to a power. Furthermore, the look direction dependence appeared sinusoidal, with a slight asymmetry in the upwind/downwind peaks. For these reasons, the general scatterometer equation has the simple form:

$$\sigma_0^p(U, \theta, \phi) = aU^\gamma(1 + b\cos\phi + c\cos 2\phi)^\beta$$
(9.28)

---

4 An upwind look means that the radar looks into the wind.

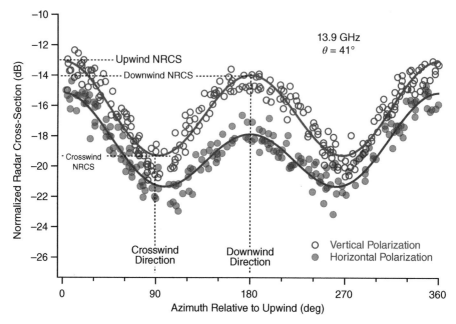

Figure 9.18 Measured NRCS as a function of angle relative to the mean wind direction. Data from Schroeder et al. (1984).

In this expression:

ϕ is the wind direction relative to radar look direction.

$a, b, c,$ and γ are empirically determined coefficients with a and γ depending on incident angle θ and coefficients b and c depending on θ and wind speed.

p is polarization (H = horizontal; V = vertical).

β = constant taken to be 1.0 for the CMOD2 algorithm; and 1.6 for CMOD4.

The upwind/crosswind asymmetry in equation (9.28) comes from the $\cos 2\phi$ term, which is much larger than the $\cos \phi$ term that gives the upwind/downwind asymmetry. If $\beta = 1$, then measurements of NRCS at 0°, 90°, and 180° could be used to determine the coefficients b and c. Coefficients a and γ could then be obtained by comparison of scatterometer measurements with data obtained at ground calibration sites.

The general problem is to estimate wind speed U and direction ϕ from measurements of NRCS made at different angles. Since there are two unknowns, at least two measurements

will be needed. Let us motivate how bad the ambiguity is with just two measurements. Begin by ignoring the smaller $\cos \phi$ term:

$$\sigma_0^p(U, \theta, \phi) = aU^\gamma(1 + c\cos 2\phi) \quad (9.29)$$

Assume measurements of the NRCS at two angles ϕ and $\phi - \pi/2$. Then recall the two trigonometric identities:

$$\cos 2\phi = \cos(-2\phi) \quad (9.30)$$

$$\cos 2(\phi - \pi/2) = \cos 2(\pi/2 - \phi) \quad (9.31)$$

Hence, with only two orthogonal measurements, there is a fourfold ambiguity [ϕ, $-\phi$, $\pi/2 - \phi$, and $\phi - \pi/2$]. To reduce the ambiguity, we at least need to add a third measurement at 45° to the first two. In this case:

$$\cos 2(\phi + \pi/4) = -\sin 2\phi = \sin(-2\phi) \quad (9.32)$$

$$\sin 2(\phi - \pi) = -\sin 2(\pi - \phi) \quad (9.33)$$

Application of more trig identities then suggests this further reduces the wind direction ambiguity to two possibilities, namely ±180°. This level of ambiguity can usually be resolved by referring to large-scale weather system information.

Just to reiterate, the analysis we just walked through illustrated the potential for wind direction ambiguities by simplifying the problem to the point where it can be solved in closed form.

While this is the general approach that will be taken, the actual geophysical model function used to estimate normalized radar cross-section from measurements of the wind speed U and direction ϕ are a bit more complex than equation (9.28). For example, the CMOD4 geophysical model function used for the vertically-polarized ERS-2 scatterometer was given by:

$$\sigma_0^V = b_0(1 + b_1 \cos \phi$$
$$+ b_3 \tanh b_2 \cos 2\phi)^{1.6} \quad (9.34)$$
$$b_0 = b_r 10^{\alpha + \gamma f_1(U_w + \beta)} \quad (9.35)$$

$$f_1(y) = \begin{cases} -10 & \text{if } y \leq 10^{-10} \\ \log y & \text{if } 10^{-10} < y \leq 5 \\ \sqrt{y}/3.2 & \text{if } y > 5 \end{cases}$$
$$(9.36)$$

where:

U_w = wind speed (m/s^{-1}).
ϕ = wind direction relative to look direction.

$\alpha, \beta, \gamma, b_1, b_2,$ and b_3 are expanded as Legendre polynomials to a total of 18 coefficients.
$b_r \approx 1.0$ (determined from a look-up table and depends on incident angle).

You will have to look in the literature to find the detailed coefficients, but you can see that the CMOD4 algorithm relies on small modifications to the simple form we described along with a large number of tuned parameters. Subsequent analysis has led to updated models including CMOD5 (Hersbach, 2003), CMOD6 (Elyouncha et al., 2015), and C-SARMOD (Mouche & Chapron, 2015). Similar geophysical model functions are also in use for other frequencies.

For the Radarsat SAR, the polarization was horizontal and the NRCS was given by:

$$\sigma_0^H(U, \theta, \phi) = \left(\frac{1 + 0.6\tan^2\theta}{1 + 2\tan^2\theta} \right)^2 \sigma_0^V(U, \theta, \phi)$$
$$(9.37)$$

Figure 9.19 plots the CMOD4 algorithms for vertical and horizontal polarization for five wind speeds. Imagine the dashed black line indicates the V-pol NRCS of −16 dB that was measured at a look angle of 45°. Based

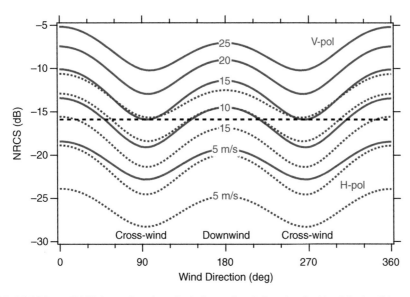

Figure 9.19 NRCS from CMOD4 as a function of wind speed and direction for V-pol (red solid curves) and H-pol (blue dotted curves). The dashed line indicates a NRCS measurement of −16 dB.

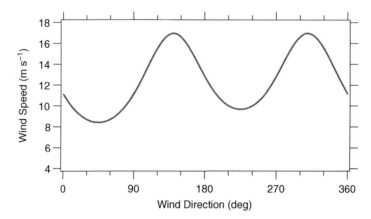

Figure 9.20 Possible wind speed and directions that could have produced a single measurement of NRCS.

on the CMOD4 algorithm, this single measurement could be the result of anything from an 8–15 m s^{-1} wind speed at a full range of possible angles.

Figure 9.20 shows all of the possible wind speed and directions that CMOD4 predicts could produce an NRCS of −16 dB.

Now suppose the radar makes a second NRCS measurement at an azimuth angle that is offset 90° from the first measurement. The second measurement then produces a second curve of possible solutions, as shown in Figure 9.21. But these solutions will only intersect at up to four points, as shown. Thus the true wind speed and direction is given by one of these four points, and wind speed and direction have not been uniquely determined.

If three observations are made, the ambiguity is further reduced. Note that these NRCS

observation must be very precise because scattering is only weakly anisotropic.

Figure 9.22 plots the wind speed and direction solutions for NRCS measurements at four different look angles. The incident angles were 45° for the two solid curves and 53° for the dashed and dotted curves, all with vertical polarization. The azimuths relative to the satellite track are 45° for the red curve, 135° for the blue curve, 55° for the dashed curve, and 145° for the dotted curve.

The four arrows marked by the number 2 show the four solutions derived from two-looks; the arrows marked by the number 3 show the two solutions derived from three looks, and the arrow marked by the number 4 shows the single solution derived from four looks.

Figure 9.23 shows wind speed measured by the Metop-B ASCAT instrument, compared

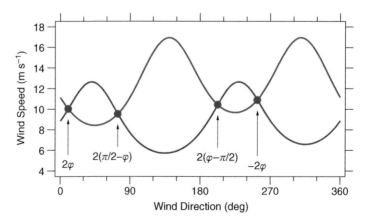

Figure 9.21 Multiple solutions for wind direction from two NRCS measurements at different look angles.

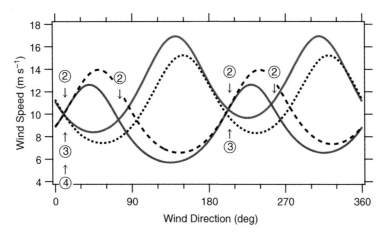

Figure 9.22 Ambiguity resolution with four NRCS measurements at different look angles. Arrows indicate the possible solutions based on two, three or four looks. Modified from Naderi et al. (1991).

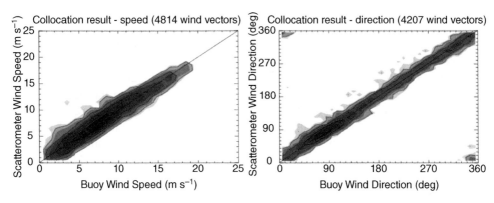

Figure 9.23 Comparison of 12.5 km Metop-B ASCAT wind product versus moored buoy winds from January–March 2017. From OSI SAF/EARS Winds Team (2018) with permission from European Space Agency.

with wind speed from buoys. These data indicate a wind speed bias of $0.04 \, \text{m s}^{-1}$ with standard deviations of the cartesian wind components of less than $1.8 \, \text{m s}^{-1}$. The standard deviation of the wind direction error is less than 20°. These data were using the 12.5 km resolution product produced by ASCAT.

9.5 Fan-Beam Scatterometers

There are two basic designs for satellite scatterometers: fan beam and conical scan. Fan-beam scatterometers use antennae that are long in the azimuth direction but narrow in the elevation direction. This creates a beam that is narrow in azimuth but broad in elevation. We call these fan beams, because they look like an open fan held vertically.

Conical-scan scatterometers are discussed in the next section.

The illuminated swath for a typical satellite scatterometer would be 25–50 km in azimuth and 500–600 km in elevation. Current fan-beam scatterometers use six antennae to project three beams to the left of the satellite track and three beams to the right, as shown in Figure 9.24. The incident angle in each beam typically varies from about 25° in the near range, to about 55° in the far range.

Fan-beam scatterometers can use pulse timing to determine range, or they can be designed to determine range using Doppler processing. In Doppler processing a long FM chirped pulse is transmitted, as shown in red in Figure 9.25. The returned signal from a scatterer comes back time delayed and hence frequency shifted relative to the original transmitted pulse.

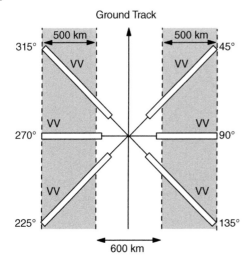

Figure 9.24 Canonical fan-beam geometry.

By mixing the return with the transmitted chirp, the power coming back from any individual range cell will occur at a different frequency. Thus a Fourier decomposition of the returned signal is equivalent to separating the returned signal into contiguous range cells.

Independent of how range resolution is obtained, the returns from each individual pulse is noisy, so multiple pulses, typically 50–100, are averaged to obtain a low-noise RCS estimate.

Table 9.1 is a list of past scatterometers that have flown since 1978, and Table 9.2 is a list of current or soon-to-be operational scatterometers.

Notice that half are C-band scatterometers and half are Ku-band. Again, about half are fan-beam systems of the type discussed in this section, and half are conical scan systems that are discussed in the next section. All have resolutions of either 25 or 50 km.

It is also worth noting that SASS, the first fan-beam scatterometer launched as a demonstration project on Seasat, only had two beams projected to each side. So it could only obtain two look angles at each point on the surface, which produced ambiguous results. All later fan beam systems used three beams on each side, to reduce measurement ambiguities.

The scatterometers deployed on the now defunct ERS-1 and ERS-2 satellites were fan-beam systems (Figure 9.26).

These were C-band systems with three antennae looking off to the right side of the flight track. One beam looked forward 45°, one looked orthogonal to the flight track, and one looked aft 45°. This provided RCS measurements of every point within the 500 km swath of the instrument from three different look directions.

The center beam antenna and sampling characteristics were different from the fore and aft beams. These differences were choices made to maximize the overlap of the data.

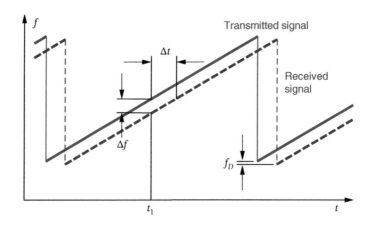

Figure 9.25 Obtaining range resolution from a linear FM chirp.

Table 9.1 Past satellite scatterometers.

Instrument / Satellite(s)	Agency	Frequency / Operation	Res. (km)	Launch date / End date
SASS	NASA	14.6 GHz, 4-ant.	50	
Seasat		Doppler bin, left & right		6/78 – 10/78
AMI	ESA	5.3 GHz, 3-ant.	50	
ERS-1		range bin, right side		7/91 – 6/96
ERS-2				4/95 – 7/11
NSCAT	NASA	14 GHz, 6-ant.	25	
ADEOS-1	JAXA	Doppler bin, left & right		8/96 – 6/97
SeaWinds	NASA	13.4 GHz	25	
QuikSCAT	NASA	two rotating pencil beams		6/99 – 10/09
ADEOS-2	JAXA			12/02 – 10/03
OSCAT	ISRO	13.5 GHz	25	
OceanSat-2		two rotating pencil beams		9/09 – 2/14
RapidSCAT	NASA	13.4 GHz	25	
ISS		two rotating pencil beams		9/14 – 8/16

Table 9.2 Current and near-future satellite scatterometers.

Instrument / Satellite	Agency	Frequency / Operation	Res. (km)	Launch date
ASCAT[1]	ESA	5.3 GHz, 6-ant.	50	
METOP-A		range bin, left & right		10/06
METOP-B				9/12
METOP-C				11/18
SCAT	CSA	13.256 GHz	25	
HY-2A		two rotating pencil beams		8/11
HY-2B				10/18
OSCAT	ISRO	13.5 GHz	25	
ScatSat-1		two rotating pencil beams		9/16
RFSCAT	CSA/France	13.256 GHz, dual pol	10	
CFOSAT		two rotating fan beams		10/18
WindRAD	CSA	5.3 & 13.256 GHz, dual pol	10–20	
FY-3E		four rotating fan beams		7/21

[1] Figa-Saldaña et al. (2002) and Gelsthorpe et al. (2000)

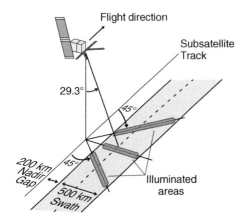

Figure 9.26 ERS-1 fan beam geometry. Adapted from Lecomte (1998), with permission from European Space Agency.

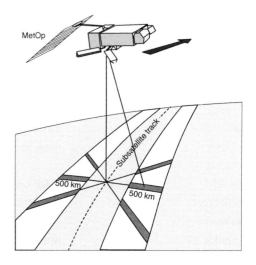

Figure 9.27 ASCAT six-beam geometry. Figure from https://earth.esa.int/web/eoportal/satellite-missions/m/metop-sg with permission from EUMETSAT.

The yaw orientation of the spacecraft was constantly adjusted to keep the returns from center beam at zero Doppler.

The ASCAT scatterometer is a current design used on the European METOP satellites. It is similar to the ERS scatterometer, but is a six-beam system with three beams looking out each side of the spacecraft (Figure 9.27). This produces two 500-km swaths separated by a holiday (that is the technical term for a gap in coverage) of 650 km centered on the satellite track.

NSCAT was the last U.S. fan-beam scatterometer that flew last century on the Japanese ADEOS-1 spacecraft. The mission lasted less than a year and ended with a structural failure to the solar panel array. As shown in Figure 9.28, the arrangement of the fan beams for NSCAT was a bit different than previous systems, with an asymmetric set of azimuth look angles.

NSCAT and its predecessor SASS used Doppler processing to obtain range resolution. This makes the measurements sensitive to Doppler shifts induced by motion. We mentioned that the attitude of the ERS-1 and 2 satellites were adjusted to remove the Doppler induced by the satellite's orbital motion from the center beams. But the Earth also rotates,

causing a Doppler shift that must be accounted for in the processing.

Figure 9.29 shows the lines of constant Doppler shift projected down onto the Earth's surface for the very first satellite scatterometer, SASS. The slight tilt of the symmetry axis in this diagram is a result of the Earth's rotation.

9.6 Conical-Scan Pencil-Beam Scatterometers

The second basic design for satellite scatterometers is the conical-scan pencil beam. Conical-scan pencil beam scatterometers are designed with a rotating dish antenna with two feeds that produces two pencil beams at different incident angles. As illustrated in the left-hand panels of Figure 9.30, the scan pattern on the surface for a single beam is a spiral, similar to the scan pattern of a passive microwave radiometer. The bottom left panel illustrates that a sensor with a single beam would only provide two looks at each point in the swath, one looking forward and one looking backward. The right-hand panels illustrate the benefits of the two beam system that can

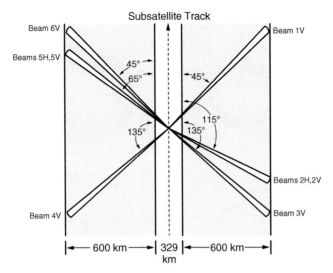

Figure 9.28 Asymmetric arrangement of the NSCAT fan beams. Source: Naderi et al. (1991) with permission from IEEE.

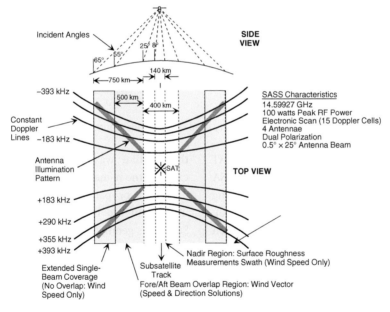

Figure 9.29 Doppler contours for SASS at the equator. Source: Grantham et al. (1975).

provide up to four looks within some regions of the swath.

The center vertical line in the right-hand panels of Figure 9.30 represents the flight track of the satellite. When the satellite reaches the point T_1 it is in a position to measure the RCS of the location labeled FOV using the outer beam. At time T_2, the satellite can measure the RCS of the same location from a different

azimuth using the inner beam. Then two more looks can be achieved at times T_3 and T_4.

Even before launching NSCAT, NASA had come to the conclusion that scatterometers were useful, but expensive. So they undertook the design of this radically new conical-scan scatterometer, which was first implemented in the SeaWinds instrument. SeaWinds was initially designed to be flown on the ADEOS-II

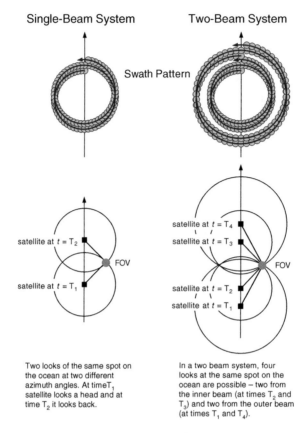

Single-Beam System

Two-Beam System

Swath Pattern

satellite at $t = T_2$

FOV

satellite at $t = T_1$

satellite at $t = T_4$

satellite at $t = T_3$

FOV

satellite at $t = T_2$

satellite at $t = T_1$

Two looks of the same spot on the ocean at two different azimuth angles. At timeT$_1$ satellite looks a head and at time T$_2$ it looks back.

In a two beam system, four looks at the same spot on the ocean are possible – two from the inner beam (at times T$_2$ and T$_3$) and two from the outer beam (at times T$_1$ and T$_4$).

Figure 9.30 Example of the conical scan geometry used by SeaWinds. Source: Freilich (2000).

satellite for launch in 2002. But NSCAT failed, leaving a coverage gap. So NASA threw together the QuikSCAT mission using a commercial satellite bus and the backup SeaWinds flight hardware. The haste in which the satellite was assembled and launched explains the name of the mission.

The major parameters for the SeaWinds instrument are listed in Table 9.3 and the SeaWinds geometry illustrated in Figure 9.31.

Table 9.3 SeaWinds parameters.

Parameter	Inner beam	Outer beam
Rotation rate (rpm)	18	
Pulse length (unchirped, ms)	1.5	
Pulse length (chirped, μs)	Programmable, > 2.7	
Polarization	HH	VV
Zenith angle (deg)	40	46
Surface incident angle (deg)	47	55
Slant range (km)	1100	1245
3-dB footprint (along- × cross-scan, km)	24 × 31	26 × 36
Along-track spacing (km)	22	22
Along-scan spacing (km)	15	19

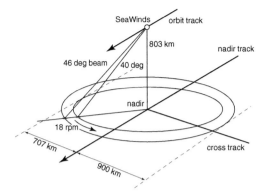

Figure 9.31 SeaWinds conceptual design and scan coverage. From Freilich (2000).

The instrument's two beams, which rotated at 18 rpm, were set to look angles of 40° and 46° relative to nadir, resulting in incident angles on the surface of 47° and 55°. Each beam had a footprint of about 25 × 35 km. The instrument was spun fast enough that all of the measurements in each beam overlapped with those made on the previous rotation.

The total ground swath of the instrument was 1800 km wide, but the geometry of having two conically-scanned beams effectively divided the swath into three zones as shown in Figure 9.32.

Every point in zone III would get four looks, two in the forward and two in the aft directions at each of the two incident angles. This is

four looks, but there is little difference in the azimuth direction of the two forward looks or two aft looks. Thus in this region, the satellite gets four looks but effectively only two azimuth angles that are nearly 180° apart, which is far from ideal.

Every point in zone II also gets four looks, but these looks occur with a useful spread of azimuth directions at each of the two incident angles. Note that at the outer edge of these zones there is effectively only three distinct looks, because the looks from the inner beam are made nearly orthogonal to the track. But with 3–4 different looks, this is the sweet spot for the instrument. Each of these zones is about 450 km wide.

Points in zone I are only accessible by the 55° beam. Thus they only get two looks, which effectively reduces to one look at the outer edges of the swath.

Because the viewing geometry changed as a function of distance from the center track, with variations in both the numbers of looks and the directional spread of those looks, the accuracy and degree of ambiguity of the measurements varied substantially across the swath. This weakness in the design was mitigated by using geophysical models of the winds to smooth the wind fields and eliminate ambiguities.

To further improve spatial resolution, the QuikSCAT footprint was processed into 12

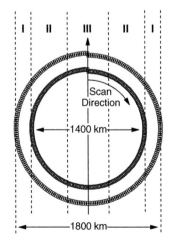

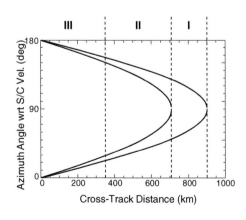

Figure 9.32 Three major zones in the SeaWinds scan pattern (left) and the azimuth look angle with respect to cross-track distance for the two beams (right). Adapted from Freilich (2000).

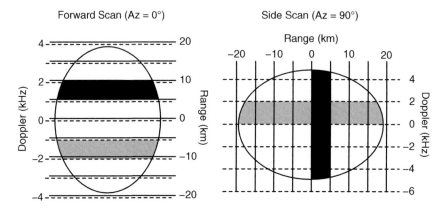

Figure 9.33 QuikSCAT range cells. Source: Spencer et al. (2000) with permission from IEEE.

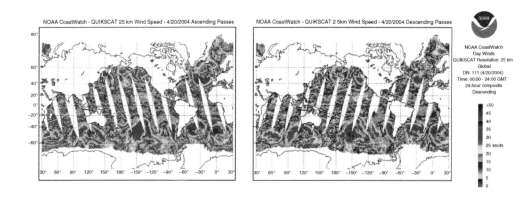

Figure 9.34 QuikSCAT wind data from one day of ascending (left) and descending (right) passes. Figures from NOAA.

range cells, as shown in Figure 9.33. The standard operational data computed the RCS over the full footprint, but the inner eight range cells were also used to provide higher resolution wind estimates for experimental purposes.

Figure 9.34 shows the wind data from one day of ascending (left) and descending (right) passes from QuikSCAT. These look a lot like some of the passive microwave radiometer maps.

But it is important to remember that the multiple looks allow active scatterometers to estimate wind speed and direction. Figure 9.35 is a map of the vector winds obtained from one pass of QuikSCAT over the waters off the Baja peninsula.

Finally, Figure 9.36 shows global maps of vector winds produced from QuikSCAT data. From all the logos it is clear this is not a regular product, but it provides a reminder of how a single satellite can provide data that may be otherwise impossible to obtain.

9.7 Conical-Scan Fan-Beam Scatterometers

There is a third, hybrid design for satellite scatterometers: the conical-scan fan-beam scatterometer, also known as a rotating fan-beam scatterometer. As illustrated in

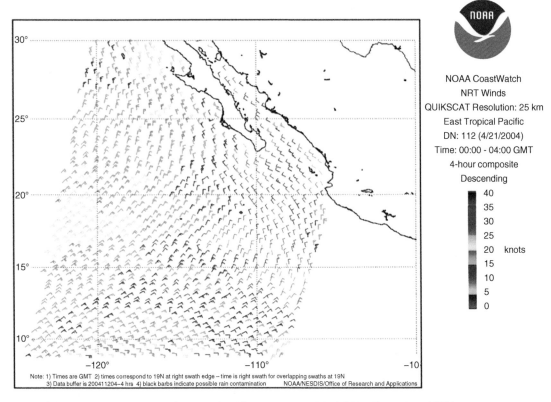

Figure 9.35 Example vector wind field obtained from one pass of QuikSCAT. Figure from NOAA.

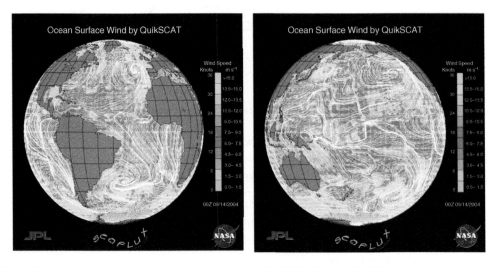

Figure 9.36 Global maps of vector winds produced from QuikSCAT. Source: JPL, NASA.

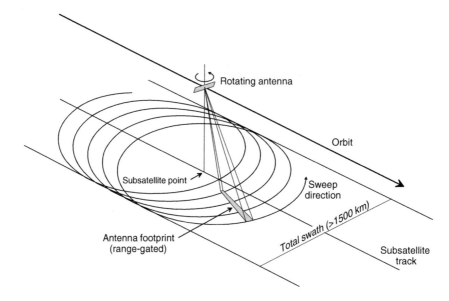

Figure 9.37 Canonical conical-scan fan-beam geometry. From Lin et al. (2000) with permission from IEEE.

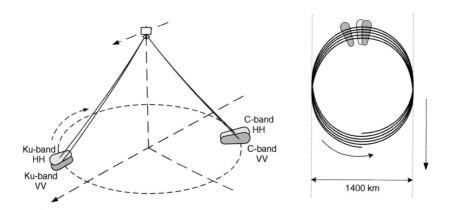

Figure 9.38 Four-beam dual-band conical-scan fan-beam geometry. From Li et al. (2019).

Table 9.4 Key parameters for RFSCAT and WindRad (Li et al. 2019).

Parameter	RFSCAT	WindRad
Altitude (km)	514	836
Swath (km)	1000	1400
Footprint (km)	280	200
Rotation rate (rpm)	3.5	3.0
Frequency (GHz)	13.256	13.256 & 5,4
Polarization	HH & VV	HH & VV
Number of beams	2	4
Incident angle (deg)	25–48	34.7–44.5
Resolution (km)	25	25

Figure 9.37, these are fan-beam scatterometers that rotate. The goal of this design is to improve performance by increasing the number of looks that contribute to each wind estimate. The basic design originally proposed by Lin et al. (2000) would have supported 10–11 looks in most cells, providing lower noise performance at a 15 km resolution.

The first operational conical-scan fan-beam scatterometer (RFSCAT) is a Ku-band sensor hosted on the Chinese/French CFOSAT that was launched in October 2018. RFSCAT has two fan beams that operate at VV and HH polarization.

China launched a second conical-scan fan beam scatterometer (WindRad) on the FY-3E satellite in July 2021. As illustrated in Figure 9.38 this is a dual-band scatterometer (Ku- and C-bands) with four beams. The C-band beams point to one side of the sensor and the Ku-band beams point to the opposite side. Table 9.4 lists the primary parameters for the RFSCAT and WindRad scatterometers.

The performance gains associated with these designs have been analyzed through simulation by Lin et al. (2003) and Li et al. (2019).

References

Bragg, W. H., & Bragg, W. L. (1913). The reflection of x-rays by crystals. *Proceedings of the Royal Society of London. Series A, Containing Papers of a Mathematical and Physical Character*, *88*(605), 428–438.

Elyouncha, A., Neyt, X., Stoffelen, A., & Verspeek, J. (2015). Assessment of the corrected CMOD6 GMF using scatterometer data. In *Remote sensing of the ocean, sea ice, coastal waters, and large water regions 2015* (Vol. 9638, p. 963803). International Society for Optics and Photonics.

Figa-Saldaãa, J., Wilson, J. J. W., Attema, E., Gelsthorpe, R., Drinkwater, M. R., & Stoffelen, A. (2002). The advanced scatterometer (ASCAT) on the meteorological operational (MetOp) platform: A follow on for European wind scatterometers. *Canadian Journal of Remote Sensing, 28*(3), 404–412.

Freilich, M. H. (2000). SeaWinds algorithm theoretical basis document. *Oregon State University Tech. Rep., Corvallis, OR*. Retrieved from https://eospso.nasa.gov/sites/default/files/atbd/atbd-sws-01.pdf

Gelsthorpe, R. V., Schied, E., & Wilson, J. J. W. (2000). ASCAT-Metop's advanced scatterometer. *ESA bulletin, 102*, 19–27.

Grantham, W. L., Bracalente, E. M., Jones, W. L., Schrader, J. H., Schroeder, L. C., & Mitchell, J. L. (1975). *An operational satellite scatterometer for wind vector measurements over the ocean*. National Aeronautics and Space Administration.

Hersbach, H. (2003). *CMOD5: An improved geophysical model function for ERS C-band scatterometry*. European Centre for Medium-Range Weather Forecasts.

Johannessen, J. A., Kudryavtsev, V., Akimov, D., Eldevik, T., Winther, N., & Chapron, B. (2005). On radar imaging of current features: 2. Mesoscale eddy and current front detection. *Journal of Geophysical Research: Oceans, 110*(C7), ID C07017.

Jones, W. L., & Schroeder, L. C. (1978). Radar backscatter from the ocean: Dependence on surface friction velocity. *Boundary-Layer Meteorology, 13*(1), 133–149.

Pierson Jr., W. J., & Moskowitz, L. (1964). A proposed spectral form for fully developed wind seas based on the similarity theory of SA Kitaigorodskii. *Journal of Geophysical Research, 69*(24), 5181–5190.

Kudryavtsev, V., Akimov, D., Johannessen, J., & Chapron, B. (2005). On radar imaging of current features: 1. Model and comparison with observations. *Journal of Geophysical Research: Oceans, 110*(C7).

Kudryavtsev, V., Hauser, D., Caudal, G., & Chapron, B. (2003a). A semiempirical model of the normalized radar cross section of the sea surface, 2. Radar modulation transfer function. *Journal of Geophysical Research: Oceans, 108*(C3).

Kudryavtsev, V., Hauser, D., Caudal, G., & Chapron, B. (2003b). A semiempirical model of the normalized radar cross-section of the sea surface 1. Background model. *Journal of Geophysical Research: Oceans, 108*(C3).

Lecomte, P. (1998). Proceedings of a joint ESA-EUMETSAT workshop on emerging scatterometer applications–from research to operations. ESA. Retrieved from https://earth.esa.int/documents/10174/1602497/WSC13.pdf

Li, Z., Stoffelen, A., & Verhoef, A. (2019). A generalized simulation capability for rotating beam scatterometers. *Journal of Atmospheric Measurement Technology*. Retrieved from https://amt.copernicus.org/articles/12/3573/2019/

Lin, C.-C., Rommen, B., Wilson, J. J. W., Impagnatiello, F., & Park, P. S. (2000). An analysis of a rotating, range-gated, fanbeam spaceborne scatterometer concept. *IEEE Transactions on Geoscience and Remote Sensing, 38*(5), 2114–2121.

Lin, C.-C., Stoffelen, A., de Kloe, J., Wismann, V. R., Bartha, S., & Schulte, H.-R. (2003). Wind retrieval capability of rotating range-gated fanbeam spaceborne scatterometer. In *Sensors, systems, and next-generation satellites VI* (Vol. *4881*, pp. 268–280). International Society for Optics and Photonics.

McAllister, M. L., Draycott, S., Adcock, T., Taylor, P. H., & Van Den Bremer, T. S. (2019). Laboratory recreation of the Draupner wave and the role of breaking in crossing seas. *Journal of Fluid Mechanics, 860*, 767–786.

Moskowitz, L. (1964). Estimates of the power spectrums for fully developed seas for wind speeds of 20 to 40 knots. *Journal of Geophysical Research, 69*(24), 5161–5179.

Mouche, A., & Chapron, B. (2015). Global C-Band Envisat, RADARSAT-2 and Sentinel-1 SAR measurements in copolarization and cross-polarization. *Journal of Geophysical Research: Oceans, 120*(11), 7195–7207.

Naderi, F. M., Freilich, M. H., & Long, D. (1991). Spaceborne radar measurement of wind velocity over the ocean – an overview of the NSCAT scatterometer system. *Proceedings of the IEEE, 79*(6), 850–866.

OSI SAF/EARS Winds Team. (2018). ASCAT wind product user manual. *EUMETSAT OSI SAF, 1.15*.

Rice, S. O. (1951). Reflection of electromagnetic waves from slightly rough surfaces. *Communications on Pure and Applied Mathematics, 4*(2-3), 351–378.

Schroeder, L. C., Jones, W. L., Schaffner, P. R., & Mitchell, J. L. (1984). Flight measurement and analysis of AAFE RADSCAT wind speed signature of the ocean. Technical Memorandum 85646. NASA.

Spencer, M. W., Wu, C., & Long, D. G. (2000). Improved resolution backscatter measurements with the SeaWinds pencil-beam scatterometer. *IEEE Transactions on Geoscience and Remote Sensing, 38*(1), 89–104.

Stewart, R. H. (1985). *Methods of satellite oceanography*. University of California Press.

Thompson, T. W., Weissman, D. E., & Gonzalez, F. I. (1983). L band radar backscatter dependence upon surface wind stress: A summary of new Seasat-1 and aircraft observations. *Journal of Geophysical Research: Oceans, 88*(C3), 1727–1735.

Unal, C. M., Snoeij, P., & Swart, P. J. F. (1991). The polarization-dependent relation between radar backscatter from the ocean surface and surface wind vector at frequencies between 1 and 18 GHz. *IEEE transactions on Geoscience and Remote Sensing, 29*(4), 621–626.

Valenzuela, G. R. (1978). Theories for the interaction of electromagnetic and oceanic waves: A review. *Boundary-Layer Meteorology, 13*(1), 61–85.

Voronovich, A. (1994). Small-slope approximation for electromagnetic wave scattering at a rough interface of two dielectric half-spaces. *Waves in Random Media, 4*(3), 337–368.

Wright, J. (1966). Backscattering from capillary waves with application to sea clutter. *IEEE Transactions on Antennas and Propagation, 14*(6), 749–754.

10

Synthetic Aperture Radar

10.1 Introduction to SAR

Synthetic aperture radars (SAR) are used to produce high-resolution imagery of the surface at microwave frequencies, even from the long ranges implied by space-based observation. Figure 10.1 contains an image of Willapa Bay, WA, made from the Radarsat-1 SAR that operated at C-band with horizontal polarization. An ocean surface wave pattern can be seen propagating generally from the upper left toward land. The refraction of the wave pattern as the waves shoal can be seen. The bright and dark linear features

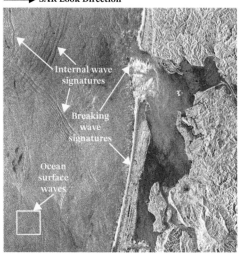

SAR Look Direction

throughout the water region are signatures from internal waves. The imaged area is 50 × 50 km.

As we will see, images like this are incredibly useful. SARs are used to study ocean waves, ocean currents, bathymetry, and sea ice. They also support a variety of land applications.

To motivate SAR, let us consider the problem of imaging from an X-band, 3-cm wavelength, real aperture radar transmitting pulses that can be compressed to the equivalent of 3.33 ns pulses. Assume the radar has a circular dish antenna with a diameter of 1.0 m pointed along a line of sight 30° from nadir. Next recall that the slant range resolution δ_{sr}, ground range resolution δ_g, and azimuth resolution δ_{az} are given by:

$$\delta_{sr} = c\tau/2 \tag{10.1}$$

$$\delta_g = \frac{c\tau}{2 \sin \theta_i} \tag{10.2}$$

$$\delta_{az} = \frac{H\lambda}{D_{az} \cos \theta_i} \tag{10.3}$$

where:

$\tau =$ pulse length
$\theta_i =$ incident angle
$H =$ antenna altitude
$\lambda =$ wavelength
$D_{az} =$ antenna length in azimuth direction.

Evaluation of these equations for a satellite platform at an altitude of 800 km yields a slant range resolution of 0.5 m and a ground resolution of 1.0 m, which seem quite reasonable.

Figure 10.1 An image of Willapa Bay, WA, made from the Radarsat-1 SAR. Source: Wackerman and Clemente-Colon (2004).

Remote Sensing Physics: An Introduction to Observing Earth from Space, Advanced Textbook 3, First Edition.
Rick Chapman and Richard Gasparovic.
© 2022 American Geophysical Union. Published 2022 by John Wiley & Sons, Inc.
Companion website: www.wiley.com/go/chapman/physicsofearthremotesensing

The problem is that the azimuth resolution of this system would be 27.7 km!

Equation 10.3 can be inverted to obtain a formula for the size of the antenna that would be needed to achieve 1-m azimuth resolution from an 800-km altitude:

$$D_{az} = \frac{H\lambda}{\delta_{az}\cos\theta_i} = 27.7 \text{ km} \qquad (10.4)$$

Obviously a 27-km long antenna is impractical! The solution is to leverage the motion of the platform to synthesize an aperture this long, which is what a SAR does.

The SAR was invented by Carl Wiley at Goodyear Aerospace in 1954. Wiley realized that as a side-looking radar flew along, the motion of the platform caused the returns to be Doppler shifted, and that these Doppler shifts could be used to separate signals arriving from different directions.

Consider Figure 10.2, which shows a radar aperture at three different positions as it moves down the page. The aperture in this case is illustrated as having a beamwidth of ±25°. So returns arising when the antenna is in the middle position will contain: positive Doppler shifts from scatterers ahead of the radar; zero Doppler shifts from scatterers broadside to the antenna; and negative Doppler shifts from scatterers behind the radar.

Wiley's idea was to sum together energy with positive Doppler from data taken early in the pass, zero Doppler from data in the middle of the pass, and negative Doppler from data taken late in the pass. The original implementations were referred to as Doppler beam sharpening, and were kept as a military secret for a number of years.

A series of three illustrations further explain the concept. The illustration in Figure 10.3 shows an airborne radar with a broad-beamed side-looking radar. The beam is indicated in yellow and the radar pulses in green.

The coordinate system for SAR imagery is often expressed in terms of range and azimuth directions, where the azimuth direction is essentially cross-range. We have adopted this nomenclature because it is so widely used.

Figure 10.4 illustrates the returned signal as the aircraft flies past a point scatterer

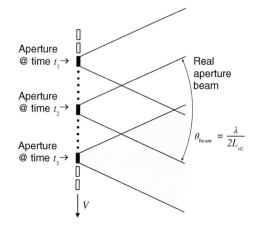

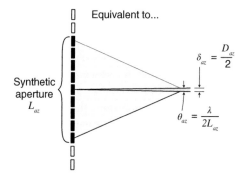

Figure 10.2 Formation of a synthetic aperture. Source: Moreira (2013) with permission of ESA.

sitting on the surface, indicated by point SA. A concrete example of such a point scatterer could be a navigation buoy sticking out of the water.

Early in time, the plane would be approaching the buoy from a distance and the returned frequency would be Doppler shifted to a higher frequency based on the aircraft speed. This is schematically illustrated in the time-frequency curve. Then as the plane passes by the location of the scatterer, the Doppler shift falls and goes through zero. The zero Doppler point occurs at the time of the plane's closest point of approach to the scatterer, or when the scatterer is broadside to the radar track. Finally, the Doppler shift goes negative as the plane continues past the scatterer.

Figure 10.5 adds one more scatterer (point SB) that is at the same range as the original

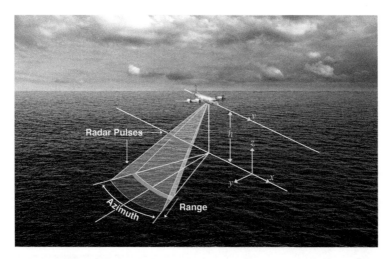

Figure 10.3 An airborne radar with a broad-beamed side-looking radar with permission from Johns Hopkins University Applied Physics Laboratory (JHU/APL).

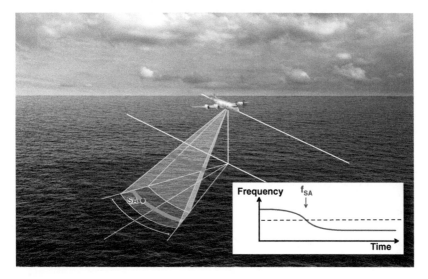

Figure 10.4 Returned Doppler signal as the aircraft flies past a point scatterer with permissions from Johns Hopkins University Applied Physics Laboratory (JHU/APL).

scatterer, but offset from it in the azimuth direction.

Note the second scatterer creates a Doppler return that is shifted in time, but otherwise identical to the first. The time shift is clearly the azimuthal distance between the two scatterers divided by the speed of the aircraft. The SAR locates these scatterers in azimuth by distinguishing between these two Doppler shifted signals. In fact, the core of SAR azimuth processing is little more than the application of a matched filter looking for these characteristic signals.

A mathematical description of SAR imaging is provided in the next section, but first consider the general geometry of SAR. Figure 10.6 illustrates the geometry of Seasat, the first space-based SAR used for Earth observations[1]. The Seasat antenna had an along-track length D_{AT} of 10.7 m and a height D_R of 2.2 m. It flew at an altitude of 800 km with a nadir angle of

---

1 Seasat was launched in 1978 and lasted for about 100 days before it died from a power supply failure. The data from Seasat are still available on the web from the Alaska SAR facility and well worth a look.

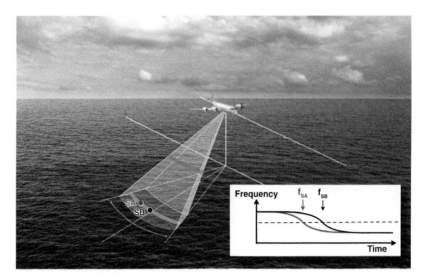

Figure 10.5 Returned Doppler signals as the aircraft flies past two point scatterers with permissions from Johns Hopkins University Applied Physics Laboratory (JHU/APL).

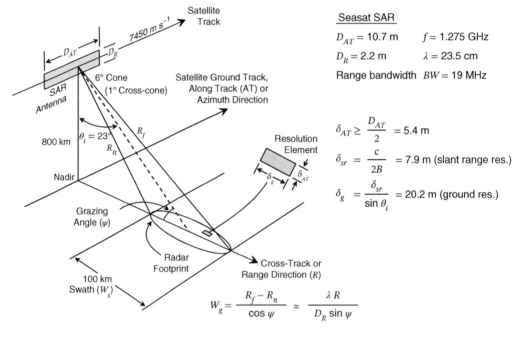

Figure 10.6 Seasat SAR imaging geometry. Source: McCandless and Jackson (2004) with permission from Chris Jackson.

23° to beam center. The range to beam center was about 860 km.

Because the antenna was long and thin, the radar beam illuminated a swath on the Earth with a range extent of 100 km and an azimuth extent of about 19 km. The azimuth extent can be derived from the range and the azimuth beamwidth λ/D_{AT}, which is about 1.25°.

We call the range extent of the footprint W_g because it represents the swath width on the ground.

The range resolution of a SAR is identical to the range resolution of a real aperture radar. The slant range resolution is given by:

$$\delta_{sr} = \frac{c\tau}{2} = \frac{c}{2\,BW} \qquad (10.5)$$

and the ground resolution is:

$$\delta_g = \frac{\delta_r}{\sin\theta_i} \qquad (10.6)$$

where:

τ = compressed pulse length (chirp pulse)

BW = bandwidth of frequency variation in chirp pulse

θ_i = incident angle of pulse.

For example, Seasat transmitted a 33.8 μs pulse with a 19 MHz bandwidth. The time-bandwidth product or pulse compression ratio was 634, and the compressed pulse length was 52.6 ns. The slant and ground range resolutions are then easily computed: $\delta_{sr} = 7.9$ m and $\delta_g = 20.2$ m.

It turns out that the azimuth resolution of Seasat was half its antenna length, or 5.4 m. The reasons for this seemingly odd result are provided in the next section.

10.2 SAR Azimuth Resolution

10.2.1 Doppler Time History

In order to derive the azimuth resolution of a SAR, we begin by carefully considering the Doppler returns that are fundamental to SAR processing.

Consider Figure 10.7 showing a SAR antenna beam at three positions, progressing to the right at a velocity v_{ant}. The antenna is illuminating a single scatterer in the middle of the dashed line that represents the ground. As the antenna moves, the scatterer first enters the beam at time $-t_{B/2}$, at a range of $R(-t_{B/2})$. At time t_0, the scatterer is broadside to the antenna at range $R(t_0)$. And the scatterer exits the beam at time $t_{B/2}$, at range $R(t_{B/2})$.

The phase of the signal scattered back to the antenna is given by the phase of the transmitted

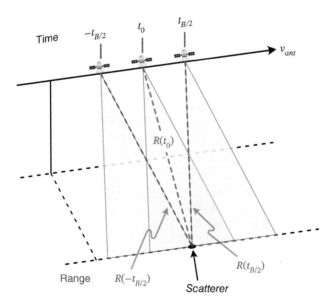

Figure 10.7 SAR antenna at three different times as it moves past a scatterer.

signals minus the extra phase associated with the propagation of the signals to and from the scatterer at range R, which is a function of time. This phase difference between transmitted and received signal is given by radar wavenumber k_{rf} times the two-way path distance $2R$. The phase of the received signal is thus:

$$\varphi(t) = \omega_0 t - 2\bar{k}_{rf} \cdot \bar{R}(t) \qquad (10.7)$$

where:

$$k_{rf} = \frac{2\pi}{\lambda_{rf}}, \quad \omega_0 = 2\pi v_0 \quad \text{and} \quad v_0 \lambda_{rf} = c \qquad (10.8)$$

and where v_0 is the transmitted pulse frequency, and λ_{rf} is the wavelength of the transmitted pulse.

The frequency of the signal received by the antenna ω is the time derivative of the phase, so it depends on dR/dt:

$$\omega = \frac{\partial \varphi}{\partial t} = \omega_0 - 2k_{rf} \frac{dR(t)}{dt} \qquad (10.9)$$

As derived below, the range is approximately given by:

$$R(t) \approx R(t_0) \left(1 + \frac{1}{2} v_{ant}^2 \frac{(t - t_0)^2}{R^2(t_0)} \right) \qquad (10.10)$$

Therefore the Doppler-shifted received frequency is approximately given by:

$$\omega = \omega_0 - \frac{2k_{rf} v_{ant}^2}{R(t_0)} (t - t_0) \qquad (10.11)$$

and linearly depends on t. Now recall how nonlinear the time history of the Doppler frequency was (as illustrated in Figure 10.5). This linear approximation turns out to be valid over just a limited region about the zero Doppler point, but that is useful for many purposes.

The other major point here is that the Doppler shift term looks like a "chirp" signal with time scale t set by the velocity of the antenna and the range to the target. Radar engineers often refer to this as slow time, to distinguish this time scale from the fast time used in range pulse compression. The terminology strikes us as a bit odd, but it is common in the literature so you have been warned.

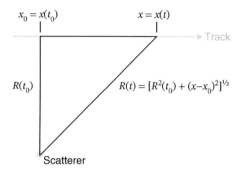

Figure 10.8 Range-time geometry.

The derivation of the range as the SAR moves past a scatterer involves all of the complexity of high school algebra. If we designate t_0 to be the time of closest point of approach, then $R(t_0)$ is a fixed value. Referring to Figure 10.8, the range is then specified by the Pythagorean formula:

$$R^2(t) = R^2(t_0) + (x - x_0)^2$$
$$= R^2(t_0) \left[1 + \frac{(x - x_0)^2}{R^2(t_0)} \right] \qquad (10.12)$$

So R is simply the square root of this formula, which when $R \gg (x - x_0)$ can be written as a Taylor series expansion about t_0:

$$R(t) = R(t_0) \sqrt{1 + \frac{(x - x_0)^2}{R^2(t_0)}}$$
$$= R(t_0) \left[1 + \frac{1}{2} \frac{(x - x_0)^2}{R^2(t_0)} + \cdots \right] \qquad (10.13)$$

The formula can be completed by noting that x and t are related by the antenna velocity: $(x - x_0) = v_{ant}(t - t_0)$, so to lowest order:

$$R(t) = R(t_0) \sqrt{1 + \frac{v_{ant}^2 (t - t_0)^2}{R^2(t_0)}}$$
$$\approx R(t_0) \left(1 + \frac{1}{2} v_{ant}^2 \frac{(t - t_0)^2}{R^2(t_0)} \right) \qquad (10.14)$$

It is the Taylor series expansion that limits the validity of the approximation to times close to the zero Doppler point. In the case of Seasat, R is about 860 km and $v_{ant} t$ turns out to be no more than 10 km. So the approximation is quite good!

Equation (10.11) is an expression for the returned frequency as a function of time that can be rewritten as the transmitted frequency plus a Doppler shift:

$$\omega = \omega_0 - \frac{2k_{rf}\, v_{ant}^2}{R(t_0)}(t - t_0) = \omega_0 + \omega_{Dopp}$$

$$(10.15)$$

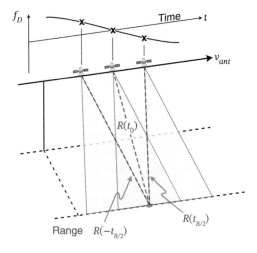

Figure 10.9 Doppler frequency of return pulses from the point P as the radar beam passes by. Adapted from Elachi (1987).

The Doppler shift is positive as the antenna approaches the scatterer, zero at broadside, and negative as the antenna recedes from the scatterer.

The total Doppler shift from the scatterer as it passes through the entire beam is then given by:

$$\omega_{Dopp}^{Total} = 2k_{rf}\frac{v_{ant}^2}{R(t_0)}T_B = 2k_{rf}\frac{v_{ant}^2}{R(t_0)}T_{int} \quad (10.16)$$

where T_B is the dwell time of the scatterer in the beam, which we assume is also the SAR processor's coherent integration time, T_{int}.

The key point of all of this is that, at any instant of time, the position of a scatterer can be determined from its range and its Doppler frequency. This is equivalent to determining its location in the surface intersection of a coordinate system utilizing range and Doppler frequency.

Figure 10.9 illustrates the time history of the frequency returns as the antenna beam passes over a scatterer. This geometry is closer to that used for satellites, so the Doppler frequency shift varies nearly linearly with time.

Figure 10.10 is a perspective plot of the SAR image coordinate system consisting of a family

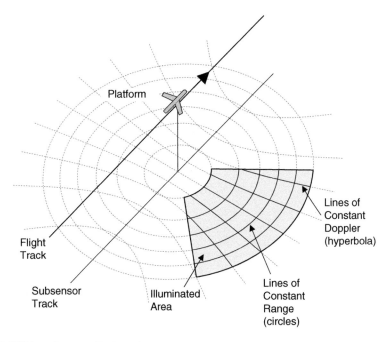

Figure 10.10 SAR imaging coordinate system.

of iso-range circles, which are the intersection of spheres with the surface, and iso-Doppler hyperbolas which are intersections of cones with the surface.

The coordinate system is not orthogonal or linear, but each x, y point on the surface can be mapped to a unique range and Doppler frequency.

10.2.2 Azimuth Extent, Integration Time, and Doppler Bandwidth

The azimuth beamwidth of a SAR, like any other radar, is given by $\theta_{beam} = \lambda/D_{az}$. As shown in Figure 10.11, the azimuth extent of the footprint is then $L = R(t_0)\theta_{beam}$. We can then derive the SAR integration time as L/v_{ant}:

$$T_{int} = \frac{L}{v_{ant}} = \frac{R(t_0)\theta_{beam}}{v_{ant}} = \frac{R(t_0)\lambda_{rf}}{v_{ant} D_{az}} \quad (10.17)$$

Note that in equation (10.17) the term λ/D_{az} is an angle that is effectively divided by the term v_{ant}/R, which is the angular speed of the SAR. It turns out the inverse of the SAR's angular speed, usually expressed as "R over V", is an important parameter, as we will see in following sections.

From before, the total Doppler shift is:

$$\begin{aligned} \omega_{Dopp}^{Total} &= 2k_{rf}\frac{v_{ant}^2}{R(t_0)}T_{int} \\ &= \frac{4\pi v_{ant}^2}{\lambda_{rf} R(t_0)}\left[\frac{R(t_0)\lambda_{rf}}{v_{ant}D_{az}}\right] \\ &= 4\pi\frac{v_{ant}}{D_{az}} = 2\pi f_{Dopp}^{Total} \quad (10.18) \end{aligned}$$

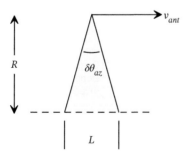

Figure 10.11 Beam geometry diagram.

so:

$$f_{Dopp}^{Total} = \frac{2v_{ant}}{D_{az}} \quad (10.19)$$

Take Seasat as an example, with key parameters: $D_{az} = 10.7$ m; $k_{rf} = 23.5$ cm; $R(t_0) = 860$ km; and $v_{ant} = 7.45$ km s^{-1}. Substitution shows the azimuth extent of the illuminated footprint was 18.9 km, the integration time was 2.56 seconds, and the Doppler bandwidth was almost 1400 Hz.

10.2.3 Azimuth Resolution

As before, the azimuth resolution is derived by considering the returns from two scatterers located at the same range, but at different azimuths. The azimuth resolution is defined to be the minimum distance that the scatterers can be separated and yet still resolved in azimuth. Resolving the scatterers essentially means that the Doppler returns from the two scatterers can be separated in time. Figure 10.12 illustrates the Doppler returns from two nearby scatterers.

The minimum resolvable time difference in a signal of a given bandwidth is just one over that bandwidth, so we get the minimum resolvable time to be:

$$\Delta t_{az} = \frac{1}{f_{Dopp}^{Total}} = \frac{D_{az}}{2v_{ant}} \quad (10.20)$$

The azimuth resolution is then how far the antenna moves in this time:

$$\delta_{az} = v_{ant}\Delta t_{az} = \frac{D_{az}}{2} \quad (10.21)$$

Think about what this formula means, namely azimuth resolution of a SAR is one-half the antenna length in the azimuth direction, independent of radar frequency or range! In other words, a SAR can achieve azimuth resolutions of a few meters from space. We will see later that this is not quite as magical as it sounds, but it is still pretty cool.

There is another, more precise way to look at SAR resolution that you may find informative. Imagine that we send out a series of pulses as the beam passes by the scatterer,

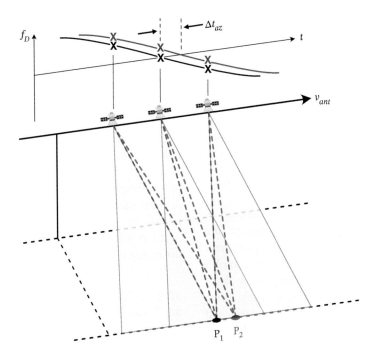

Figure 10.12 Doppler returns from two scatterers. Adapted from Elachi (1987).

sampling the amplitude and phase of the returned signals. The frequency of these samples will vary linearly with time in a known fashion:

$$\omega = \omega_0 - \frac{2k_{rf}\, v_{ant}^2}{R(t_0)}(t - t_0) \qquad (10.22)$$

As illustrated in Figure 10.13, this linear FM signal can be compressed in time, just like a single chirped pulse can be compressed in the range dimension. In this case, the azimuth-compressed signal will have a waveform with a time resolution $\Delta t_{az} = 1/f_{Dopp}^{Total}$.

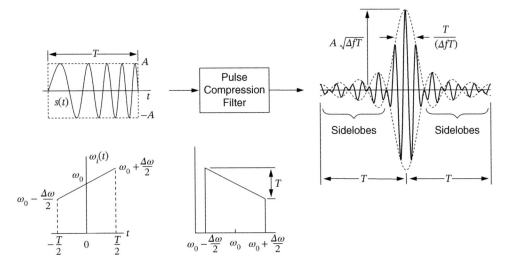

Figure 10.13 Pulse compression using a matched or dispersive filter. Source: P. Z. Peebles (2007) with permission from John Wiley & Sons.

Thus azimuth compression is analogous to range compression, but while range compression occurs within a pulse (on that fast time scale talked about before), azimuth compression occurs on the slow time scale required for the antenna beam to pass by the scatterer.

All this magic does not come for free. The total Doppler bandwidth of the returned signal is:

$$BW_{Dopp} = f_{Dopp}^{Total} = \frac{2v_{ant}}{D_{az}} \qquad (10.23)$$

So if we want to avoid aliasing of this signal, we need to sample it at a high enough rate to meet the Nyquist sampling requirement. It turns out that SARs usually use complex sampling, meaning they sample an in-phase and quadrature phase signal at each range cell of each pulse. This means we can avoid aliasing if we sample at a rate equal to the Doppler bandwidth[2].

We sample in azimuth by sending out pulses, and this sampling requirement means that we need to send out at least two pulses in the time that the antenna moves a distance equal to its length in the azimuth direction.

For the Seasat SAR, this meant a minimum pulse repetition frequency of 1394 pulses s^{-1}, giving about 3600 pulses while an individual scatterer is in the beam. In practice, the Seasat SAR pulse rate was usually 1640 pulses s^{-1}, and there were about 4200 pulses per beamwidth. Faster pulsing is almost always used in combination with spectral filters to better avoid aliasing. The typical PRF values used in space-based radars range from 1 to 3 kHz.

We said before that it is more precise to formulate the azimuth resolution of a SAR in terms of compression of the coherent time series formed by the returned pulses. This is because azimuth resolution fundamentally depends on the total bandwidth of the signal

returned during the coherent integration time. The previous "proof" that the azimuth resolution of a SAR is at best 1/2 of the antenna length is an approximation derived for a stripmap SAR following a linear flight trajectory. Satellite SARs follow a curved trajectory. Furthermore the antenna of a SAR can rotate during the imaging process to keep its beam focused on a spot on the surface in order to extend the coherent integration time and hence form a larger synthetic aperture. This is called spotlight SAR and will be discussed further in a following section. Ultimately it is the bandwidth of the returned signal that drives resolution.

10.2.4 SAR Timing, Resolution, and Swath Limits

Most SARs use the same antenna for transmitting and receiving. The transmitter emits far too much power for the receiver to handle, so most radars blank their receivers during transmissions. Thus it was not unusual that the Seasat receiver could only listen between pulse transmissions. The time gap between Seasat pulses is the time interval between successive pulses minus the length of an individual pulse. Seasat had a time gap of 576 µs during which it could receive signals.

This time gap corresponds to a two-way range of 173 km or a one-way range of 86 km. This means that the Seasat ground swath could not have been any more than $86 \div \sin(23°)$ or 221 km in extent. The actual Seasat swath was set to 100 km in order to limit the data rates transmitted to the ground. This also allowed the antenna beam to be narrowed a bit in the range direction, improving the signal-to-noise ratio.

We have told this story in terms of Seasat, but there is a general trade-off on all single-aperture SARs between azimuth resolution and swath size. So what is the impact of azimuth resolution on swath?

SAR azimuth resolution is given by $\delta_{az} = D_{az}/2$. This means that increased resolution

---

2 Note that if we were only sampling one component of this signal, then we would need to sample twice this fast, which is the more common form of the Nyquist sampling critera.

requires a shorter antenna. Now we can walk down the remainder of a logic tree, leaving out details to make the major points. The equations are all written down so you can follow along algebraically on your own. But the big picture is:

- SAR azimuth resolution is $\delta_{az} = D_{az}/2$, so increased resolution requires a shorter antenna.
- Antenna azimuth beamwidth is $\theta_{beam} = \lambda/D_{az}$, so a shorter antenna means a broader beam.
- Total Doppler bandwidth is $BW_{Dopp} = 2v/D_{az}$, so a broader beam means the returns have higher bandwidth.
- Minimum PRF to avoid aliasing is $f_p = 1/BW_{Dopp}$, so a higher bandwidth means increased PRF.
- Maximum time available for receiving returns is $\Delta t_p = 1/f_p$, so an increased PRF means shorter gaps between transmissions to receive signal.
- Maximum swath width is $W_g = c\Delta t_p/2\sin\theta_i$, so shorter gaps for receipt of the signal means decreased swath width.

Thus higher resolution comes with the penalty of narrower swath width.

In the end we are left with two fundamental design constraints for the operation of a SAR:

- The time between pulses must be shorter than the time it takes to move 1/2 of antenna length.
- The return from the Nth pulse must be over before the return from the $(N + 1)^{th}$ pulse arrives.

This implies that the pulse timing must satisfy these constraints:

$$\frac{2W_g\sin\theta_i}{c} \leq \Delta t_p \leq \frac{\delta_{az}}{v_{ant}} \qquad (10.24)$$

Now there are a dozen other important constraints and design considerations for a SAR having to do with power and timing and so forth, but these two constraints are fundamental and key to SAR operation.

10.2.5 The Magic of SAR Exposed

It was previously shown that azimuth resolution of a SAR is one-half the antenna length in the azimuth direction, independent of radar frequency or range. This result is true, but it can be misleading in real applications. So let us be a bit more complete. And in doing so, we hope to convince you SAR is not quite as magical as it seems.

The prior result on SAR azimuth resolution was based on the implicit assumption that the synthetic aperture was formed over the entire time that the main beam illuminated a point on the surface at range R. This is called the Coherent Processing Interval (CPI). This was the parameter T_{int} in our calculation and the CPI for Seasat was 2.56 s.

Achieving the same azimuth resolution at greater ranges would have required a longer synthetic aperture and, hence, a longer CPI. The top right panel in Figure 10.14 shows the antenna beam from a SAR at two locations representing when a scatterer at range R might first enter the beam and leave the beam. Note that the distance traveled by the SAR between these two extremes is both the extent of the synthetic aperture that was formed by the SAR processing and also the azimuth footprint of the real beam. Thus the azimuth angular resolution in this case is given by λ/L.

The bottom right panel is the same drawing, but now with the scatterer moved to a range of $2R$. Yes the SAR can achieve the same azimuth resolution in this case, but it requires a synthetic aperture that is twice as long, which means that the coherent processing interval must also be doubled. So getting the same resolution at increasing range is not free, but requires processing an increasing amount of data.

It is important to note that the azimuth resolution of a SAR will grow with range if a constant CPI is used in the processing, and this is often the case. We will also discuss going the other way with this relationship, because it will turn out that processing of partial apertures,

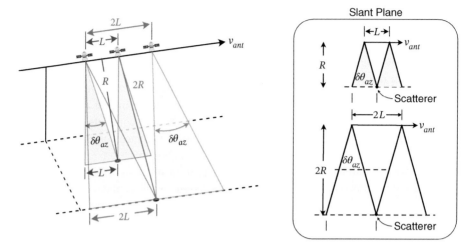

Figure 10.14 Achieving the same resolution at longer ranges requires a longer synthetic aperture.

which means shorter CPIs, is an approach often used to reduce imaging noise.

10.3 SAR Image Formation and Image Quality

A "single-look" SAR image is formed by processing the return signal over the total Doppler bandwidth – equivalent to keeping track of the phase shift from each scatterer as it passes through the entire beamwidth. As we have discussed, for Seasat SAR, a single-look image had a ground range resolution of 20.2 m and an azimuth resolution of 5.4 m.

The left-hand panel of Figure 10.15 is not from Seasat, but it is a single-look image of a famous landmark in Paris. If you look carefully at any of the bright portions of this image, you'll notice that it looks like it was painted

by a pointalist. This is speckle arising from the random summation of coherent signals. Sometimes the returns sum together constructively, in which case the return is bright, and sometimes they sum destructively, in which case the return is dark. It turns out that the speckle noise is chi-squared distributed with two-degrees of freedom. This means that the standard deviation of the speckle in a single-look image is equal to its mean.

To reduce the "noise" (speckle) in a single-look image, the Doppler spectrum can be divided into N equal segments, and then each segment can be processed separately to form N images. The N images are then summed incoherently to produce what is called an "N-look" image.

The right-hand panel in Figure 10.15 is a multilook image with far less speckle. But there is no free lunch.

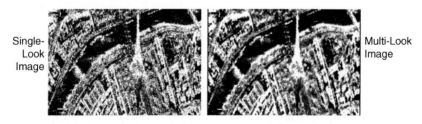

Single-Look Image

Multi-Look Image

Figure 10.15 Comparison of single-look versus multilook imagery. Source: Sarti (2012), ESA.

The azimuth resolution of an N-look image will be degraded by a factor of N because the effective Doppler bandwidth of each segment is reduced by a factor of N:

$$\delta_{az}^N = ND_{az}/2 \qquad (10.25)$$

An N-look image is equivalent to tracking the phase variation of the return signal over $1/N$-th of the total azimuth beamwidth. Figure 10.16 illustrates four-look processing.

Most of the Seasat SAR images were processed to four looks, giving a final image with a ground range resolution of 20.2 m and an azimuth resolution of 21.6 m. In practice, slightly less than full bandwidth was used in range and azimuth to produce images with 25×25 m pixels.

The quantitative analysis of SAR imagery is badly contaminated by speckle noise in single-look imagery, so multilook processing is often required to produce usable estimates of geophysical parameters. This multilook processing comes at the expense of resolution. So SARs with higher single-look resolution inherently contain more information than lower-resolution SARs. While resolution may not matter much for the detection of a large feature, many SAR techniques, such as interferometry or measurement of compact polarization, require significant speckle noise reduction in order to work effectively.

The image formation process consists of applying a two-dimensional compression to the complex (amplitude and phase) return signal:

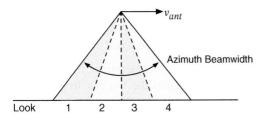

Figure 10.16 Multilook processing effectively splits the beam into multiple sub-beams.

- Range resolution is obtained by applying a standard pulse compression algorithm for frequency-modulated pulses.
- Azimuth resolution is obtained by compressing the Doppler frequency imposed on the return signal as a result of the antenna velocity.

The image formation process is generally implemented in the frequency domain, and is equivalent to applying matched filters in both the range and azimuth dimensions.

SAR images are typically large. For example, standard ERS-1/2 images were 100×100 km, sampled at 12.5×12.5 m pixels = 64 Megapixels per image. It took approximately 13.3 seconds to acquire one standard image. The number of pixels processed at a reduced resolution of 25×25 m was 1.2×10^6 per second.

Raw SAR data often have low dynamic range because the antenna beam is so broad that the returns are averages over large regions. Thus digitization of raw SAR data requires only a few bits[3]. Today this is typically eight bits per in-phase (I) and quadrature (Q) channel.

SAR processing involves weighted sums of thousands of pulses to perform the matched filtering that sharpens the spatial response of the data. This processing significantly increases dynamic range. Thus processed SAR data are usually distributed as floating point values.

You may find it interesting to note that SAR data are almost always transmitted from the satellite to the ground in RAW format, because of the reduced dynamic range of the raw data.

We have made SAR processing seem like a trivial exercise, but this is far from the truth. For example, we have completely ignored that the target's range location constantly changes as the SAR moves past a target. This is called range migration and there are clever algorithmic approaches for taking this into account. In addition, there are many other complicating effects that, while not discussed here, must

---

3 Researchers at The Johns Hopkins University Applied Physics Laboratory once built a space-based SAR processor for looking at ocean waves that worked with I and Q channels each digitized to 1-bit!

be taken into account in the image formation process. All the gory details can be found in any number of texts on SAR processing. Three that are particularly useful are Curlander and McDonough (1991), Jakowatz et al. (2012), and Jansing (2021). We think the Jakowitz formulation of SAR processing as a tomographic process is particularly informative.

Now while we are not going to discuss SAR processing in detail, the images in Figure 10.17 showing results from several key steps in the SAR image formation process are informative.

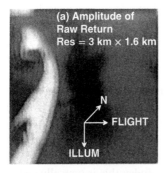

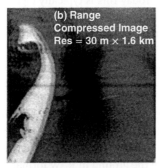

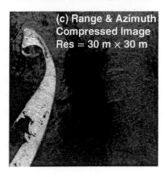

Figure 10.17 Key steps in the SAR image formation process illustrated with SIR-C data acquired over Cape Cod. Source: McCandless and Jackson (2004) NOAA.

These are images of the tip of Cape Cod obtained from the SIR-C mission[4]. Each of these images covers an extent of approximately 38 km in range by 43 km in azimuth.

The upper image is the amplitude of the raw radar return. The radar energy is spread out over the SIR-C chirp length of about 10 µs, which corresponds to a range resolution of about 3 km. Furthermore, the azimuth resolution for this image is set by the real aperture and is about 1.6 km. So yes you can see Cape Cod in the image, but it is a blurry view.

Each pulse in the SIR-C data was range compressed to form the middle image. This image has the same poor azimuth resolution, but a range resolution of 30 m. The lower image has been fully processed, including both range and azimuth compression. This image has a resolution of 30 m in both range and azimuth.

It may be interesting to note that the shuttle was a challenging platform from which to make scientific measurements because astronauts moving around on the shuttle cause platform motions that can disturb the measurements. SAR processing is based on maintaining phase coherence over the integration period, which means maintaining knowledge of the location of the antenna to within a fraction of the radar wavelength during the few seconds it takes to form the image.

10.4 SAR Imaging of Moving Scatterers

Now let us discuss another fact of SAR imaging, the effects of scatterer motion. We saw before that the Doppler frequency from a stationary scatterer is zero when the antenna is broadside to the scatterer. This tells us where to position the scatterer in the processed image:

- Azimuth location determined by where the antenna is when the Doppler frequency is zero.

4 SIR stands for Spaceborne Imaging Radar. The SIR-C mission had L-band (1.25 GHz) and C-band (5.3 GHz) radars from JPL and a German X-band (9.6 GHz) radar.

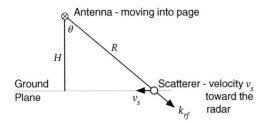

Figure 10.18 A SAR antenna moving into the page at an altitude H flies past a scatterer moving along the ground toward the radar.

- Range location determined by the arrival time of the signal when the Doppler frequency is zero.

So it is worth asking the question of what happens when the scatterer is moving either toward or away from the radar at the time it is illuminated? The answer is one of those things that makes life interesting: The motion of the scatterer toward or away from the radar introduces an additional Doppler shift to the backscattered signal.

Consider Figure 10.18, illustrating a SAR antenna moving into the page at an altitude H above the ground. The SAR is looking off to the side as usual at a scatterer that is moving along the ground with a velocity v_s toward the radar.

The Doppler frequency caused by motion of the scatterer toward the radar is the radar wavenumber times twice the component of the scatter velocity in the look direction:

$$\omega^M_{Dopp} = 2\bar{k}_{rf} \cdot \bar{v}_s = 2k_{rf} v_s \cos(\pi/2 + \theta)$$
$$= -2k_{rf} v_s \sin\theta \qquad (10.26)$$

The factor of two comes from the fact that a change in the scatterer range to the radar of ΔR means a change in the radar's two-way path length of $2\Delta R$.

We previously derived an expression for the Doppler frequency resulting from the motion of the antenna over the position increment $(x - x_0)$:

$$\omega_{Dopp} = -\frac{2k_{rf} v^2_{ant}}{R(t_0)}(t - t_0)$$

$$= -\frac{2k_{rf} v_{ant}}{R(t_0)}(x - x_0) \qquad (10.27)$$

Hence, we can relate the Doppler frequency caused by the scatterer motion to an azimuth displacement:

$$\omega^M_{Dopp} = -2k_{rf} v_s \sin\theta = -\frac{2k_{rf} v_{ant}}{R(t_0)}(x - x_0)$$
$$(10.28)$$

Thus the Doppler frequency introduced by the velocity component toward the radar from a moving object is equivalent to an apparent azimuth displacement of the target in the SAR image. This displacement is given by the component of the scatterer velocity projected into the radar line of sight times a factor of R/V where R is the range to the target and V is the antenna velocity:

$$x - x_0 = \begin{cases} +\dfrac{R(t_0)}{v_{ant}} v_s \sin\theta & \text{if object moving toward radar} \\[2ex] -\dfrac{R(t_0)}{v_{ant}} v_s \sin\theta & \text{if object moving away from radar} \end{cases} \qquad (10.29)$$

Since the target return is positioned in the processed image at the azimuth location where the Doppler frequency is zero, the location of a moving object will be shifted in the image by a distance $(x - x_0)$ in azimuth from its true ground position. Hence, the image location of an object moving toward the radar appears at a later time than the true location of the object on the ground. And a target moving away from the radar will appear at an earlier time than the true location. This is colloquially known as the "train off the tracks" phenomenon.

Instead of a train, suppose a ship is moving at 25 knots ($12.9\,\mathrm{m\,s^{-1}}$) toward the Seasat

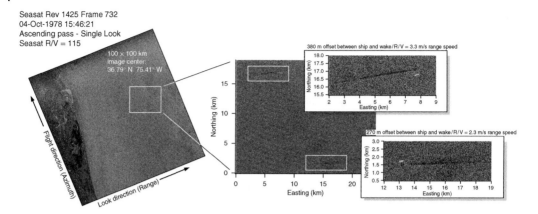

Figure 10.19 SAR image of two ships displaced azimuthally from their wakes. Credit: NASA 1978, processed by ASF DAAC 2013.

SAR at the time it is imaged. The range is $R = 860$ km and $v_{ant} = 7.45$ km s^{-1}, so R/V is 115 seconds. The ship's velocity projected into the look direction of the radar is about 5 m s^{-1}. So the azimuth displacement in the image will be 580 m:

$$x - x_0 = \frac{R(t_0)}{v_{ant}} v_s \sin \theta = 580 \text{ m} \quad (10.30)$$

given $\theta = 23°$.

The factor of R/V turns out to be a key parameter for SAR interpretation, because it reflects the degree of motion blurring that will occur in a SAR image. As previously mentioned, V/R is actually the angular velocity of the SAR, which may help you understand why the parameter is important.

Note that there is no Doppler shift from the azimuth component of the scatterer velocity. Instead, this component gives rise to a smearing of the target return in the azimuth direction within the processed image. The blur is equivalent to SAR processing with an error in the antenna velocity equal to the azimuthal speed of the target.

Such motion effects are especially important for SAR imaging of ocean waves, where they result in highly nonlinear imaging effects,

complicating the interpretation of SAR images of ocean wave systems.

Figure 10.19 is a SAR image of two ships and their wakes. One ship is traveling toward the radar and the other traveling away. The figure illustrates that the SAR misregisters both ships off their wakes because they are moving in a way that the wake is not. Note the ship traveling toward the radar look direction is shifted in the flight direction, and the ship traveling away from the radar look direction is shifted opposite to the flight direction.

The analysis of target motion effects involved a Taylor-series approximation to the range as a function of time. The exact formula for range is shown first in equation (10.31), followed by the Taylor-series expansion:

$$R(t) = R(t_0) \left[1 + \frac{(x - x_0)^2}{R^2(t_0)} \right]^{1/2} \quad (10.31)$$

$$\approx R(t_0) \left[1 + \frac{1}{2} \frac{(x - x_0)^2}{R^2(t_0)} + \cdots \right] \quad (10.32)$$

In fact much of SAR literature has been based on the usually good approximation provided by truncation of this Taylor-series expansion, in exactly the form we have used

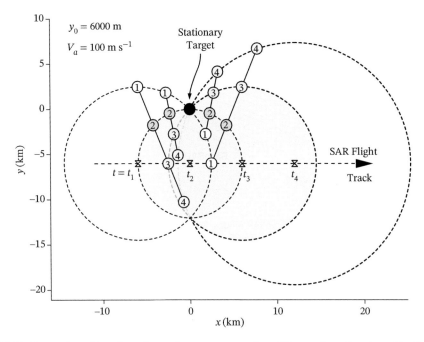

Figure 10.20 Illustration of four linearly-moving scatterers at four different times (numbered 1–4) that are indistinguishable from a stationary scatterer (black circle). From Chapman et al. (2010).

here. As a result, a significant number of papers have appeared in the literature showing how to estimate the motion of moving targets from an analysis of the higher-order terms in this Taylor series expansion. Many of these papers were wrong.

Figure 10.20 illustrates that there is an entire family of moving targets that are absolutely indistinguishable from a stationary target in a single-aperture SAR.

The dashed line indicates the track of a SAR, the locations of which are represented at four different times by the bow tie symbols. The SAR is assumed to be moving at a constant speed over a flat earth. The dark black circle is a stationary scatterer being observed by the SAR as it passes. As we have just learned, the SAR is able to image this scatterer because of its ostensibly unique phase time history. The four dotted circles are drawn to be centered on

each of the four SAR locations. All four pass through the stationary scatterer. Four other lines were added, each representing a possible moving scatterer that would have exactly the same phase time history as the stationary scatterer.

It turns out that there are an infinite number of moving scatters that are indistinguishable from each stationary scatterer so motion ambiguities are fundamental.

10.5 Multimode SARs

As we have described, SARs are interesting, but fundamentally simple devices that provide high-resolution images of radar cross-section. And truth be told, the initial SARs were sort of simple devices. But as SARs have become more widely used, their complexity has been

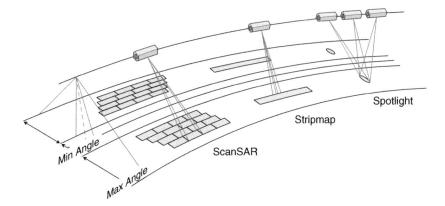

Figure 10.21 Stripmap, spotlight, and ScanSAR modes. Source: Airbus Defence and Space Geo-Intelligence, TerraSAR-X Image Product Guide with permission from Airbus.

increased by adding multiple modes to support novel capabilities.

Almost all modern SARs support multiple modes by combining flexible, programmable timing with a phased-array antenna to produce electronically-steered beams. As illustrated in Figure 10.21, the most common three modes for a modern SAR are stripmap, spotlight, and ScanSAR.

In stripmap mode the antenna beam is fixed in elevation and azimuth, producing a continuous image strip. This is conventional SAR, although modern stripmap modes often allow tradeoffs between azimuth resolution and swath by adjustments to timing and, in some instruments, antenna beamwidth.

In spotlight mode the antenna beam is steered in the azimuth direction to spotlight one region as the satellite passes over. In modern systems this is often accomplished with a steerable beam created by a phased array antenna, but other systems achieve the same effect by changing the attitude of the entire spacecraft to hold the beam on a fixed location. Spotlight mode is attractive, especially for military applications, because it can be used to significantly increase the length of the synthetic aperture and therefore improve resolution. Submeter azimuth resolution is possible with spotlight modes.

In ScanSAR mode, the antenna elevation is steered to acquire adjacent, slightly overlapping coverage areas with different incident angles. These tiles are then processed into images and merged into a single scene with a large swath.

It is important to realize that all of these modes acquire data at rates no greater than the antenna velocity times the swath width, a product that can be shown to equal $c/2$ times the azimuth resolution. Thus the ScanSAR mode creates wide swath imagery, but can only do so at a reduced azimuth resolution.

10.6 Polarimetric SAR

Most modern SARs support multiple polarizations using multiple antenna feeds and interleaved pulses. Polarization is a particularly useful parameter to measure for an active sensor like a SAR, because the SAR can control the polarization of both the transmit and receive channels. Thus the SAR can measure the complex response of scatterers to various polarizations. This makes SAR polarimetry particularly sensitive to the shape, orientation, and dielectric properties of the scatterers. This sensitivity provides leverage to make qualitative and quantitative estimates of geophysical properties of the scattering surface, whether it be crop land, snow and ice, the ocean or an urban scene.

10.6.1 Polarimetric Response of Canonical Targets

Up to this point the radar cross-section of a target has been presented as a real-valued parameter that may be polarization dependent. In fact, the radar cross-section is more completely represented as a complex scattering matrix, with some targets producing a scattered field with a different polarization than the incoming field. The reader may want to review Section 6.4 on polarization before proceeding.

Mathematically, given the Jones vector for the incident $\underline{E_i}$ and scattered fields $\underline{E_s}$, the complex scattering matrix $\mathbf{S}$ is defined:

$$\underline{E_s} = \mathbf{S}\,\underline{E_i} = \begin{bmatrix} S_{\perp\perp} & S_{\perp\parallel} \\ S_{\parallel\perp} & S_{\parallel\parallel} \end{bmatrix} \underline{E_i} \qquad (10.33)$$

where the indices ($\perp$ and $\parallel$) correspond to an orthogonal coordinate system relative to

the orientation of the target instead of the radar. The diagonal elements of this matrix are referred to as the copolar terms and the off-diagonal elements are the cross-polar terms. The radar cross-section is related to the scattering matrix through the formula:

$$\sigma = 4\pi \mathbf{S}^2 \qquad (10.34)$$

The scattering matrix, and hence the response of the radar to a scatterer, varies depending on the type of scatterer. Table 10.1 shows the complex scatterer matrix for several basic types of scatterers. Spheres, flat plates, and trihedrals all exhibit similar scattering characteristics. If the incident field is linear, then the scattered field has the same polarization but with a 180° phase change. If the incident field is circular, the 180° phase change on reflection causes the scattered field to have the opposite polarization of the incident field.

Table 10.1 Complex scattering matrix for several types of canonical scatterers.

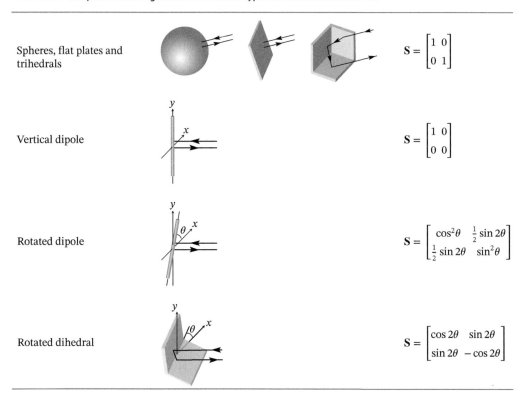

Spheres, flat plates and trihedrals	$\mathbf{S} = \begin{bmatrix} 1 & 0 \\ 0 & 1 \end{bmatrix}$
Vertical dipole	$\mathbf{S} = \begin{bmatrix} 1 & 0 \\ 0 & 0 \end{bmatrix}$
Rotated dipole	$\mathbf{S} = \begin{bmatrix} \cos^2\theta & \frac{1}{2}\sin 2\theta \\ \frac{1}{2}\sin 2\theta & \sin^2\theta \end{bmatrix}$
Rotated dihedral	$\mathbf{S} = \begin{bmatrix} \cos 2\theta & \sin 2\theta \\ \sin 2\theta & -\cos 2\theta \end{bmatrix}$

For example, left-hand circular incident radiation scatters as right-hand circular radiation.

Vertical dipoles respond only to the vertically-polarized component of the incident radiation that induce currents in the dipole. These currents then produce vertically-polarized scattered radiation. A rotated dipole is only sensitive to the component of the incident radiation oriented along the axis of the dipole, producing scattered radiation that is linearly polarized parallel to the axis of the dipole.

A dihedral is a two-sided corner. When aligned horizontally or vertically, the copolar response is maximized for horizontal or vertical linear polarization, and horizontally or vertically-oriented elliptical polarization, including circular polarization. This copolar response falls to zero when the dihedral is rotated 45° about the radar line of sight, and the cross-polar return is maximized. This means that dihedrals are effective at creating a cross-polarized response.

10.6.2 Decompositions

So the complex scattering matrix contains information about the general types of scatterers that contribute to the signal within each pixel. Decompositions are mathematical combinations of the complex scattering matrix elements that are designed to separate the contributions from different scatterer types into individual channels. While dozens of decompositions have been developed for various purposes, the concept is best illustrated by considering two common decompositions, the Pauli decomposition and the Freeman–Durden decomposition.

Path reciprocity applies to a monostatic radar, so $S_{hv} = S_{vh}$. Note that we have switched back to a horizontal and vertical polarization basis, but one defined by the orientation of the scatterer instead of the radar. In this case, the Pauli decomposition produces three channels given by:

$$\alpha = \frac{S_{hh} + S_{vv}}{\sqrt{2}}$$

$$\beta = \frac{S_{hh} - S_{vv}}{\sqrt{2}} \tag{10.35}$$

$$\gamma = \sqrt{2}\, S_{hv}$$

The value of this decomposition is that:

- α^2 corresponds to the power scattered by targets characterized by either a single bounce (e.g. a sphere or flat plate), or an odd number of bounces (e.g. a trihedral).
- β^2 corresponds to the power scattered by targets characterized by either a double bounce (e.g. a dihedral), or an even number of bounces.
- γ^2 corresponds to the power scattered by targets that can scatter linearly-polarized incident radiation into the orthogonal polarization. Targets of this type typically consist of multiple scatterers, such as occurs with volume scattering.

The Freeman–Durden decomposition takes a different approach to estimate the contributions from three different scattering mechanisms: the scatter from a volume of randomly oriented dipoles (e.g. a forest canopy), the even- or double-bounce scatter from a pair of orthogonal surfaces with different dielectric constants (e.g. ground and tree trunks), and Bragg scatter from a moderately rough surface (e.g. a plowed field or water waves).

These parameterizations represent aggregates of scatterers that are best characterized by their statistical characteristics. So the general approach taken was to formulate models for the second-order statistics of the measured signals based on the characteristics of each scatterer type. The second-order statistics derived from the measured data are computed from terms like $\langle |S_{hh}|^2 \rangle$, $\langle |S_{vv}|^2 \rangle$, $\langle |S_{hv}|^2 \rangle$, and $\langle |S_{hh}S_{vv}^*| \rangle$. A procedure is then followed to estimate the power arising from each of the three scatterer types in each pixel. The details are available in Freeman and Durden (1998).

The left-hand panel in Figure 10.22 shows a scene obtained with a quad-pol SAR, displayed using the Pauli decomposition. The blue, red, and green channels of this image correspond to the single-bounce (α), double-bounce (β), and volume scattering (γ) channels, respectively.

The ocean is blue in this image indicating the ocean return is predominantly single scattering. The multiple scattering in trees causes depolarization, which makes the wooded areas green. And the urban areas have lots of multiple scattering, but less depolarization, and so in the end they are a combination of red and green.

The right-hand panel in Figure 10.22 is based on the Freeman and Durden decomposition. Blue, red and green in this image correspond to the Bragg scattering (surface), double-bounce, and volume scattering channels, respectively. The net result is a means for classifying key physical parameters associated with the different types of scattering surfaces in the scene.

Table 10.2 indicates some of the applications of specific combinations of polarization channels.

10.6.3 Compact Polarimetry

Some of the polarimetric parameters described in Table 10.2 depend on a full characterization of the complex scattering matrix. Such a characterization requires separate transmission of at least two polarized signals, and receipt of either two coherent channels that include phase, or four amplitude-only channels based on the Stokes matrix. Support for full or quad-polarization (two transmit + two receive channels) significantly increases SAR hardware complexity and quadruples the amounts of data that need to be transmitted and processed. Compact polarimetry is an approach for obtaining as much information as possible with reduced hardware and data transmission requirements.

Table 10.3 lists the three major approaches that have been investigated for compact polarimetry.

The $\pi/4$ or *slant* mode was one of the first to be analyzed, but is blind to scatterers that

Pauli Decomposition

$|S_{hh} + S_{vv}| \ |S_{hh} - S_{vv}| \ |S_{hv}|$

Freeman–Durden Decomposition

$P_{surface}(\text{dB}) \ P_{double\text{-}bounce}(\text{dB}) \ P_{volume}(\text{dB})$

Figure 10.22 Examples of SAR polarimetry. Data were originally taken by the NASA JPL aircraft, but the imagery were obtained from https://earth.esa.int/documents/653194/656796/Single_Multi_Polarization_SAR_data.pdf.

Table 10.2 Some polarimetric observables and applications. From Moreira et al. (2013).

2 × 2 Sinclair matrix	Radar observables	Application examples
Scattering amplitude (complex) and scattering power	S_{ij} $\sigma_{ij}^0 = 4\pi \lvert S_{ij} S_{ij}^* \rvert$	Classification/segmentation Change detection Glacier velocities (feature tracking) Ocean wave and wind mapping Coherent scattering
Total power	$TP = \lvert S_{HH} \rvert^2 + \lvert S_{XX} \rvert^2 + \lvert S_{VV} \rvert^2$	Classification/segmentation Feature tracking
Amplitude ratios	$\sigma_{HH}^0 / \sigma_{VV}^0$ $\sigma_{XX}^0 / \sigma_{VV}^0$ $\sigma_{XX}^0 / (\sigma_{HH}^0 + \sigma_{VV}^0)$	Dry/wet snow discrimination Soil moisture Surface roughness
Polarimetric phase differences	$\delta_{HH} - \delta_{VV}$	Sea ice thickness Crop types Forest/nonforest classification
Helicity[1]	$\lvert S_{LL} - S_{RR} \rvert$	Man-made target classification

[1]Subscripts L and R refer to left- and right-circular polarization, respectively.

Table 10.3 Different approaches to compact polarimetry.

Name	Transmit signal	Receive channels
$\pi/4$	linear 45°	coherent H + V
Circular	R or L circular	coherent R + L circular
Hybrid CP	R or L circular	coherent H + V

are oriented orthogonal to the transmitted signal. This mode is widely used in meteorological radars, a suitable application because of the random orientation of atmospheric scatterers.

A circularly polarized transmission eliminates the blind spot of slant mode, and was initially used with coherent dual-circular polarized receive channels. Yet there is no fundamental reason that the type of polarization received (linear or circular) must equal the type of polarization transmitted.

(Measurements made at any two orthogonal polarizations can always be converted to any other orthogonal basis.) This led to the concept of hybrid compact polarimetry. Raney (2007) argues that the hybrid compact polarimetry is superior because it nominally assures similar received signal levels in both receive channels, which reduces the effect of noise on the derived products. According to a more recent review article (Raney, 2019), hybrid compact polarimetry has emerged as the preferred method for operational systems.

10.7 SAR Systems

This section provides a brief review of some of the specific SAR systems that have been used, or are currently in use, for Earth remote sensing.

Table 10.4 lists the major satellite SARs used for Earth remote sensing that are no longer

Table 10.4 SARs used for Earth remote sensing that are no longer operational.

Satellite	Country/ Year	Band	Freq. (GHz)	Incident angle (deg)	Polariz.	Resolution (m)
Seasat (Failed 10/1978)	U.S.A. 1978	L	1.275	23	HH	25
ERS-1/2 (Off 2000/2011)	Europe 1991/95	C	5.25	23	VV	25
Almaz (Failed 10/1992)	Russia 1991	S	3.0	30–60	HH	15
JERS-1 (Off 10/1998)	Japan 1992	L	1.275	39	HH	30
Radarsat-1 (Off 5/2008)	Canada 1995	C	5.3	20–50	HH	Multimode 9–100
Envisat (Failed 4/2012)	Europe 2002	C	5.25	15–45	HH, HV VV, VH	Multimode 28–150
ALOS-PALSAR (Failed 4/2011)	Japan 2006	L	1.27	8–60	HH, HV VV, VH	Multimode 7–100

operational. Seasat was the first, a U.S. satellite launched in 1978. Seasat lasted only 100 days, but the data it returned were revolutionary. Seasat was also the last free-flying SAR flown by NASA for Earth remote sensing. NASA did fly a SAR on the space shuttle, but all other unmanned SARs have been flown by other nations.

Table 10.5 lists currently operating SARs that are used for Earth remote sensing. There is an interesting trend to be noted. First, while all of the original SARs were developed and operated by national governments, the second-generation systems were largely built as a public/private partnership. The systems were expensive to develop and operate, while the data were thought to be useful to both government and commercial entities. So most of the western nations decided to attempt some form of cost sharing.

This funding model has not been entirely successful. It turned out that governmental users ordered more than a factor of 10 more images than commercial users, so the companies set up to commercialize the data struggled. This economic experiment is largely being abandoned in the design of the next generation systems that is currently underway.

The design of the Italian COSMO-SkyMed SARs illustrates other trends. These SARs were designed to be flown in a coordinated constellation of four satellites with a sharing of data between scientific and military users. The desire to provide frequent revisits and timely information for military targets is the primary driver behind the use of a constellation, although it certainly also has benefits for science. The German TerraSAR is another example of a SAR designed to support both scientific and military users.

TanDEM-X is interesting – it is nearly a duplicate of TerraSAR-X designed to fly in close formation with its twin to support a

Table 10.5 SARs used for Earth remote sensing that are currently operational.

Satellite	Year	Country	Band Freq. (GHz)	Incident angle (deg)	Polariz.[1]	Resolution[2] (m)
Radarsat-2	2007	Canada	C (5.3)	10–60	Quad	3–100
TerraSAR-X TanDEM-X[4]	2007 2010	Germany	X (9.65)	20–55	Quad	1–16
COSMO-SkyMed[3]	2007[1,2] 2008[3] 2010[4]	Italy	X (9.6)	20–60	Dual	1–30
Risat-1	2012	India	C (5.35)	12–55	Compact	1–50
HJ-1C	2012	China	S (3.13)	25–47	HH or VV	5–25
Kompsat-5	2013	South Korea	X (9.66)	20–55	Single	1–20
Sentinel-1A Sentinel-1B	2014(A) 2016(B)	Europe	C (5.405)	20–45	Dual	5–40
Alos-2	2014	Japan	L (1.2)	20–65	Quad	1–100
PAZ	2018	Spain	X (9.65)	15–60	Quad	1–16
SAOCOM-1A	2018	Argentina	L (1.275)	20–50	Compact	10–100
Radarsat Constellation (3 satellites)	2019	Canada	C (5.405)	19–53	Compact	1–100

[1] Polarimetry: Single is 1 Co- or Cross-pol channel. Dual is either HH+HV or VH+VV. Quad is four linear channels (HH,HV,VH,VV). Compact includes modes beyond linear.
[2] 1-m resolution only in spotlight mode with spot sizes limited to 5–10 km.
[3] Italian Space Agency (2007).
[4] TanDEM-X is companion to TerraSAR-X for land surface digital elevation mapping.

bistatic SAR configuration. The utility of this configuration is discussed in Section 10.8.1.

Table 10.6 provides the key parameters for a selected set of SARs.

10.7.1 Radarsat-1

Radarsat-1 was a Canadian C-band SAR designed for a variety of missions, including ice monitoring. Radarsat-1 was Canada's first SAR and operated from 1995 to 2013.

Figure 10.23 shows the seven modes supported by Radarsat-1. The table lists the swath widths and azimuth resolutions for each of the modes. All of these modes were either stripmap or ScanSAR. Radarsat-1 did not support a spotlight mode.

The Radarsat-1 operating modes are detailed in Table 10.7. Note that the standard products for six of the seven modes were two-look to eight-look images. And note that because the antenna is steerable, the field of regard for the sensor, the extent of ground range that it could make a measurement, exceeds the instantaneous field of view of the sensor for all but the Wide ScanSAR mode.

Table 10.6 Key parameters for selected SARs.

	Radarsat-1	Envisat	PALSAR	Radarsat-2	TerraSAR-X	COSMO-SkyMed	Sentinel-1
Launched	4-Nov-95	1-Mar-02	Jan-06	Dec-07	Jul-07	Jun-07–2009	Apr-14–Apr-16
Orbit	Sun-synch	Sun-sync	Sun-sync	Sun-sync	Sun-sync	Sun-sync	
Inclination (deg)	98.6	98.55	98.16	98.6	97.4		98.18
Altitude (km)	798	800	692	798	514	619	693
Period (min)	100.7	100.6	98.7	100.7	95	97.86	98.6
Repeat cycle (days)	24	35	46	24	11	16	12
Subcycles (days)	3, 7, & 17		2	3, 7, and 17			
Orbits/day	14	14	14	14		14.7	14.6
Equatorial crossing (LST)	1800	1000	1030	1800	1800	0600	1800
Node	ascending	descending	ascending	ascending	ascending	ascending	ascending
Radar band	C	C	L	C	X	X	C
Radar frequency (GHz)	5.3	5.33	1.27	5.3	9.65	9.6	5.405
Wavelength (cm)	5.6	5.6	23.6	5.6	3.1	3.1	5.6
Polarization	Horizontal	HH,HV VH,VV	HH,HV VH,VV	HH,HV VH,VV			HH+HV,HH VV+VH,VV
RF bandwidth (MHz)	11.6–30.0	28–150	7–100	11.6–100.0	150–300	180–400	300
Range resolution (m)	8–100	28–150	7–100			1–100	5–20
Azimuth resolution (m)	8–100	28–150	7–100			1–100	5–40
Transmit pulse length (μs)	42			42		70–80	5–100
PRF	1270–1390 Hz			1270–1390 Hz	3–6.5 kHz	2.9–4.1 kHz	1–3 kHz
Peak power	5 kW	713 – 1395 W		1650–2280 W	2260 W		4400 W
Average power (W)	300						200–500
Antenna size (az × el) (m)	15.0 × 1.5	10.0 × 1.3	3.1 × 8.8	15.0 × 1.5	4.8 × 0.8	5.7 × 1.4	12.3 × 0.8
Look direction	Right[1]	Right	Right	R or L	R or L	R or L	Right
Antenna apertures	1	1	1	2			
Sampling rate (MHz)	12.9–30.0				110–330	82–176	300
I/Q Sample size (bits)	4	8	5		8		10
Average data rate (Mb s⁻¹)	74–100	100	240				25
SAR on time (min/orbit)	28	30					
Recorder time/capacity	10 min	12 min	18 min	256 Gbits	256 Gbits	300 Gbits	1443 Gbits

[1] Entire satellite could be rotated 180° to look on the other side.

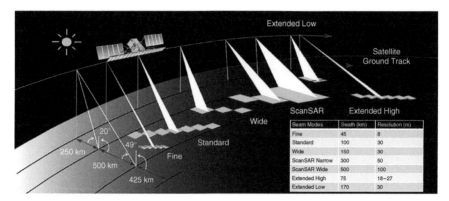

Figure 10.23 Radarsat-1 imaging modes. Source: Canadian Space Agency, http://www.asc-csa.gc.ca/eng/satellites/radarsat/radarsat-tableau.asp with permission from Canadian Space Agency.

Table 10.7 Radarsat-1 operating modes.

	Standard mode	Wide mode	Fine Res. mode	SCANSAR narrow mode	SCANSAR wide mode	Extended high mode	Extended low mode
Number of beams	7	3	5	N/A	N/A	6	N/A
Nadir offset (km)	250	250	500	240 or 400	250	500	125
Swath width (km)	100	150	50	300	500 or 440	75	75
Range resolution (m)	25	23, 27, 35	8–9	50	100	25	25
Azimuth resolution (m)	28	28	9	50	100	28	28
Number of looks	4	4	1	2–4	4–8	4	4
Incident angle range (deg)	20–49	20–45	35–49	20–46	20–50	50–60	10–20

This satellite, like many of the Earth remote sensing SARs, was placed in a dawn/dusk, sun-synchronous orbit. That means that all of the satellite overpasses occur near either dawn or dusk in local time. Placing a satellite in this orbit minimizes the time the satellite is in the Earth's shadow and hence maximizes power generation from the solar cells.

Radarsat-1 was designed to acquire no more than 28 minutes of data during each orbit. Such restrictions are also not uncommon because of either limitations of power, data storage, or download capabilities. This is not a trivial amount of data given the SAR can produce data rates of up to $100\,\mathrm{Mb\ s^{-1}}$. The Radarsat-1 system could acquire up to two terabytes per orbit.

10.7.2 Envisat

Envisat was the third European SAR after the highly successful ERS-1 and 2. Envisat produced data for a decade before it failed in 2012.

10.7.3 PALSAR

The Japanese PALSAR was an L-band radar that ran for five years. It supported five polarimetric modes, a fine resolution mode available at 18 grazing angles, and a ScanSAR mode.

10.7.4 Radarsat-2

Radarsat-2 is the second generation SAR from Canada launched in 2007. It is still operational.

You may not have noticed but the few SARs discussed up to this point could only look off to one side of their track. Radarsat-2 is the first SAR we have discussed that could look left or right of track, although changing look direction requires a spacecraft maneuver.

Figure 10.24 illustrates Radarsat-2, which added a wide variety of additional modes, including spotlight and multiple polarizations.

Table 10.8 contains the details of some of the Radarsat-2 imaging modes.

10.7.5 TerraSAR-X

TerraSAR-X is a German SAR launched in 2007 that produces data for both scientific and military applications. In many ways this SAR represents the current state of the art in operational SARs.

The design of TerraSAR-X is being shared with other nations in Europe. Spain is building a version for launch in a few years. The goal is to operate these SARs along with the Italian COSMO-SkyMed SARs as a super constellation with coordinated tasking.

10.7.6 COSMO-SkyMed

COSMO-SkyMed (COnstellation of small Satellites for the Mediterranean basin Observation) is an Italian constellation of four X-band (9.6 GHz) SAR satellites designed for both civilian and military applications. All four satellites are in the same orbit plane and each repeats the same ground track every 16 days.

Each radar can operate in stripmap, ScanSAR, and spotlight modes, with two options available in each mode. One stripmap mode has a swath of 40 km and 3–5 m resolution, and the other can collect-dual polarization data over a 30-km swath at 15-m resolution. ScanSAR options provide 30-m resolution imagery over a 100-km swath, or 100-m resolution data over a 200-km

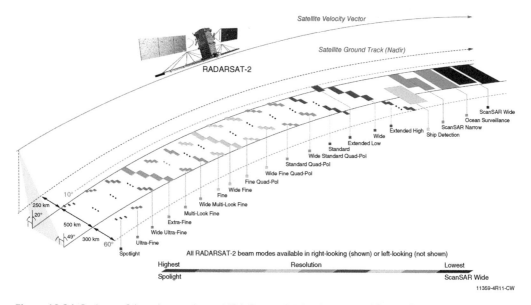

Figure 10.24 Radarsat-2 imaging modes. © MDA Geospatial Services Inc. All Rights Reserved.

Table 10.8 Some of the Radarsat-2 imaging modes.

Beam modes	Scene size (km) [Rng × Az] (km)	Resolution [Rng × Az] (m)	Incident angle (deg)	No. looks [Rng × Az]
Selective Single or Dual Polarization Transmit H and/or V, receive H and/or V				
Fine	50 × 50	10.4 – 6.8 × 7.7	30–50	1 × 1
Wide Fine	150 × 150	14.9 – 7.3 × 7.7	20–45	1 × 1
Standard	100 × 100	26.8 – 18.0 × 24.7	20–49	1 × 4
Wide	150 × 150	40.0 – 19.2 × 24.7	20–45	1 × 4
ScanSAR Narrow	300 × 300	79.9 – 37.7 × 60	20–46	2 × 2
ScanSAR Wide	500 × 500	160 – 72.1 × 100	20–49	4 × 2
Ocean Surveillance	530 × 530	Variable		
Polarimetric Transmit H and V on alternate pulses / receive H and V on any pulse				
Fine Quad-Pol	25 × 25	16.5 – 6.8 × 7.6	18–49	1 × 1
Wide Fine Quad-Pol	50 × 25	17.3 – 7.8 × 7.6	18–42	1 × 1
Standard Quad-Pol	25 × 25	28.6 – 17.7 × 7.6	18–49	1 × 1
Wide Standard Quad-Pol	50 × 25	30.0 – 16.7 × 7.6	18–42	1 × 1
Single Polarization HH Transmit H, receive H				
Extended High	75 × 75	18.2 – 15.9 × 24.7	49–60	1 × 4
Extended Low	170 × 170	52.7 – 23.3 × 24.7	10–23	1 × 4
Selective Single Polarization Transmit H or V, receive H or V				
Spotlight	18 × 8	4.6 – 2.1 × 0.8	20–49	1 × 1
Ultra Fine	20 × 20	4.6 – 2.1 × 2.8	20–49	1 × 1
Wide Ultra Fine	50 × 50	3.3 – 2.1 × 2.8	29–50	1 × 1
Extra Fine (best resolution)	125 × 125	3.1 × 4.6	22–49	1 × 1
Extra Fine (most looks)	125 × 125	24 – 12 × 23.5	22–49	4 × 7
Multilook Fine	50 × 50	10.4 – 6.8 × 7.6	30–50	2 × 2
Wide Multilook Fine	90 × 90	10.8 – 6.8 × 7.6	29–50	2 × 2
Ship Detection F	450 × 500	103 – 71 × 40 – 81	35–56	16 × 2
Ship Detection S	450 × 500	33 – 23 × 19 – 77	35–56	5 × 1

swath. Spotlight mode operation for civilian applications provides 1-m resolution over an area of 10 km on a side. The other spotlight mode is restricted to military-only use. Spotlight operation is limited to either HH or VV polarization.

10.7.7 Sentinel-1

With the individual European states building their own or collaborating on the development of military satellites, ESA remains in the lead for the development of Earth remote sensing satellites. The latest SARs are on the Sentinel-1A and 1B spacecraft. The Sentinel-1 SAR supports four modes: Stripmap with pixel sizes as small as 1.5 × 3.6 m; Interferometric Wide swath with pixel sizes as small as 2.3 × 14.1 m; Extra-Wide swath with pixel sizes as small as 5.9 × 19.9 m; and a Wave mode that produces imagettes for ocean surface wave analysis with pixel sizes as small as 1.7 × 4.1 m, all at one-look. The Sentinel-1 SAR can transmit either H or V pulses but has two parallel receivers for receiving H and V.

Both Sentinel-1A and Sentinel-1B share the same orbital plane with a 180° orbital phasing difference, which reduces revisit times by a factor of 2.

10.7.8 Radarsat Constellation Mission (RCM)

The Radarsat Constellation Mission or RCM is Canada's next generation space-based SAR system. This system is designed to be deployed in a three-satellite constellation.

Figure 10.25 shows the modes for the RCM, which are fewer in number than for Radarsat-2. The detailed mode parameters are listed in Table 10.9. Note that either compact polarimetry or full quad-polarization are supported in all of the RCM modes.

10.7.9 Military SARs

It should be no surprise that military organizations have an interest in exploiting the surveillance capabilities of SAR systems – all-weather, day and night imaging with meter-scale spatial resolution. Starting in the first decade of the 21st century, several countries with radar spacecraft development capabilities have orbited SARs for intelligence, surveillance, and reconnaissance applications. Multimode operation is a common characteristic of these SARs, affording an ability to change the swath size, and to redirect the illumination to areas of particular interest with a high-resolution spotlight mode. As might be expected, many of the design parameters and capabilities of these radars are not readily available.

Countries with known military SARs include Germany, Italy, Israel, India, China, Japan, Russia, and the United States.

Germany's SAR-Lupe system consists of five identical satellites with X-band radars. The first member of the constellation was launched in December 2006; the full system was completed in July 2008. The five satellites are distributed in three orbital planes approximately 60 degrees apart. In stripmap mode, a SAR image frame covers an area of 8 × 60 km at about 1-m resolution. In spotlight mode, the resolution is about 0.5 m with a frame size of 5.5 km on a side. Agreements exist to

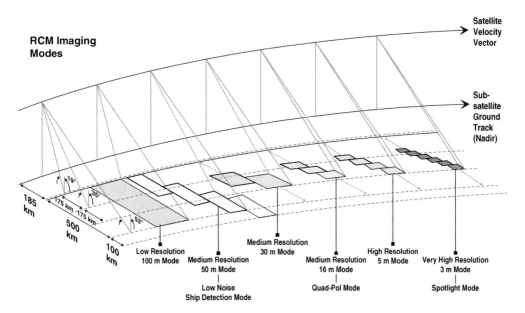

Figure 10.25 Radarsat Constellation Mission imaging modes. Source: Canadian Space Agency, http://www.asc-csa.gc.ca/eng/satellites/radarsat/radarsat-tableau.asp with permission from Canadian Space Agency.

Table 10.9 Radarsat Constellation Mission imaging modes. All modes support either compact polarimetry or quad-polarization.

Beam modes	Swath width (accessible) (km)	Resolution (m)	Looks [Rng × Az]
Low Resolution 100 m	500 (500)	100 × 100	8 × 1
Medium Resolution 50 m	350 (600)	50 × 50	4 × 1
Medium Resolution 30 m	125 (350)	30 × 30	2 × 2
Medium Resolution 16 m	30 (350)	16 × 16	1 × 4
High Resolution 5 m	30 (500)	5 × 5	1 × 1
Very High Resolution 3 m	20 (500)	3 × 3	1 × 1
Low Noise	350 (600)	100 × 100	4 × 2
Ship Detection	350 (350)	Variable	5 × 1
Spotlight	20 (350) × 5	1 × 3	1 × 1
Quad-Polarization	20 (250)	9 × 9	1 × 1

share SAR data with France and possibly other European nations.

The SAR-Lupe system is scheduled to be replaced in 2021 with a second generation three-satellite system called SARah. Two of these radars will use the same passive-reflector antenna technology used in SAR-Lupe, while the third system will have a phased array antenna as used on the TerraSAR-X radar.

The Italian COSMO-SkyMed radar system discussed previously is a dual purpose system for both environmental monitoring and military surveillance applications. In particular, this X-band SAR has a high-resolution spotlight mode reserved for military-only use.

In January 2008, Israel orbited a SAR technology demonstration satellite known as TecSAR using launch facilities provided by India. The satellite is also known as Ofek-8. This X-band radar is a multimode, multipolarization (HH, HV, VH, or VV) system with multibeam electronic steering. It has four operating modes:

- Stripmap mode with 3-m resolution.
- ScanSAR mode providing 8-m resolution imagery and wide area coverage by electronic beam steering.

- Spotlight mode using mechanical steering for spot imagery with resolution of less than 1 m.
- Mosaic mode providing 1.8-m resolution images of large areas by using both mechanical and electronic steering.

Additional TecSAR radars are reported to have been launched on Ofek-10 (April 2014) and Ofek-11 (September 2016).

India launched RISAT-2 (Radar Imaging Satellite 2) in April 2009. This X-band SAR is similar to TecSAR, having been obtained from Israel Aerospace Industries, the builder of Israel's TecSAR. The system is reported to be used for both civil and military applications. Capabilities include a 30-km stripmap mode with 3-m resolution, and a spotlight mode to image a 10 × 10 km spot with 2-m resolution.

The Chinese military has launched several SAR-equipped satellites as part of its Yaogan series. The military satellites also carry the designation Jianbing (JB). The first generation SAR satellites, known as JB-5, were launched in April 2006 (Yaogan-1), November 2007 (Yaogan-3), and August 2010 (Yaogan-10). Yaogan-1 was reported to be an L-band radar with 5-m resolution. Yaogan-1 appears to have broken up in February 2010. No information is available for the other JB-5 radars but they are likely to operate at L-band also.

JB-7 is the second generation radar spacecraft (Yaogan -6, -13, -18), having been launched between April 2009 and October 2013. No information regarding the characteristics of these SARs is available.

Japanese spy satellites are known as Information Gathering Satellites (IGS). The first two SAR satellites were launched in 2003 and 2007, respectively; both are now retired. Four newer generation spacecraft and one on-orbit spare were launched between December 2011 and June 2018. No details such as radar frequency, operation modes, and swath size have been released, although the resolution is believed to be about 1 m for those deployed between 2011 and 2015; the two most recent systems are thought to have 0.5-m resolution.

The Russian Kondor satellite is a military system with an S-band (3.13 GHz) SAR that can image a 10-km wide stripmap with 1–3 m resolution. It also has a spotlight mode with 1–2 m resolution, and a ScanSAR mode with 5–30 m resolution. Polarization modes include HH and VV, depending on imaging mode. A 6 × 6 m parabolic dish antenna allows a cross-track pointing capability (55 degrees to either side of the satellite) without changing the spacecraft attitude. Information on the number of Kondor satellites and their launch dates is not available.

Kondor-E is an export version reported to have been built for the military of South Africa and launched in 2013.

Finally, in 2008, the United States National Reconnaissance Office declassified the existence of its SAR satellite constellation known as Onyx. No additional information has been released.

10.8 Advanced SARs

Advanced synthetic aperture radar capabilities have been developed, including interferometric SARs that utilize two or more antennae to obtain additional information. Interferometric SARs have been designed with cross-track antennae for terrain mapping and along-track antennae for measurement of ocean currents and moving targets. We will also briefly discuss the coming generation of high-resolution wide-swath SARs that break the resolution/swath constraint that exists for single aperture radars.

SARs are coherent radars. Up to this point we have described how the phase time history is used to provide azimuth resolution, but have ignored the phase of the final result. It turns out that the final product of SAR processing is a complex image of the radar scene. Specifically, each pixel in the scene has an amplitude proportional to the measured RCS in that pixel, and a phase proportional to the mean distance to the scattering phase center for that pixel, modulo the radar wavelength.

SAR interferometry measures the phase differences between two or more complex SAR images that have been acquired from antennae in slightly different positions or at slightly different times. This is done because the phase in each pixel is an exquisitely sensitive measure of distance even if measured from orbit.

Four basic types of SAR interferometry have been developed, each with its own specific uses:

- Cross-track interferometry is used for measuring land topography.
- Along-track interferometry is used for measuring ocean currents and moving target detection.
- Differential interferometry is used to precisely measure topographic changes.
- Tomographic interferometry is used to measure the vertical separation of volume scatterers, a useful capability for separating levels within a forest canopy.

10.8.1 Cross-Track Interferometry

A cross-track SAR interferometer transmits a broad beam from one antenna and receives on

3D Geometry

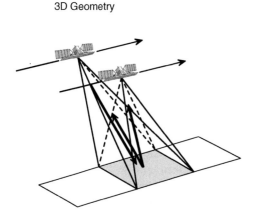

Along-Track View

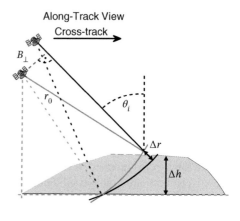

Figure 10.26 Geometry of cross-track interferometry. Modified from Moreira et al. (2013).

two antennae, one separated in the cross-track direction from the other (Figure 10.26). The two SAR images that are formed from the received pulses will have nearly identical amplitude, but the phases will differ according to the height of the scatterers in each pixel.

For example, if the antennae are separated perpendicular to the line of sight by $B_\perp$ then the difference in the slant range between the two paths is given by:

$$\Delta r \cong \frac{B_\perp}{r_0 \sin \theta_i} \Delta h \qquad (10.36)$$

where r_0 is the slant range, θ_i is the local incident angle, and Δh is the surface elevation difference.

The phase difference is then just 2π times the slant range difference divided by the radar wavelength:

$$\Delta \varphi \cong \frac{2\pi}{\lambda} \Delta r \qquad (10.37)$$

By measuring the phase difference between the two channels we can measure the height on the surface.

Techniques have been developed to allow these interferometric measurements to be made from a single platform on different passes or even different radars.

While phase depends on scene elevation, the phase measured by the InSAR wraps every 2π radians. Thus interferometric phase data need

to be unwrapped in order to derive relative elevation.

The top panel in Figure 10.27 illustrates the phase centers of two InSAR antennae viewing an artificial hill. The bottom left panel shows the interferometric phase data that might be acquired from observation of this hill. The phase wrapping is clear on this image. Some of this phase difference is due to the viewing geometry of the surface and some is due to the hill. The next panel to the right shows the phase differences that would have occurred if the surface was flat. Subtracting the predicted phases for a flat Earth from the measured phases leaves just the phase differences due to the hill as shown in the third panel. The unwrapped phase image is shown on the right, providing an accurate representation of the original hill.

Figure 10.27 was based on simulated data, while Figure 10.28 is based on real data. In real data, the topography is far higher than a radar wavelength, so there are thousands of fringes in the scene. The raw phase data shown on the left is so wrapped it almost looks like noise. The middle image shows these radar phase data minus a phase ramp due to the mean height of the Earth. This subtraction reduces the phase gradients, significantly reducing the number of phase wraps in the image. This image can then be unwrapped to find the deviations of

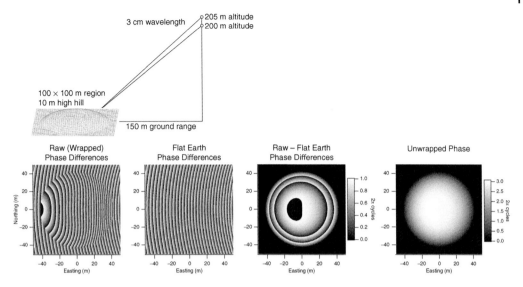

Figure 10.27 Modeled topography, phase map from a notional interferometric SAR, and the unwrapped phase used to measure the topography.

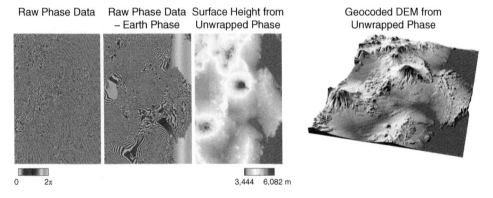

Figure 10.28 Example raw and unwrapped phase maps from real data. Source: Moreira et al. (2013) with permission of Institute of Electrical and Electronics Engineers.

the topography from the mean height. The result is shown on the right.

10.8.2 Along-Track Interferometry

An along-track SAR interferometer transmits a broad beam from one antenna and receives on two antennae, one separated in the along-track direction from the other (Figure 10.29).

If two SAR images are formed by selecting the received pulses from the two receivers corresponding to the same locations along the track, then the two SAR images will be formed at different times. The effective time difference between the two images is given by one half of the distance between the antennas divided by the platform speed:

$$\Delta t = \frac{D_{offset}}{2V_{platform}} \tag{10.38}$$

The factor of one-half arises from the fact that the effective phase center of the SAR formed by transmitting from one antenna and receiving from the second is halfway between the two physical antennae.

Along-track interferometric SAR is sensitive to any scatterer motions in the time

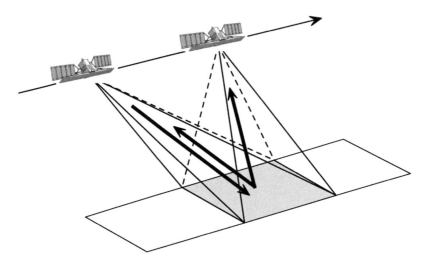

Figure 10.29 Geometry of along-track interferometry.

between the images. For example, consider two images from an airborne along-track SAR interferometer flying at 100 m s^{-1} with an antenna separation of 0.6 m. This corresponds to an effective time difference Δt of 3 ms, recalling the factor of two from equation (10.38).

This turns out to be a typical time lag for measuring ocean currents with an X-band ATI SAR. In 3 ms, the scatterers arising from a 1 m s^{-1} current will move 3 mm, which equals 1/10 of a wavelength or 0.6 radians, at least if the SAR was looking near the horizon. At an incident angle of 60°, the phase shift would be half this value.

Because we are measuring phase, and phase wraps, there is an ambiguity velocity that equals the wavelength divided by the time between the images:

$$v_{amb} \, \Delta t = \lambda_{radar}$$
$$v_{amb} = \frac{2V_{platform} \, \lambda_{radar}}{D_{offset}} \quad (10.39)$$

In the example aircraft case, this ambiguity velocity would equal 10 m s^{-1}.

Figure 10.30 is one of the first along-track interferometric SAR images measuring ocean currents. This image was taken by the JPL AirSAR at L-band from 20 kft. The scene is a ship-generated internal wave wake in

Loch Linnhe, Scotland with surrounding mountains. The black and white portions of the image are from the single-channel SAR, while the color inlays are from the along-track interferometric SAR estimates of surface currents.

The original JPL analysis of these data suggested that currents in the ship wake had a magnitude of ± 50 cm s^{-1}, while in situ measurements made with current meters showed a magnitude of ± 5 cm s^{-1}. It took a year of work, but Thompson and Jensen (1993) ultimately showed that the ATI SAR actually measured the sum of the ± 5 cm s^{-1} current and the ± 45 cm s^{-1} phase velocity of the L-band Bragg scatterers!

One difficulty with along-track SAR interferometery is that everything on the ocean is moving, and so the surface decorrelates with time. Figure 10.31 shows plots depicting the distribution of phase measured from ocean backscatter at 35 GHz as a function of time. The left-hand plot shows the distribution of phase differences measured with separations of 0.33, 1.0, 2.0, and 3.0 ms. The right-hand curve is a contour plot of probability density as a function of phase difference and temporal lag, which is a more complete representation of the same data. These data show the phase at 35 GHz decorrelates completely in about 3 ms.

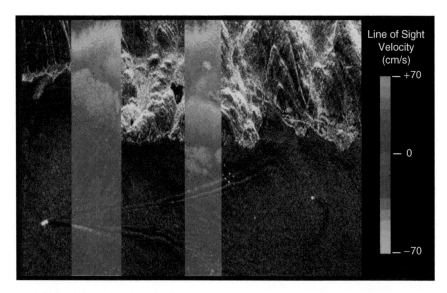

Figure 10.30 A SAR image of a ship's internal wave wake in Loch Linnhe, Scotland (grey scale) with a color inlays from an along-track interferometric measurements of Doppler velocities. Source: NASA JPL.

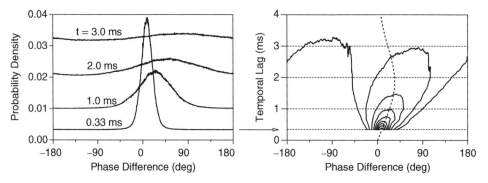

Figure 10.31 Measured probability density for ocean backscatter phase differences indicates decorrelation in 3 ms at 35 GHz (Ka-band). From Chapman et al., (1994).

Similar measurements made at a range of frequencies from L-band to Ka-band (Chapman et al., 1994) suggest that ocean backscatter decorrelates according to a simple rule of thumb:

$$\tau_{decorrelate}(\text{ms}) = \frac{100}{f(\text{GHz})} \qquad (10.40)$$

To make a good ATI SAR current measurement, the time spacing should be about 1/3 of the decorrelation time. For a satellite in low-Earth orbit, the satellite speed is about 7 km s^{-1}. Using the rule of thumb, we can compute the decorrelation times and optimal antenna spacing as shown in Table 10.10.

Table 10.10 Decorrelation times and optimal antenna spacing.

Band/ Freq	Decorrelation time (ms)	Optimal antenna spacing (m)
L (1.5 GHz)	66 ms	311 m
C (6 GHz)	16 ms	78 m
X (10 GHz)	10 ms	47 m

This clearly indicates that the optimal antenna spacing is too large to be deployed on a single satellite.

Romeiser et al. (2014) have demonstrated satellite-based ATI-SAR measurement of ocean currents using two satellites flying in close formation: TerraSAR-X and Tandem-X.

10.8.3 Differential Interferometry

Differential interferometry combines cross-track interferometric measurements made at different times with an existing digital elevation model (DEM). Differencing each set of cross-track interferometric measurements made with an existing DEM nearly flattens the phase field, allowing for very small changes in phase (fractions of a wavelength) to be detected.

Figure 10.32 contains two images of Mexico City taken six months apart. These data were used with a DEM to measure subsidence, and the rate of subsidence over this period of time. This technique is capable of measuring subsidence rates on the order of 1 cm per month. Differential interferometry is also being used to regularly monitor the movement of glaciers and volcanoes.

10.8.4 Tomographic Interferometry

InSAR measurements can be used to separate scatterer motions relative to the platform within the same range cell. The relative motion of tree canopies is different than that of the ground because of height. In fact each level in the canopy presents a slightly different apparent velocity to the SAR. Thus InSAR data can be used to estimate the vertical structure of a forest – this is called SAR tomography.

The panel on the right in Figure 10.33 shows an example of this technique. The bottom panel is a SAR image of a forested area, with a yellow line indicating those data acquired from a single ground range. The top panel represents the density of the scatterers as a function of height at the specific ground range indicated by the yellow line.

10.8.5 High-Resolution, Wide-Swath SAR

Our earlier analysis showed that the swath width of a high-resolution single-aperture radar is limited based on azimuth resolution.

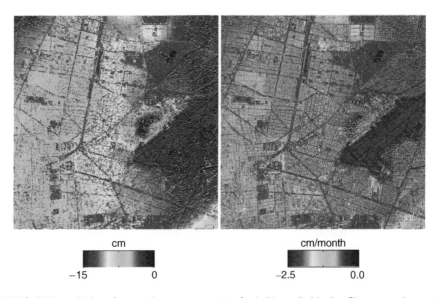

Figure 10.32 Differential interferometric measurements of subsidence in Mexico City over a six-month period. Source: Moreira et al. (2013) with permission of IEEE.

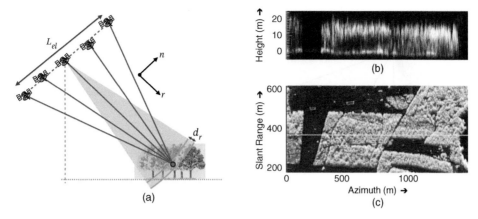

Figure 10.33 Examples of tomographic interferometry. From Moreira et al. (2013) with permission from IEEE.

A new class of high-resolution wide-swath SARs is currently under development. In order to see how these systems are designed to work, let us review the swath limitation for a standard, single-aperture SAR.

Figure 10.34 illustrates how a narrow-swath SAR avoids range ambiguities. The light blue region indicates the range extent of the beam. When a pulse hits the surface it starts to scatter in all directions. The initial scattering location is at the closest point that the beam intersects the surface and the scattered energy from multiple pulses are represented by multiple red arcs.

The one arc that is solid red represents the energy that is currently being received by the SAR. The small red circle on the surface indicates the location from which that received energy was scattered. If time were to advance in this diagram, the scattered location on the surface being received by the SAR would race across the footprint. The key point is that the beam is small enough that the location of scattering would slide beyond the radar's footprint before the signals from the next pulse start arriving from the nearest portions of the footprint. So there are multiple pulses in the air at once, but the swath is narrow enough that returns from only a single pulse are being received at the SAR at any one time.

Figure 10.35 illustrates the range ambiguities that would occur if a wide beamwidth is used for a conventional SAR. Again we have multiple pulses in the air at any one time.

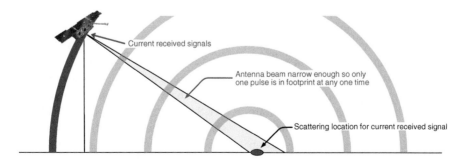

Figure 10.34 Conventional high-resolution narrow-swath SAR.

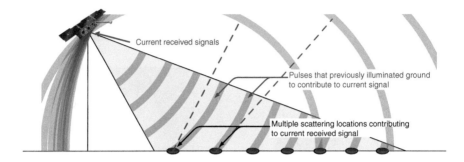

Figure 10.35 Ambiguous returns prevent a wide-swath from SAR being implemented with a single broad beam.

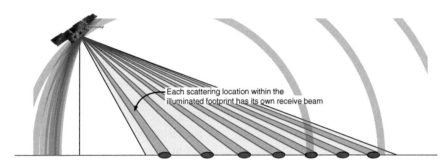

Figure 10.36 A phased array antenna can produce multiple receive beams to separate multiple received pulses, allowing for high resolution and wide swath.

But now we have multiple scattering locations contributing to the return at the same time. A single SAR receiver has no way of separating these returns from each other. Thus we will not be able to identify where the returns came from on the surface.

Figure 10.36 illustrates how a new generation of high-resolution wide-swath SARs will use multiple beams to eliminate range ambiguities.

Again the SAR transmits pulses in a broad beam. But this system uses a phased array antenna to form multiple independent receive beams, one beam per pulse. The beams are swept across the surface at the same rate as the pulses. Each of these beams feeds an independent receiver. Thus these receivers can separate the returns from individual pulses

achieving wide swath at the same time as high resolution.

High-resolution wide-swath SARs are currently under development in Germany, Canada, China, India, and the U.S.A. The German system to be launched in 2026 will have 400-km swath at 2-m resolution.

10.9 SAR Applications

SARs are expensive, power-hungry instruments that produce huge amounts of data. Yet SARs are also the most flexible remote sensor yet devised, with many applications for remote sensing of the oceans, land, and even the atmosphere. This section focuses on applications of SAR data.

10.9.1   SAR Ocean Surface Waves

This subsection discusses the use of SARs to measure ocean surface waves. Figure 10.37 is a typical SAR image of ocean waves, showing their refraction as they approach the coast near Peniche, Portugal. This image clearly demonstrates the capability of SARs to image a complex and varying pattern of ocean surface waves.

Figure 10.38 is a second image of ocean waves under low wind conditions. The variable winds cause large variations in the image intensity across the scene, yet the image also clearly shows a primary surface gravity wave field as well as some atmospheric waves that are propagating in different directions.

These images lead to the obvious question of how to use such data to measure geophysical

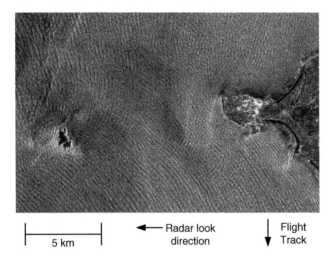

Figure 10.37 Seasat image of refracting surface waves near Peniche, Portugal. NASA figure acquired 20 August 1978, processed by ASF DAAC 2013.

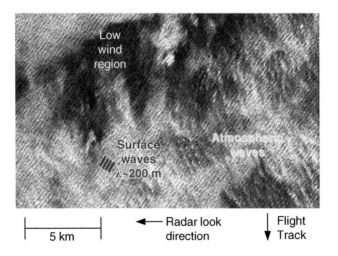

Figure 10.38 Another example of Seasat imaging of ocean waves under low wind conditions. NASA figure acquired 29 September 1978, processed by ASF DAAC 2013.

quantities such as the directional spectrum of the ocean waves. As always, in order to understand how to interpret such data, we need to have a detailed understanding of exactly how the SAR is imaging these waves.

We know that an X-band radar operated at mid-incident angles is sensitive to Bragg scattering from waves of about the radar's wavelength. But the surface waves imaged here have wavelengths of about 200 m. So how is the radar sensitive to these waves? Specifically, what ocean wave parameter or combination of parameters is the SAR sensing: height, slope, velocity, or some mix of these? As usual, the answer to this question is interestingly complicated.

There are three primary mechanisms by which a SAR images ocean waves. The first mechanism is one we discussed in Section 9.2, namely tilt modulation.

Figure 10.39 is a not-to-scale cartoon illustrating that short waves ride on top of the long waves. The radar responds to those waves matching the Bragg scattering condition, which depends on the local surface slope. And this local surface slope is driven by the slope of the propagating long waves. As the surface tilts toward the radar, the RCS increases, and as the slope tilts away, the RCS decreases.

This is a simple mechanism that in some ways is like the variations in radiance that allow us to see ocean waves.

The second mechanism, referred to as hydrodynamic modulation, arises because the long waves actually modulate the amplitude of short waves. As illustrated, the short wave amplitudes increase between the crests and front face of the long waves and decrease a bit in the troughs and back face of the long waves.

This is called one mechanism, but is actually driven by several physical effects. The first physical effect is that the orbital currents strain the surface, compressing the short waves on the leading face of the long wave, and stretching them out in the troughs. This stretching does have a small dynamic impact on the amplitude of a given wave, but more importantly it changes the wavenumber of the waves, with the phase fronts compressing or stretching as if coils on a spring. When the waves are compressed, a wave that may have started out as a 3-cm Bragg wave is compressed so it no longer matches the Bragg condition. But the wave that was previously a 4-cm wave may be compressed to a 3-cm wavelength, and now it becomes the Bragg wave. The important point is that the wave spectrum indicates that long waves in the vicinity of the Bragg wave have higher amplitude than shorter waves, so compression makes a higher amplitude wave meet the Bragg condition. Another effect is that the wind stress varies along the phase of the long wave, and wind stress has a direct impact on the amplitude of the short waves.

The third mechanism is called velocity bunching, which sounds like something that happens in a traffic jam, but is actually caused

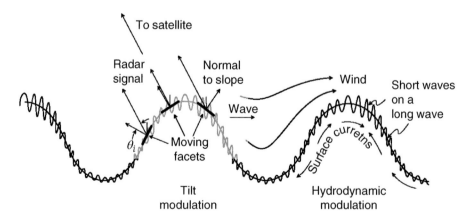

Figure 10.39 Two SAR imaging mechanisms: tilted Bragg scattering and hydrodynamic modulations. Adapted from Stewart (1985).

by SAR velocity effects. Section 10.4 describes how the SAR imaging mechanism will misregister range-moving objects by shifting their apparent location in azimuth. The impact of misregistration on hard targets is easy to visualize but is more complex for the ocean surface where everything on the surface is moving all the time. For example, while this constantly changing motion can act to blur an object sitting on the surface, it can also lead to an important wave imaging mechanism.

It turns out that these three mechanisms: tilt modulation, hydrodynamic modulation, and velocity bunching, contribute to wave imaging in different proportions depending on the direction the wave is traveling relative to the radar look direction. The mechanisms can cancel in some cases, leading to a complete loss of the visibility of waves traveling in some directions.

Figure 10.40 is an image of what are called range-traveling waves. These are waves that are predominantly propagating toward or away from the radar look direction. In this image, the waves are propagating toward the right, the spacecraft track is down the page, and the Seasat SAR is looking to its right (left on the page). Thus these waves are propagating toward the radar, nearly along the line of sight.

The two dominant mechanisms for imaging range-traveling waves are tilt and hydrodynamic modulation, with the tilt modulation usually being the larger of the two effects. As will be explained, the velocity-bunching mechanism turns out to be negligible for range-traveling waves.

Figure 10.41 is an image of azimuth-traveling waves, which are waves that are propagating orthogonal to the radar line of sight. In this case the satellite track is down the page, which is toward the south southwest. The radar looks to the right relative to its track and the waves are propagating up the page which is to the north.

Tilt modulation is not a significant effect for azimuth-traveling waves, because the wave slopes are not tilting toward or away from the radar. So there is a null in the tilt modulation for azimuth-traveling waves. Hydrodynamic modulation is still occurring, but it is relatively weak. Instead the imaging of these waves is primarily due to velocity bunching.

In order to understand velocity bunching, consider in some detail the SAR imaging of range-traveling waves. Figure 10.42 illustrates a single cycle of a long wave propagating toward the radar flying along the bottom of the page looking up the page. This single cycle consist of two wave crests and one trough.

The short vertical arrows indicate the direction of the peak horizontal component of the orbital currents for the wave. The current at the crests is in the same direction as the wave propagation, and the current in the trough is

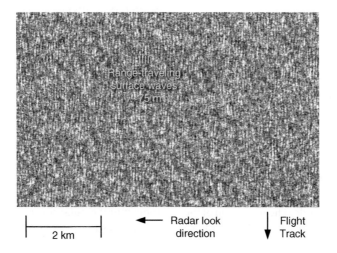

Figure 10.40 An example of range-traveling waves west-northwest of Scotland. NASA figure acquired 5 September 1978, processed by ASF DAAC 2013.

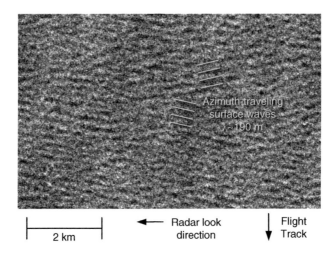

2 km   Radar look direction   Flight Track

Figure 10.41 An example of azimuth-traveling waves. NASA figure acquired 1978-07-18, processed by ASF DAAC 2013.

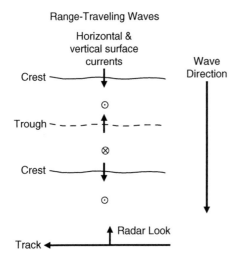

Figure 10.42 Surface currents for a range-traveling wave relative to crests and troughs. Adapted from Holt (2004).

in the opposite direction. The circles indicate the locations of the peak vertical component of the orbital currents, with the center dot indicating the current is moving up out of the page, and the cross indicating the current is down into the page.

To first order these orbital currents are circular, so the horizontal and vertical components have the same magnitude. The projection of the orbital currents into the radar will then have

a peak value toward the radar somewhere on the leading face of the waves, and a peak value away from the radar on the leading edge of the trough.

Tilt and hydrodynamic modulation will mean that the RCS varies along the phase of the long wave. But the range velocity also varies with the phase of the long wave, so let us consider the effect of that varying velocity on the SAR image.

Figure 10.43 is a cartoon representation of SAR misregistration effects applied to this range-traveling wave. Consider the imaging of scatterers in the upper right square that happen to be on the leading face of the wave. Scatterers in this box are being advected by the long waves toward the radar, so the energy will be shifted in the SAR image to a later time, which is to the left. This shift is indicated by the arrow. In fact, all of the image cells along the leading face of the wave will have a similar velocity toward the radar and so will be shifted to the left in the image.

The squares on the leading edge of the trough are all moving away from the radar, so they will all be imaged to the right of their true location. The projection of the long wave orbital currents into the radar line of sight is zero for the squares on the trailing edges of the crests or

Range-Traveling Waves

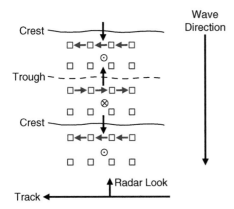

Azimuth shifts occur along wave phase fronts

Figure 10.43 Velocity bunching has no effect on range-traveling waves because azimuthal shifts are parallel to wave crests and troughs. Adapted from Holt (2004).

Azimuth-Traveling Waves

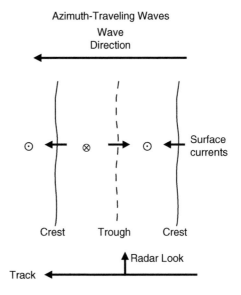

Figure 10.44 Surface currents for an azimuth-traveling wave relative to crests and troughs. Adapted from Holt (2004).

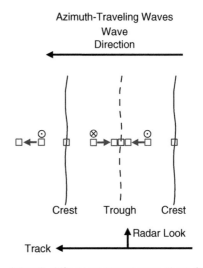

Azimuth shifts occur across wave phase fronts

Figure 10.45 Velocity bunching is the primary imaging mechanism for azimuth-traveling waves. Adapted from Holt (2004).

troughs, so they will be imaged in the correct location.

The key point is that while SAR imaging effects shift energy around for range-traveling waves, these shifts are all along lines of constant phase of the long waves, so it has no effect on the distribution of energy across the wave. The leading edges of the crests are still bright and the troughs are still dark.

Next consider an azimuth-traveling wave as illustrated in Figure 10.44. This time the horizontal components of the orbital currents are orthogonal to the radar look, so they cause no Doppler shift in the returned signal. But the vertical components of the orbital currents that peak in the locations shown by the $\otimes$ and $\odot$ symbols (indicating flow into and out of the page, respectively) do have nonzero projection into the radar line of sight that causes Doppler shifts in the scattered signal.

Figure 10.45 illustrates the effect SAR misregistration has on this azimuth-traveling wave. On the leading edge of the crests the vertical velocity is up and hence toward the radar. This will cause the energy in these pixels to be shifted to the left. On the leading edge of

the trough, the vertical velocity is down and hence away from the radar. This will cause the energy in these pixels to be shifted to the right. The vertical velocity on the crests or on the troughs is zero, so the energy in these locations is not shifted in azimuth.

The net effect of this is to bunch energy in the troughs and to spread energy away from the crests. We call this effect velocity bunching and it is the primary reason that azimuth-traveling waves can be imaged in a SAR.

Figure 10.46 further illustrates velocity bunching. The vertical component of the velocities cause Doppler shifts in the returned signal that leads to SAR misregistration. And this misregistration shifts energy away from the crests and toward the troughs. This is the diagram you will find in many texts on SAR imaging.

Figure 10.47 is another diagram showing the same effect. But let us dig a bit deeper than the cartoon. The azimuth shift that occurs will equal the radar factor R/V times the velocity in the radar look direction. So the size of the azimuth shift depends on the amplitude of the wave.

When the amplitude of the long wave is small, the azimuth shift of scatterers is less than the long-wave wavelength, as shown in all of the figures here. This is a linear imaging situation where the image wavelength is approximately the same as the long-wave wavelength. For large amplitude waves, the azimuth displacement can be greater than the long-wave wavelength, and the image wavelength no longer matches the long-wave wavelength. Here wave imaging becomes nonlinear.

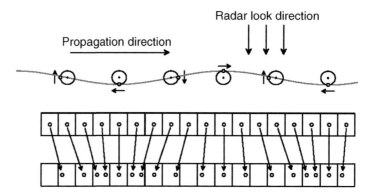

Figure 10.46 Azimuth shifts due to velocity bunching. Modified from Alpers (1983).

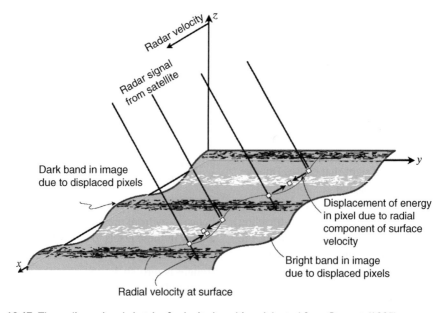

Figure 10.47 Three-dimensional sketch of velocity bunching. Adapted from Stewart (1985).

Real ocean wave systems are not as simple as these cartoons. They have many wave components of different amplitudes and wavelengths. In general, SAR wave images will have a mixture of linear and nonlinear effects. The net effect is that the image spectrum is nonlinearly related to the surface wave spectrum.

The best estimate of a wave spectrum can be computed by an iterative inversion process, starting with a best guess of the wave spectrum, from which the SAR spectrum is computed. Adjustments are then made until the computed SAR spectrum matches the observed spectrum (Hasselmann et al., 1996). But this process is far from perfect and can work over only a limited range of wavelengths.

10.9.2 SAR Winds

Section 9.3 in the chapter on scatterometry describes how the radar cross-section of the ocean surface depends on wind speed and direction. Radar scatterometers are designed to measure the RCS from multiple directions to solve for wind speed and direction. But if the wind direction was available from some other source, then a calibrated measurement of RCS from a single look direction can be used to estimate wind speed. This is the central concept behind estimating wind speeds from SAR imagery.

The rationale for such measurements is simple. While scatterometers measure ocean winds with resolutions of 25–50 km, a SAR can measure wind speed at a resolution of 25 m. So the SAR can be used to study phenomenon on scales that scatterometers cannot detect.

Figure 10.48 contains high resolution SAR wind speed estimates obtained in a region off the coast of Alaska. Colors in this figure indicate wind speed. The arrows in this figure indicate the wind speed and directions obtained from a weather forecast model that

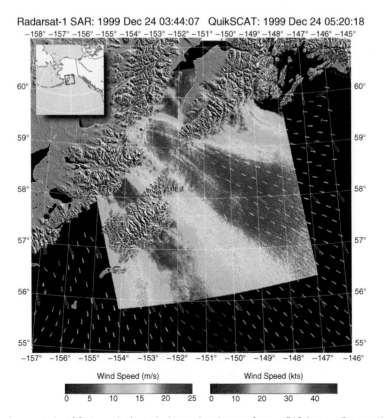

Radarsat-1 SAR: 1999 Dec 24 03:44:07   QuikSCAT: 1999 Dec 24 05:20:18

Figure 10.48 An example of fine resolution wind speed estimates from a SAR image. Source: JHU/APL, http://fermi.jhuapl.edu.

was driven in large part by scatterometer wind measurements.

The SAR image was then processed to estimate the wind speed at every location that was consistent with both the measured RCS and the predicted wind directions interpolated to the finer resolution of the SAR image.

The winds at this time were flowing offshore. The low wind regions were due to blockage from mountains, while the high wind region is what is called a gap flow, a region of enhanced winds flowing though gaps in the topography.

Where the model predicted wind speed is the same as the SAR estimated wind field, the arrows have the same color as the image and disappear into the image. But the contrast of the arrows in the regions of the highest and lowest winds indicates that the model had significant errors in predicting some of the finer scale features.

Figure 10.49 is an example of a fine-scale feature called a barrier jet, a region of enhanced winds caused by flows constrained by the coastal mountains.

Figure 10.50 is an example that shows wave-like oscillations in the wind speed over the ocean. These oscillations are due to internal waves in the atmosphere that are caused by the mean wind flowing over the peninsula.

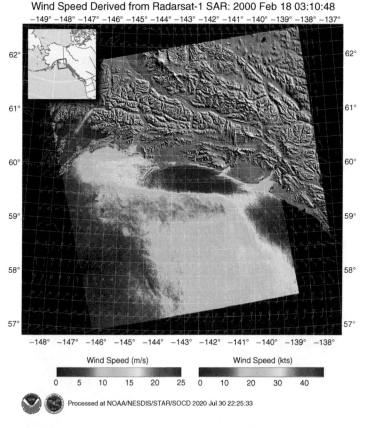

Figure 10.49 An example of a barrier jet created by winds near topography. Source: NOAA, National Environmental Satellite, Data, and Information Service (NESDIS), Center for Satellite Applications and Research (STAR), Satellite Oceanography and Climatology Division (SOCD).

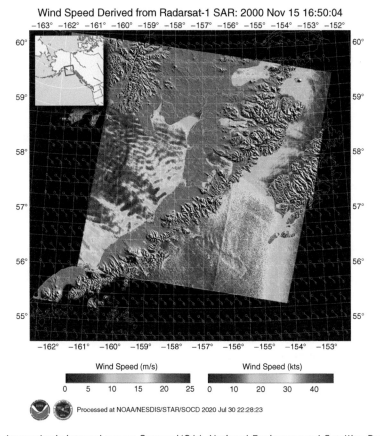

Figure 10.50 Atmospheric internal waves. Source: NOAA, National Environmental Satellite, Data, and Information Service (NESDIS), Center for Satellite Applications and Research (STAR), Satellite Oceanography and Climatology Division (SOCD).

In Figure 10.51, an instability in the atmospheric flow south of the Aleutian Islands is made visible. This oscillatory instability is referred to as von Karman Sheet Vortices, which are shed by flow past an obstacle. The SAR image was taken over a few seconds of time, but the figure on the right, obtained from Wikipedia, illustrates a theoretical model of such a flow.

Figure 10.52 is an example of gap flows at a variety of spatial scales.

Figure 10.53 is a finer-scale example, showing gap flows into the Chesapeake Bay through various creeks and inlets.

SAR winds is a simple concept that requires the SAR data to be accurately calibrated. Given

this calibration and a good wind direction estimate, SAR winds can be as accurate as a scatterometer. Figure 10.54 shows a comparison of SAR winds versus QuikSCAT winds, with an RMS error of 1.4 m s^{-1}.

Note that some, and possibly most, of the differences of the wind estimates from the SAR and the scatterometer are real. The scatterometer measures a spatially-averaged wind field that differs significantly from the true winds at high spatial resolution.

Unstable atmospheric stratification often exists in coastal regions of the world's oceans. For example, along the east coast of the United States, westerly winds often blow cold air from the land out over the warmer ocean. The

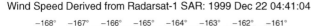

Wind Speed Derived from Radarsat-1 SAR: 1999 Dec 22 04:41:04

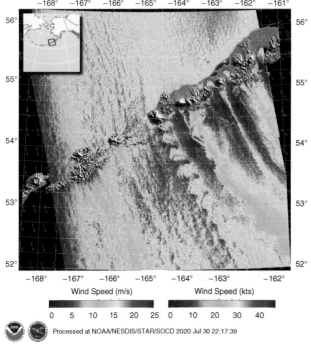

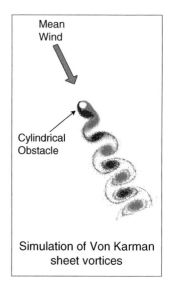

Mean Wind

Cylindrical Obstacle

Simulation of Von Karman sheet vortices

Wind Speed (m/s) Wind Speed (kts)

0 5 10 15 20 25 0 10 20 30 40

Processed at NOAA/NESDIS/STAR/SOCD 2020 Jul 30 22:17:39

Figure 10.51 Von Karman sheet vortices in the atmosphere. Left-hand panel from NOAA, National Environmental Satellite, Data, and Information Service (NESDIS), Center for Satellite Applications and Research (STAR), Satellite Oceanography and Climatology Division (SOCD). Right-hand panel includes simulation modified from https://en.wikipedia.org/wiki/Karman_vortex_street\LY1\textbackslash#/media/File:Vortex-street-animation.gif.

presence of cold air over warm water creates atmospheric convection cells. These convection cells cause wind speed variations at the ocean surface, as shown in Figure 10.55, producing variations in the surface roughness that show up as a mottled effect in the SAR imagery.

Notice that the largest size of the atmospheric cells is capped by the height of the marine atmospheric boundary layer or MABL. Thus analysis of the texture in a SAR image can be used to infer the degree of instability and the height of the marine atmospheric boundary layer.

Figure 10.56 is a SAR image acquired from Radarsat-1 (C-band, HH-pol) off the U.S. east coast when the MABL was unstable. The image contains the mottled SAR-signature of convection throughout. The imaged area is 300 × 300 km. The scale size of the mottling grows larger to the lower right of the image, an indication of the growing height of the boundary layer.

Figure 10.57 illustrates another application of SAR winds, namely the high-resolution measurement of wind speed to support studies of where to locate wind farms. One of the wind

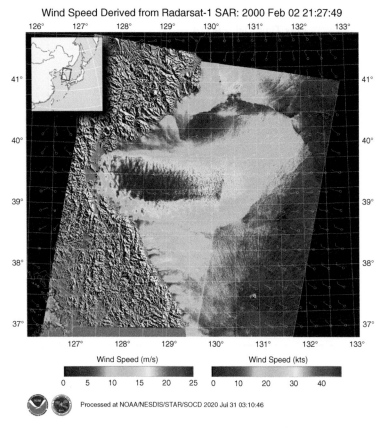

Figure 10.52 Example gap flows at multiple spatial scales. Source: NOAA, National Environmental Satellite, Data, and Information Service (NESDIS), Center for Satellite Applications and Research (STAR), Satellite Oceanography and Climatology Division (SOCD).

Figure 10.53 Gap flows into the Chesapeake Bay through various creeks and inlets. Source: Winstead and Young (2000) © American Meteorological Society. Used with permission.

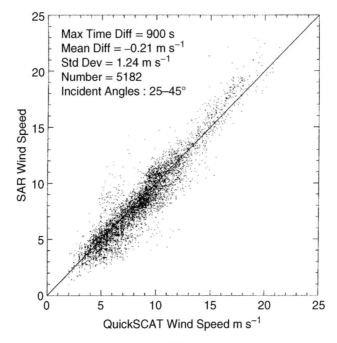

Figure 10.54 Comparison of SAR winds versus QuikSCAT winds. From Monaldo et al. (2004) with permission from IEEE.

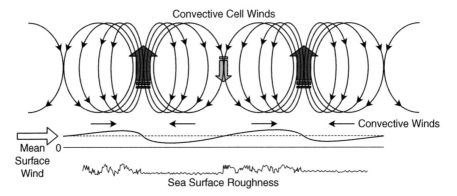

Figure 10.55 Schematic of convective cells in the marine atmospheric boundary layer. Source: Sikora and Ufermann (2004).

farms off the coast of Denmark is clearly visible in this image.

Rain cells are another atmospheric phenomena that can be imaged by a SAR. Figure 10.58 shows images of a pair of large rain cells acquired with multiple polarizations at L-, C- and X-bands.

The differences in these images are due to variations in the scattering and absorptive properties of rain, wind, and surface waves.

There are at least three factors that make rain cells visible in SAR images:

• There is scattering and absorption from drops in the atmosphere. Scattering from

Figure 10.56 SAR image with mottled texture associated with active convective cells in the marine atmospheric boundary layer. Source: Sikora and Ufermann (2004) NOAA.

rain drops is often weak at the lowest SAR frequencies. Absorption in the heaviest regions of rain often lead to dark patches in the SAR imagery.

- The rain drops hitting the surface change the surface roughness. Moderate rain can dampen surface waves by increasing turbulence at the surface. The heaviest drops can create splashes that effectively increase the roughness. The heaviest rain often comes at the leading edges of the rain cell, causing an enhanced RCS.
- Rain cells are usually associated with wind variations around the cell that cause surface roughness changes.

The differences in the RCS in Figure 10.58 can be explained in terms of the factors illustrated in Figure 10.59. The heaviest rain is dampening the L-band Bragg waves, but the small splashes from the drops may be increasing the RCS at C-and X-bands. The C-band SAR may also be more sensitive to rain drop scattering, leading to the bright central regions, while the X-band is more sensitive to absorption causing dark regions. All wavelengths indicate a strong gust front ahead of the cell. These images make it clear that measurements at multiple frequencies and polarizations can be useful for estimating the characteristics of individual rain cells.

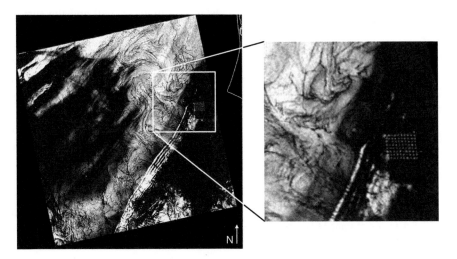

Figure 10.57 SAR wind measurements near an ocean wind farm. Source: Christiansen (2006).

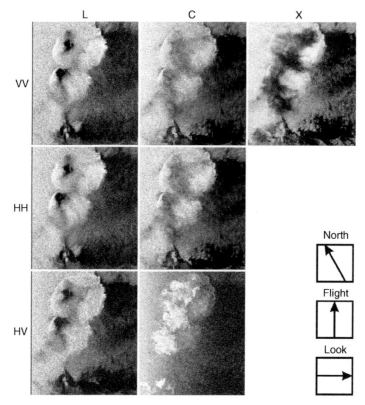

Figure 10.58 Rain cell observed with multiple polarizations at L-, C- and X-bands. The size of each image is approximately 16 × 18 km. Source: Alpers and Melsheimer (2004).

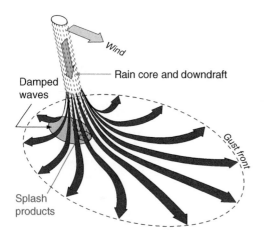

Figure 10.59 Schematic of some key processes in a rain cell. Source: Alpers and Melsheimer (2004).

10.9.3 SAR Bathymetry

Synthetic aperture radars can also be used to measure ocean bathymetry in coastal regions.

Figure 10.60 is a SAR image of Nantucket Island (outlined in red) and the surrounding shoals. The complexity of the bottom topography surrounding the island is indicated by the depth contours that are overlaid on the image in color. It is clear from this picture that despite the fact that the L-band energy from the SAR does not penetrate into the water, somehow the SAR is sensitive to the complex bottom topography. Exactly how this works is the topic of this section.

Figure 10.61 is a second SAR image taken from the C-band SAR on ERS-2 on 7 October 1996 at 15:25 UTC. Again we see that the SAR image includes some sort of distorted information about the bottom topography.

These sorts of variations in radar cross-section are observed in virtually all coastal areas with complex bottom topography.

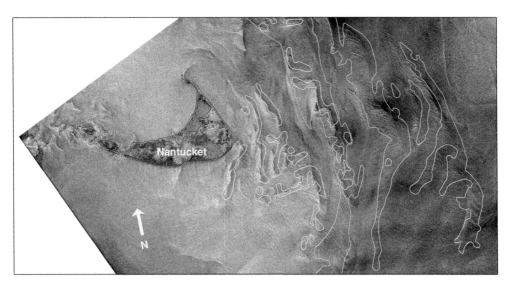

Figure 10.60 A Seasat SAR image of Nantucket shoals with overlaid bathymetric contours. Data acquired 27 August 1978 12:34 UTC. Similar to McCandless and Jackson (2004), but with data downloaded from ASF DAAC, courtesy of NASA.

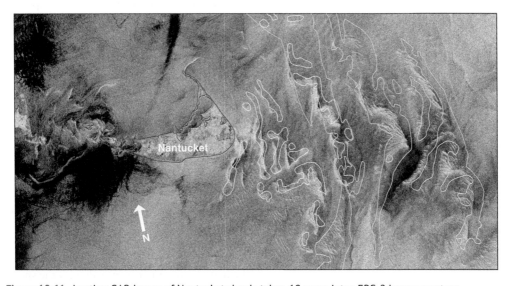

Figure 10.61 Another SAR image of Nantucket shoals taken 18 years later. ERS-2 image courtesy of ESA.

Figure 10.62 is a Seasat image over sandbanks in the English channel.

So how is it that a SAR can be sensitive to bottom topography? First, there is no magic here. The radar image intensity is proportional to backscatter cross-section. And what we will show is that the bottom topography affects the surface currents which then affect the surface roughness.

As an aside, the SAR misregistration effects in this case are small in comparison with the scales of the bottom topography, so velocity misregistration effects are generally not important.

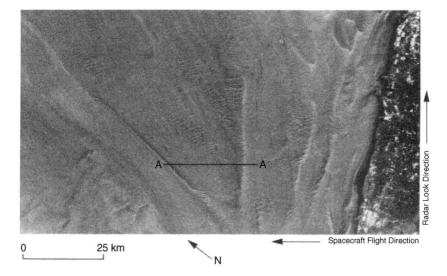

Radar Look Direction

Spacecraft Flight Direction

0   25 km

N

Figure 10.62 Seasat image over sandbanks in the English channel. Source: Fu and Holt (1982).

A simple Bragg scattering model assumes that the radar cross-section is proportional to the surface wave spectrum at the Bragg wavenumber. Thus the variation in intensity is proportional to the variation in RCS which is proportional to the variation of the height spectrum at the Bragg wavenumber:

$$\frac{\delta I}{I} = \frac{\delta \sigma}{\sigma_0} = \frac{\delta \Psi(k_B)}{\Psi_{eq}} \qquad (10.41)$$

where:

$$k_B = 2k_{RF} \sin \theta \qquad (10.42)$$

Surface strain is defined as the horizontal derivative of a varying surface current, for example du/dx. We discussed before how surface strain modulates the wavenumber of ambient waves, therefore modulating the waves that meet the Bragg condition. At the same time the surface strain is perturbing the waves away from their equilibrium state, the wind and wave–wave interactions begin to work to restore the spectrum to its equilibrium state. This restoring force is broadly characterized by a wave relaxation time $\tau(k, U_w)$ that depends on wavenumber and wind speed.

It turns out that for small strain, the change in wavenumber to which the radar responds is proportional to $-\tau$ times the strain, where τ is the relaxation time of the Bragg wave[5]. Furthermore at these wavenumbers, wave spectral models exhibit a $k^{-4.5}$ wavenumber dependence. Thus in this limit of small variations in strain, the image intensity variations are roughly proportional to $-4.5 \, \tau \, du/dx$:

$$\frac{\delta I}{I} = \frac{\delta \sigma}{\sigma_0} = -4.5 \, \tau \left(\frac{\partial u}{\partial x} \right)_{z=0} \qquad (10.43)$$

While this formula is not precise, it does serve as a reasonable rule of thumb for modulations observed at L-band and lower frequencies.

Actual cross-section modulation is more complex and depends on the modulation of the meter-scale gravity waves that tilt the short waves that contribute to the cross-section. At higher frequencies, two-scale cross-section models can predict the observed image modulations for most situations, although the

---

5 To get the signs straight, note that positive strain compresses the waves, decreasing the wavelength and increasing the wavenumber.

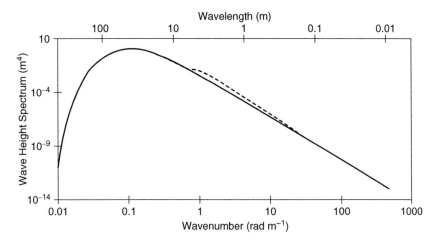

Figure 10.63 Example of spectral modulations due to a varying current. From Thompson (1988) with permission from John Wiley & Sons.

section on internal waves includes an example where the two-scale model starts to fail.

The general solution for the interaction of a wave propagating through a varying current is formulated in terms of wave action, which is the wave energy divided by the intrinsic frequency. This solution arises because wave action is conserved as a wave propagates through a slowly-varying current.

The spectral perturbations due to the varying current can then be computed, although we will not go through the details here. The spectrum in Figure 10.63 shows such a calculation, with the perturbed spectrum shown as a dashed line just above the ambient spectrum. The results indicate that long waves ($\lambda > 8$ m corresponding to $k < 0.8$ rad m^{-1}) are not modulated very much by passing through a current – they are energetic and move fast enough that they spend little time interacting with the current field. The shortest waves of less than 6 cm are also not modulated very much but for a different reason – the restoring force of the wind is too strong. So only waves from 10 cm to a few meters are effectively modulated by a current.

Note that while these modulations have a small direct effect on the Bragg waves

for lower-frequency SARs (e.g. L-band), they will have little to no impact at higher frequencies (e.g. X-band). Thus the modulations we are interested in are not due to amplitude variations in the spectrum, but changes in the wavenumber as we have discussed.

So surface currents can modulate the RCS. How does the bottom topography affect the surface currents?

Figure 10.64 shows an uneven wedge on the bottom with the mean tidal current above. We begin by assuming the current is to the right. The figure is two dimensional, but imagine the bottom topography extends a long distance into the page. The variations in the currents in the x-direction are then easy to compute.

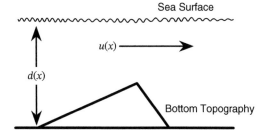

Figure 10.64 Flow over bottom topography.

Bernoulli's law states that the current at each location, $u(x)$, times the depth at each location, $d(x)$, is a constant:

$$u(x)d(x) = \text{constant} = C \qquad (10.44)$$

And according to equation (10.43), the image intensity is proportional to $-du/dx$:

$$\frac{\delta I}{I} = \frac{\delta \sigma}{\sigma_0} = -4.5\tau \left(\frac{\partial u}{\partial x} \right)_{z=0} \qquad (10.45)$$

We then compute du/dx to be $-C/d^2$ times the derivative of the depth in the x-direction, d':

$$\frac{\partial u(x)}{\partial x} = \frac{-C}{d^2(x)} \frac{\partial d(x)}{\partial x} = -C \frac{d'(x)}{d^2(x)} \qquad (10.46)$$

In the region over the gentle leading slope in Figure 10.64, $d' < 0$ so $du/dx > 0$ and the surface is smoother, which makes the image darker than ambient. In the region over the steep trailing slope in Figure 10.64, $d' > 0$ so $du/dx < 0$ and the surface is rougher, which makes the image brighter than ambient. Later in the tidal cycle, the current will reverse direction, and the SAR image features will change polarity.

Figure 10.65 is a more detailed plot of what we just walked through, with the addition of some comments regarding the amplitude of the modulations. The left-hand panel shows the surface roughness and image intensity arising from a tidal flow to the left, indicated by the small arrow at the bottom of the figure. The right-hand panel is for a tidal flow to the right. Note the image contrast reverses as the flow reverses. Also note that in both cases the highest contrast intensity signal (positive or negative) arises from flow over the steepest part of the topography.

10.9.4 SAR Ocean Internal Waves

The ocean is generally stratified with warmer, less-dense water near the surface and colder, denser water below. Salinity can also play a role in stratification, especially in coastal regions in the vicinity of strong fresh water outflows, but near-surface stratification in most parts of the ocean is thermally driven.

Imagine a small parcel of fluid at rest 100 m deep in a stratified ocean. Now imagine reaching in, moving that parcel up to 90 m and then releasing it. The parcel would find itself surrounded by less dense fluid, so its mass would exceed the buoyancy of the water it displaced and it would thus be subject to a

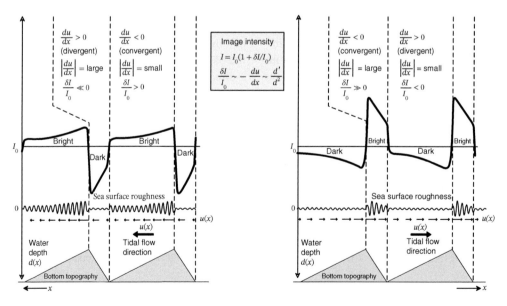

Figure 10.65 Wave current interactions over sandbanks. From Alpers and Hennings (1984) with permissions from John Wiley & Sons.

Figure 10.66 Photo taken from the space shuttle of the surface roughness expression caused by internal waves. Source: Jackson (2004).

downward force which would cause it to sink. That parcel would begin to accelerate until it reached some terminal fall velocity. When it reached its original depth of 100 m, the buoyancy forcing it downwards would disappear, but the fluid parcel would have some momentum so it would overshoot and end up deeper than it started. Then it would be surrounded by denser fluid and be forced upward. The natural period of this oscillation is called the Brunt–Väisälä frequency and is proportional to the gradient of the density stratification.

Any vertical displacement of fluid in the stratified ocean can produce such oscillations, with each parcel affecting the adjacent fluid parcels leading to a propagating wave. We call these internal gravity waves. For example, internal waves are periodically generated by tidal flow past a sill that exists in the Strait of Gibraltar. Figure 10.66 is an optical photo taken from the space shuttle of internal waves emanating from the Strait of Gibraltar.

Figure 10.67 is a comparable SAR image of internal waves emanating from the Strait of Gibraltar. Actually to be precise, we should

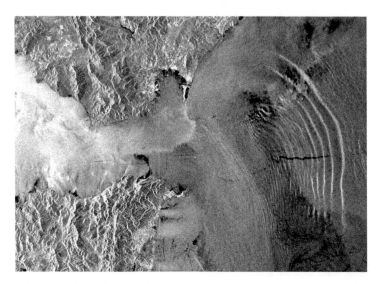

Figure 10.67 SAR image of the surface roughness expression caused by internal waves. See Alpers et al. (2008) for a more complete description. Source: ESA (2019).

say that this is an image of the surface effects caused by internal waves, since the radar is only sensitive to the surface of the ocean.

Internal waves exist on buoyancy gradients within the ocean interior. And while there are both horizontal and vertical currents in the interior of the fluid, the vertical motions decay as you approach the bottom or the free surface. Thus an internal wave may have a vertical displacement of 100 m at depth, but only create horizontal currents at the ocean surface.

Small-amplitude linear internal waves that can be modeled as sinusoids are nearly ubiquitous in the ocean. They can be caused by a number of processes, including interactions of the tides with the deep bottom topography and wind stress variations at the surface. Yet these deep water internal waves are not the internal waves that are usually imaged by SARs.

Instead, SARs typically detect the largest nonlinear internal waves that can be created by tidal flows over coastal bottom topography, e.g. the slope at the edge of the continental shelf. These large-amplitude waves often begin life as hydraulic jumps arising from the outgoing tide. Initially the propagation of that hydraulic jump is arrested by the tidal outflow, but when the outflow relaxes the jump begins to propagate. As illustrated in Figure 10.68 the hydraulic jump eventually disperses into a rank ordered set of solitons, with the longest wavelengths having the highest amplitude at the front of the packet. Ultimately these solitons may decay into more linear, sinusoidal forms.

A packet of these waves can be created once per tidal cycle, which is 12.5 hours in most locations.

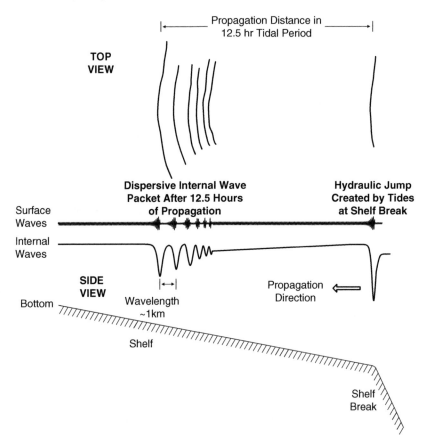

Figure 10.68 Internal wave creation and dispersion.

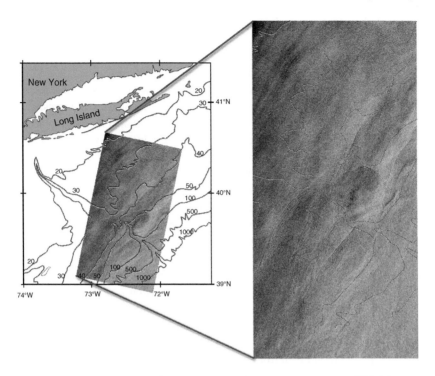

Figure 10.69 Internal waves created by tidal interactions with bottom topography. ERS-1 images acquired on 18 July 1992 courtesy of European Space Agency. Bathymetry contours derived from Liu (1988) with permission from John Wiley & Sons

Figure 10.69 shows an ERS-1 SAR image taken in July 1992 over the Hudson Canyon off the coast from Long Island, NY. The map on the left is actually from a similar experiment conducted in conjunction with Seasat, and it indicates the topography in the region. This has been a popular place to perform internal wave experiments because internal waves are regularly created by tidal flow past the shelf and the canyon. These waves propagate toward shallower water.

Figure 10.70 is a rotated blow up of one portion of the previous image. The short packet of 1 or 2 waves on the right-hand side of this

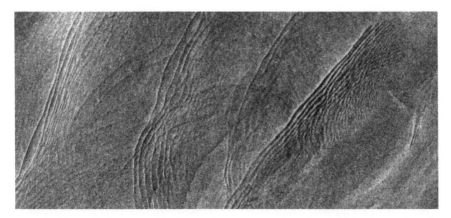

Figure 10.70 Internal wave packets created by tidal interactions with the shelf break near the Hudson Canyon. ERS-1 images acquired on 18 July 1992 courtesy of European Space Agency.

image are the initial surface expressions of a new packet of waves being generated.

Each subsequent packet to the left was generated 12½ hours earlier than the last based on the tidal cycle. So the waves on the left-hand side of the image were likely created 50 hours before this image was acquired. You can see that the longest and strongest waves are at the leading edge of each packet. The packets weaken and become more linear as they propagate toward shore. To get some sense of scale, the leading phase front of each packet is probably 1–2 km ahead of the second phase front. So these are long wavelength waves with very long crests.

During the 1992 experiment, current meters were deployed on moorings to measure the currents of these waves and a Russian research vessel was towing instruments through these waves to measure their characteristics.

Figure 10.71 is a photograph taken from the research vessel of a dark band of roughened water due to one of these internal waves.

That a SAR can image these waves should not be a surprise at this point. The internal waves induce a varying surface current field. The surface strain modulates the ambient surface gravity waves, creating rough and smooth bands on the ocean surface. The surface RCS

then depends on the surface roughness. The entire process is illustrated in Figure 10.72. The same formula for the modulation of the Bragg waves by surface strain that was derived in the last section: $-4.5\,\tau\,du/dx$, also applies to the imaging of internal waves.

Figure 10.72 also illustrates a unique aspect of solitary internal waves in that they can change polarity as they propagate up an incline. This is one of many odd properties that solitons have, making them a fascinating phenomenon to study.

Figure 10.73 is another SAR image of internal waves and some rain cells in the Gulf of Mexico.

The images and phenomena in this section have been described in general terms, yet remote sensing is a quantitative science with a goal of extracting geophysical parameters. In that quantitative sense, the measurement of internal waves from SAR is in its infancy. Most of the papers in the literature talk about the locations of the internal waves, and little else.

One exception is a paper by Hogan et al. (1996). This paper compares in situ data with SAR data acquired during an internal wave experiment conducted in Loch Linnhe Scotland in 1989. The images on the right in Figure 10.74 are nearly space- and time-aligned

Figure 10.71 Photograph of the surface expression of an internal wave.

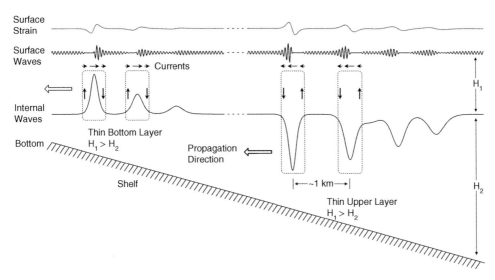

Figure 10.72 Internal wave imaging process. Adapted from Liu et al. (1998) with permission from John Wiley & Sons.

Figure 10.73 Seasat SAR image of internal waves and some rain cells in the Gulf of Mexico. Data acquired on 24 August 1978 14:01 UTC. Similar to Holt (2004), but with data downloaded from ASF DAAC, courtesy of NASA.

SAR imagery obtained from three different aircraft. The SARs were operated at C-, X-, Ku-, and Ka-bands. The scene is dominated by the internal wave wake created by a deep-draft vessel that was employed to create these internal waves on demand. Fresh water runoff and saline deep water makes Loch Linnhe a nearly perfect two-layer system in which to conduct internal wave experiments.

The white dots on the left of these images are moored research vessels and buoys with instruments making in situ measurements of the water column and the internal waves. These instruments measured the peak internal wave currents to be about ± 5 cm s^{-1}. The ambient surface wave spectrum was also measured. The strain rates produced by these waves were then estimated and used with a two-scale Bragg model to predict the observed SAR signals. The results are shown in the plots on the left at each of the four frequencies.

These results indicate that the theory works extremely well at the lowest frequency of

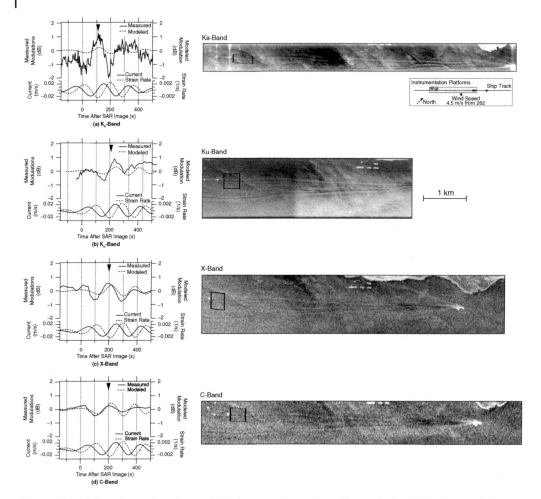

Figure 10.74 Internal wave imaging model/data comparisons. From Hogan et al. (1996) with permissions from IEEE.

C-band, but the errors become increasingly larger at higher frequencies, with the two-scale model under predicting the signal. From the work of Johannessen et al. (2005) and Kudryavtsev et al. (2003a, 2003b, 2005), it now seems clear that the deficiency was due to not incorporating surface wave microscale breaking into the formulation, a concept that was developed over a decade after the Hogan paper was written.

10.9.5 SAR Sea Ice

As discussed earlier in Chapter 6, passive microwave radiometers can be used to monitor large-scale sea ice coverage and type, but their resolution is poor. While optical measurements provide high resolution to monitor sea ice coverage and type, cloud cover significantly limits their usefulness in the polar regions. The primary solution is SAR, which can provide excellent resolution with continuous coverage, working at night as well as through clouds. An example is shown in Figure 10.75.

Figure 10.76 shows that RCS varies for different ice types as a function of frequency, polarization, and incident angle. The plots are for L-, C-, X-, and Ku-bands. For example, X-band measurements at moderate incident angles can easily distinguish ice from calm

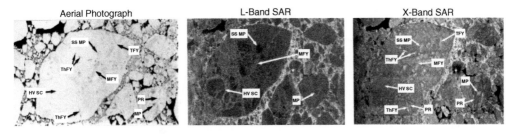

Figure 10.75 Optical photo and SAR images of sea ice. Labeled features include heavy-snow covered ice (HV SC); thick (TFY), medium (MFY), and thin (ThFY) first-year ice; pressure ridges (PR); subsurface melt ponds (SS MP); and melt ponds (MP). Source: Cavalieri et al. (1990) with permission of John Wiley & Sons.

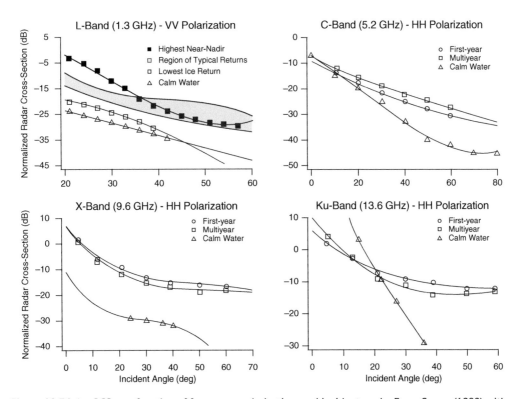

Figure 10.76 Ice RCS as a function of frequency, polarization, and incident angle. From Carsey (1992) with permissions from John Wiley & Sons.

water, but differentiating between first-year and multiyear ice is more challenging. The separation of first-year and multiyear ice increases in Ku-band measurements at 40° incident angles.

In fact, the contrast between first-year and multiyear ice varies depending on the time of year. Figure 10.77 illustrates contrast reversals that have been observed to occur between first-year and multiyear ice for C-, X-, and Ku-band frequencies.

Despite the fact that RCS variations are complex, they can be understood, modeled and hence used to estimate ice characteristics.

Figure 10.78 is an image from a C-band, VV-pol SAR of winter pack ice. The image was taken from an aircraft SAR so the incident angles vary from 20° to 70° across the image,

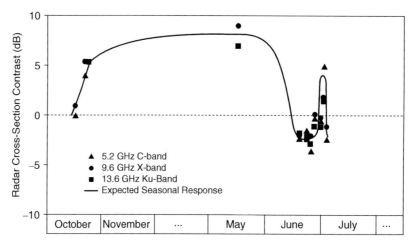

Figure 10.77 Contrast reversals between first-year and multiyear ice for C-, X, and Ku-band frequencies. Source: Onstott and Gogineni (1985) with permissions from John Wiley & Sons.

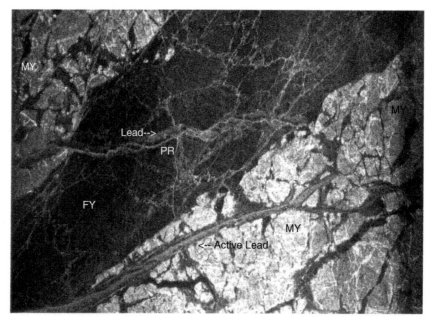

Figure 10.78 SAR image of winter pack ice. Source: Onstott and Shuchman (2004).

which is just 10 km across. The large spatial variations within this scene reinforce the difficulty in using lower resolution sensors like passive microwave radiometers to characterize ice types.

Figure 10.79 is an ERS-1 SAR image of pack ice with active leads. Such high-resolution maps can be particularly important for operation of ice breakers.

Figure 10.80 shows a three-day mosaic of all Envisat ASAR coverage of the Arctic Ocean. SARs are regularly used to monitor sea ice because of their high resolution and ability to operate under all weather conditions.

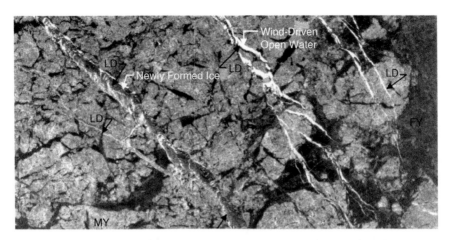

Figure 10.79 ERS-1 SAR image of pack ice with active leads. Source: Onstott and Shuchman (2004).

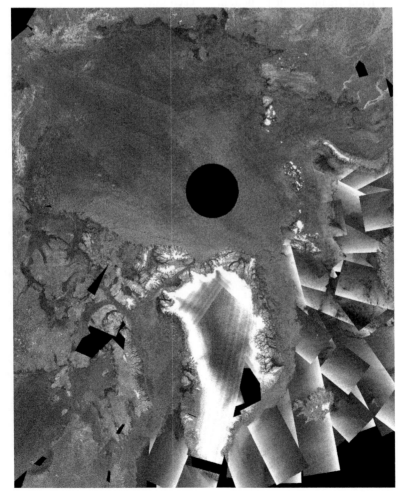

Figure 10.80 A three-day mosaic of Envisat ASAR coverage of the Arctic Ocean. From Technical University of Denmark and Sensing (2009).

10.9.6 SAR Oil Slicks and Ship Detection

SARs are excellent sensors for monitoring coastal waters for illegal oil discharges from ships. Figure 10.81 is a SAR image showing a high RCS ship traveling northward while discharging oil, creating a dark wake that is more than 80 km in length. The oil disperses with time, causing the oil trail to widen. This ERS-1 image was acquired in 1994 over the Pacific, east of Taiwan. The imaged area is 100×100 km.

Algorithms have been developed to detect illegally dumped oil, but these algorithms have to compete against natural or biogenic oils that are created by living organisms or by bottom seeps. Many of the oils are natural surfactants, meaning they can be trapped on the surface, where the molecules arrange themselves into thin surface films that dampen the smallest surface waves.

Figure 10.82 is an image of biogenic slicks in the Caspian Sea south of the Volga Estuary. This is an area of eutrophication because of high nutrient inputs from the Volga river. The biogenic oils are formed into massive slicks that are then advected by surface currents. The surface features visible in this image are likely a result of weak wind-induced surface currents. This is an ERS-2 image with a 100 km extent on each side.

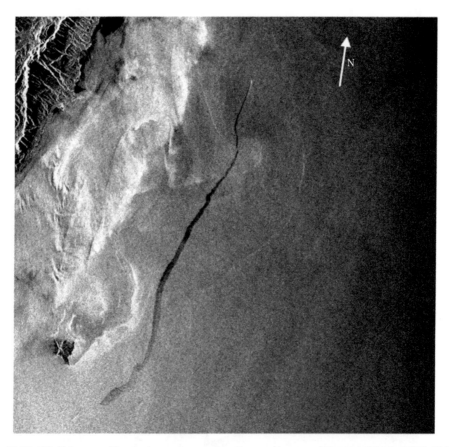

Figure 10.81 Oil slick created by discharge from a ship. Source: Alpers and Espedal (2004), © ESA 1994.

Figure 10.82 SAR image of natural (biogenic) oils on the Caspian Sea. Source: Alpers and Espedal (2004), © ESA 1993.

Figure 10.83 shows some other slicks created by dumping in the North Sea, just west of Sylt. The European Union uses every pass of ESA SARs to look for man-made oil slicks in European waters. The ground receiving station for most of these passes is on Svalbard, a Norwegian island north of the Arctic Circle. The data are then transmitted via fiber optic link to a processing center in Tromsø, Norway.

SARs can also be used to detect and track surface ships. Many nations employ this capability to detect and track ships in their economic zones. This is especially important for so-called dark ships, ships that, for one reason or another, are not emitting AIS messages as required by international maritime law.

Figure 10.84 is a particularly interesting image of the fish-rich waters in the Bering Sea. A small portion of Siberia can be seen on the left side of the full image, and the tip of St Lawrence Island is in the upper right corner. The white line indicates the maritime border between the U.S.A. and Russia. The white dots in the zoomed in region are likely fishing vessels on either side of the border. This fishery has large economic value, so authorities on both sides of the border monitor the fishing fleet using such imagery.

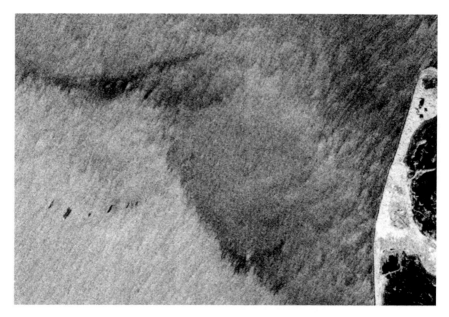

Figure 10.83 North sea oil slicks observed by SIR-C. Source: NASA (1999).

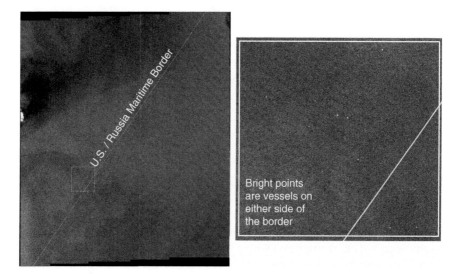

Figure 10.84 Fishing ships on either side of the maritime border in the Bering Sea. These data were taken by Sentinel-1B on 17 August 2018, and downloaded from ESA's Copernicus Open Access Hub.

In general, ship detection depends on ship size, SAR resolution and incident angle. Figure 10.85 is a plot of a ship detection Figure of Merit for the beam modes of Radarsat-1 as a function of SAR angle of incidence. The ship detection Figure of Merit is defined as the minimum detectable ship length for a wind speed of $10\,\mathrm{m\,s^{-1}}$, with wind blowing in the direction of the radar. Ship detectability is reduced in higher sea states due to increased

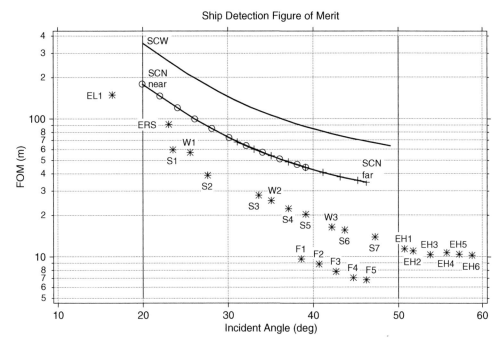

Figure 10.85 Ship detection figure of merit for Radarsat-1. Refer to Table 10.7 for designations of Radarsat-1 operating modes. From Pichel et al. (2004).

surface wave clutter, but increased for lower winds speeds and wind directions other than toward the radar due to decreased surface wave clutter.

It is not just the high RCS of the ship hull that is a potential target – ship wakes are also detectable. Figure 10.86 shows three components of a ship wake:

- The centerline turbulent wake which is the dark region extending maybe 3 km behind the ship.
- The vortex wake indicated by the two dark lines extending well beyond the turbulent wake.
- At least one bright arm of the Kelvin wake extending aft of the ship at an angle less than 19.5° from the centerline turbulent wake.

The centerline turbulent wake is caused by turbulence created by the ship's propeller. This turbulence dampens ambient surface waves, causing the turbulent wake to be dark at most radar wavelengths.

The ship's hull has lift, so it creates a pair of vortices like those coming off airplane wing tips. These vortices can collect surfactants that act to dampen short waves, creating a pair of dark lines arising from the edges of the turbulent wake. These vortices are organized flows and can be longer lived than the turbulent wake.

The Kelvin wake is simply the pattern of surface waves created by an object moving on the surface of the water at a steady speed. It consists of all the surface gravity wave components that have an effective phase speed, projected into the direction of the ship's travel, that equals the ship's speed. Wake data such as these can be used to estimate the ship's direction and speed. A far more complete description of ship wakes and how they are imaged in radars is provided in Reed and Milgram (2002).

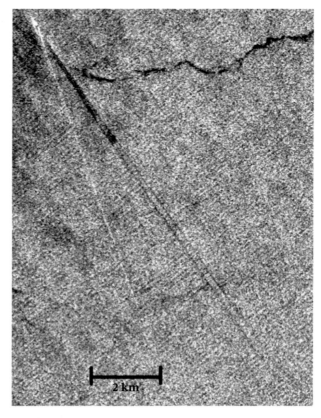

2 km

Figure 10.86 SAR image of ship wake showing centerline turbulent wake, vortex wake, and Kelvin wake. Source: Pichel et al. (2004), © ESA 1996.

In 1996, Tony Liu published a SAR image from the South China Sea (Figure 10.87) showing an obvious ship wake in the big image and a second ship wake in the inset (Liu et al., 1996).

This second wake was interesting because there was no detectable ship at the front of the wake. In his paper, Tony commented:

> "The wake structures could have been formed by a small ship, a ship of low radar-reflective material such as wood or fiberglass, or perhaps a submarine traveling close to the surface?"

We instead think this could be a large ship with a large flat deck. Similar SAR images have been obtained of the wake from a large tanker

with a flat deck. The RCS of ships with flat decks can be very small when observed from above because the radar energy reflects off the deck and away from the radar.

Figure 10.88 is a Seasat image from Revolution 407 that caught a ship making a narrow-V wake off the coast of Florida. This wake caused quite an argument in oceanographic circles because the bright line is well within the outer cusp of the Kelvin wake. In fact, the narrow-V wake looks a bit like an internal wave wake. Yet these waters were not stratified, so it was not initially understood how a ship could create a narrow-V wake.

The answer took a while to develop, but was actually quite simple. The Kelvin wake consists of many wave components. Figure 10.88 is an L-band image, and that V-wake looks just like

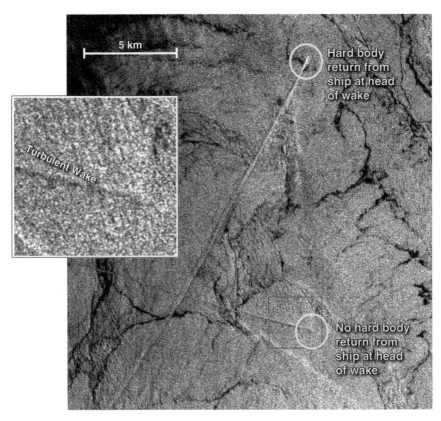

Figure 10.87 Ship wake from the South China Sea with no apparent ship. ERS-1 data acquired on 13 May 1995. Source: European Space Agency.

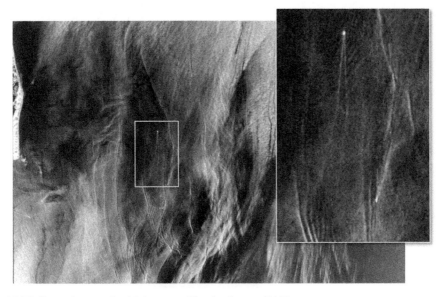

Figure 10.88 Seasat image of a ship's narrow-V wake. Source: NASA.

a map of the locations of the L-band Bragg scatterers in the interior of the Kelvin wake. Years later, SAR data from the Loch Linnhe experiments showed multiple narrow-V ship wakes, each with an opening angle that was entirely predictable based on the frequency of the radar (Stapleton, 1995).

10.9.7 SAR Land Mapping Applications and Distortions

SAR imaging of land surfaces suffers from several types of distortions that arise because SAR does not work geometrically like an optical camera.

Slant Plane Distortion

The first distortion is called Slant Range distortion. It arises because the slant plane is not linearly related to the ground plane. As illustrated in Figure 10.89, scatterers that are evenly distributed on a horizontal surface (points $x_1 - x_5$) will appear with nonlinear

spacing in a slant range image (points $s_1 - s_5$). The opposite is also true with scatterers evenly distributed in the slant plane being nonlinearly spaced on a horizontal surface.

Figure 10.90 further illustrates the nonlinear mapping between the slant and ground planes. The left-most picture is a slant plane radar image of Lake Geneva. The middle panel is the radar image after a nonlinear mapping has been applied to convert it into ground range coordinates. An optical image from space is shown on the right for comparison.

Foreshortening

The second distortion is foreshortening, which is the effect of elevated scatterers appearing closer than they truly are. As illustrated in Figure 10.91, the surface at x_3 is elevated. Thus the point s_3 in the radar image will appear closer to point s_2 than it truly is.

Foreshortening also produces a radiometric effect. All of the scattered energy from

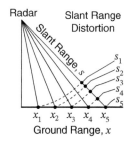

Figure 10.89 Slant range distortion.

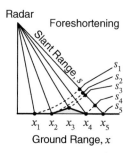

Figure 10.91 Foreshortening.

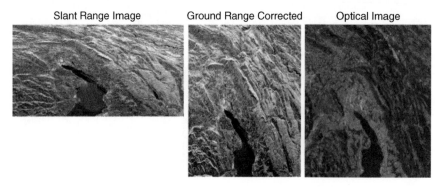

Figure 10.90 Examples of slant range distortion. Source: ERS Radar Course (ESA, 2018), courtesy ESA.

the $x_2 - x_3$ region in the image will be compressed into a smaller area of the ground mapped image, making that region excessively bright. Likewise all of the energy from the x_3-x_4 region will be spread over a larger region in the ground mapped image, making that region dark. This would be true even if the RCS was absolutely uniform across the entire scene.

So foreshortening affects both radiometric and geometric calibration.

Foreshortening is particularly significant in mountainous areas, making the mountains appear to "lean" toward the sensor. The bright and dark regions in the mountainous upper left corner of Figure 10.92 are examples of foreshortening.

Layover

The third distortion is layover, the extreme case of foreshortening where the peak at location x_3 is so high it appears to be located between s_1 and s_2 (Figure 10.93). So the peak at location s_3 is said to "lay over" the base s_2. Radiometrically,

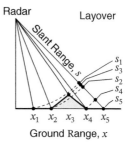

Figure 10.93 Layover.

layover looks similar to foreshortening with bright and dark bands.

As illustrated in Figure 10.94, layover zones face the radar illumination and generally appear as bright features on the image.

Shadowing

The final distortion is shadowing. This is just what it sounds like. The peak at location x_3 in this case is high enough that it shadows the point at location x_4 behind it, causing a dark return (Figure 10.95).

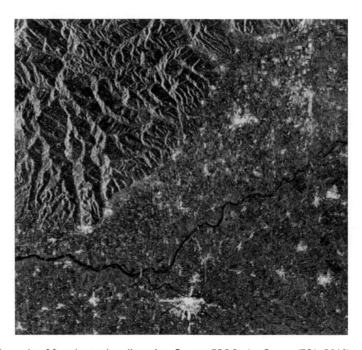

Figure 10.92 Example of foreshortening distortion. Source: ERS Radar Course (ESA, 2018).

Optical Image   SAR Image

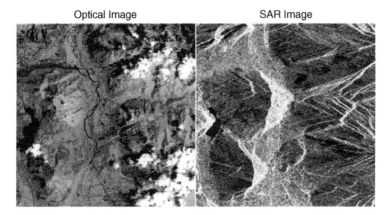

Figure 10.94 Examples of layover distortion. Source: ERS Radar Course (ESA, 2018).

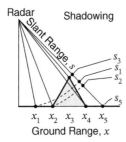

Figure 10.95 Shadowing.

Figure 10.96 further illustrates shadowing. The bright region (1) in the figure is an example of layover, while the dark region (2) in the image is an example of shadowing.

Examples

This section contains a few examples of SAR image distortions. Figure 10.97 is a radar image of the French and Italian Alps, illustrating foreshortening and layover that cause the mountains to appear to lean toward the radar.

Figure 10.98 shows Ku-band images of the Washington Monument and the Pentagon. The most prominent feature in the image of the Washington Monument appears to be the shadow of the monument, as there appears to be no direct return from the monument itself. There are substantial returns from the Pentagon, but you can see the impact of layover and the shadow. These make it easy to identify the SAR's look direction.

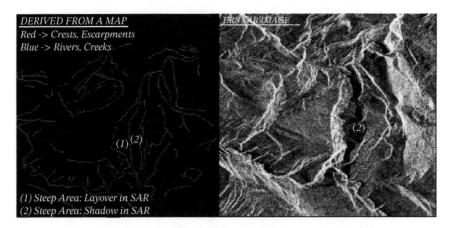

Figure 10.96 Examples of shadowing distortion. Source: ERS Radar Course (ESA, 2018).

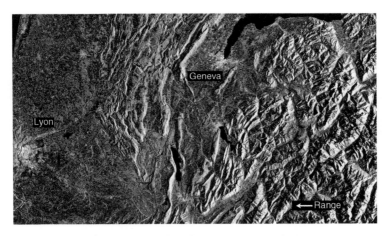

Figure 10.97 Sentinel-1 image of the French and Italian Alps. Figure contains modified Copernicus Sentinel data (2017), processed by ESA, CC BY-SA 3.0 IGO.

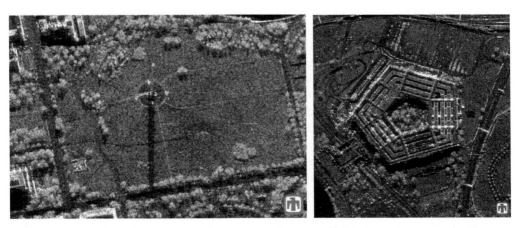

Figure 10.98 Ku-Band SAR images of the Washington Monument (left) and the Pentagon (right). Source: Sandia National Laboratory.

Figure 10.99 is a SAR image of the U.S. Capitol, taken with the radar looking down the page. The Capitol dome and the edges of the roof are all suffering from layover.

Of course, synthetic aperture radars are also used for military applications such as reconnaissance, surveillance, and targeting. These applications are driven by the military's need for all-weather, day-and-night imaging sensors. SAR can provide sufficiently high resolution to distinguish terrain features and to recognize and identify selected man-made

Figure 10.99 U.S. Capitol Building – Ku-Band SAR. Source: Sandia National Laboratory.

Figure 10.100 SAR image of M-47 Tanks, 30-cm resolution. Source: Sandia National Laboratory.

targets. Figure 10.100 shows a SAR image with 30-cm resolution of M-47 tanks along with an optical photo of the same tanks.

Figure 10.101 is a pair of images that illustrates coherent change detection (CCD), a common land measurement technique that offers the capability for detecting changes between imaging passes. The right-hand panel is an optical image of what looks like the end of a runway. The left-hand panel is constructed from a pair of SAR images taken of the same scene, but at different times. The dark loop that is not evident in the optical scene is evidence of active vehicle tracks.

To form this CCD image, the two images of the same scene are geometrically registered so that the same target pixels in each image align. After the images are registered, they are cross correlated pixel by pixel. Where a change has not occurred between the imaging passes, the pixels remain correlated. For example, there is no change in the paved surface so it appears uniformly bright which means correlated. In the areas where change has occurred, the pixels are uncorrelated and appear dark. Vehicles running along a dark track between the two imaging events apparently compressed the surface enough to decorrelate the pixels.

Of course, targets that are not fixed or rigid, such as trees blowing in the wind, will naturally decorrelate and show as having "changed." While this technique is useful for detecting change, it does not measure the direction or magnitude of the change.

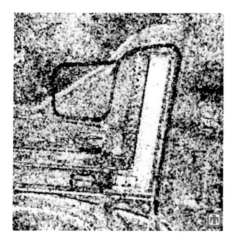

Figure 10.101 Example of coherent change detection (left: SAR CCD image, right: optical image). Source: Sandia National Laboratory.

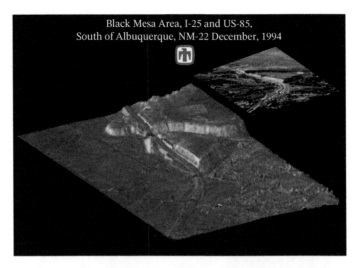

Figure 10.102 Elevation variations measured from interferometric SAR processing. Source: Sandia National Laboratory.

Interferometric synthetic aperture radar data can be acquired using two antennae on one aircraft or by flying two slightly offset passes of an aircraft with a single antenna. As illustrated in Figure 10.102, interferometric SAR can be used to generate very accurate surface profile maps of the terrain.

Figure 10.103 shows a photograph and SAR image of the Albuquerque airport. SAR is obviously a great all-weather sensor for mapping and target identification.

All the SAR imagery we have looked at so far have been derived from a single channel. Figure 10.104 is an image of New York City obtained from three separate SAR channels,

each rendered in this single false color image. The three channels were L-band H-pol (red), L-band cross-pol (green), and C-band cross-pol (blue). It is clear that this scheme has some ability to separate out various land types.

Figure 10.105 shows two SAR images of San Francisco. The left-hand panel is a conventional single channel L-band SAR. The right-hand panel is almost the same image, but this time from a multichannel SAR with the same false colors as for New York City. The differences in RCS between these channels can be used to help differentiate between various land types.

Figure 10.103 A photograph and Ku-band SAR image of Albuquerque airport. Source: Sandia National Laboratory.

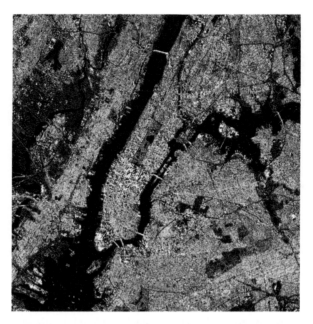

Figure 10.104 Multiband SAR image of New York City with three separate SAR channels rendered as a false color image. NASA SIR-C mission, 10 October 1994. https://www.jpl.nasa.gov/images/space-radar-image-of-new-york-city.

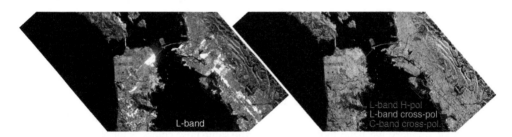

Figure 10.105 L-band (left) and multichannel (right) image of San Francisco. Left-hand panel from: https://www.jpl.nasa.gov/images/space-radar-image-of-san-francisco-california. Right-hand panel from: https://www.jpl.nasa.gov/images/space-radar-image-of-san-francisco-california-2. Both images were acquired on 3 October 1994.

10.9.8 SAR Agricultural Applications

Agricultural applications of SAR are of particular economic value and have become increasingly widespread over the past 20 years. Such applications include the measurement and mapping of:

- Soil condition (tilled vs uncultivated).
- Soil moisture.
- Crop type.
- Crop condition assessment.

For example, beginning in 2009 the Canadian government started producing digital maps of crop type based on a combination of optical imagery and Radarsat data. Field data suggest that this national crop inventory achieves a classification accuracy of greater than 85% at a spatial resolution of 30 m. An interactive version of the map is available at https://www.agr.gc.ca/atlas/aci.

Radar measurements over land depend on variations in the dielectric constant, roughness, and scatterer orientations associated with the soil and the plants growing above it. These variations usually occur in combination, so multiple observations are usually needed to infer useful parameters from the measurements. These multiple observations typically span one or more polarizations, incidence angles, radar frequencies, or apertures. Multitemporal data are also often used. The structures of plants change in predictable ways as they grow, so multiple images acquired at various times during the growing season can provide strong evidence of vegetation type and growth rates.

Soil condition measurements are based on the difference in scattering between tilled and untilled bare ground. Tilling changes the surface roughness which can be detected in any polarization. The direction of radar look relative to the direction of the tilled rows can have a significant impact on RCS for like-polarization, but less so for cross-polarization (Bradley and Ulaby, 1981; Mcnairn et al., 1996).

The earliest research on using radars to measure **soil moisture** indicated that the measurements should be made at near-nadir angles (7° – 15°) using C-band in order to minimize sensitivity to foliage (Ulaby & Batlivala, 1976). Subsequent work found it difficult to disambiguate soil moisture, surface roughness, and vegetative cover (McNairn & Brisco, 2004). One proposed algorithm utilized multitemporal data because soil moisture, vegetative growth, and surface roughness can be expected to change on different time scales (Paloscia et al., 2013).

While the accurate measurement of soil moisture with only SAR data remains a research challenge, operational soil moisture estimates are being made by combining SAR data with other data sources. The measurement of soil moisture at resolutions of tens of kilometers have historically been made with passive microwave sensors. The NASA Soil Moisture Active Passive (SMAP) mission was designed to utilize coincident passive microwave radiometry and active radar measurements to accurately measure soil moisture at a higher resolution than possible with passive microwave measurements alone. The SMAP radar operated as an unfocused SAR, using Doppler sharpening to provide a 3-km resolution. Unfortunately the SMAP radar failed after seven months, so new algorithms were developed to combine the SMAP L-band radiometer data with the polarimetric C-band data from Sentinel-1A/1B (Das et al., 2019; Lievens et al., 2017). Other alternatives to what is referred to as spatial downscaling of remotely sensed data are also being investigated (Alexakis et al., 2017; Peng et al., 2017).

Discrimination between **crop types** depends on unique dielectric and structural characteristics of each crop type. Some crops grow low to the ground and some grow tall with long stalks. Figure 10.106 illustrates a simple structural model of a crop field consisting of a ground plane, the vertically-oriented stalks of the crop, and the randomly-oriented leaves. Scattering can occur directly from the ground, stalks or leaves in this model or it can occur via a double bounce mechanism from any one component to another component and then back to the radar. The geometry of this simple model suggests why the returns can be expected to be polarization dependent.

Crop type mapping most commonly utilizes the time variations of multiple polarizations to distinguish between crop types. Modern classifiers, such as the aptly named "random forest" classifier, are typically used with classification boundaries determined from ground-truthed data. So while the classification algorithms are not physics-based, the physics of the scattering from the soil and crop do determine the variations in the polarimetric returns.

Figure 10.107 illustrates an example land cover map derived from compact polarimetric images obtained over 12 days during the growing cycle. The confusion matrix for these

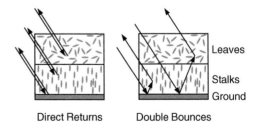

Figure 10.106 Simple geometric model of crop scattering including single and double-bounce scattering from ground, stalks and leaves. Adapted from Lopez-Sanchez and Ballester-Berman (2009).

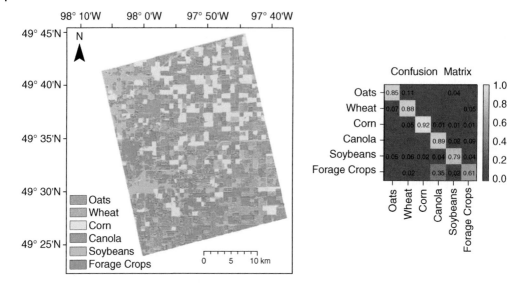

Figure 10.107 Land cover map derived from multiple compact-polarimetry images obtained over a growing season, and associated classification confusion matrix. Adapted from Mahdianpari et al. (2019).

data indicates that other than some difficulty separating forage crops from canola, SAR data alone could identify these crops at accuracies exceeding 79%.

The use of polarimetric interferometric SAR is being investigated to provide even more information on crop type, as discussed in Lopez-Sanchez and Ballester-Berman (2009).

Crop condition assessment has also been a topic of some research. Kim and van Zyl (2009) proposed a Radar Vegetation Index (RVI) as an intermediate product in their work to estimate soil moisture. The RVI was defined to be $RVI = \frac{8\sigma_{hv}}{\sigma_{hh} + \sigma_{vv} + 2\sigma_{hv}}$ and shown to correlate with biomass. Subsequent work has shown that the RVI is correlated with both vegetation water content and the leaf area index, both of which can be indicators of plant health (Szigarski et al., 2018).

The interested reader is referred to McNairn and Brisco (2004) and McNairn and Shang (2016) for a more comprehensive review of the agricultural uses of SAR.

References

Alexakis, D. D., Mexis, F.-D. K., Vozinaki, A.-E. K., Daliakopoulos, I. N., & Tsanis, I. K. (2017). Soil moisture content estimation based on Sentinel-1 and auxiliary earth observation products. A hydrological approach. *Sensors, 17*(6), 1455.

Alpers, W. (1983). Monte Carlo simulations for studying the relationship between ocean wave and synthetic aperture radar image spectra. *Journal of Geophysical Research: Oceans, 88*(C3), 1745–1759.

Alpers, W., Brandt, P., & Rubino, A. (2008). Internal waves generated in the Straits of Gibraltar and Messina: Observations from space. In V. Barale & M. Gade (eds.), *Remote sensing of the European seas* (pp. 319–330). Springer.

Alpers, W., & Espedal, H. A. (2004). Oils and surfactants. In C. R. Jackson and J. R. Apel (Eds.), *Synthetic aperture radar: Marine user's manual* (pp. 263–272). NOAA/NESDIS.

Alpers, W., & Hennings, I. (1984). A theory of the imaging mechanism of underwater bottom topography by real and synthetic aperture radar. *Journal of Geophysical Research: Oceans, 89*(C6), 10529–10546.

Alpers, W. G., & Melsheimer, C. (2004). Rainfall. In C. R. Jackson and J. R. Apel (Eds.), *Synthetic aperture radar: Marine user's manual* (pp. 355–372). NOAA/NESDIS.

Bradley, G. A., & Ulaby, F. T. (1981). Aircraft radar response to soil moisture. *Remote Sensing of Environment, 11,* 419–438.

Carsey, F. D. (Ed.) (1992). *Microwave remote sensing of sea ice. Geophysical Monograph Series* (Vol. 68). Washington, DC: American Geophysical Union.

Cavalieri, D. J., Burns, B. A., & Onstott, R. G. (1990). Investigation of the effects of summer melt on the calculation of sea ice concentration using active and passive microwave data. *Journal of Geophysical Research: Oceans, 95*(C4), 5359–5369.

Chapman, R. D., Gotwols, B. L., & Sterner, R. E. (1994). On the statistics of the phase of microwave backscatter from the ocean surface. *Journal of Geophysical Research: Oceans, 99*(C8), 16293–16301.

Chapman, R. D., Hawes, C. M., & Nord, M. E. (2010). Target motion ambiguities in single-aperture synthetic aperture radar. *IEEE Transactions on Aerospace and Electronic Systems, 46*(1).

Christiansen, M. B. (2006). *Wind energy applications of synthetic aperture radar* (Doctoral dissertation). Institute of Geography, Faculty of Science, University of Copenhagen.

Curlander, J. C., & McDonough, R. N. (1991). *Synthetic aperture radar*. New York, NY: John Wiley & Sons, Inc. USA.

Das, N. N., Dunbar, S., Colliander, A., Chaubell, M., Yueh, S., Entekhabi, D., al. (2019). Soil Moisture Active Passive (SMAP) – algorithm theoretical basis document SMAP-Sentinel L2 radar/radiometer (active/passive) soil moisture data products. *NASA JPL*. Retrieved from https://smap.jpl.nasa.gov/system/

internal_resources/details/original/522_L2_ SM_SP_Release_Version3.pdf.

Elachi, C. (1987). *Introduction to the physics and technology of remote sensing*. John New York, NY: John Wiley & Sons, Inc.

ESA (European Space Agency). (2018). ERS radar course 3 - slant range / ground range. Retrieved from https://earth.esa.int/web/ guest/missions/esa-operational-eo-missions/ ers/instruments/sar/applications/ radarcourses/content-2/-/asset_publisher/ qIBc6NYRXfnG/content/radar-course-2-slant-range-groundrange

ESA (European Space Agency). (2019). Oceanic internal waves generated at the Strait of Gibraltar. ESA. Retrieved from https://earth .esa.int/web/guest/missions/esa-operational-eo-missions/ers/instruments/sar/ applications/tropical/-/asset_publisher/ tZ7pAG6SCnM8/content/oceanic-internal-waves-strait-of-gibraltar

Freeman, A., & Durden, S. L. (1998). A three-component scattering model for polarimetric SAR data. *IEEE Transactions on Geoscience and Remote Sensing, 36*(3), 963–973.

Fu, L.-L., & Holt, B. (1982). *Seasat views oceans and sea ice with synthetic-aperture radar*. California Institute of Technology, Jet Propulsion Laboratory.

Hasselmann, S., Brüning, C., Hasselmann, K., & Heimbach, P. (1996). An improved algorithm for the retrieval of ocean wave spectra from synthetic aperture radar image spectra. *Journal of Geophysical Research: Oceans, 101*(C7), 16615–16629.

Hogan, G. G., Chapman, R. D., Watson, G., & Thompson, D. R. (1996). Observations of ship-generated internal waves in SAR images from Loch Linnhe, Scotland, and comparison with theory and in situ internal wave measurements. *IEEE Transactions on Geoscience and Remote Sensing, 34*(2), 532–542.

Holt, B. (2004). SAR imaging of the ocean surface. In C. R. Jackson and J. R. Apel (Eds.), *Synthetic aperture radar: Marine user's manual* (pp. 25–80). NOAA/NESDIS.

Italian Space Agency. (2007). *COSMO-SkyMed system description & user guide*. Rep. ASI-CSM-ENG-RS-093-A.

Jackson, C. R. (2004). An atlas of internal solitary-like waves and their properties. US Office of Naval Research, Code 322PO. Retrieved from http://www.internalwaveatlas.com/Atlas2_index.html

Jakowatz, C. V., Wahl, D. E., Eichel, P. H., Ghiglia, D. C., & Thompson, P. A. (2012). *Spotlight-mode synthetic aperture radar: A signal processing approach: A signal processing approach*. Springer Science & Business Media.

Jansing, E. D. (2021). *Introduction to synthetic aperture radar: Concepts and practice*. McGraw-Hill Education.

Johannessen, J. A., Kudryavtsev, V., Akimov, D., Eldevik, T., Winther, N., & Chapron, B. (2005). On radar imaging of current features: 2. Mesoscale eddy and current front detection. *Journal of Geophysical Research: Oceans*, *110*(C7).

Kim, Y., & van Zyl, J. J. (2009). A time-series approach to estimate soil moisture using polarimetric radar data. *IEEE Transactions on Geoscience and Remote Sensing*, *47*(8), 2519–2527.

Kudryavtsev, V., Akimov, D., Johannessen, J., & Chapron, B. (2005). On radar imaging of current features: 1. Model and comparison with observations. *Journal of Geophysical Research: Oceans*, *110*(C7).

Kudryavtsev, V., Hauser, D., Caudal, G., & Chapron, B. (2003a). A semiempirical model of the normalized radar cross section of the sea surface, 2. Radar modulation transfer function. *Journal of Geophysical Research: Oceans*, *108*(C3).

Kudryavtsev, V., Hauser, D., Caudal, G., & Chapron, B. (2003b). A semiempirical model of the normalized radar cross-section of the sea surface 1. Background model. *Journal of Geophysical Research: Oceans*, *108*(C3).

Lievens, H., Reichle, R. H., Liu, Q., De Lannoy, G. J., Dunbar, R. S., Kim, S. B., et al. (2017). Joint Sentinel-1 and SMAP data assimilation to improve soil moisture estimates.

Geophysical Research Letters, *44*(12), 6145–6153.

Liu, A., Peng, C., & Chang, Y.-S. (1996). Mystery ship detected in SAR image. *Eos, Transactions American Geophysical Union*, *77*(3), 17–17.

Liu, A. K. (1988). Analysis of nonlinear internal waves in the New York Bight. *Journal of Geophysical Research: Oceans*, *93*(C10), 12317–12329.

Liu, A. K., Chang, Y. S., Hsu, M.-K., & Liang, N. K. (1998). Evolution of nonlinear internal waves in the East and South China Seas. *Journal of Geophysical Research: Oceans*, *103*(C4), 7995–8008.

Lopez-Sanchez, J. M., & Ballester-Berman, J. D. (2009). Potentials of polarimetric SAR interferometry for agriculture monitoring. *Radio Science*, *44*(02), 1–20.

Mahdianpari, M., Mohammadimanesh, F., McNairn, H., Davidson, A., Rezaee, M., Salehi, B., & Homayouni, S. (2019). Mid-season crop classification using dual-, compact-, and full-polarization in preparation for the Radarsat Constellation Mission (RCM). *Remote Sensing*, *11*(13), 1582.

McCandless Jr., S. W. & Jackson, C. R. (2004). Principles of synthetic aperture radar. In C. R. Jackson and J. R. Apel (Eds.), *Synthetic aperture radar: Marine user's manual* (pp. 1–24). NOAA/NESDIS.

McNairn, H., Boisvert, J. B., Major, D. J., Gwyn, Q. H. J., Brown, R. J., & Smith, A. M. (1996). Identification of agricultural tillage practices from C-band radar backscatter. *Canadian Journal of Remote Sensing*, *22*(2), 154–162.

McNairn, H., & Brisco, B. (2004). The application of C-band polarimetric SAR for agriculture: A review. *Canadian Journal of Remote Sensing*, *30*(3), 525–542.

McNairn, H., & Shang, J. (2016). A review of multitemporal synthetic aperture radar (SAR) for crop monitoring. In Y. Ban (Ed.), *Multitemporal remote sensing: Methods and Applications* (pp. 317–340). Springer.

Monaldo, F. M., Thompson, D. R., Pichel, W. G., & Clemente-Colón, P. (2004). A systematic comparison of QuikSCAT and SAR ocean surface wind speeds. *IEEE Transactions on Geoscience and Remote Sensing*, *42*(2), 283–291.

Moreira, A. (2013). Synthetic aperture radar SAR: Principles and applications. ESA 4th Advanced Course in Land Remote Sensing. Retrieved from https://earth.esa.int/ documents/10174/642943/6-LTC2013-SAR-Moreira.pdf

Moreira, A., Prats-Iraola, P., Younis, M., Krieger, G., Hajnsek, I., & Papathanassiou, K. P. (2013). A tutorial on synthetic aperture radar. *IEEE Geoscience and remote sensing magazine*, *1*(1), 6–43.

NASA. (1999). North Sea oil slicks observed by SIR-C. Retrieved from https://www.jpl.nasa .gov/spaceimages/details.php?id=PIA01748

Onstott, R. G., & Gogineni, S. P. (1985). Active microwave measurements of Arctic sea ice under summer conditions. *Journal of Geophysical Research: Oceans*, *90*(C3), 5035–5044.

Onstott, R. G., & Shuchman, R. A. (2004). SAR measurements of sea ice. In C. R. Jackson and J. R. Apel (Eds.), *Synthetic aperture radar: Marine user's manual* (pp. 81–116). NOAA/NESDIS.

Paloscia, S., Pettinato, S., Santi, E., Notarnicola, C., Pasolli, L., & Reppucci, A. (2013). Soil moisture mapping using Sentinel-1 images: Algorithm and preliminary validation. *Remote Sensing of Environment*, *134*, 234–248.

Peng, J., Loew, A., Merlin, O., & Verhoest, N. E. C. (2017). A review of spatial downscaling of satellite remotely sensed soil moisture. *Reviews of Geophysics*, *55*(2), 341–366.

Pichel, W. G., Clemente-Colon, P., Wackerman, C. C., & Friedman, K. S. (2004). Ship and wake detection. In C. R. Jackson and J. R. Apel (Eds.), *Synthetic aperture radar: Marine user's manual* (pp. 277–304). NOAA/NESDIS.

Raney, R. K. (2007). Hybrid-polarity SAR architecture. *IEEE Transactions on Geoscience and Remote Sensing*, *45*(11), 3397–3404.

Raney, R. K. (2019). Hybrid dual-polarization synthetic aperture radar. *Remote Sensing*, *11*(13), 1521.

Reed, A. M., & Milgram, J. H. (2002). Ship wakes and their radar images. *Annual Review of Fluid Mechanics*, *34*(1), 469–502.

Romeiser, R., Runge, H., Suchandt, S., Kahle, R., Rossi, C., & Bell, P. S. (2014). Quality assessment of surface current fields from TerraSAR-X and TanDEM-X along-track interferometry and Doppler centroid analysis. *IEEE Transactions on Geoscience and Remote Sensing*, *52*(5), 2759–2772.

Sarti, F. (2012). Remote sensing and SAR images processing – characterization and speckle filtering in radar images. ESA Radar Remote Sensing Course Tartu, Estonia. Retrieved from https://earth.esa.int/c/document_library/get_ file?folderId=226458\LY1\textbackslash& name=DLFE-2125.pdf

Sikora, T. D., & Ufermann, S. (2004). Marine atmospheric boundary layer cellular convection and longitudinal roll vortices. In C. R. Jackson and J. R. Apel (Eds.), *Synthetic aperture radar: Marine user's manual* (pp. 321–330). NOAA/NESDIS.

Stapleton, N. R. (1995). Bright narrow V-shaped wakes observed by the NASA/JPL AIRSAR during the Loch Linnhe experiment 1989. In *Geoscience and remote sensing symposium, 1995. IGARSS'95.'quantitative remote sensing for science and applications', international* (Vol. 3, pp. 1646–1648). IEEE.

Stewart, R. H. (1985). *Methods of satellite oceanography*. University of California Press.

Szigarski, C., Jagdhuber, T., Baur, M., Thiel, C., Parrens, M., Wigneron, J.-P., et al. (2018). Analysis of the radar vegetation index and potential improvements. *Remote Sensing*, *10*(11), 1776.

Technical University of Denmark, D. o. M., DTU Space, & Sensing, R. (2009). Envisat ASAR 3-day mosaic of the Arctic Ocean. Retrieved from http://north.seaice.dk/2009/02/15/ 20090215.envisat.n.GMM3d.jpg

Thompson, D. R. (1988). Calculation of radar backscatter modulations from internal waves.

Journal of Geophysical Research: Oceans, *93*(C10), 12371–12380.

Thompson, D. R., & Jensen, J. R. (1993). Synthetic aperture radar interferometry applied to ship-generated internal waves in the 1989 Loch Linnhe experiment. *Journal of Geophysical Research: Oceans*, *98*(C6), 10259–10269.

Ulaby, F. T., & Batlivala, P. P. (1976). Optimum radar parameters for mapping soil moisture. *IEEE Transactions on Geoscience Electronics,* *14*(2), 81–93.

Wackerman, C. C., & Clemente-Colon, P. (2004). Wave refraction, breaking and other near-shore processes. In C. R. Jackson and J. R. Apel (Eds.), *Synthetic aperture radar: Marine user's manual* (pp. 171–188). NOAA/NESDIS.

Winstead, N. S., & Young, G. S. (2000). An analysis of exit-flow drainage jets over the Chesapeake Bay. *Journal of Applied Meteorology*, *39*(8), 1269–1281.

11

Lidar

11.1 Introduction

Lidar is a type of remote sensor that emits pulses of light that are then reflected or scattered back into the sensor. Characteristics of the returned signal are measured, recorded, and analyzed to extract a variety of geophysical parameters. Lidar sensors are in many ways similar to radar, but operate at much shorter wavelengths, ranging from the ultraviolet through the visible and into the infrared. The energy at these wavelengths can interact with the atmosphere, land or ocean in a variety of complex ways, providing the opportunity to measure a wide range of phenomena.

The term lidar was originally an acronym that stood for **L**ight **D**etection **a**nd **R**anging, although like radar, the term is now most commonly written in all lower case. You will also occasionally see the term ladar, based on the acronym for **La**ser **D**etection **a**nd **R**anging. While lasers are the most common source of illumination for all lidar systems, the term ladar is preferred by some when referring to sensors designed to detect man-made targets. The term lidar is predominantly used in the field of remote sensing.

11.2 Types of Lidar

A wide variety of lidar types have been developed that exploit a range of physical interactions of light and matter for a number of applications (Figure 11.1). The most common parameter measured by lidars is the time of

Figure 11.1 Lidar types and applications.

Remote Sensing Physics: An Introduction to Observing Earth from Space, Advanced Textbook 3, First Edition.
Rick Chapman and Richard Gasparovic.
© 2022 American Geophysical Union. Published 2022 by John Wiley & Sons, Inc.
Companion website: www.wiley.com/go/chapman/physicsofearthremotesensing

flight of the transmitted pulse. Time-of-flight measurements can provide a very accurate estimate of range to the target for altimeter applications including ground mapping or measurement of sea surface or ice topography. Lidars operating at blue–green wavelengths can penetrate into the water and scatter from the bottom to measure bathymetry, while multiple returns from lidars operated over forests can measure canopy heights and vegetation densities relative to the forest floor.

In addition to time of flight, lidars can also measure the range profile of the returned power for each pulse. Such power profile measurements can be made at the same single wavelength that was transmitted, or at multiple wavelengths or polarizations to estimate the spectral and polarization characteristics of the returned signal. Such approaches are used in a variety of lidars designed to measure scattering and absorption profiles in the atmosphere. Differential scatter and absorption lidars transmit pulses at two or more closely-spaced wavelengths using the additional data to separate wavelength-dependent parameters. Some lidars measure Rayleigh scattering and some transmit at one wavelength and measure returns at different wavelengths to measure Raman scattering, Brillouin scattering, or fluorescence (discussed in the next section). Such lidars are used to measure temperature, humidity, and constituent concentration profiles in both the atmosphere and the oceans. Other noncoherent lidars measure power at multiple adjacent wavelengths in order to detect a Doppler shift in the spectral returns due to wind (the double-edge detection lidar).

Finally, coherent lidars, described in the following section, are used in remote sensing, primarily for the measurement of wind speed.

11.2.1 Direct vs Coherent Detection

Lidars are classified as coherent or noncoherent depending on whether they transmit a coherent signal and then maintain and utilize the phase of the returned signal in measuring the returned signal. Optical sensors, such as photodiodes, are sensitive to irradiance but not phase. These sensors are fundamentally noncoherent devices and cannot by themselves utilize the phase coherence that is a characteristics of most lasers. The receiver for what is called a direct detection lidar (top panel of Figure 11.2) simply responds to the radiant flux received by the sensor for each lidar pulse.

The bottom panel of Figure 11.2 illustrates the simplest additional optical elements needed to form a coherent receiver, one that responds to the phase of the returned signal. For a coherent lidar, some light that is phase coherent with the transmitted light is mixed

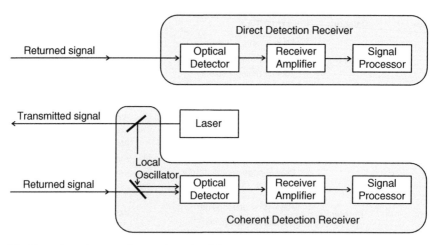

Figure 11.2 Direct detection vs. coherent detection lidar.

with the returned light at the face of the sensor. The two beams coherently combine to produce a phase-dependent flux at the face of the optical sensor. The illustration shows some of the original transmitted beam as the source of what is referred to as the local oscillator[1], but other stable sources can also be used. Use of the original transmitted beam is what is called a homodyne receiver, while using a different stable source is called a heterodyne receiver. The wavelengths of light are small, so the physical alignment and stability of the optical elements are critical to making such a system work.

Coherent lidar systems can be extremely sensitive to phase, making them the most sensitive detectors of Doppler velocity. This technique is used in some lidars designed to measure wind speed. Yet wind speed can also be measured using one or more direct detection receivers with narrow band filters placed in front of the sensors. A noncoherent wind lidar transmits a pulse with a relatively narrow bandwidth. The pulse scatters off of aerosols, and the Doppler of the returned signal is shifted by the mean velocity of the aerosols, and broadened by the spread in the velocity of the aerosols. The output of a single narrowband filter located on the edge of the returned Doppler spectrum will increase or decrease in amplitude depending on the mean Doppler shift. A pair of narrowband filters located on opposite edges of the returned Doppler spectrum provides more sensitivity to mean Doppler shifts, since one output will increase and the other will decrease in amplitude for a given shift.

While coherent lidars can achieve exquisite sensitivity to Doppler-induced phase shifts, coherence of returned signals is reduced by a wide variety of physical processes and it can be an engineering challenge to maintain coherence in the lidar receiver. For these reasons, most remote sensing lidars are noncoherent.

---

1 This term originates from the design of heterodyne radio receivers.

11.3 Processes Driving Lidar Returns

There are basically three processes involving the interaction of light with molecules (or in some cases atoms) that produce the lidar returned signal:

- Elastic scattering – light is absorbed and immediately reradiated by the molecule at the same wavelength.
- Inelastic scattering – light is absorbed by the molecule, affecting the energy state of the molecule, which then reradiates at a different wavelength.
- Fluorescence – light is absorbed by the molecule, increasing the energy state of the molecule for some period of time before the molecule eventually reradiates at a different wavelength.

Each of these processes is discussed in more detail in the following sections.

11.3.1 Elastic Scattering

In elastic scattering, photons excite molecules which then reradiate at the same frequency – the molecules neither gain or lose vibrational, rotational or electronic energy. In general, the scattering cross-section of a particle depends on its size, shape, and refractive index.

For particles much smaller than the wavelength of the incident radiation, the reradiation takes the form of a dipole-current induced by the light's electric field. This is called Rayleigh scattering. The intensity distribution for Rayleigh scattering from a single particle is:

$$I = I_0 \frac{8\pi^4\alpha^2}{\lambda^4 R^2}(1 + \cos^2\theta) \quad (11.1)$$

where I_0 is the intensity of the incident radiation, α is the molecular polarizability of the scatterer, λ is the wavelength, R is the distance to the scattering particle, and θ is the scattering angle relative to the incoming photon.

Mie scattering applies to larger particles, whose circumference is greater than the incident wavelength. In the Mie limit, particles have a cross-section that approaches the physical cross-section of the particle. Mie scattering is strongly peaked in the forward direction, defined to be the same direction as the incoming photon, with a weaker secondary peak in the backwards direction.

Scattering can be polarization dependent, making polarization of the returned signal a useful measurement.

While scattering technically occurs at the same encounter frequency as the incident radiation, motion of the scatters does impart a Doppler shift on the scattered radiation. Mean scatterer motions toward or away from the lidar cause a mean Doppler shift while random motions toward or away from the lidar cause Doppler broadening. Spectral measurements of these effects can be used to infer the mean and standard deviation of particle velocities.

11.3.2 Inelastic Scattering

All molecules exist in varying states of vibrational, rotational or electronic energy. In inelastic scattering, the incident light excites the molecule into a higher energy state. The molecule then reradiates at a different wavelength, falling back into a lower energy state. There are generally two forms of inelastic

scattering: Raman scattering and Brillouin scattering.

Raman Scattering

Raman scattering occurs when a photon incident on a material causes it to either gain (Stokes Raman) or lose (anti-Stokes Raman) vibrational, rotational or electronic energy. Figure 11.3 illustrates the energy states and changes in photon frequency arising from elastic and inelastic Raman scattering. The molecule does not change energy state in elastic scattering, so the frequency of the incident and emitted photons are identical (left panel). In Stokes Raman scattering, the molecule ends up in a higher energy state than it started, and the emitted photon has a lower frequency than the incident photon (middle panel). In anti-Stokes Raman scattering, the molecule started in an excited state and ends up in a lower energy state, with the excess energy being carried off by the emitted photon which has a higher frequency than the incident photon (right panel).

It is important to note that Raman scattering is not a resonant phenomena – it can be excited by any wavelength. In each case, the returned signal has a fixed offset from the excitation

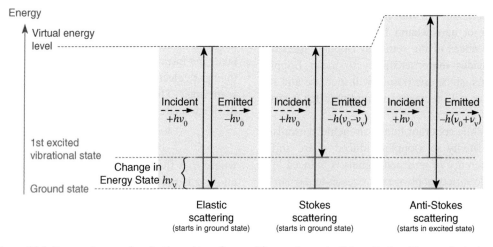

Figure 11.3 Energy changes for elastic and two-forms of Raman (non-elastic) scattering. Figure adapted from original https://en.wikipedia.org/wiki/Raman_scattering\LY1\textbackslash#/media/File:Ramanscattering.svg.

wavelength based on which energy levels have been excited or depleted. The frequencies of molecular vibrational modes typically are in the infrared (10^{12}–10^{14} Hz, corresponding to 30–3000 cm^{-1}). In general, only a small fraction of photons (1 in 10 million) scatter inelastically, but the scattering spectrum contains valuable information about the scattering molecules.

In the most common form of Raman scattering, the lidar transmits one wavelength and measures the returns at two or more different wavelengths, with one wavelength corresponding to a Stokes return and one to an anti-Stokes return. The ratio of the Stokes and anti-Stokes returned power depends on the relative populations of the illuminated molecules in the ground and excited states, which in turn depends on temperature. Thus the ratio of the Stokes and anti-Stokes returned power as a function of range can be used to measure atmospheric temperature profiles.

Brillouin Scattering

Brillouin scattering occurs when incident photons excite acoustic phonons in a material which then reradiates at a lower frequency. This is similar to Raman scattering, but in Brillouin scattering the material gains energy in the form of acoustic waves. The frequency shift imparted by acoustic waves is far smaller than Raman scattering, typically 3–180 GHz. This makes Brillouin scattering a more challenging measurement to make than Raman scattering. Because the material is excited acoustically, the return spectrum depends on elasticity, temperature and pressure of the material.

11.3.3 Fluorescence

Fluorescence is similar to Raman scattering in that energy is absorbed at one wavelength and reradiated at a lower energy. In fluorescence this reradiation does not occur immediately. The fluorescence lifetime characterizes how

long it takes before a molecule or atom emits a photon after excitation, and is typically 1–10 ns. Fluorescence is a resonant phenomena so fluorescent absorption and reemission occur at specific, material-dependent frequencies. Thus both the lifetime and spectral characteristics are parameters that can be measured to characterize the illuminated material.

11.4 Lidar Range Equation

To develop the lidar equation, consider a laser that emits a pulse with power P_0, duration τ_{pulse}, and energy $Q_0 = P_0\,\tau_{pulse}$. The laser optics focuses this energy into a solid angle Ω_T, which at a distance R produces an illuminated area $A = \Omega_T R^2$. In a vacuum this beam would create an irradiance at distance R of $E_{inc} = P_0/\Omega_T R^2$. More typically, the lidar is being operated through the atmosphere, which attenuates the beam through a combination of scattering and absorption. The irradiance at range R is thus given by:

$$E_{inc} = \frac{P_0 T_{atm}}{\Omega_T R^2}$$

$$= \frac{P_0}{\Omega_T R^2} \exp\left[-\int_0^R \alpha(u)\,du\right] \quad (11.2)$$

where T_{atm} is the transmittance through the atmosphere computed by integrating the range-dependent atmospheric attenuation coefficient α over the atmospheric path.

11.4.1 Point Scattering Target

We are going to consider multiple types of targets to develop different forms of the lidar equation. We begin by assuming the lidar illuminates a target that is smaller than the incident area of the beam, and that can be characterized by an effective backscatter cross-section σ. Then the scattered power is:

$$P_{scat} = E_{inc}\,\sigma$$

$$= \frac{P_0\,\sigma}{\Omega_T R^2} \exp\left[-\int_0^R \alpha(u)\,du\right] \quad (11.3)$$

The power density at the lidar receiver is then:

$$E_{rcv} = \frac{P_{scat}}{4\pi R^2} \exp\left[-\int_0^R \alpha(u)\,du\right]$$

$$= \frac{P_0\sigma}{4\pi\Omega_T R^4} \exp\left[-2\int_0^R \alpha(u)\,du\right] \tag{11.4}$$

and the received power becomes:

$$P_{rcv} = \frac{P_0 A_{rcv}\,\sigma}{4\pi\Omega_T R^4} \exp\left[-2\int_0^R \alpha(u)\,du\right] \tag{11.5}$$

where A_{rcv} is the area of the lidar receiver optics. In a transparent atmosphere, application of the definition of gain $(G_T = 4\pi/\Omega_T)$ makes this equivalent to the standard radar range equation. This form of the standard lidar equation is applicable to the detection of targets smaller than the illuminating beam.

11.4.2 Lambertian Surface

A second form of the lidar equation can be formulated assuming the target is a Lambertian surface with reflectivity ρ. In this case, we will further assume that the field of view for the receive telescope is broad enough to encompass the entire illuminated area A on the surface. The total flux incident on the surface is:

$$\Phi_{inc} = E_{inc} A = P_0 \exp\left[-\int_0^R \alpha(u)\,du\right] \tag{11.6}$$

Then the radiance of the illuminated portion of the surface is:

$$L = \frac{\rho\,\Phi_{inc}}{\pi A}$$

$$= \frac{\rho P_0}{\pi\Omega_T R^2} \exp\left[-\int_0^R \alpha(u)\,du\right] \tag{11.7}$$

and the received power is:

$$P_{rcv} = LT_{atm}^2\,A_{rcv}\,\Omega_{rcv}$$

$$= \frac{\rho P_0 A_{rcv}}{\pi R^2} \exp\left[-2\int_0^R \alpha(u)\,du\right] \tag{11.8}$$

where we have assumed the receiver optics has the same solid angle as the transmit optics,

$\Omega_{rcv} = \Omega_T$. Unlike the lidar equation for a point target, the power falls off as R^2 because all of the transmitted power falls on the surface, and is presumed to be within the receive beam. This form of the standard lidar equation is applicable to the detection of surfaces, as in a laser altimeter.

11.4.3 Elastic Volume Scattering

A third form of the lidar equation can be formulated assuming the returned signal arises from elastic volume scattering. In this case, the cell-backscattered intensity at range R is $I_{scat} = \beta(R)E_{inc}\,V$, where β is the range-dependent volume scattering coefficient evaluated in the backscatter direction with units of $m^{-1}sr^{-1}$ and V is the volume of the cell being interrogated. Given a range resolution of $\delta R = c\tau_{pulse}/2$, the cell volume becomes $V = \Omega_T R^2 \delta R = \Omega_T R^2 c\tau_{pulse}/2$. The scattered intensity from the cell is thus:

$$I_{scat} = \beta(R)E_{inc}V = \frac{1}{2}P_0\,\beta(R)\,c\,\tau_{pulse}$$

$$\times \exp\left[-\int_0^R \alpha(u)\,du\right] \tag{11.9}$$

Given that the solid angle from the scattering volume to the receive telescope is given by $\Omega = A_{rcv}/R^2$, the backscattered power collected by the receive telescope is:

$$P_{rcv} = I_{scat}\,\Omega \exp\left[-\int_0^R \alpha(u)\,du\right]$$

$$= \frac{cQ_0 A_{rcv}}{2R^2}\beta(R)$$

$$\times \exp\left[-2\int_0^R \alpha(u)\,du\right] \tag{11.10}$$

where we have substituted the pulse energy Q_0 for the product of the pulse power P_0 and duration τ_{pulse}. This is the conventional form of the lidar equation applicable to elastic volume scattering, as in an aerosol profiling lidar.

11.4.4 Bathymetric Lidar

With the introduction of a few additional factors, this same model can be extended to predict the power received from a bathymetric

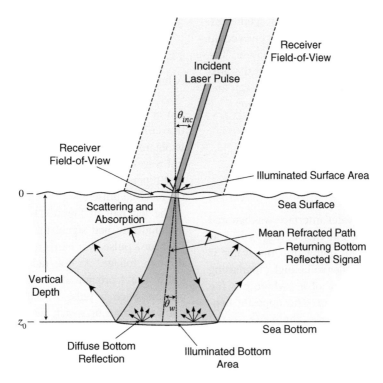

Figure 11.4 Geometry for a bathymetric lidar. Adapted from Guenther (1985).

lidar[2]. Figure 11.4 illustrates a bathymetric lidar pointing a narrow beam at the ocean surface from an altitude of h. In the following discussion, we assume the beam is pointing straight down, normal to the surface[3].

The laser pulse power just below the ocean surface $P_w(0)$ is given by:

$$P_w(0) = P_0 T_{atm} T_s \qquad (11.11)$$

where T_s is the transmission coefficient at the air–sea interface. Once below the surface, the narrow beam is attenuated as it propagates and it spreads out because of scattering. Both attenuation and spreading are included in a single term called the beam spread function $BSF(z, \theta)$, which is a function of depth and angle. The downwelling irradiance at a given depth z is then given by:

$$E_i(z) = P_w(0) \, BSF(z, 0°) \qquad (11.12)$$

The upwelling irradiance from the bottom with reflectivity ρ is simply $\rho E_i(z_0)$. The bottom is assumed Lambertian, so the fraction of the upwelling irradiance E_u scattered towards the receiver from the bottom is then:

$$E_u(z_0) = E_i(z_0) \, \rho \, \frac{\Omega}{\pi} \qquad (11.13)$$

where Ω is the solid angle of all rays originating underwater within the illuminated volume that can intersect the receive aperture. Temporarily ignoring refraction at the air water interface, $\Omega = A_{rcv}/(h + z_0)^2$.

The upwelling irradiance will again be attenuated on its path to and through the surface so:

$$E_u(0^+) = E_u(z_0) \exp(-K_{eff} z_0) \, T_s \qquad (11.14)$$

where 0^+ indicates just above the surface and K_{eff} is an effective attenuation coefficient. For a typical lidar with a large receive solid angle, the effective attenuation coefficient will

---

2 This derivation follows the general approach taken in Guenther (1985) and the Ocean Optics Web Book, http://www.oceanopticsbook.info/view/radiative_transfer_theory/level_2/the_lidar_equation

3 In practice, the laser beam is usually pointed slightly off nadir to reduce the surface flash and to keep the waveform within the dynamic range of the receiver.

approach the diffuse attenuation coefficient, $K_{eff} \approx K_d$, but for a narrow-beam receiver could approach the beam coefficient, $K_{eff} \approx c$. The received power will then depend on the area viewed by the receiver $A_{fov} = \Omega_{rcv}(h + z_0)^2$ and the attenuation through the atmosphere:

$$
\begin{aligned}
P_{rcv} &= E_u(0^+) \, A_{fov} T_{atm} \\
&= P_0 T_{atm}^2 T_s^2 \, A_{rcv} \, \Omega_{rcv} \\
&\quad \times BSF(z_0, 0°) \, \frac{\rho}{\pi} \, \exp(-K_d \, z_0)
\end{aligned}
$$

(11.15)

Refraction at the water interface was ignored in this derivation, but this is not as bad as it sounds. The light going into the water is refracted into a smaller solid angle, increasing its radiance by a factor of n^2, where n is the refractive index of water. The opposite happens when the light leaves the water, so the refractive effects cancel out.

It should seem remarkable that according to equation (11.15) the returned lidar power is independent of the lidar's altitude, at least ignoring atmospheric attenuation. This is an artifact of the assumptions made in deriving this equation, but it turns out to be an informative artifact. The derivation assumes all of the power is delivered to the surface in a small beam. The beam spreads once it gets into the water, producing a large spot on the ocean surface. The field of view of the lidar receiver covers just a portion of the illuminated spot on the surface.

The derivation in Chapter 3.3.1 shows that the response of a sensor looking at a surface with a constant radiance does not vary with distance. So the result should not be surprising, and under some range of conditions is correct. But if either the illuminated area on the surface becomes large with respect to the underwater spreading, or the receiver field of view exceeds the area on the surface illuminated from below, then equation (11.15) must be modified to include an R^2 term in the denominator. And if both conditions are met, then equation (11.15) would include an R^4 term in the denominator.

11.5 Lidar Receiver Types

11.5.1 Linear (full waveform) Lidar

A linear lidar uses a conventional photodiode or other linear optical sensor as a receiver. The sensor converts incoming photons into an electron current that is amplified and then sampled to obtain range resolution. This is also called a full waveform lidar, because the sampled current represents the full returned waveform of the received pulse[4].

The left half of Figure 11.5 illustrates the notional returned signal for a full waveform lidar for a pulse over bare ground, and a pulse through foliage. As for radar, a 1-ns wide pulse would provide about 15 cm range resolution. Such a pulse would contain many photons, some of which would be reflected back to the lidar by leaves, and some would find paths in between the leaves, ultimately hitting the ground before returning to the

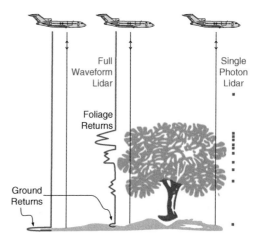

Figure 11.5 Example signals from a full waveform lidar (left) and singe photon lidar (right). Figure uses elements from https://commons.wikimedia .org/wiki/File:Forces-acting-on-an-airplane.svg and https://commons.wikimedia.org/wiki/File: CherokeePrimer-p5-tree.svg.

---

4 Some early lidar altimeters recorded only the time of flight to the leading edge of the returned pulse. A full waveform lidar reports returned power at multiple range cells that can provide useful information about the surface.

lidar. The received signal for such a lidar would be appropriately integrated for a 1-ns period before sampling. It is typical for a full waveform lidar to require the accumulation of at least hundreds of photons during each sample period to make a reasonable estimate of range to a target and/or intensity of the returned signal.

11.5.2 Single Photon Lidar

A new class of lidars has been developed that uses receive sensors sensitive enough to detect single photons. Such a lidar measures the time of flight to each returned photon to generate information about the scene. The returned signal on the right-hand side of Figure 11.5 is meant to illustrate the notional returns from such a lidar, with the red dots graphically indicating the times of reception of individual photons. Multiple pulses in quick succession then produce a point cloud that can be used to estimate the characteristics of the scene, which in this example would be canopy height, foliage density, and ground location. The increased sensitivity of such lidars allows the lidar to be operated with either reduced laser power or with broad or multiple beams to improve scan rates.

Single photon lidars are well suited for altimetry and ground mapping applications that require time-of-flight measurements. The advantage of single photon lidars is reduced for measurements that require accurate characterization of the intensity of the returned signal. Returned intensity depends on the rate of photons being received, so measurement of the intensity of the returned signal requires integration of multiple photons in order to reduce the statistical variations caused by shot noise. Single photon lidars can measure intensity by temporal or spatial integration of single or multiple returns, but this reduces some of their advantage over low-noise, full waveform lidars. Detailed analysis is needed in each case to determine the best approach for each individual measurement.

There are at least two competing technologies for such sensitive receivers: SPL and Geiger-Mode Lidar (GML). While the term "single photon lidar" really describes a class of lidars, a number of papers apply the term to a specific implementation. We will refer to that implementation as SPL.

The SPL, one specific type of single photon lidar, divides each transmitted pulse into multiple beamlets that are received by an array of microchannel plate detectors. Each channel in the microchannel plate acts as a miniature photomultiplier, producing a cascade of photoelectrons in response to each single photon hitting the photocathode. The current pulse generated by each photon is time tagged when it is received. Microchannel plates are good for this application because they produce pulses with low jitter and have a very fast (1–2 ns) recovery time. This fast recovery time allows the sensor to detect multiple photon events per pixel per pulse. The IceSat-2 lidar is an SPL design.

Geiger-Mode Lidars are an alternative form of single photon lidar. GMLs use a divergent laser pulse to produce a large laser footprint on the ground. The reflected signal is captured by an array of Geiger-mode Avalanche Photo Diodes (GmAPD) that are each sensitive to a single photon. Time of flight and the location on the array indicating the angle of arrival are measured for each received photon.

These sensors are solid state devices that have high detection efficiency, fast response and their operation requires only low bias voltages. A good reference on avalanche photodiodes can be found in Aull et al. (2002). GmAPD arrays as large as 32 × 128 elements have been fabricated. Itzler et al. (2014) describe a camera built with such an array that can be operated at a 115 kHz frame rate. When coupled with a laser source that illuminates the entire footprint of the receive array, photons can be received from the entire imaged field of view using a single pulse.

The primary issue with GmAPDs is that when an avalanche photodiode receives a photon, the avalanche diode changes state, preventing it from responding to any additional photons until it is reset. Furthermore, the cells in current GmAPD arrays can only be reactivated by resetting the entire GmAPD array. Thus only one photon can be detected in each array cell from a single pulse. The impact of this limitation varies depending on the specific application.

11.6 Lidar Altimetry

Lidar altimetry offers some advantages over radar altimetry. Lidar altimeters can provide exceptional accuracy and can provide measurements over some cross-track extent. They can be designed to have much finer spatial resolution than radar altimeters – the illuminated beams can be smaller, and less pulse averaging is needed to obtain a good measurement. Lidars altimeters can also be used to measure the profiles of thin clouds, vegetation canopy height and density, among other parameters.

There are also disadvantages. Lidar altimeters cannot make measurements through dense clouds, whereas radar altimeters work all the time. This is particularly important in regions with significant amounts of cloud cover, such as high-latitude ocean areas. Lidar altimeters making measurements over ice determine the height of the snow sitting on the ice, and not the ice itself. In contrast, sea ice measurements with radar altimeters can in some cases see the transitions from air to snow, snow to ice, and ice to water.

Lidar altimeters are usually designed to be beam limited, so precise pointing measurements are required to determine where the measurement was made. Radar altimeters are usually pulse limited, which means pointing is not critical, but they measure the distance to the closest point on the ground, which is not always directly below the radar when the surface is sloped. The delay-Doppler, phase-monopulse altimeter was designed to at least partly compensate for this issue.

11.6.1 NASA Airborne Topographic Mapper

The NASA Airborne Topographic Mapper (ATM), illustrated in Figure 11.6, is a scanning lidar altimeter primarily used to measure sea ice, continental ice sheets, and glaciers. Repeated yearly surveys of Greenland have contributed to our understanding of changes to Greenland's ice sheet. Operated at an altitude of 400–800 meters, topography is measured by combining the lidar range measurements with high-quality GPS, inertial navigation, and attitude sensors. The ATM is also used to

Figure 11.6 ATM overview and example 2-m-resolution digital elevation map produced by ATM of Mt. Erebus, Antarctica. Source: NASA https://atm.wff.nasa.gov.

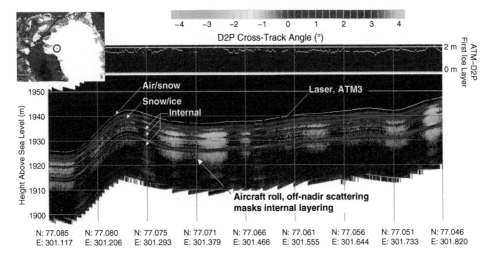

Figure 11.7 Comparison of data over Greenland from the ATM lidar altimeter and D2P radar altimeter. Source: Leuschen and Raney (2005) with permission from Johns Hopkins University Applied Physics Laboratory (JHU/APL).

acquire ground truth data for both radar and lidar satellite altimeters.

Figure 11.7 is a comparison of data from the ATM lidar altimeter and the D2P radar altimeter obtained during a NASA-sponsored flight over Greenland. The agreement between the ATM data (thin white line) and the first returns from the D2P is excellent, but the D2P also observes that the first layer of ice is 1.5–2 m below the snow surface observed by the ATM. The color in the bottom figure indicates the cross-track angle from which the D2P returns originated. Thus each instrument provides a unique and complementary measurement of the surface.

11.6.2 Space-Based Lidar Altimeters (IceSat-1 & 2)

Two examples of satellite lidar altimeters are IceSat-1 and IceSat-2. Figure 11.8a is a digital elevation map produced by the Geoscience Laser Altimeter System (GLAS) carried on IceSat-1, which was launched in 2003. GLAS was a high-power full waveform lidar. Unfortunately, the instrument had a design flaw that severely limited the lifetime of the three onboard lasers. The first laser failed after 38 days of operation. The decision was then made

to use the remaining lasers only one month out of every four. The system was operated for seven years in this mode before the third laser failed in 2010.

The GLAS lasers produced 1064-nm pulses for altimetry and frequency-doubled 532-nm pulses to measure clouds and aerosols (Abshire et al., 2005). The 1064-nm pulses had an energy of about 70 mJ and were produced at a pulse-repetition frequency of 40 Hz. The pulses illuminated a 65-m diameter spot on the Earth's surface. Successive spots were separated on the surface by 172 m.

IceSat-2 was launched in 2018 carrying the more sophisticated Advanced Topographic Laser Altimeter System (ATLAS) (Markus et al., 2017). ATLAS is a single photon lidar specifically designed to improve measurements over sloped ice. ATLAS utilizes a novel 6-beam design (Figure 11.8b). The beams are organized into pairs, with a nominal 90-m cross-track separation within a pair, and 3.3 km between each pair. The paired beams provide the capability to estimate cross-track slope, in addition to height.

The ATLAS laser produces only 2 mJ per pulse but at a 10-kHz pulse repetition rate. The ground footprint of each beam is about 10 m,

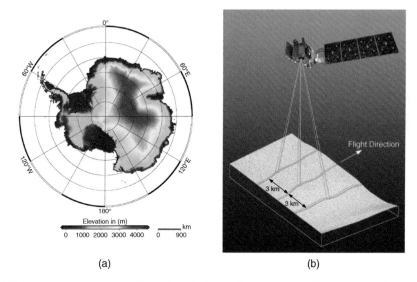

(a) (b)

Figure 11.8 (a) Antarctic topography determined by IceSat-1 and (b) IceSat-2 beam layout. IceSat-1 & 2 figures adapted from NASA.

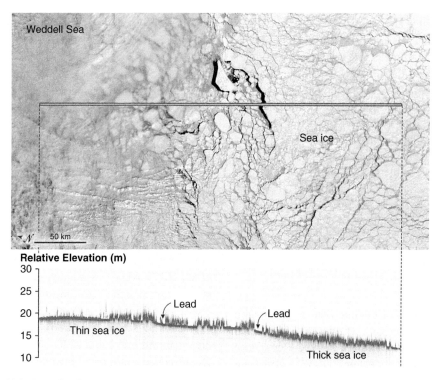

Figure 11.9 An IceSat-2 track over sea ice in the Weddell Sea. Source: NASA https://www.nasa.gov/feature/goddard/2018/icesat-2-reveals-profile-of-ice-sheets-sea-ice-forests.

with individual measurements being made with a 0.7-m along-track separation. Because this is a single photon detector, there is substantial noise in each measurement, so along-track averaging is needed for an accurate solution.

Figure 11.9 illustrates early results from IceSat-2, showing data from a track over sea ice in the Weddell Sea. The differences between thick sea ice, thin sea ice, and the leads (open water) between the floes is clearly evident.

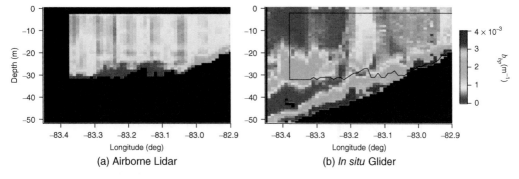

Figure 11.10 Example lidar remote sensing of underwater scattering layers (a) compared to in situ measurements (b). Based on data from Churnside et al. (2017).

11.6.3 Bathymetric Lidar

Figure 11.10 is an example of underwater scattering profiles measured from an airborne lidar. Figure 11.10a is the lidar-inferred profile of the particulate backscatter coefficient b_{bp} along a track off the Florida coast. Figure 11.10b shows the ground truth data obtained by in situ instrumentation deployed from a glider. To achieve this level of comparison, the particulate backscatter coefficient inferred from the lidar was multiplied by 0.75, as if the inversion algorithm was a bit uncalibrated. Note the lidar penetration is limited, and while there are similarities between the remote measurement and in situ data, significant differences remain. This is partly due to ambiguities in the inversion of the lidar equation. A good review of bathymetric lidars is provided in Churnside (2013).

11.7 Lidar Atmospheric Sensing

There are multiple approaches for lidars to sense the atmosphere. They can utilize backscatter from the atmosphere to measure profiles of clouds and aerosols. They can utilize the Doppler shift of the returned signal to measure profiles of the wind velocity component in the look direction. Or they can utilize spectral measurements of radiation reflected off the Earth's surface to estimate absorption profiles in the atmosphere. One example of this latter approach is the Differential Absorption Lidar (DIAL), which measures absorption lines in the atmosphere using multiple closely-spaced (<1 nm) wavelengths to eliminate surface reflectivity and other common confounding factors.

While a general review of lidars for atmospheric sensing is beyond the scope of this text, a few specific satellite-based sensors are worth highlighting.

11.7.1 ADM-Aeolus

The Atmospheric Dynamics Mission - Aeolus, is an ESA satellite launched in 2018. The primary sensor on the satellite is the Atmospheric LAser Doppler INstrument (ALADIN). This is a direct-detection Doppler lidar operating in the ultraviolet (355 nm). ALADIN is designed to provide global measurements of profiles of one component of the wind velocity. The sensor measures the Doppler shift of the returned signals to estimate a single component of wind velocity in the direction of look. Measuring only one component of velocity may seem limiting, but such measurements can significantly improve numerical weather predictions.

The ALADIN instrument utilizes two noncoherent detection approaches to sense the wind-induced Doppler shifts in the returned signals. The ALADIN instrument transmits 60-mJ pulses of ultraviolet light with an inherent linewidth of about 30 MHz (Figure 11.11). Some of that transmitted radiation is scattered by aerosols (the Mie signal) and some of it is scattered by gas molecules (the Rayleigh signal). Each individual scatterer shifts the mean Doppler of the transmitted pulse by an amount proportional to its line-of-sight velocity with

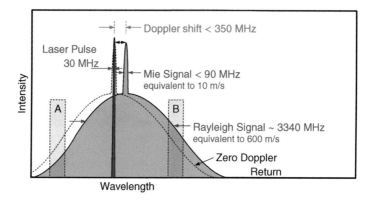

Figure 11.11 ALADIN transmit pulse (red), backscattered signal for zero wind (dashed black curve), and Doppler-shifted backscattered signal (blue). Adapted from Drinkwater et al. (2016) and Andersson et al. (2008), courtesy of ESA.

respect to the lidar. The returned signal then includes the Doppler shifts induced by all of the scatterers within the volume sampled by the lidar. ALADIN utilizes one sensor design to measure the mean Doppler shifts of the Mie signals and a different design for the Rayleigh signals.

In the case of the Mie signal, the standard deviation of the aerosol velocities within a sampled volume are typically about 10 m s^{-1}. Thus the returned signal is Doppler shifted, but it is also broadened, returning with a bandwidth of about 90 MHz. A Fizeau spectrometer in ALADIN is then used to measure the mean frequency of the Mie return. The spectrometer, like a prism, bends different frequencies of light into different angles. The Mie signal is then measured as the location of the peak signal on the CCD sensor.

In the case of the Rayleigh signal, the standard deviation of the molecular velocities within a sampled volume are typically about 600 m s^{-1}, a result of thermal vibrations with a small contribution from the wind. Thus the returned Rayleigh signal is Doppler shifted, but has been broadened to a bandwidth of about 3340 MHz. The Rayleigh spectrometer in ALADIN includes two narrow-band filters located on opposite sides of the spectrum of the transmitted pulse (filters A and B in Figure 11.11). Sensors measure the total power in each band, then use the ratio of the signals to estimate the mean frequency of the Rayleigh return.

This technique is referred to as double-edge detection.

In operation, the ALADIN lidar is aimed 35° from nadir and 90° to the satellite track on the side away from the Sun (Figure 11.12). The lidar is pointed orthogonal to the spacecraft track to minimize the component of the sensed velocity that could be caused by the satellite motion. The system is designed to measure from a height of 30 km down to either the top of thick clouds or the surface, with a horizontal resolution of 200 km.

The ALADIN instrument consists of four major components: a transmitter, a 1.5-m diameter Cassegrain telescope, and the Mie and Rayleigh spectrometers (Figure 11.13). The transmitter is a 150-mJ diode-pumped Nd:YAG laser, which is frequency-tripled to produce 60-mJ pulses at 355 nm with a pulse repetition frequency of 50 Hz. A UV source was chosen because it is relatively eye safe and produces a larger Rayleigh scattering signal. The Mie spectrometer is a Fizeau interferometer that produces a linear fringe at a position determined by the wind velocity component in the line of sight. The spectrometer has a resolution of 100 MHz, corresponding to a speed of 18 m s^{-1}, but the centroid of the fringe is estimated to higher precision, resulting in a measurement precision of 1.8 m s^{-1}. The Rayleigh receiver utilizes narrow band filters (etalons) constructed from Fabry–Perot interferometers. The power in the edges of the Rayleigh spectrum is then measured with a CCD.

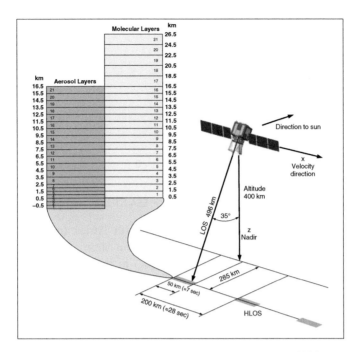

Figure 11.12 Baseline Aeolus measurement geometry. From Andersson et al. (2008), courtesy of ESA.

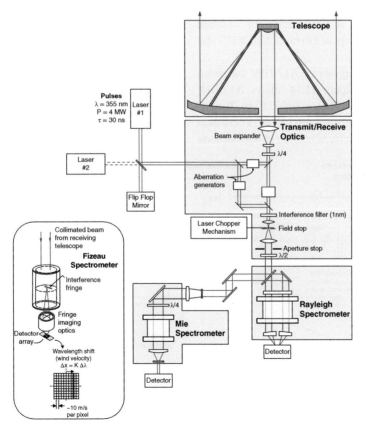

Figure 11.13 ALADIN optical design. Adapted from ESA (2005), courtesy of ESA.

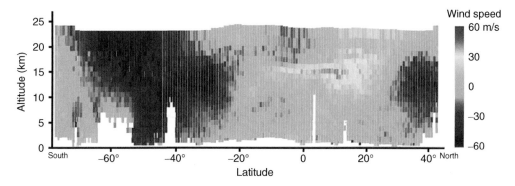

Figure 11.14 Example wind speed data from Aeolus. Source: ESA (2019).

Development of the ALADIN instrument was challenging. Laboratory testing showed that the high-power laser was damaging the optical surfaces in the instrument, not an uncommon issue for high-power lidars. Once on orbit, the energy of the UV laser declined significantly in the first nine months of operation, forcing ESA to switch to the backup laser in July 2019. According to news reports, the first-of-its-kind design issues with ALADIN caused a 11-year schedule delay (the satellite was originally supposed to launch in 2007) and a 50% cost overrun.

Since switching lasers, ALADIN has been working well. Figure 11.14 shows ALADIN wind speed data from a track from Turkey, over Africa, to the Southern Ocean. These data show strong easterly winds at 15 km altitude over north Africa (orange between 0° and 20°N) and strong westerly winds (blue and purple) in the upper troposphere and lower stratosphere over the Southern Ocean. The white areas are regions where thick clouds prevent measurements by blocking the laser signal.

11.7.2 NASA CALIOP

The Cloud-Aerosol Lidar with Orthogonal Polarization (CALIOP) is deployed onboard the Cloud-Aerosol Lidar and Infrared Pathfinder Satellite Observation (CALIPSO) satellite. CALIOP is a spaceborne lidar system designed to measure the vertical structure of aerosol and cloud distributions on a global scale (Figure 11.15).

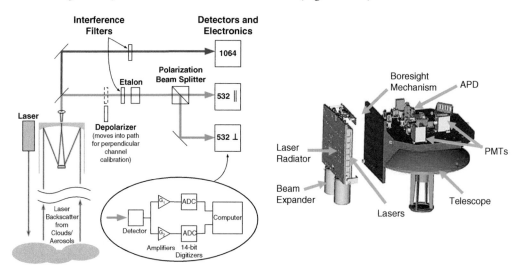

Figure 11.15 CALIOP instrument design. Adapted from Winker et al. (2006).

CALIOP uses a Nd:YAG laser to generate 1064-nm pulses and includes a frequency doubler to produce coaligned pulses at 532 nm (Winker et al., 2006). The 532-nm pulses are linearly polarized while the 1064-nm pulses are not. The lidar receiver consists of three channels: unpolarized power at 1064 nm, and like- and cross-polarized power at 532 nm. Returns from the different wavelengths provide information about aerosol vs molecular scattering, while the orthogonal polarizations of back-scattered signals at 532 nm provide information about the shape (spherical or nonspherical) of aerosol and cloud particles.

Both the 1064-nm and 532-nm pulses have an energy of 110 mJ. The pulse lengths are about 20 ns. The resulting signals are filtered and then sampled every 30 m from the surface to 8.2 km above mean sea level. At a pulse repetition frequency of 20.16 Hz, CALIOP produces a vertical profile every 335 m along the ground track.

The Hybrid Extinction Retrieval Algorithm (HERA) is an iterative algorithm that retrieves aerosol extinction profiles for aerosol layers, using the lidar ratio along with estimates of the aerosol types for each layer. HERA averages profiles to reduce noise, varying the number of profiles averaged to balance noise and resolution. HERA then estimates the layer-by-layer attenuation, correcting the underlying signals for the attenuation of the intervening atmosphere. The algorithm results in profiles of extinction and backscatter at both 532 nm and 1064 nm.

Figure 11.16 illustrates several of the derived products produced by CALIOP. These data were acquired during a descending track from the Sahara into the South Atlantic. Figure 11.16a plots the total attenuated backscatter signal computed from sum of the two 532-nm channels after calibration, range scaling, and gain normalization to account for the effects of near-range attenuation on far-range signals. The color scale for this plot was selected so that the blues correspond to molecular scattering, the yellows and reds to strong aerosol returns, and the grays to strong cloud returns.

Figure 11.16b illustrates the 532-nm depolarization channel computed from the ratio of the cross-polarization signal to the like-polarization signal. The signal backscattered from spherical particles (e.g., water droplets or spherical aerosols) has the same polarization as the transmitted signal, while nonspherical particles (e.g., dust and ice) tend to depolarize the signal. The depolarization ratio for cirrus clouds consisting of ice particles is usually about 0.25–0.40. Dust aerosols usually have a lower depolarization ratio of about 0.15. Generally depolarization increases with penetration depth because of multiple scattering; in other words, depolarization is a cumulative effect.

Figure 11.16c illustrates the attenuated color ratio channel computed from the ratio of the 1064-nm signal to the 532-nm signal, after accounting for the two-way, range-dependent transmittance due to molecules, including ozone, but not aerosols. This channel varies with the size of the particles in the scattering volume, increasing as the aerosols become larger.

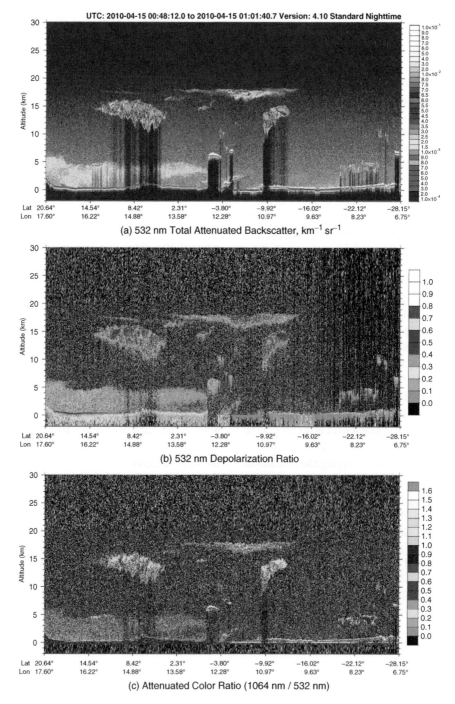

Figure 11.16 CALIOP attenuated backscatter at 532 nm. Source: NASA https://www-calipso.larc.nasa.gov/resources/calipso_users_guide/browse/index.php.

References

Abshire, J. B., Sun, X., Riris, H., Sirota, J. M., McGarry, J. F., Palm, S., et al. (2005). Geoscience laser altimeter system (GLAS) on the ICESat mission: On-orbit measurement performance. *Geophysical Research Letters, 32*(21).

Andersson, E., Dabas, A., Endemann, M., Ingmann, P., Källén, E., Offiler, D., & Stoffelen, A. (2008). *ADMAeolus Science report* (tech. rep. No. SP-1311). ESA. Retrieved from https://earth.esa.int/documents/10174/1590943/AEOL002.pdf.

Aull, B. F., Loomis, A. H., Young, D. J., Heinrichs, R. M., Felton, B. J., Daniels, P. J., & Landers, D. J. (2002). Geiger-mode avalanche photodiodes for three-dimensional imaging. *Lincoln Laboratory Journal, 13*(2), 335–349.

Churnside, J., Marchbanks, R., Lembke, C., & Beckler, J. (2017). Optical backscattering measured by airborne lidar and underwater glider. *Remote Sensing, 9*(4), 379.

Churnside, J. H. (2013). Review of profiling oceanographic lidar. *Optical Engineering, 53*(5), 051405.

Drinkwater, M., Borgeaud, M., Elfving, A., Goudy, P., & Lengert, W. (2016). ADM-Aeolus Mission requirements document. ESA. Retrieved from https://earth.esa.int/pi/esa?type=file&table=aotarget&cmd=image&alias=Aeolus_MRD

ESA. (2005). The Atmospheric Dynamics Mission, ADM-Aeolus draft. ESA. Retrieved from https://earth.esa.int/pi/esa?id=1740&sideExpandedNavigationBoxId=Aos&cmd=image&topSelectedNavigation NodeId=AOS&targetIFramePage=%2Fweb%2Fguest%2Fpi-community%2Fapply-for-data%2Fao-s&ts=1362822330221&type=file&colorTheme=03&sideNavigationType=AO&table=aotarget

ESA. (2019). Example wind measurements from Aeolus second laser. ESA Space in Images. Retrieved from http://www.esa.int/spaceinimages/Images/2019/07/Wind_measurements_from_Aeolus_second_laser

Guenther, G. C. (1985). *Airborne laser hydrography: System design and performance factors*. NOAA, Rockville MD.

Itzler, M. A., Entwistle, M., Jiang, X., Owens, M., Slomkowski, K., & Rangwala, S. (2014). Geiger-mode APD single-photon cameras for 3D laser radar imaging. In *2014 IEEE aerospace conference* (pp. 1–12). IEEE.

Leuschen, C. J., & Raney, R. K. (2005). Initial results of data collected by the APL D2P radar altimeter over land and sea ice. *Johns Hopkins APL Technical Digest, 26*(2), 114–122.

Markus, T., Neumann, T., Martino, A., Abdalati, W., Brunt, K., Csatho, B., et al. (2017). The Ice, Cloud, and land Elevation Satellite-2 (ICESat-2): Science requirements, concept, and implementation. *Remote Sensing of Environment, 190*, 260–273.

Winker, D. M., Hostetler, C. A., Vaughan, M. A., & Omar, A. H. (2006). *CALIOP algorithm theoretical basis document, part 1: CALIOP instrument, and algorithms overview* (tech. rep. No. PC-SCI-202 Part 1). NASA.

12

Other Remote Sensing and Future Missions

12.1 Other Types of Remote Sensing

There are still more types of remote sensing missions that we have not previously covered as well as some future missions. We will begin by providing a brief overview of two types of remote sensors that have yet to be discussed.

12.1.1 GRACE

The first is GRACE, a pair of satellites that were placed in the same orbit, but separated by about 220 km (Figure 12.1). These satellites were designed to produce high-resolution maps of gravity gradients around the globe by continuously measuring the distance between each other. As the lead satellite approached an area of higher gravity, it accelerated and pulled away from the trailing satellite. As the

satellites straddled the area of higher gravity, the lead satellite slowed down and the trailing satellite sped up. As the trailing satellite passed the area of higher gravity, it slowed down and the lead satellite was not affected.

The monthly gravity anomaly maps generated by GRACE were up to 1000 times more accurate than previous maps, with higher spatial resolutions (Tapley et al. 2004). Tapley et al., (2019) summarized the impact of GRACE measurements on understanding climate change.

GRACE was the first Earth-monitoring mission in the history of space flight whose key measurement is not derived from electromagnetic waves either reflected off, emitted by, or transmitted through Earth's surface and/or atmosphere. GRACE was launched in March 2002 and ceased operations in October 2017. The two GRACE Follow-On satellites were launched in May 2018.

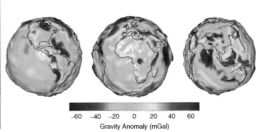

-60 -40 -20   0   20   40   60
Gravity Anomaly (mGal)

Figure 12.1 Gravity Recovery and Climate Experiment (GRACE). Source: NASA Right-hand panel from https://earthobservatory.nasa.gov/Features/GRACE/page3.php. Left-hand panel from https://www.jpl.nasa.gov/news/press_kits/grace-fo/mission/why-more-grace/\LY1\textbackslash#gallery-7.

Remote Sensing Physics: An Introduction to Observing Earth from Space, Advanced Textbook 3, First Edition.
Rick Chapman and Richard Gasparovic.
© 2022 American Geophysical Union. Published 2022 by John Wiley & Sons, Inc.
Companion website: www.wiley.com/go/chapman/physicsofearthremotesensing

12.1.2 Limb Sounding

We previously talked about atmospheric sounding, but entirely in the context of nadir-looking instruments. A variety of atmospheric sounders have been developed that work by looking at low grazing angles through the atmosphere. This is called limb sounding and is used to measure trace constituents in the upper atmosphere. As illustrated in Figure 12.2, there are generally three approaches:

- *Occultation* measures a refracted source, such as the sun, the moon, a star or a transmitter from another satellite such as GPS. In this case the sensor is measuring variations in absorption as the source-to-receiver path passes through the atmosphere. This is a often a high SNR measurement, but requires a very specific geometry that may only occur a few times a day.
- *Emissions* measures emissions from the atmosphere, much like the nadir-looking measurements. This approach usually has a lower SNR but the measurements can be made continuously.
- *Scattering* measures scattered sunlight along the refracted path.

12.2 Future Missions

Now let us discuss the increasing need for Earth observations, and future Earth remote sensing missions.

The world faces significant environmental challenges:

- We need to monitor the substantial changes that are occurring in our climate.
- Human activity has caused changes in the chemistry of the atmosphere.
- Worldwide there are shortages of clean and accessible freshwater.
- There is widespread degradation of terrestrial and aquatic ecosystems.
- Many places have experienced increases in soil erosion.
- Virtually all of the world's fisheries are in decline.

Addressing these societal challenges requires that we confront key scientific questions related to:

- Ice sheets and sea level change.
- Large-scale and persistent shifts in precipitation and water availability.

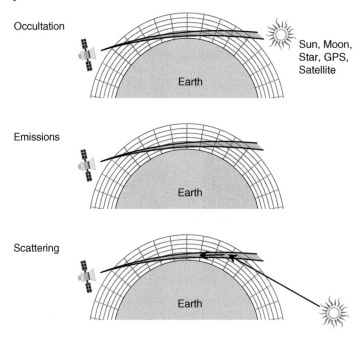

Figure 12.2 Three major modes for limb sounding: occultation, emissions, and scattering. Adapted from Carlotti and Magnani (2009).

- Transcontinental air pollution.
- Shifts in ecosystem structure and function in response to climate change.
- Impacts of climate change on human health.
- Occurrence of extreme events, such as severe storms, heat waves, wild fires, earthquakes, and volcanic eruptions.

We believe there should be no debate about the need for such scientific measurements and research. We also think that questions of how we respond to these societal challenges should be a matter of serious political debate, and we can only hope that such debate would be informed by our best science.

12.2.1 NASA Missions

Figure 12.3 shows NASA's current and future Earth science missions. We will briefly discuss each of these missions.

We begin with those extended missions that are currently operating beyond their originally planned lifetime. The names of these satellites are colored blue in Figure 12.3.

- SORCE measures solar radiation at wavelengths from X-ray to near-infrared to study long-term climate change.
- Landsat 7 is an EO/IR imager that is mostly functional, but its scan line corrector failed in 2003.
- Terra and Aqua are satellites with multiple sensors including MODIS.
- Cloudsat is a 94-GHz cloud radar.
- Calipso, which stands for Cloud-Aerosol Lidar and Infrared Pathfinder Satellite Observations, combines lidar with passive IR and visual imagers to measure clouds and aerosols.
- Aura includes IR and microwave sounders.
- OSTM/JASON-2 is the Ocean Surface Topography Mission, which is a radar altimeter.
- GPM is a microwave imager and dual-frequency radar designed to measure global precipitation.

Active United States missions operating within their planned lifetime include:

- ICESat-2 launched in September 2018.
- The GRACE follow-on (FO) mission launched in March 2018.

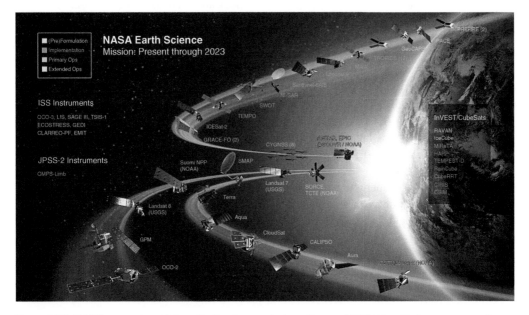

Figure 12.3 NASA's current and future Earth science missions. Source: NASA https://science.nasa.gov/earth-science.

- CYGNSS stands for Cyclone Global Navigation Satellite System, a set of eight low-cost microsatellites designed to measure the ocean surface wind field using a bistatic scatterometry technique based on GPS signals. The system was launched in 2016.
- The DSCOVR satellite orbits around the L_1 Earth–Sun Lagrange point, primarily to provide space weather information.
- NISTAR is a radiometer to measure the spectral irradiance from the Earth.
- EPIC is a whole Earth imager. The satellite was launched in February 2015.
- TCTE is the Total Solar Irradiance Calibration Transfer Experiment, launched in November 2013. This is a gap-filler designed to continue measurements of the total energy output of the sun until the launch of the next such instrument.
- SMAP stands for Soil Moisture Active Passive. This is a microwave radiometer plus an L-band SAR launched in January 2015. Unfortunately, the radar failed in July 2015.
- The Suomi NPP, launched in October 2011, is another gap-filler between the old NOAA POES and the replacement JPSS to provide weather and climate measurements from low Earth polar orbit.
- Landsat 8 launched to provide continuity with the failing Landsat 7 failing. It is primarily an EO sensor operated by the USGS.
- Landsat 9 launched in September 2021.
- The Orbiting Carbon Observatory, OCO-2, is an EO/IR radiometer designed to measure atmospheric CO_2. It was launched in 2014 as a replacement for OCO-1 which failed to reach orbit.
- There are a few instruments on the International Space Station. These include the Lightning Imaging Sensor (LIS); the Stratospheric Aerosol and Gas Experiment III (SAGE-III), which is an atmospheric limb sounder; the Total and Spectral Solar Irradiance Sensor (TSIS-1); ECOSTRESS is a six-band scanning infrared radiometer; and GEDI is a high-resolution lidar for measuring forests and topography.

Future United States missions currently under development include:

- TEMPO, which is an ultraviolet and visible spectrometer in geostationary orbit to monitor daily variations in ozone, nitrogen dioxide, and other key elements of air pollution.
- SWOT, which stands for Surface Water & Ocean Topography. This is a nadir altimeter plus a high-resolution radar interferometer.
- NI-SAR is a Synthetic Aperture Radar being designed jointly with the Indian space agency. It is a dual-frequency (L- and S-band) polarimetric SAR that will have some wide-swath capability.
- Multi-Angle Imager for Aerosols (MAIA) will use twin cameras to measure atmospheric aerosols.
- TROPICS is a constellation of 12 3U-CubeSats with scanning microwave radiometers to measure temperature, humidity, precipitation, and cloud profiles.
- GeoCarb is a follow-on to OCO-2.
- PACE, which stands for Plankton, Aerosol, Cloud, ocean Ecosystem, is an advanced multispectral imager.
- PREFIRE, which stands for Polar Radiant Energy in the Far-InfraRed Experiment, will use infrared spectral radiometers to measure long-wave radiation over the arctic.

12.2.2 ESA Missions

ESA also has a robust Earth observation program. Its operational observations are organized within the ESA Copernicus program with the stated goal to "provide accurate, timely and easily accessible information to

Table 12.1 Current and planned ESA sentinel missions.

Satellite	Launch date	Description
Sentinel-1A	Apr 2014	
Sentinel-1B	Apr 2016	SAR
Sentinel-1C	2022	
Sentinel-1D	2024	
Sentinel-2A	Jun 2015	
Sentinel-2B	Mar 2017	High-resolution optical imager for land services
Sentinel-2C	2023	
Sentinel-2D	2025	
Sentinel-3A	Feb 2016	
Sentinel-3B	Apr 2018	Altimeter and multispectral imager for ocean and land
Sentinel-3C	2023	
Sentinel-3D	2026	
Sentinel-4A	2023	Atmosphere radiometer, sounder, and imager to be
Sentinel-4B	2027	flown on Meteosat Third Generation (MTG)
Sentinel-5P	Oct 2017	Sounder for trace gases and aerosols
Sentinel-5A	2023	
Sentinel-5B	2024	Spectral radiometer to be flown on MetOp-SG A
Sentinel-5C	2038	
Sentinel-6A	Nov 2020	Radar altimeter
Sentinel-6B	2025	

improve the management of the environment, understand and mitigate the effects of climate change and ensure civil security."[1] The primary instruments for the Copernicus program are carried on the Sentinel satellites, as shown in Table 12.1.

Table 12.2 lists the current and planned EUMETSAT satellites.

In addition, ESA supports a variety of other satellite remote sensors:

- Aeolus – The Atmospheric Dynamics Mission launched in August 2018 to provide global observations of wind profiles from space.
- Swarm – A constellation of three satellites designed to survey the Earth's geomagnetic field.
- Proba-V – Provides multispectral images to study the evolution of the vegetation cover on a daily and global basis.
- CryoSat – The radar altimeter to measure continental ice thickness and marine ice cover.

1 https://www.esa.int/Our_Activities/Observing_the_Earth/Copernicus/Overview3

Table 12.2 Current and planned EUMETSAT satellites.

Satellite	Launch date	Description
Meteosat-8	Aug 2002	Geostationary imagers
Meteosat-9	Dec 2005	
Meteosat-10	Jul 2012	
Meteosat-11	Jul 2015	
Metop-A	Oct 2006	Microwave and infrared sounders and imager
Metop-B	Sep 2012	
Metop-C	Nov 2018	
Jason-2	Jun 2008	Altimeter
Jason-3	Jan 2016	
MetOp-SG A-1	2021	Sentinel-5 spectral radiometer, sounder and imager, LEO orbit
MetOp-SG A-2	-	
MetOp-SG A-3	-	
MetOp-SG B-1	2022	Passive microwave imager and scatterometer
MetOp-SG B-2	-	
MetOp-SG B-3	-	

- SMOS – Designed to observe soil moisture over the land and salinity over the oceans.

12.2.3 Summary

This may seem like a lot of activity, but budgets for Earth remote sensing in the U.S.A. have been falling in real terms for years. While small satellites offer some opportunities for more cost effective remote sensing, there are many measurements that require larger satellites. In any case, monitoring climate change requires time series measured in decades, which in turn requires sustained investment in satellites, sensors, ground stations, and the scientific community needed to analyze and interpret the data.

A report by the National Academy of Sciences (National Research Council, 2012) made the following argument for the value of remote sensing:

"Understanding the complex, changing planet on which we live, how it supports life, and how human activities affect its ability to do so in the future is one of the greatest intellectual challenges facing humanity. It is also one of the most important challenges for society as it seeks to achieve prosperity, health, and sustainability."

References

Carlotti, M., & Magnani, L. (2009). Two-dimensional sensitivity analysis of MIPAS observations. *Optics Express, 17*(7), 5340–5357.

National Research Council. (2012). *Earth science and applications from space: A midterm assessment of NASA's implementation of the decadal survey.* Washington, DC: National Academies Press.

Tapley, B. D., Bettadpur, S., Watkins, M., & Reigber, C. (2004). The gravity recovery and climate experiment: Mission overview and early results. *Geophysical Research Letters, 31*(9).

Tapley, B. D., Watkins, M. M., Flechtner, F., Reigber, C., Bettadpur, S., Rodell, M., et al. (2019). Contributions of GRACE to understanding climate change. *Nature Climate Change, 9*(5), 358–369.

Appendix A

Constants

c	2.9979×10^8 m s^{-1}	speed of light
G	6.67×10^{-11} N m^2 kg^{-2}	gravitational constant
g	9.807 m/s$^2 = Gm_e/R^2$	acceleration of gravity at Earth's surface
h	6.62607×10^{-34} J $\cdot$ s	Planck's constant
J_2	1.0826×10^{-3}	Earth quadrapole coefficient
k_B	1.38065×10^{-23} J/K	Boltzmann's constant
m_e	5.976×10^{24} kg	mass of the Earth
R	6371 km	mean radius of the Earth
R_{eq}	6378 km	equatorial radius of the Earth
R_m	1737 km	Moon radius
R_{me}	3.844×10^5 km	mean Moon/Earth distance
R_{po}	6357 km	polar radius of the Earth
R_s	6.955×10^5 km	Sun radius
R_{se}	1.496×10^8 km	mean Sun/Earth distance
T_{sid}	86164 s	sidereal day
σ	1.80498×10^{-8} W m^{-2} $\cdot$ sr $\cdot$ K^4	Stefan–Boltzmann radiance constant
σ_M	5.6705×10^{-8} W m^{-2} $\cdot$ K^4	Stefan–Boltzmann total emittance constant

Remote Sensing Physics: An Introduction to Observing Earth from Space, Advanced Textbook 3, First Edition.
Rick Chapman and Richard Gasparovic.
© 2022 American Geophysical Union. Published 2022 by John Wiley & Sons, Inc.
Companion website: www.wiley.com/go/chapman/physicsofearthremotesensing

The right of .... to be identified as the author of this work has been asserted in accordance with law.

Registered Offices ....

Editorial Office ....

For details of our global editorial offices, customer services, and more information about Wiley products visit us at www.wiley.com.

Appendix B

Definitions of Common Angles

A variety of angles are used to describe the geometry of remote sensing. The most common of these angles are defined in this appendix.

Figure B.1 illustrates angles that are defined with respect to the vertical on a flat Earth. In this case, vertical is defined in terms of two directions: zenith is straight up and nadir is straight down. Both of these directions are usually described relative to the mean surface of the Earth at the point being sensed.

We draw a ray from the sensor to the sensed location on the surface. The angle this ray makes with the surface normal is called the incident angle θ_i. This angle is also called the sensor zenith angle because it is the angle between the direction of the sensor as viewed from the ground and the zenith.

Similarly, the solar zenith angle θ_s is the angle from the sun to the zenith.

Most, but not all, remote sensors look down. So the sensor look angle θ_n or θ_{look} is usually measured with respect to nadir. Note that on a flat Earth, the sensor look angle equals the incident angle. *This is not true for the spherical Earth.* The flat Earth approximation may seem odd, but it is often used for low-altitude airborne remote sensing when the curvature of the Earth can be ignored.

Sometimes angles are measured with respect to the horizontal, instead of the vertical. In this case we talk about the sensor elevation angle θ_e, which is the angle between the sensor as viewed from the surface and the horizon (Figure B.2). Likewise the solar elevation angle θ_{se} is the angle of the sun above the horizon. From the sensor viewpoint, we define the sensor grazing angle θ_g to be the angle from the sensor look direction relative to the horizon. This is also occasionally referred to as the sensor depression angle.

Things become slightly more complicated on a spherical Earth. The incident angle is still measured with respect to the surface normal measured at the location being sensed. And again the sensor look angle is measured

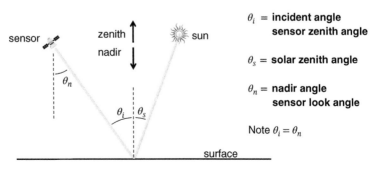

Figure B.1 Common angles defined relative to the vertical over a flat surface.

θ_i = incident angle
    sensor zenith angle

θ_s = solar zenith angle

θ_n = nadir angle
    sensor look angle

Note $\theta_i = \theta_n$

Remote Sensing Physics: An Introduction to Observing Earth from Space, Advanced Textbook 3, First Edition.
Rick Chapman and Richard Gasparovic.
© 2022 American Geophysical Union. Published 2022 by John Wiley & Sons, Inc.
Companion website: www.wiley.com/go/chapman/physicsofearthremotesensing

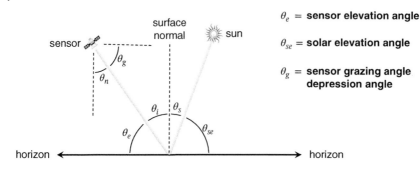

θ_e = **sensor elevation angle**

θ_{se} = **solar elevation angle**

θ_g = **sensor grazing angle**
depression angle

Figure B.2
Common angles
defined relative
to the horizontal
over a flat surface.

relative to nadir, but nadir for a satellite sensor is the direction from the satellite towards the center of the Earth. So the surface normal at the location being sensed is not in general parallel with the sensor nadir angle. Thus the incident angle does not necessarily equal the sensor look angle (Figure B.3).

Let us again come back down close to the surface, but consider looking at a surface with varying height. The angle of incidence θ_i is always measured with respect to the local surface normal, even as this normal varies from location to location (Figure B.4). For this reason we will sometimes refer to this as the local incident angle. The surface tilt ψ is measured with respect to the local vertical, usually defined by the local gravity vector. And now the sensor look angle equals the sum of the incident angle and surface tilt, at least in those cases where the sensor is close enough to the location being sensed that the local vertical is parallel to the sensor nadir direction.

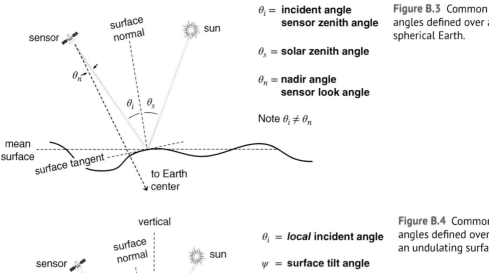

θ_i = **incident angle**
sensor zenith angle

θ_s = **solar zenith angle**

θ_n = **nadir angle**
sensor look angle

Note $\theta_i \neq \theta_n$

Figure B.3 Common
angles defined over a
spherical Earth.

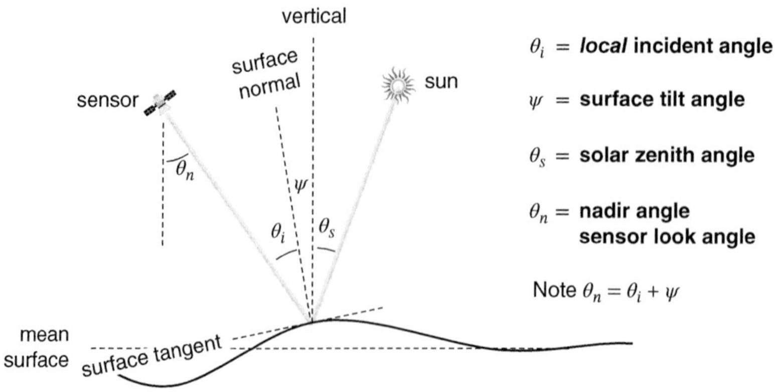

θ_i = *local* **incident angle**

ψ = **surface tilt angle**

θ_s = **solar zenith angle**

θ_n = **nadir angle**
sensor look angle

Note $\theta_n = \theta_i + \psi$

Figure B.4 Common
angles defined over
an undulating surface.

Figure B.5 Bidec
angle.

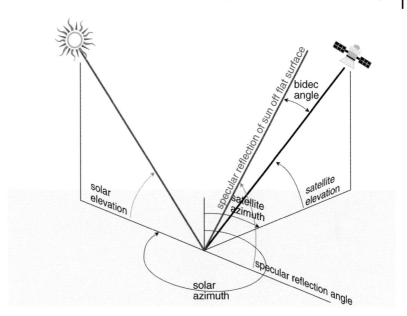

While these definitions have been presented in two-dimensional diagrams, remote sensing is performed in a three-dimensional world. The extra dimension is usually expressed in terms of an azimuth angle, with those azimuths being referenced to either true north or sometimes relative to the solar azimuth.

One important parameter for optical remote sensing is the angular distance between the sun's glitter pattern and the sensor look angle. This is commonly known as the bidec angle, as illustrated in Figure B.5.

Appendix C

Example Radiometric Calculations

Radiometry can be a confusing topic when first encountered. This appendix provides some additional hints and some example worked problems in order to better understand radiometry.

General hints:

- Visualize the flow and spread of energy before writing down equations. You may find it helpful to sketch the problem to make the flows clear.
- Radiance is defined as $L = \frac{d\Phi}{d\Omega dA \cos\theta}$. When calculating the radiance emitted by a source, $A\cos\theta$ is an area on the surface of the source projected into the direction the light is emitted, and Ω is the solid angle into which the light is emitted. When calculating radiance received by a detector, $A\cos\theta$ is an area on the surface of the detector projected into the direction the light is coming from, and Ω is the solid angle subtended by the source as viewed from that detector.
- Be careful about units. Make sure the units agree with the property you are asked to compute. If the problem calls for radiance, do not produce a result with units of $W\,sr^{-1}$. Always use mks and deg K in calculations.

Description for problems in radiometry 1–10:
Consider a room with a 5-m high ceiling that is illuminated by a single small flashlight. The flashlight is sitting on the floor, pointed straight up. The flashlight draws 3 W of power from its batteries, with an LED bulb that is 50% efficient. The reflector of the flashlight has been carefully designed to create a uniform circular beam with a half angle of 10°. Consider the ceiling to be a gray Lambertian surface with a reflectivity of 0.8. The floor and walls of the room are painted flat black with a reflectivity of 0. Next to the flashlight on the floor there is a 20 × 20 cm solar cell with 10% efficiency connected to a power meter. There is a second solar cell, identical to the first, also sitting on the floor looking up, but 5 m away from the flashlight. Assume the flashlight and the solar cells sit flush to the floor.

1. What is the radiant intensity of the flashlight?
2. What is the irradiance illuminating the ceiling within the flashlight beam?
3. What is the emittance of the ceiling within the flashlight beam?
4. What is the total flux emitted from the ceiling?
5. What is the radiance of the ceiling just above the flashlight as viewed from the location of the first solar cell?
6. What is the radiance of the ceiling just above the flashlight as viewed from the location of the second solar cell?
7. What is the radiance of the ceiling outside of the flashlight beam?
8. What is the irradiance on the floor at the location of the first solar cell?
9. What is the irradiance on the floor at the location of the second solar cell?

Remote Sensing Physics: An Introduction to Observing Earth from Space, Advanced Textbook 3, First Edition.
Rick Chapman and Richard Gasparovic.
© 2022 American Geophysical Union. Published 2022 by John Wiley & Sons, Inc.
Companion website: www.wiley.com/go/chapman/physicsofearthremotesensing

10. If all of the energy falling on both of the solar cells for an hour was stored in a battery, how long would it run the flashlight at full output?

Solutions to problems 1–10:
Begin by sketching the problem.

The light energy flows from the flashlight towards the ceiling. The beam from the flashlight forms a 10° cone. This creates a spot on the ceiling with diameter $d_{beam} = 2(5 \text{ m}) \tan(10°) = 1.76$ m. The area of the beam is $A_{beam} = \pi(d_{beam}/2)^2 = 2.44 \text{ m}^2$.

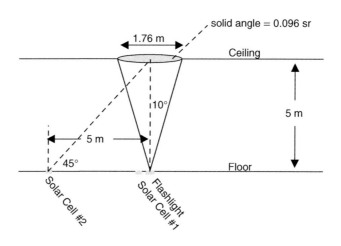

1. What is the radiant intensity of the flashlight?

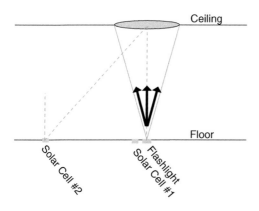

The units of radiant intensity are W sr^{-1}. Thus we need the emitted power and the solid angle over which the power is spread.

The input power to the flashlight is 3 W but it only has a 50% efficiency. So the flashlight emits $\Phi_{rad} = 1.5$ W of power as light and the rest is lost as heat.

At least three expressions can be used to compute the solid angle subtended by the beam, one exact formula and two approximations. For the approximate forms we note that the full cone angle $\alpha = 20° = 0.349$ rad.

- Exact: $\Omega_{beam} = 4\pi \sin^2(\alpha/4) = 4\pi \sin^2(20°/4) = 0.0955$ sr.
- Small angle: $\Omega_{beam} = \pi\alpha^2/4 = \pi(0.349 \text{ rad})^2/4 = 0.0957$ sr.
- Area over distance squared: $\Omega_{beam} = A_{beam}/R^2 = 2.44 \text{ m}^2/(5/\cos 10°)^2 = 0.0947$ sr.

Note that even the crudest of these approximations yields an estimate accurate to better than 1%.

The radiant intensity is then given by $\boxed{I = \Phi_{rad}/\Omega_{beam} = 15.7 \text{ W sr}^{-1}}$. In real life the flashlight has a finite aperture, but for this calculation we are appropriately treating it as a point source, albeit one that is focused into a particular span of directions.

2. What is the irradiance illuminating the ceiling within the flashlight beam?

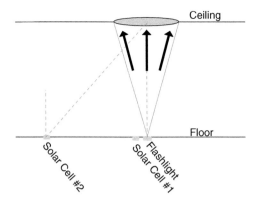

Irradiance is a measure of how much power is falling onto each unit area of a surface. The units of irradiance are W m^{-2}, so we need the total power falling on the surface and the area over which the power is spread.

Over these distances there is nothing absorbed in the atmosphere, so the total power falling on the ceiling is 1.5 W. We previously computed the beam area to be 2.44 m. Thus the irradiance is $\boxed{E_{onto} = \Phi_{rad}/A_{beam} = 0.61 \text{ W m}^{-2}}$.

3. What is the emittance of the ceiling within the flashlight beam?

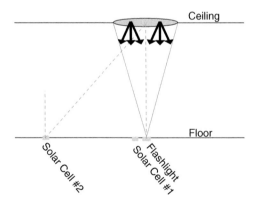

Emittance is a measure of how much power is being emitted from each unit area of a surface. The units of emittance are also W m^{-2}, so we need the total power leaving the surface and the area over which the power is spread.

This power leaving the surface is simply the power falling on the surface times the ceiling reflectivity, where $R_{ceiling} = 0.8$. The emitted

area is the beam area. So the ceiling emittance is $\boxed{E_{emit} = R_{ceiling}E_{onto} = 0.49 \text{ W m}^{-2}}$.

4. What is the total flux emitted from the ceiling?

Total flux is a power, so the units are simply W.

There are two equivalent approaches that could be used to do this calculation. One could multiply the ceiling emittance times the beam area, $\boxed{\Phi_{ceiling} = E_{emit} A_{beam} = 1.2W}$. Alternatively you can simply multiply the total incident power times the reflection coefficient, $\Phi_{ceiling} = \Phi_{rad} R_{ceiling} = 1.2W$.

Note the missing 0.3 W is being absorbed into the surface, heating the surface. Once the ceiling reaches equilibrium, all of this power will be reradiated as heat.

5. What is the radiance of the ceiling just above the flashlight as viewed from the location of the first solar cell?

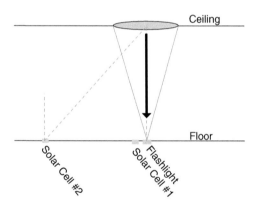

This question is now asking about the radiance that has been scattered off the ceiling and back down to the floor. The units of radiance are W m$^{-2} \cdot$ sr.

The other critical information is that the surface is Lambertian, which means the radiant intensity leaving the surface is proportional to $\cos\theta$ and the radiance leaving the surface is independent of θ, where θ is the viewing angle measured relative to the surface normal. This is a common property of surfaces that consists of fine scatterers. Good examples are a sheet of white matte paper or snow. Note that the directionality of the light illuminating a Lambertian surface has no affect on the uniformity of the outgoing radiation. In many ways a mirror is the opposite of a Lambertian surface, because the radiance leaving a perfectly mirrored surface is completely dependent on the directionality of the illumination.

Equation (3.12) is a formula relating the radiance of a Lambertian surface to the total flux leaving that surface, $\Phi = \pi L A_s$. Solving for the radiance yields: $L = \Phi/\pi A_s$. We know the flux leaving the ceiling is $\Phi = 1.2$ W, and the effective area of the surface is $A_s = A_{beam} = 2.44 \text{ m}^2$. Thus the radiance of the ceiling looking up from the solar cell immediately next to the flashlight is $\boxed{L_1 = \Phi/\pi A_s = 0.16 \text{ W m}^{-2} \cdot \text{sr}}$.

6. What is the radiance of the ceiling just above the flashlight as viewed from the location of the second solar cell?

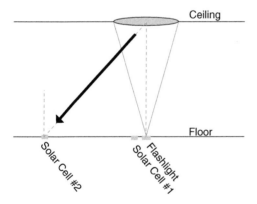

No calculation is needed here. A key property of a Lambertian surface is that the radiance of the surface is independent of view angle. Thus the radiance looking from a 45° angle is exactly the same as looking straight up. Thus the radiance of the ceiling just above the flashlight as viewed from the second solar cell is $\boxed{L_2 = L_1 = 0.16 \text{ W m}^{-2} \cdot \text{sr}}$.

7. What is the radiance of the ceiling outside of the flashlight beam?

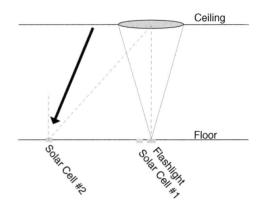

Here is another calculation that is easy. The answer is $\boxed{\text{zero}}$. The problem specifies the floor and walls are perfect absorbers. Very little scattering will occur in the atmosphere. So the room will be dark except for the ceiling within the beam of the flashlight.

8. What is the irradiance on the floor at the location of the first solar cell?

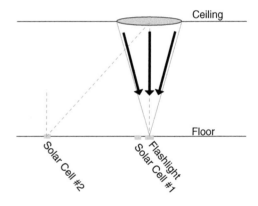

Again we begin with units. Irradiance has units of W m^{-2}. We just computed (in Example 5) the radiance of the ceiling L which is the radiance of the ceiling in every direction subtended by the beam. To get the irradiance falling on the floor we need to integrate the radiance of the ceiling over all directions looking up from the floor. But this is easy. The ceiling radiance is a constant, so we can factor that out of the integral as long as we limit the integration to the angular extent of the beam.

The integral over the angular extent of the beam then becomes the solid angle subtended by the beam, which we had previously computed. Thus the irradiance at solar cell #1 is $\boxed{E_1 = L_1\Omega_{beam} = (0.16\text{ W m}^{-2}\cdot\text{sr})(0.095\text{ sr}) = 0.015\text{ W m}^{-2}}$.

9. What is the irradiance on the floor at the location of the second solar cell?

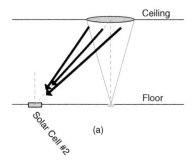

(a)

This calculation is similar to the last, but with additional factors to account for the reduced solid angle that the source on the ceiling subtends when viewed from solar cell #2 and the fact that the solar cell is not perpendicular to the incoming radiation. As usual the solution is most easily seen by conceptually breaking it up into individual steps.

Referring to panel (a), first we will note that the distance from the center of the illuminated area on the circle to solar cell #2 is $R = \sqrt{5^2 + 5^2} = 7.07$ m. Note this is further away than the distance to solar cell #1, so we expect the irradiance to be lower.

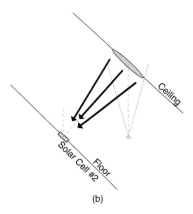

(b)

Now we begin by first solving a problem we know how to solve: the irradiance if the illuminated spot is a circle of the same size as the illuminated beam on the ceiling, with that circle located directly above the solar cell at a distance $R = 7.07$ m. As illustrated in panel (b) this is the same problem we solved earlier, just with a greater distance. We previously computed the radiance to be $L_2 = 0.16$ W m^{-2} sr. The solid angle subtended by the source is approximately given by $\Omega_{2b} \approx A_{beam}/R^2 = 2.44\text{ m}^2/(7.07\text{ m})^2 = 0.049$ sr. Note the subscripts "2" and "b" in the expression Ω_{2b} are meant to indicate the variable is as seen from the position of solar cell #2, as shown in panel (b).

The irradiance for this geometry is thus $E_{2b} = L_2\Omega_{2b} \approx L_2 A_{beam}/R^2 = 0.0076\text{ W/m}^2$. But this is not yet the problem we were asked to solve.

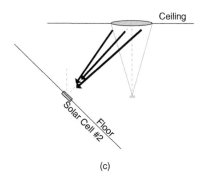

(c)

As the next step, estimate the solid angle that the illuminated region on the ceiling actually subtends as viewed from solar cell #2. As illustrated in panel (c) this is nearly equivalent to rotating the illuminated circle from the problem we just solved back to the horizontal. Doing so will have no impact on the apparent width of the circle, but the apparent height of the circle will shrink by a factor of about $\cos\theta$. Thus the illuminated circle appears as an ellipse with area $A_{beam2} \approx A_{beam}\cos\theta$. In this geometry, $\theta = 45°$, so $A_{beam2} \approx (2.44\text{ m}^2)\cos 45° = 1.73\text{ m}^2$. This reduces the solid angle by a similar factor $\Omega_{2c} \approx \Omega_{2b}\cos\theta = 0.035$ sr.

Thus the irradiance on solar cell #2 in panel (c) is $E_{2c} = L_2\Omega_{2c} \approx 0.0054$ W m^{-2}

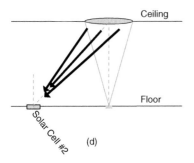

(d)

The last step in solving the original problem is to rotate the solar cell in panel (c) back to the horizontal, as shown in panel (d). This reduces the apparent area of the solar cell by another factor of $\cos\theta$. So the irradiance at solar cell #2, as shown in panel (d) is $\boxed{E_{2a} \approx E_{2c}\cos\theta \approx L_2 A_{beam}\cos^2\theta/R^2 = 0.0038 \text{ W m}^{-2}}$

10. If all of the energy falling on both of the solar cells for an hour was stored in a battery, how long would it run the flashlight at full output?

The total power falling on each solar cell is simply the irradiance falling on that cell times the area of the cell. The total energy is then the sum of the power from the two cells times the solar cell efficiency times 3600 seconds in an hour, with units of W s, or equivalently joules. The total energy falling on both cells in an hour is 0.27 J. The flashlight requires 3 W of power to run at full strength. The energy stored by these cells in a full hour of operation would run this flashlight for a grand total of time equal to the total energy in one hour, divided by the power required for the flashlight = $\boxed{0.27 \text{ W s} / 3 \text{ W} = 0.090 \text{ s}}$.

Thus we have proven beyond a shadow of a doubt that powering a flashlight from solar cells that are indirectly illuminated by the light from that very same flashlight is not a very bright idea.

The other takeaway is that you can solve radiometry problems by following the flow of energy, thinking about the units, and applying the formula according to whether the energy is spreading or being accumulated (integrated) over area or angle.

Appendix D

Optical Sensors

Optical sensors can be categorized as either thermal or photon detectors.

Thermal sensors work by absorbing incoming radiation, causing a change in temperature of the detector. This change in temperature then impacts some measurable parameter – for example, resistance. Thermal sensors are often used in the infrared because they are typically sensitive across a wide range of incident wavelengths.

Photon or quantum detectors depend on the direct interaction of the incoming light with the detector materials, for example a photon hitting a semiconductor sensor creates an electron-hole pair. Such photon-generated carriers can then be measured from either:

- The charge collected during an integration period.
- A photocurrent.
- A change in resistance.
- By voltage generation across a junction.

No matter how the incoming radiation is converted to an electrical response, all optical sensors can be characterized by certain parameters. The first of these is responsivity, which is defined as the ratio of the photocurrent output by the sensor to the radiant power incident on the sensor:

$$R_\lambda = \frac{I_p}{P_{inc}} = \frac{\text{photocurrent (A)}}{\text{power incident on sensor (W)}}$$
$$\text{(D.1)}$$

The units for responsivity are amps per watt. Figure D.1 shows the sensitivity of various sensor types to irradiance, which equals the sensor responsivity per unit area.

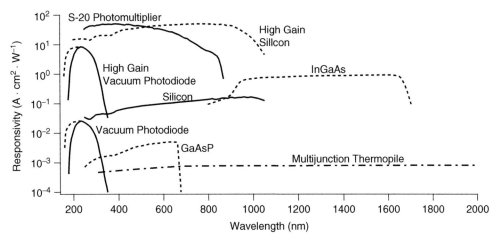

Figure D.1 Area responsivity of various sensor types. From Ryer (1997).

Remote Sensing Physics: An Introduction to Observing Earth from Space, Advanced Textbook 3, First Edition.
Rick Chapman and Richard Gasparovic.
© 2022 American Geophysical Union. Published 2022 by John Wiley & Sons, Inc.
Companion website: www.wiley.com/go/chapman/physicsofearthremotesensing

The responsivity curves show:

- Silicon is generally the best detector at optical wavelengths.
- Silicon detectors are generally more sensitive in the red than they are in the blue.
- Indium Gallium Arsenide is the best sensor at long wavelength IR.
- Thermopiles are not very sensitive, but have a very broad response.

Quantum efficiency η measures the probability of converting photons to electrons. It is the one of the key factors in determining the performance of low-light-level imagers.

$$\eta = \frac{\text{\# of electrons collected}}{\text{\# of incident photons}} \quad (D.2)$$

Quantum efficiency depends on wavelength, the material's absorption coefficient, the device geometry and, doping, among other factors. Modern devices can have quantum efficiencies in the range of 70–90%.

Figure D.2 shows there is a relationship between responsivity and quantum efficiency – a relationship that is easy to derive.

The interaction of electromagnetic radiation with a material is quantized into photons. The energy of a photon is given by Planck's constant h times the frequency of the photon v, or h times the speed of light divided by the wavelength:

$$E_{ph} = hv = \frac{hc}{\lambda} \quad (D.3)$$

The power incident on a sensor is then the energy in each photon times the rate at which the photons are incident on the detector r_p, which is the photon flux:

$$P_{inc} = r_p hv \quad (D.4)$$

The rate of electron production r_e is the quantum efficiency times the rate of photons, which can be related to the incident power:

$$r_e = \eta r_p = \frac{\eta P_{inc}}{hv} \quad (D.5)$$

The photocurrent is then given by the charge of each electron, e, times the electron flux, which can be expressed in terms of the incident power, frequency, and quantum efficiency:

$$I_p = e r_e = \frac{e \eta P_{inc}}{hv} \quad (D.6)$$

The responsivity can then be expressed in terms of the quantum efficiency η and the wavelength λ:

$$R_\lambda = \frac{I_p}{P_{inc}} = \frac{e\eta}{hv} = \frac{e\eta\lambda}{hv} = \frac{\eta\lambda}{1.24}(A/W) \quad (D.7)$$

This is the basis for the curves in Figure D.2.

Unfortunately, optical detectors are not perfect. Specifically there are a number of sources of noise in any measurement made with an optical sensor. The six major noise sources are:

Shot noise – Arrival of photons is a statistical process that is related to the Poisson distribution. In general, the standard deviation of the shot noise, measured in photon counts, equals the square root of the mean photon count. If a mean value of one million photons hits a detector in any given second,

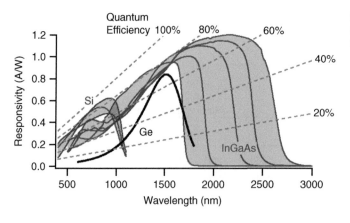

Figure D.2 Responsivity and quantum efficiency of various materials.

then the photon counts will vary from second to second with a standard deviation of 1000 photons, or 0.1% of the mean. Shot noise is often the dominant noise source in modern visible-wavelength sensors.

Dark current – Real detectors exhibit some current even in the absence of illumination. This is a significant issue for low-light-level imaging. Improvements in dark current are the reason cell phones can take pictures at night better than they could a decade ago.

Readout noises – Due to clock and other electronic noise in the readout circuitry of the sensor. Readout noise is minimized by good design.

Johnson noise – Due to the quantum nature of electrical currents. Can be expressed in terms of the square root of four times Boltzmann's constant times the bandwidth of the circuit divided by the resistance of the circuit. It is essentially shot noise for electrons.

Flicker noise – Due to quantum resistance fluctuations. It is usually expressed as a noise current I_p times the square root of the bandwidth divided by a flicker frequency f. The noise current and flicker frequency are characteristics of the detector material and how it is processed.

Digitization noise – Due to the finite steps used in digitizing the current output of the detector.

It is easy to derive the formula for the standard deviation of digitization noise.

Figure D.3 illustrates the digitization of a continuous signal. Assume the error δ in each digitized measurement is uniformly distributed between $+\Delta/2$ and $-\Delta/2$ where Δ is the digitization step size.

The variance of the noise is thus given by an integral of the square of the error times the probability that the error occurs:

$$\sigma_{dig}^2 = \int \delta^2 p(\delta) d\delta$$

$$= \frac{\int_{-\Delta/2}^{+\Delta/2} x^2 dx}{\int_{-\Delta/2}^{+\Delta/2} dx} = \frac{x^3|_{-\Delta/2}^{+\Delta/2}}{3\Delta} = \frac{\Delta^2}{12}$$

$$(D.8)$$

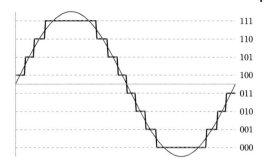

Figure D.3 Digitization of a continuous signal.

The result is that the noise variance is one-twelfth the step size squared and the standard deviation is the step size divided by the square root of 12:

$$\sigma_{dig} = \frac{\Delta}{\sqrt{12}} \tag{D.9}$$

D.1 Example Optical Sensors

Optical sensors can be categorized in multiple ways. This section reviews a variety of sensor types. The discussion begins with photodiodes, a fundamental type of photon detector that can be made from different semiconductors. The two different approaches used to form photodiodes into imaging devices are then described: the charge-coupled device and CMOS imagers. Finally, bolometers and microbolometers, common forms of thermal sensors are described.

Optical sensors can be configured as a single pixel, a line array or a two-dimensional image array. Single-pixel sensors would include photodiodes, no matter what material they are made from, and bolometers. Line or imaging sensors would include CCDs, CMOS arrays and microbolometers.

D.1.1 Photodiodes

A photodiode is a semiconductor device that converts light into current. The current is generated when photons are absorbed in the photodiode.

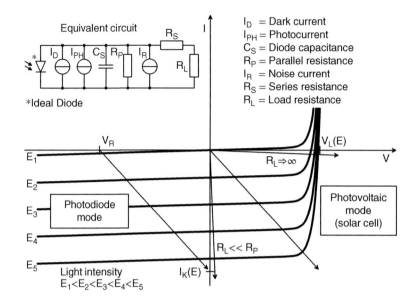

Figure D.4 Voltage–current curves for a photodiode. Figure from https://en.wikipedia.org/wiki/Photodiode#/media/File:Photodiode_operation.png.

The voltage–current curves in Figure D.4 show that photodiodes can be operated in various modes depending on what biasing voltage is applied to the device. Without a bias voltage, they operate in photovoltaic mode, producing a current directly from the incoming photons like a solar cell. But photodiodes are more commonly operated with a reverse bias, which is called the photodiode mode.

As illustrated in Figure D.5, the output current in photodiode mode is linearly proportional to irradiance incident on the sensor over a wide range of values. Yet photodiodes are not perfect sensors. When operated with reverse bias, a small amount of current, called the dark current, is also produced when no light is present. This dark current is responsible for the deviation from linearity at the lowest values of irradiance.

Silicon photodiodes are widely used at visible frequencies because their responsivity and quantum efficiency are reasonably large at visible wavelengths (Figure D.6).

Other semiconductor materials such as indium gallium arsenide or germanium are used for infrared sensing because that is where they have peak responses.

Indium gallium arsenide is a semiconductor alloy with a narrow direct bandgap in the near infrared (Figure D.7). The sensor works when an infrared photon of sufficient energy kicks an electron from the valence band to the conduction band. Such an electron is collected by a suitable external readout integrated circuits (ROIC) and transformed into an electric signal. InGaAs is the preferred material to detect infrared radiation in the 1 to 2 μm (NIR/SWIR) atmospheric windows.

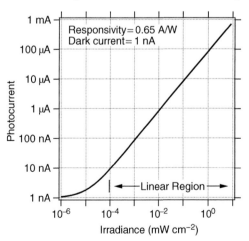

Figure D.5 Output current in photodiode mode.

Figure D.6 Quantum efficiency and responsivity of silicon, germanium and indium gallium arsenide.

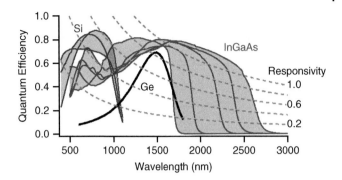

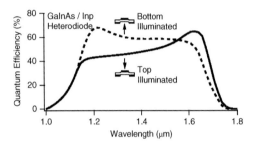

Figure D.7 Quantum efficiency of indium gallium arsenide (InGaAs) detectors. Figure adapted from https://commons.wikimedia.org/wiki/File:GaInAs_and_Ge_Photodiodes.jpg.

Mercury cadmium telluride is an alloy with a narrow direct bandgap spanning the NIR to LWIR regions. Mercury cadmium telluride is the only common material that can detect infrared radiation in both the 3–5 μm (MWIR) and 8–12 μm (LWIR) atmospheric windows (Figure D.8).

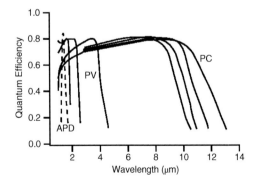

Figure D.8 Quantum efficiency of mercury cadmium telluride (HgCdTe) detectors. Based on data from Norton (1999).

LWIR mercury cadmium telluride detectors need cooling to temperatures near 77 K to reduce noise due to thermally excited current carriers. The main competitors of mercury cadmium telluride detectors are less sensitive silicon based bolometers.

Figure D.9 illustrates the spectral coverage in the UV, visible and infrared of various types of photon sensors. Again the silicon sensors are the preferred choice in the UV and visible, with some overlap into the near IR. Indium gallium arsenide is preferred in the short-wave IR. And mercury cadmium telluride is preferred at mid-wave to long-wave IR. There are other devices listed on this chart that are sometimes used for special applications, but we've discussed the most common forms of the photodiode.

D.1.2 Charge-Coupled Devices

Charge-coupled devices are one type of design used to form photodiodes into line or image arrays. The problem is not one of creating a 2D array of millions of photodiodes. The problem is how to accumulate charge and get the tiny currents off the chip. The pixels on a CCD image sensor are metal oxide semiconductor capacitors that store the electrons created by the incoming photons.

These capacitors are essentially storage wells that are formed by the application of small bias voltages. The key design element of the CCD is that alternating fields are used to transfer the charges from one charge well to the next. This process is illustrated in Figure D.10.

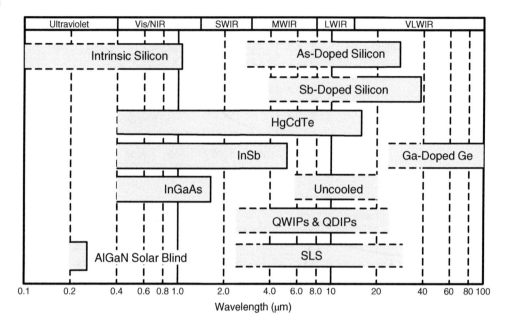

Figure D.9 Spectral coverage of various types of photon sensors. Figure from National Research Council (2010).

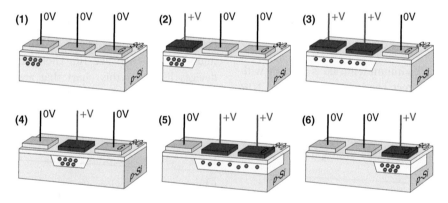

Figure D.10 CCD sensor phases. (1) charge accumulates, (2) well 1 is established, (3) transfer begins to well 2, (4) well 2 established, (5) transfer begins to well 3, (6) well 3 established. Figures adapted from Wikipedia, original by Michael Schmid, https://en.wikipedia.org/wiki/File:CCD_charge_transfer_animation.gif.

CCD image sensors are more expensive than CMOS sensors, but widely used in professional applications because of their lower noise, although CMOS sensors have been improving.

CCDs can be configured as 1D (line scanner) or 2D (imager) arrays.

Color imaging with a 2D imager can be performed by tripling the number of pixels and placing colored filters over each and every pixel in the scene, as illustrated in Figure D.11.

The pattern of blue, green and red filters shown in the diagram is called a Bayer filter. It has fewer red pixels than blue and green because of the poorer resolution of red in the human eye.

Although this is the way most cameras are built, most scientific instruments do not use this scheme because usually the measurements made in each color channel need to be made at precisely aligned locations, and this

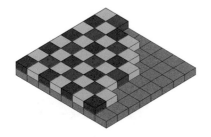

Figure D.11 Bayer patterned color filter. Figure from Wikipedia: https://commons.wikimedia.org/wiki/File:Bayer_pattern_on_sensor.svg.

cannot be done if different pixels are sensitive to different colors.

Figure D.12 illustrates that there are multiple architectures for getting data off of CCD imaging devices.

The Full Frame CCD is the simplest structure. In it the light falls on the sensor elements, which fill the area of the sensor. To read out the signals, the data columns are shifted down one row, then the readout row at the bottom of the array is shifted out to the right. The process is repeated until all of the rows have been read out. Note that the entire sensor, other than the readout row, is light sensitive, so the design has a 100% fill factor. The downside of this design is that it requires a mechanical shutter to be placed in front of the sensor. Otherwise the image will be blurred by light that continues to shine on the sensor face while the image data are being shifted and read out.

The Frame Transfer CCD was developed to alleviate some of the issues with the full frame CCD for applications like TV cameras that require continuous imaging. This design doubles the number of pixels, but covers up half of them under an opaque mask. That way only half the chip is exposed to light. During readout, all of the rows are shifted down into the bottom half of the chip, where the data are preserved during readout. There is still some blurring with this design, but the blurring occurs only over the time it takes to shift 1000 rows, instead of the time it take to shift 1 million pixels. This design offers higher frame rates, but at the expense of larger chip size.

The next design is the Interline Transfer CCD, which is quite common today. This design interleaves photosensitive columns with readout columns that have an opaque covering. The data are transferred into the interline channels in a single clock sequence, before being read out. This design produces the highest frame rates, but at the expense that only 50% of the sensor area is actually photosensitive. This loss of fill factor is often compensated for by inclusion of a microlens above each and every pixel.

D.1.3 CMOS Image Sensors

Pixels on a CMOS image sensor are also photodiodes, but each has an active circuit

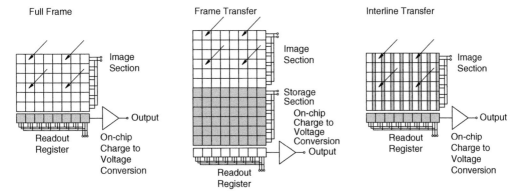

Figure D.12 CCD readout architectures. Figure from http://www.andor.com/learning-academy/ccd-sensor-architectures-architectures-commonly-used-for-high-performance-cameras with permission from Oxford Instruments.

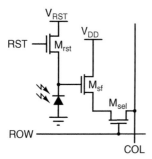

Figure D.13 Design of a CMOS cell. Figure from https://commons.wikimedia.org/wiki/File:Aps_pd_pixel_schematic.svg.

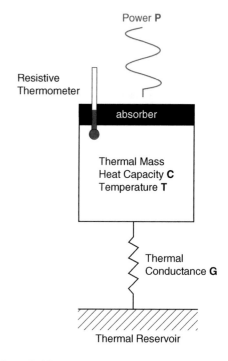

Figure D.14 Conceptual design of a bolometer. Figure from Wikipedia by D.F. Santavicca, https://commons.wikimedia.org/wiki/File:Bolometer_conceptual_schematic.svg.

of transistors to reset the diode, select the diode for readout, and act as a source-follower amplifier for signal readout (Figure D.13). These are essentially the same circuit elements used in the design of dynamic random access memory circuits. In fact, early experimenters created cheap cameras by prying the tops off of RAM chips.

CMOS image sensors are cheaper to manufacture than CCD sensors, and are widely used in consumer products such as cell phones. For a variety of technical reasons, CMOS sensors have historically suffered from higher noise levels than CCD imagers. But the desirability of cell phone cameras that work well in dimly-lit bars has led manufacturers to make significant reductions in CMOS sensor noise.

D.1.4 Bolometers and Microbolometers

A bolometer is a device for measuring the power of incident electromagnetic radiation via the heating of a material with a temperature-dependent electrical resistance (Figure D.14). Bolometers can be used to measure radiation of any frequency, but more sensitive devices exist for many wavelengths. Bolometers are the most sensitive detectors for wavelengths from around 200 μm to 1 mm, namely the far-infrared into the terahertz bands.

To achieve the best sensitivity, the bolometer must be cooled to a fraction of a degree above absolute zero, typically from 50 to 300 mK.

A microbolometer is a kind of integrated circuit bolometer used as a detector in thermal cameras (Figure D.15). Microbolometers are arranged into large arrays and combined with readout electronics for thermal imaging. Commercial units are available today with focal planes containing 1K × 1K pixels, although larger arrays could be built.

Microbolometers are typically sensitive to wavelengths between 7.5 and 14 μm. The most commonly used absorbing materials are amorphous silicon and vanadium oxide. Unlike other types of infrared detecting equipment, microbolometers do not require cooling. Microbolometer sensors are small, low power

Figure D.15 Design of a microbolometer cell. Source: Vincent et al. (2015) with permission from John Wiley & Sons.

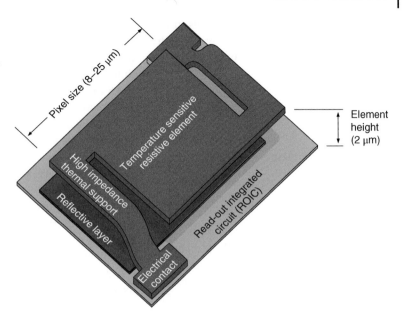

and lightweight, but less sensitive than other sensors.

Figure D.16 illustrates how the resolution of IR imaging sensors has historically lagged behind visible wavelength imaging sensors. This is an old plot showing the growth of pixel counts in focal planes as a function of year. The top curve shows the number of cells in a dynamic RAM as a surrogate for CCD or CMOS imagers operating in the visible. The bottom curves show the number of pixels achieved with various infrared sensing imagers. Notice

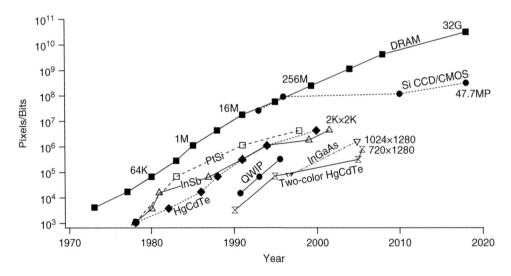

Figure D.16 Improving resolution of IR and visible wavelength imaging sensors. Adapted from Norton (2005).

that, at any time, the number of pixels in IR sensors typically lag the visible wavelength sensors by 1–2 orders of magnitude.

D.2 Optical Sensor Design Examples

D.2.1 Computing Exposure Times

Table D.1 lists the parameters for a notional satellite-based optical sensor[1]. It is instructive to step through some elements of this design.

The ground resolution of the sensor is computed by noting that the ratio of the pixel pitch δ_{pixel} to focal length f is the same as the ratio of the ground sample distance δ_g to the range R. Thus $\delta_g = \delta_{pixel} \times \frac{R}{f} = 1.63$ m. The area of a single pixel projected onto the ground is thus $A_g = \delta_g^2 = 2.65$ m^2.

Note that diffraction can also limit the achievable ground resolution. The diffraction limit of the ground resolution for this lens at 710 nm is about $\delta_g = 1.4 \frac{\lambda}{d} R = 0.8$ m. So diffraction does not limit the resolution for this design. Diffraction limits can become a limiting factor for smaller satellites. For example, a 3U

Table D.1 Example sensor specifications.

Aperture diameter (m)	d	0.60
Focal length (m)	f	8.8
Quantum efficiency	q	0.5
Pixel pitch (μm)	δ_{pixel}	21
Pixel well depth	N_{well}	89,000
TDI stages	N_{tdi}	64
Panchromatic band (nm)		450–800
Blue band (nm)		450–510
Red band (nm)		650–710
Altitude (km)	R	682
Platform ground velocity (km s^{-1})	v	6.8
Nadir sensor pointing	θ	90°

---
1 While all of these values are realistic, they are not meant to represent any individual system.

cubesat must fit within a $10 \times 10 \times 30$ cm volume, limiting the optical aperture diameter to something less than 10 cm. The diffraction limit at 710 nm for a 10-cm-diameter aperture is 4.8 m, so this would be the best achievable resolution at the edge of the red band.

The solid angle Ω that the sensor intercepts is given by the sensor aperture area A_o divided by the range squared R^2. The aperture area is $A_o = \pi d^2/4 = 0.28$ m^2. Thus the solid angle is $\Omega = 6.08 \times 10^{-13}$ sr.

Next we will compute the integration time required to fill the wells of each pixel. To perform this calculation we need to compute (1) the radiance of the scene, (2) the flux at the sensor, (3) the total flux on a pixel, and (4) the photoelectron production rate. We will perform the calculations for two channels: a panchromatic channel and a blue channel.

For quick calculations we note that the spectral radiance at the top of the atmosphere from 450–800 nm in clear sky conditions is approximately given by $2 \times 10^9 \frac{W \cdot nm^3}{m^2 \cdot sr^1} \times \lambda^{-4}$. Likewise, the radiance within a band specified by λ_1 and λ_2 is approximately: $6.7 \times 10^8 \frac{W \cdot nm^3}{m^2 \cdot sr^1} \times (\lambda_1^{-3} - \lambda_2^{-3})$. We will use these formulas to compute the band-averaged radiance for the panchromatic, blue and red channels to be 6.0, 2.3, and 0.6 W m^{-2} sr^{-1}. respectively. These correspond to spectral radiances of 0.017, 0.038, and 0.0095 W m^{-2} sr^{-1} nm^{-1}, values that are fully consistent with the top of the atmosphere curves in Figures 4.3 and 4.28. Note that while the spectral radiance is almost a factor of two higher in the blue channel than the panchromatic channel, the blue channel is 1/6 as wide, so the radiance in the blue channel is almost 1/3 of the radiance in the panchromatic channel. Furthermore, the radiance in the red channel is 1/10 of the radiance in the panchromatic channel, so it will be slowest to accumulate photons.

To compute the well fill time, we first consider just the panchromatic channel. The calculations for the blue and red channels are similar, and results for all three channels are shown at the end. The flux at the face of the sensor originating at each elemental area dA_s on the surface is: $d\Phi = L\cos\theta\,\Omega\,dA_s$. For a nadir looking sensor, $\cos\theta = 1$, so $d\Phi = 3.7 \times 10^{-12}$ W m^{-2}. Multiplying this spectral flux times the area of the ground pixel yields the total flux incident on a single pixel: $\Phi = 9.7 \times 10^{-12}$ W. This flux corresponds to a photon rate: $r_p = \frac{\lambda\Phi}{hc} = 3.1 \times 10^7$ photons/s. At a quantum efficiency of 0.5, this corresponds to a photoelectron production rate of $r_{pe} = q\,r_p = 1.5 \times 10^7$ photoelectrons per second.

At this rate, the well associated with a single pixel will fill in: $T_{well} = N_{well}/r_{pe} = 5.9$ ms. This value may seem reasonable until one considers that the ground velocity of the sensor is 6.8 km s^{-1}, so during this integration period the pixel moves a distance $\delta_{well} = vT_{well} = 39.8$ m, a factor 24 larger than the optical ground resolution. In other words, if we fill the wells in each pixel then motion blur will reduce the along-track image resolution from 1.6 to 39.8 m.

The solution is to incorporate time delay and integration into the design. The sensor focal plane is built with multiple rows that are each aligned in the cross-track direction, but adjacent to each other in the along-track direction. The data from one row are then shifted to the adjacent row at the same rate that the optical image moves across the rows. For example, in this design the pixels are squares 21 µm on a side, which produce ground pixels that are 1.63 m on a side. The time it takes the spacecraft to move one pixel on the ground is $T_{pix} = \delta_g/v = 0.24$ ms. If the sensor shifts the photoelectrons accumulated in each row into the adjacent row every 0.24 ms, then the accumulated photoelectrons are being moved to exactly keep pace with the motion of the optical image in the focal plane. The accumulated count in the last row then represents an integration over N_{tdi} periods without reducing the resolution of the image. This particular design would require 24 stages of Time-Delay and Integration (TDI) in order to provide the integration time to fill the wells while maintaining the optical resolution of the system.

Table D.2 summarizes these values, along with the comparable values for the blue and red channel. Note that well fill in the blue and red channels requires 3–10 times longer integration time than for the panchromatic channel. One solution is to use more stages of TDI.

Table D.2 Example sensor performance.

Ground resolution (GSD) (m)	δ	1.63		
Ground area of a pixel (m^2)	A_g	2.65		
Aperture area (m^2)	A_o	0.28		
Sensor solid angle (sr)	Ω	6.08×10^{-13}		
		Panchromatic	Blue	Red
Average spectral radiance (W·m^{-2}·sr^{-1}·nm^{-1})	L_λ	0.02	0.04	0.01
Band-averaged radiance (W·m^{-2}·sr^{-1})	L	6.0	2.3	0.6
Flux @ sensor (W m^{-2})	$d\Phi$	3.7×10^{-12}	1.4×10^{-12}	3.5×10^{-13}
Total flux on pixel (W)	Φ	9.7×10^{-12}	3.7×10^{-12}	9.1×10^{-13}
Photon rate (photons s^{-1})	r_p	3.1×10^7	8.9×10^6	3.1×10^6
Photoelectron production rate (pe s^{-1})	r_e	1.5×10^7	4.5×10^6	1.6×10^6
Time to fill well (ms)	T_{well}	5.9	20.0	57.3
Motion during time to fill well (m)	Δ_{well}	39.8	136	389
Time to move one pixel (ms)	T_{pix}	0.24	0.24	0.24
TDI stages to fill well	T_{tdi}	24	83	239

Another solution is to simply use bigger pixels for the multispectral measurements. This is the reason many of the high-resolution EO sensors have a factor of four lower resolution for multispectral imaging than for panchromatic imaging.

D.2.2 Impact of Digitization and Shot Noise on Contrast Detection

Consider an airborne remote sensing system utilizing a CCD camera with a well depth of 40,000 photoelectrons. That means that each pixel in the CCD camera could accumulate 40,000 photoelectrons before the pixel saturated.

The table in Figure D.17 compares the relative contributions of the shot noise and digitization noise for this CCD camera. The top row of the table indicates that the calculations were performed for three representative illumination levels: 100%, 25% and 2.5% well fills, corresponding to 40,000, 10,000, and 1000 photoelectrons in each well. The standard deviation of the shot noise is just the square root of the number of photoelectrons in the well, which corresponds to 200, 100, and 32 photoelectrons, respectively.

The first column indicates the three bit depths: 8-, 10-, and 12-bits. The standard deviation of the digitization noise is then the step size, in photoelectrons, divided by the square root of 12. The second column contains the values of the digitization noise, which were 45, 11, and 3 photoelectrons, respectively.

The two noises are uncorrelated, so they sum in quadrature, which means we compute the square root of the sum of squares. The numbers in the gray boxes indicate the total noise in photoelectrons for the specified mean photon count and number of bits. The table shows that when the pixels are well filled, the shot noise dominates and the digitization noise is negligible. But the digitization noise becomes more important under low light conditions when the wells are far from filled. In the worst cases, with low bit depth, the digitization noise dominates and the shot noise is negligible.

Now this table does not account for other noise sources, but it does illustrate the trade-offs between two of the most important noise sources.

Next consider the ability of this sensor to detect the contrast of one pixel relative to the rest. In this case the contrast signal is $S = C_{det}N_{well}$, and the signal-to-noise ratio is $SNR = \frac{C_{det}^2 N_{well}^2}{\sigma_{total}^2}$. Note that SNR is always expressed as a power ratio, hence the squares. Assuming an SNR of 10 is required to detect the contrast of a single pixel, the minimum detectable contrast can then be computed, as shown in Figure D.18.

In this example, the minimum detectable contrast in a single pixel is 1.6% when the wells are filled, almost independent of the digitization bit depth. If the wells are not filled then the minimum detectable contrast increases significantly, and the noise associated with digitization at low bit depths can come to dominate.

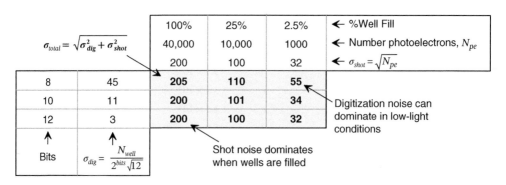

Figure D.17 Relative contributions of shot and digitization noise for a CCD camera.

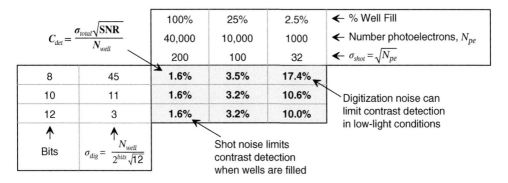

Figure D.18 Relative contributions of shot and digitization noise for a CCD camera.

References

National Research Council (Committee on Developments in Detector Technologies) (2010). *Seeing photons: Progress and limits of visible and infrared sensor arrays.* Washington, DC: National Academies Press.

Norton, P. R. (1999). HgCdTe infrared detectors. In *Proceedings of the sixth international symposium on long wavelength infrared detectors and arrays: Physics and applications* (Vol. *98-21*, pp. 49–70). The Electrochemical Society.

Norton, P. R. (2005). Third-generation sensors for night vision. In *Infrared photoelectronics* (Vol. *5957*, 59571Z). International Society for Optics and Photonics.

Ryer, A. (1997). *Light measurement handbook.* International Light, Inc.

Vincent, J. D., Hodges, S., Vampola, J., Stegall, M., & Pierce, G. (2015). *Fundamentals of infrared and visible detector operation and testing.* John Wiley & Sons.

Appendix E

Radar Design Example

Radar design can appear to be a complex topic when first encountered. This appendix provides a series of example problems in order to better understand the basics of radar design.

General hints:

- Visualize the flow and spread of energy when writing down equations. You may find it valuable to rederive the radar equation instead of remembering it because it will remind you of the simplicity of the concepts.
- Be careful about units. Make sure the units agree with the property you are asked to compute. Make sure the units are the identical on both sides of an equation.
- Never mix logarithmic units (dB) with linear units in the same calculation.

Consider the specifications for an airborne radar altimeter designed to measure the altitude of an aircraft above the ground shown in Table E.1.

Questions:

1. What is the peak power density falling on the ground?
2. What is the size of the antenna?
3. What is the area of the region on the ground illuminated by the radar?
4. What is the total peak power incident on the ground?
5. What is the RCS of the ground?
6. What is the peak power density returned to the radar antenna?
7. What is the peak power received by the radar?
8. What is the energy in a single pulse received by the radar?
9. What is the average transmit power?
10. What is the receiver noise power?
11. What is the peak or per pulse SNR?
12. What would the single-pulse SNR be of an equivalent pulsed radar with no modulation and an optimal receiver bandwidth?
13. What is the range resolution of the system without any pulse compression?
14. What pulse compression ratio can this radar support?
15. What is the range resolution after pulse compression?
16. What is the received energy in a single pulse after pulse compression?
17. What is the received power in a single pulse after pulse compression?
18. What is the single pulse SNR after pulse compression?

Solutions to questions 1–18:

We suggest that whenever you encounter a problem like this, you begin by computing a new table of input parameters that are all expressed in linear or dB units. An example starting table is shown as Table E.2.

Let us talk about linear and dB parameters and units. Given a linear parameter X, the logarithmic parameter $Y = 10 \log_{10} X$ is referred to dB. The inverse relation is $X = 10^{Y/10}$. If the units of X are cubits, then the units of Y are

Remote Sensing Physics: An Introduction to Observing Earth from Space, Advanced Textbook 3, First Edition.
Rick Chapman and Richard Gasparovic.
© 2022 American Geophysical Union. Published 2022 by John Wiley & Sons, Inc.
Companion website: www.wiley.com/go/chapman/physicsofearthremotesensing

Table E.1 Specifications for an airborne radar altimeter.

Frequency, f	4.3 GHz
Peak transmit power, P_t	+23 dBm
Pulse length, τ	50 μs
Pulse repetition frequency, *PRF*	50 Hz
Bandwidth, B	150 Mhz
Receiver noise figure, F_N	7 dB
Altitude, R	500 m
Antenna gain, G	13 dBi
Antenna type	separate transmit and receive, square horn
NRCS of surface, σ_0	2

dBcubits. (This rule also works for things other than cubits, which makes it generally useful.)

Without units, dB are always a power ratio. The only oddity is that you will often hear people say that to convert a voltage V to dB you take $20 \log_{10} V$. Actually $Y = 10 \log_{10} V$, where V is in volts and Y is in dBV. It is just that we are normally computing power in dB, and power is proportional to V^2, hence $Y_{power} = 10 \log_{10} V^2 = 20 \log_{10} V$, and the units are actually dBV². Unfortunately engineers never say this and we are probably breaking some unwritten rule of radar designers by even

mentioning it. So please don't tell anyone we told you about $10 \log_{10}$ vs $20 \log_{10}$.

Finally, note that radar engineers will sometimes use dBW, but will more often use dBm, which means dB relative to one milliwatt. Note that 0 dBW = 30 dBm.

1. What is the power density falling on the ground?

In answering these questions we will rederive the radar equation. This may seem pedantic, but it is often helpful to rederive the equation each time.

Start by assuming the peak transmitted power, P_t, is radiated isotropically, which means equally in all directions. The energy then would spread spherically. The illuminated surface area of the sphere at range R is $4\pi R^2$, so the power density, I_{s0}, at this range would be:

$$I_{s0} = \frac{P_t}{4\pi R^2} \quad \text{W m}^{-2} \quad \text{(E.1)}$$

Yet the specified radar utilizes an antenna designed to preferentially focus energy in one direction. For such an antenna, with power gain, G_t, the power density, I_s, along the beam at range R is:

$$I_s = \frac{P_t G_t}{4\pi R^2} \quad \text{W m}^2$$
$$= \frac{0.2 \times 20}{4\pi 500^2} = 1.27 \times 10^{-6} \text{ W m}^{-2} \quad \text{(E.2)}$$

Table E.2 Typical table of starting input parameters.

Parameter	Specified	Derived Linear	Derived Log
Frequency, f	4.3 GHz	4.3×10^9	96.3 dBHz
Peak transmit power, P_t	+23 dBm	0.2 W	−7 dBW
Pulse length, τ	50 μs	5×10^{-5} s	−43 dBs
Pulse repetition frequency, *PRF*	50 Hz	50 Hz	17 dBHz
Bandwidth, B	150 Mhz	1.5×10^8 Hz	81.8 dBHz
Receiver noise figure, *NF*	7 dB	5.01	7 dB
Altitude (which here = range), R	500 m	500 m	27 dBmeters
Antenna gain, G	13 dBi	20	13 dBi
NRCS, σ_0	2	$2 \text{ m}^2/\text{m}^2$	3 dB
Scale factor	4π	12.56	11 dB

We discussed dB above, but you may well be wondering why anyone would bother with dB. The answer is that radar designers find adding and subtracting easier than multiplication and division. So years ago they decided to take 10 times the logarithm of all these equations, and work in dB space. For example, equation (E.2) becomes:

$$10 \log_{10} I_s = 10 \log_{10} \left[\frac{P_t G_t}{4\pi R^2} \right]$$

$$= 10 \log_{10} P_t + 10 \log_{10} G_t$$
$$\quad - 10 \log_{10}(4\pi) - 10 \log_{10}(R^2)$$

$$= 10 \log_{10} P_t + 10 \log_{10} G_t$$
$$\quad - 10 \log_{10}(4\pi) - 20 \log_{10}(R)$$

$$= -7\,\text{dBW} + 13\,\text{dBi} - 11\,\text{dB}$$

$$\quad - 2 \times 27\,\text{dB} = -59\,\text{dBW} \qquad \text{(E.3)}$$

It is not an accident that $10^{(-59/10)} = 1.27 \times 10^{-6}$, the same answer we got before. We will do the rest of the calculations in linear units, but they all could be performed equally accurately in dB units.

2. What is the size of the antenna?

Equation (7.2) defines the gain of an antenna to be $G = 4\pi/\Omega$ where Ω is the solid angle of the antenna's main beam. The solid angle encompassed by the main beam is approximated by:

$$\Omega = \delta\theta_{az}\delta\theta_{el} \approx \left(\frac{\lambda}{d_{az}}\right)\left(\frac{\lambda}{d_{el}}\right) = \frac{\lambda^2}{A_e} \qquad \text{(E.4)}$$

This leads to a formula for the effective area of an antenna:

$$A_e = \frac{\lambda^2}{4\pi} G \qquad \text{(E.5)}$$

In this case $\lambda = c/f = 0.07$ m and $G = 20$, so $A_e = 7.73 \times 10^{-3}$ m^2. The problem says that the horn antenna is square, so the antenna dimensions are $d_{az} = d_{el} = \sqrt{A_e} = 0.088$ m.

This brings up an interesting question. When we are dealing with either a dish or horn antenna, the e-field across the aperture of the antenna is usually relatively constant. This makes the effective area of the antenna equal to the physical area of the antenna. But

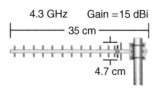

Figure E.1 Yagi antenna.

what is the effective area of a different type of antenna, for example the yagi antenna shown in Figure E.1?

The yagi's gain of 15 dBi equals a linear gain of 31.6. Equation (E.5) can then be used to compute the effective area of this antenna to be $A_e = 0.07^2 \times 31.6/(4\pi) = 0.0122$ m$^2 = 122$ cm^2. If this was a square aperture, it would correspond to an area of about 11 cm on each side, which is much larger than the cross-sectional dimensions of the antenna. There is no conflict here because the gain of this antenna is due to the length of the antenna. It is just important to realize that there is not always a simple relation between the effective area of an antenna and its physical size.

3. What is the area of the region on the ground illuminated by the radar?

One answer can be derived directly from the computation we just performed. Namely, we divide the transmitted power by the power density in the illuminated area:

$$dA = \frac{P_t}{I_s} = \frac{0.2}{1.27 \times 10^{-6}} = 1.57 \times 10^5 \text{ m}^2$$

$$\text{(E.6)}$$

But this is not the best answer. The area we just derived is not the area on the ground, but the illuminated area of the sphere centered on the radar with a radius of R. For a narrow beam, this would be nearly the same as the area on the ground, but that is not the case for a wide beam.

The alternative approach is to compute the beamwidth of the antenna, and then project that onto the ground. The antenna beamwidth is $\lambda/d = 0.07/0.088 = 0.794$ rad $= 45.5°$. The dimensions on the ground are then $R\delta\theta = 400$ m on a side for a total area of $dA = 1.57 \times 10^5$ m^2.

But wait a minute, this is virtually the same answer as we got before. Well we've led you down a false path. Geometrically, the factor we have computed should not be identical – the spherical cap is not equal to the flat base of the cone corresponding to a projection onto the flat ground. But the approximate form we used in equation (E.4) uses this exact approximation.

4. What is the total peak power incident on the ground?

This is an easy one. We transmit 0.2 W, and it is all directed towards the ground by the antenna, so the total peak power incident on the ground, $P_i = 0.2$ W.

This calculation is only slightly more difficult for an isolated scatterer. For example, the power falling on a target with radar cross-section (RCS) of σ_t with units of m^2 is:

$$P_i = I_s\sigma_t = \frac{P_tG_t\sigma_t}{4\pi R^2} \quad \text{W} \qquad (E.7)$$

At its core, this is the definition of radar cross-section. It specifies the effective intercept area of the target.

5. What is the RCS of the ground?

The problem statement gives the normalized radar cross-section (NRCS) of the ground to be 2, which although unit-less is taken to mean 2 m^2 of RCS per m^2 of illuminated area. The RCS is then the NRCS times the illuminated area:

$$\sigma_t = \sigma_0 \, dA = 2(1.57 \times 10^5) = 3.15 \times 10^5 \text{ m}^2$$
$$(E.8)$$

Notice the RCS increases with the illuminated area, which makes sense since the scattering object is effectively bigger.

It is worth pointing out that the only other area we have in the problem is the effective area of the antenna, but the antenna size does not directly enter into this calculation. So do not make the mistake of multiplying σ_0 times A_e. While this is dimensionally correct, the result is nonsense.

6. What is the peak power density returned to the radar antenna?

The surface is assumed to reradiate the energy incident on it isotropically, so the power density back at the radar, I_r, is:

$$I_r = \frac{P_tG_t\sigma_t}{(4\pi R^2)^2} \quad \text{W m}^{-2}$$
$$= \frac{P_tG_t\sigma_0 \, dA}{(4\pi R^2)^2} \quad \text{W m}^{-2}$$
$$= \frac{0.2 \cdot 20 \cdot 2 \cdot 1.57 \times 10^5}{(4\pi 500^2)^2}$$
$$= 1.27 \times 10^{-7} \text{ W m}^{-2} \qquad (E.9)$$

Note that a realistic target will actually reradiate energy with varying efficiency in different directions. The radar cross-section for such a target in the direction of the radar is actually defined as the cross-sectional area required to produce the power density observed at the radar according to equation (E.9).

7. What is the peak power received by the radar?

The radar receives the reflected energy with an antenna with an effective area aperture, A_e, that can be related to the directive receive gain, G_r. Usually this is the same antenna as the transmit antenna, but not always. Separate antennas are often used in radars that transmit and receive at the same time. The power received by the radar, P_r, is then:

$$P_r = \frac{P_tG_t\sigma_t}{(4\pi R^2)^2}A_e \quad \text{W}$$
$$= \frac{P_tG_tG_r\lambda^2\sigma_t}{(4\pi)^3R^4} \quad \text{W}$$
$$= 0.2 \cdot 20 \cdot 3.15 \times 10^5$$
$$\cdot 7.73 \times 10^{-3}/(4\pi 500^2)^2$$
$$= 9.82 \times 10^{-10} \text{ W} \qquad (E.10)$$

Two things to note. First we started with the peak transmitted power, so this is the peak received power. Second, this is about 1 nanowatt, a small but measurable level.

So we have done the calculation, but consider the dependence of the received power on range. A naive reading of equation (E.10) suggests that the received power falls off as R^4, but this fails to take into account the fact that the target RCS depends on area, which in turn depends on range. We can make the relationship explicit by making two substitutions: $\sigma_t = \sigma_0 \, dA = \sigma_0 R^2 \delta\theta^2$ and $G_r = 4\pi/\Omega = 4\pi/\delta\theta^2$, yielding:

$$P_r = \frac{P_t G_t (4\pi/\delta\theta^2) \lambda^2 (\sigma_0 R^2 \delta\theta^2)}{(4\pi)^3 R^4} \quad \text{W}$$

$$= \frac{P_t G_t \lambda^2 \sigma_0}{(4\pi)^2 R^2} \quad \text{W} \qquad \text{(E.11)}$$

This may seem surprising, because you usually hear that radar received power falls off as R^4, but this is actually only true for an isolated target. For a beam-limited system with a narrow beamwidth looking straight down at a rough surface, the falloff is only R^2. That this is the case should be easy to see from the answer to question 4, namely all of the transmitted power makes it to the surface within the main beam. So there is no power lost on the way to the surface, it is just spread out. The only real propagation loss is the $4\pi R^2$ associated with the path from the surface back to the radar.

Scatterometers are often beam-limited in the azimuth direction and pulse-length limited in the range direction. In this case, the area of each range cell increases linearly with range. So the fall-off of the signal returned from each range cell is R^3. Thus the rate of fall off of the signal depends on the specific configuration of the radar and the scene it is viewing.

Several additional effects will enter into this equation when the beam gets wide enough: the NRCS may no longer be constant over a wide range of incident angles, and the range to the surface as measured at the edge and the center of the illuminated beam may become significantly different, meaning a varying path loss on the return path. But the approach we have taken is often adequate for narrow-beam systems.

8. What is the energy in a single pulse received by the radar?

The per pulse received energy for this radar, with pulse length $\tau = 50$ μs, is:

$$E_{r1} = P_r \tau = \frac{P_t G_t G_r \lambda^2 \sigma_t \tau}{(4\pi)^3 R^4} \quad \text{W} \cdot \text{s}$$

$$= 9.82 \times 10^{-10} \cdot 50 \times 10^{-6}$$

$$= 4.91 \times 10^{-14} \quad \text{W} \cdot \text{s} \qquad \text{(E.12)}$$

and the received energy for a train of n pulses would be:

$$E_{rn} = \frac{P_t G_t G_r \lambda^2 \sigma_t n \tau}{(4\pi)^3 R^4} \quad \text{W} \cdot \text{s} \qquad \text{(E.13)}$$

9. What is the average transmit power?

The statement of the problem says that the pulse length is $\tau = 50$ μs and the pulse repetition frequency, $PRF = 50$ Hz. The pulse repetition interval can then be computed from the inverse of the PRF, $PRI = 1/PRF = 20$ ms. The average transmitted power, P_{av} is then:

$$P_{av} = \frac{\tau}{PRI} P_t = \frac{50 \times 10^{-6}}{20 \times 10^{-3}} 0.2 = 0.5 \text{ mW} \qquad \text{(E.14)}$$

The ratio of τ/PRI is also known as the duty cycle, the fraction of the time the radar is transmitting. For this radar, the duty cycle is $50 \times 10^{-6}/20 \times 10^{-3} = 0.25\%$.

10. What is the receiver noise power?

In general, thermal noise = kTB. For Earth remote sensing it is common to take T to be 300 K, although the actual scene temperature should be used if known. We have the Boltzmann constant $= k = 1.38 \times 10^{-23}$ J/K. The bandwidth of this receiver was given to be 150 MHz. If the receiver was perfect, the thermal noise floor would be:

$$P_n = kTB = N_0 B \quad \text{W}$$

$$= 1.38 \times 10^{-23} \cdot 300 \cdot 1.5 \times 10^6$$

$$= 6.21 \times 10^{-13} \quad \text{W} \qquad \text{(E.15)}$$

Note we have defined $N_0 = kT$ which is then nominally a value of $N_0 = 4.14 \times 10^{-21}$ W Hz^{-1} = -204 dBW Hz^{-1} = -174 dBm Hz^{-1}. These are common values used in radar engineering to designate the thermal noise floor for an ideal receiver.

Of course, the actual receiver front end adds an additional noise factor that was specified to be 7 dB, equivalent to a factor of five. So the received noise power is:

$$P_n = kTB(NF)$$

$$= 1.38 \times 10^{-23} \cdot 300 \cdot 1.5 \times 10^6 \cdot 5$$

$$= 3.11 \times 10^{-12} \quad \text{W} \qquad \text{(E.16)}$$

11. What is the peak or per pulse SNR?

The signal-to-noise ratio (SNR) of a single received pulse, SNR_1 is:

$$SNR_1 = \frac{P_r}{P_n} = \frac{P_t G_t G_r \lambda^2 \sigma_t}{(4\pi)^3 R^4 kTB(NF)}$$

$$= \frac{9.82 \times 10^{-10}}{3.11 \times 10^{-12}} = 315 = 25 \text{ dB}.$$

(E.17)

Notice that SNRs are defined to be ratios of power, so they are unit-less quantities. This is also a single pulse SNR prior to any gains resulting from pulse compression. That is the reason the SNR does not depend on the pulse length. Still, this is a substantial SNR, meaning the system would have no problem detecting the incoming pulses.

12. What is the single-pulse SNR of an equivalent pulsed radar with no modulation and an optimal receiver bandwidth?

This may seem an odd question, but it actually helps illustrate the value of modulation and pulse compression. Consider an unmodulated pulse of duration $\tau = 50$ μs. The bandwidth of that pulse is $B = 1/\tau = 20$ kHz. The optimal receiver bandwidth for this pulse width is then just 20 kHz – a wider bandwidth would not capture any additional transmitted energy, but would admit more noise.

This unmodulated pulsed radar would have a single-pulse SNR of:

Notice that when the receiver bandwidth is matched to the pulse length, we can rewrite the SNR in terms of pulse length τ instead of receiver bandwidth B:

$$SNR_{unmodulated} = \frac{P_t G_t G_r \lambda^2 \sigma_t}{(4\pi)^3 R^4 kTB(NF)}$$

(E.19a)

$$= \frac{P_t G_t G_r \lambda^2 \sigma_t \tau}{(4\pi)^3 R^4 kT(NF)}$$

(E.19b)

This formulation shows that the SNR increases for longer, unmodulated pulses, as long as the receiver bandwidth decreases to matches the pulse bandwidth.

13. What is the range resolution of the system without any pulse compression?

The range resolution for a simple pulse radar is $c\tau/2 = 3 \times 10^8 \cdot 5 \times 10^{-5}/2 = 7500$ m!!! Which brings up an interesting question: how long does it take for the leading edge of the pulse to travel from the radar to the surface and back? The total distance down and back is 1000 m, which at 3×10^8 m/s only takes 3.3 μs. The transmitted pulse lasts a full 50 μs, so for almost this entire period the receiver and transmitter have to be operating at the same time. This is the reason we specified separate transmit and receive antennas.

14. What pulse compression ratio can this radar support?

$$SNR_{unmodulated} = \frac{P_t G_t G_r \lambda^2 \sigma_t}{(4\pi)^3 R^4 kTB(NF)}$$

$$= \frac{0.2 \cdot 20^2 \cdot 0.07^2 \cdot 3.15 \times 10^5}{4\pi \cdot 500^4 \cdot 1.38 \times 10^{-23} \cdot 300 \cdot 2 \times 10^5 \cdot 5} = 2.37 \times 10^6$$

(E.18)

Here we have kept all of the radar parameters the same, except for the bandwidth B which has changed from 150 MHz in the originally specified radar to 20 kHz, a factor of 7500. Thus the SNR of the optimal receiver without pulse compression is 7500 times the SNR of the originally specified receiver without pulse compression, as computed in equation (E.17).

The pulse length after pulse compression is $\tau_c = 1/B = 6.67 \times 10^{-9}$ s. The pulse compression ratio (or factor) is simply given by the ratio of the pulse length before and after compression, $PCR = \tau/\tau_c = 5 \times 10^{-5}/6.67 \times 10^{-9} = 7500$.

15. What is the range resolution after pulse compression?

The range resolution after pulse compression can be computed from the compressed pulse length $c\tau_c/2 = 3 \times 10^8 \cdot 6.67 \times 10^{-9}/2 = 1$ m! Equivalently you can compute $c/2B = 3 \times 10^8/(2 \cdot 1.5 \times 10^8) = 1$ m, the difference being you should be able do that math in your head.

16. What is the received energy in a single pulse after pulse compression?

This is a trick question. Pulse compression does nothing to the energy in the pulse. It is the same as before.

17. What is the received power in a single pulse after pulse compression?

This question is more difficult than it seems. The answer is best understood in two parts. First, the energy of the pulse has stayed the same, but now the pulse is 7500 times shorter. So the power must be 7500 times larger. Again there is no magic, because the energy in the returned pulse is the same. Mathematically we have:

$$P_r = \frac{P_t G_t G_r \lambda^2 (PCR)\sigma_t}{(4\pi)^3 R^4} \quad \text{W} \qquad (E.20)$$

Looking at it another way, the radar sends out a 50-μs-long pulse and, for the case computed here, receives back a 50 μs-long pulse with a peak power of about 1 nW. This is equally true if the pulse is a 50-ms transmission of a single frequency or if the pulse is frequency modulated as specified in this radar. This is the pulse before pulse compression.

The question is not asking about the raw received power, but is instead asking about the power at the output of the correlation stage within the receiver. So now the receiver is measuring the power, which is the square of the signal amplitude, at a rate high enough to capture the full bandwidth of the transmitted pulse. It is this output that quickly peaks at a much larger power than the original signal because many samples are being summed together coherently. So at the output of the correlator, the signal looks like a 6.67 ns pulse with a power of 7500 nW. The combination of the modulation and correlation receiver allows

the radar to have the energy of the long pulse, with the resolution of a short pulse.

So part 1 of the solution is to recognize that the received power is multiplied by the pulse compression ratio. Part 2 is to realize that pulse compression also changes the value of one of the other terms in equation (E.20). Namely, the short pulse effectively changes the illuminated area on the surface dA, reducing the radar cross-section σ_t.

The effective illuminated area at the peak of the correlator output was associated with a pulse-limited circle under the altimeter. We have already computed the pulse length to be 6.67 ns, which corresponds to a 2-m-long pulse. The center of the circle is 500 m below the radar, while the edge of the circle is 502 m away from the radar. The radius of the illuminated circle is thus $r = \sqrt{502^2 - 500^2} = 44.8$ m. The illuminated area is $dA = \pi r^2 = \pi \cdot 44.8 = 6300$ m^2. This is just 4% of the beam-limited area we previously computed. The radar cross-section is thus $dA\sigma_0 = 6300 \cdot 2 = 1.26 \times 10^4$ m^2.

Combining the pulse compression gain with the smaller RCS yields an estimate for the peak received power:

$$P_{rc} = \frac{P_t G_t G_r \lambda^2 (PCR)\sigma_t}{(4\pi)^3 R^4} \quad \text{W}$$

$$= \frac{0.2 \cdot 20^2 \cdot 0.07^2 \cdot 7500 \cdot 1.26 \times 10^4}{1984 \cdot 500^4}$$

$$= 2.94 \times 10^{-7} \text{W} \qquad (E.21)$$

18. What is the single-pulse SNR after pulse compression?

Since the received power goes up by a factor of 7500, the SNR increases by the same factor:

$$SNR_c = \frac{P_{rc}}{P_n} = \frac{2.94 \times 10^{-7}}{3.11 \times 10^{-12}}$$

$$= 9.46 \times 10^4 = 49.8 \text{ dB}. \qquad (E.22)$$

Definitions

A_e	Antenna effective area, m^2
B	Receiver bandwidth, Hz
c	Speed of light, 3×10^8 m s^{-1}
dA	Area of resolution cell, m^2

d_{az}	Antenna size in azimuth direction, m	P_{av}	Average transmit power, W
d_{el}	Antenna size in elevation direction, m	P_i	Peak power incident on ground, W
$\delta\theta_{az}$	Angular extent of main beam in azimuth, rad	P_n	Received noise power, W
$\delta\theta_{el}$	Angular extent of main beam in elevation, rad	P_r	Peak received power, W
E_{r1}	Single pulse received energy, J	P_{rc}	Pulse-compressed peak received power, W
f	Radar frequency, MHz	P_t	Peak transmit power, W
G_t	Transmit antenna gain	PRF	Pulse repetition frequency, Hz
G_r	Receive antenna gain	PRI	Pulse repetition interval, Hz
I_{s0}	Power density for isotropic antenna, W m^{-2}	R	Range to target, m
		σ_t	Target radar cross-section, m^2
I_r	Power density at receive antenna, W m^{-2}	σ_0	Clutter RCS/per unit area
I_s	Power density at surface, W/m^2	SNR_1	Single pulse, uncompressed signal-to-noise ratio
k	Boltzmann constant = 1.38×10^{-23} J/K	SNR_c	Compressed pulse signal-to-noise ratio
λ	Radar wavelength, m	T	Radiometric temperature = 300K
Ω	Solid angle of main beam, sr	τ	Pulse length, s
NF	External noise figure	τ_c	Compressed pulse length, s
PCR	Pulse compression ratio		

Appendix F

Remote Sensing Resources on the Internet

The following are some of the key remote sensing resources available on the Internet.

F.1 Information and Tutorials

- https://sentinel.esa.int/web/sentinel/home – Information on Sentinel missions
- http://www.altimetry.info/radar-altimetry-tutorial/ - Radar altimetry tutorial
- https://earth.esa.int/web/polsarpro/polarimetry-tutorial – SAR polarimetry tutorial
- https://earth.esa.int/documents/10174/2700124/sar_land_apps_1_theory.pdf – SAR land applications tutorial
- http://www.oceanopticsbook.info – Ocean Optics Web book – Terrific web-based textbook on optical oceanography
- https://www.oceanopticsbook.info/packages/iws_l2h/conversion/files/LightandWater.zip – Mobley, C.D. (1994). *Light and water: Radiative transfer in natural waters.* Academic Press (entire book is available for free download).
- http://www.radartutorial.eu/index.en.html – Radar tutorial
- https://www.nrcan.gc.ca/earth-sciences/geomatics/satellite-imagery-air-photos/satellite-imagery-products/educational-resources/9309 – High school level tutorial on remote sensing
- https://www.nrcan.gc.ca/earth-sciences/geomatics/satellite-imagery-air-photos/satellite-imagery-products/educational-resources/9579 – Radar polarimetry tutorial
- http://www.sarusersmanual.com – Synthetic Aperture Radar Marine User's Manual
- http://www.internalwaveatlas.com/Atlas2_index.html – An Atlas of Internal Solitary-like Waves and their Properties.

F.2 Data

- https://eosweb.larc.nasa.gov – Atmospheric Science Data Center
- https://www.asf.alaska.edu – Alaska Satellite Facility - Primarily SAR data
- https://oceancolor.gsfc.nasa.gov – NASA Ocean Color Data Center
- https://seabass.gsfc.nasa.gov – NASA archive of in situ oceanographic and ocean color data
- https://cddis.nasa.gov – NASA Space Geodesy Data Center
- https://disc.gsfc.nasa.gov – NASA Data and Information Services Center for atmospheric composition, water & energy cycles and climate variability
- https://ladsweb.modaps.eosdis.nasa.gov – NASA Level-1 and Atmosphere Archive & Distribution System Distributed Active Archive Center
- https://lpdaac.usgs.gov – NASA Land Processes Data Center

Remote Sensing Physics: An Introduction to Observing Earth from Space, Advanced Textbook 3, First Edition.
Rick Chapman and Richard Gasparovic.
© 2022 American Geophysical Union. Published 2022 by John Wiley & Sons, Inc.
Companion website: www.wiley.com/go/chapman/physicsofearthremotesensing

- https://nsidc.org/daac/ – National Snow and Ice Data Center
- https://daac.ornl.gov – NASA Biogeochemical Dynamics Data Center
- https://podaac.jpl.nasa.gov – NASA Physical Oceanography Data Center
- http://sedac.ciesin.columbia.edu – NASA Socioeconomic Data Center
- http://www.remss.com – Company that analyzes and distributes data from satellite microwave sensors, including radiometers, sounders and scatterometers
- https://scihub.copernicus.eu – ESA open access to Sentinel-1, Sentinel-2, Sentinel-3 and Sentinel-5P user products.

F.3   Data Processing Tools

- https://earth.esa.int/web/polsarpro – ESA's PolSARPro Polarimetric SAR data processing and educational tool
- https://seadas.gsfc.nasa.gov – NASA's SeaDAS is a comprehensive software package for the processing, display, analysis, and quality control of ocean color data
- http://www.altimetry.info/toolbox/ – ESA/CNES Radar altimetry toolbox
- http://step.esa.int/main/toolboxes/ – Toolboxes for Sentinel 1, Sentinel 2, Sentinel 3, SMOS, and Proba-V
- https://earth.esa.int/eogateway/search?text=&category=Tools%20and%20toolboxes – Variety of ESA data analysis and viewing tools.

F.4   Satellite and Sensor Databases

- http://ceos.org – Committee on Earth Observation Satellites – up-to-date database of earth observing satellites and sensors
- https://space.skyrocket.de – Gunter's space page – up-to-date database of spacecraft
- https://www.wmo-sat.info/oscar/spacecapabilities – World Meteorological Organization database of environmental satellite missions and instruments

F.5   Other

- https://www.nasa.gov – NASA home page
- https://www.esa.int/ESA – ESA home page
- https://earth.esa.int/web/guest/home – ESA Earth Observation home page
- https://neo.sci.gsfc.nasa.gov – NASA Earth Observation home page
- https://svs.gsfc.nasa.gov – NASA scientific visualization studio - great source for remote sensing imagery and animations
- https://science.nasa.gov/toolkits/spacecraft-icons – Collection of NASA scientific spacecraft icons.

Appendix G

Useful Trigonometric Identities

$$\tan\theta = \frac{\sin\theta}{\cos\theta} \qquad \sec\theta = \frac{1}{\cos\theta}$$

$$\cot\theta = \frac{1}{\tan\theta} = \frac{\cos\theta}{\sin\theta} \qquad \csc\theta = \frac{1}{\sin\theta} \qquad \text{(G.1)}$$

$$\sin^2\theta + \cos^2\theta = 1 \qquad \text{(G.2)}$$

$$\cos\theta = \sin\left(\frac{\pi}{2} - \theta\right) \quad \sin\theta = \cos\left(\frac{\pi}{2} - \theta\right)$$

$$\cot\theta = \tan\left(\frac{\pi}{2} - \theta\right) \quad \tan\theta = \cot\left(\frac{\pi}{2} - \theta\right) \qquad \text{(G.3)}$$

$$\csc\theta = \sec\left(\frac{\pi}{2} - \theta\right) \quad \sec\theta = \csc\left(\frac{\pi}{2} - \theta\right)$$

$$\sin(\theta + 2\pi) = \sin\theta \quad \sin(\pi - \theta) = \sin\theta \quad \sin(-\theta) = -\sin\theta$$

$$\cos(\theta + 2\pi) = \cos\theta \quad \cos(\pi - \theta) = -\cos\theta \quad \cos(-\theta) = \cos\theta \qquad \text{(G.4)}$$

$$\tan(\theta + \pi) = \tan\theta \quad \tan(\pi - \theta) = -\tan\theta \quad \tan(-\theta) = -\tan\theta$$

$$\sin(\alpha + \beta) = \sin\alpha\cos\beta + \cos\alpha\sin\beta \qquad \text{(G.5)}$$

$$\cos(\alpha + \beta) = \cos\alpha\cos\beta - \sin\alpha\sin\beta \qquad \text{(G.6)}$$

$$\sin(\alpha - \beta) = \sin\alpha\cos\beta - \cos\alpha\sin\beta \qquad \text{(G.7)}$$

$$\cos(\alpha - \beta) = \cos\alpha\cos\beta + \sin\alpha\sin\beta \qquad \text{(G.8)}$$

$$\sin(2\theta) = 2\sin\theta\cos\theta \qquad \text{(G.9)}$$

$$\cos(2\theta) = \cos^2\theta - \sin^2\theta$$
$$= 2\cos^2\theta - 1 \qquad \text{(G.10)}$$
$$= 1 - 2\sin^2\theta$$

Law of Sines

$$\frac{\sin\alpha}{a} = \frac{\sin\beta}{b} = \frac{\sin\gamma}{c} \qquad \text{(G.11)}$$

Law of Cosines

$$a^2 = b^2 + c^2 - 2bc\cos\alpha$$
$$b^2 = a^2 + c^2 - 2ac\cos\beta \qquad \text{(G.12)}$$
$$c^2 = a^2 + b^2 - 2ab\cos\gamma$$

Remote Sensing Physics: An Introduction to Observing Earth from Space, Advanced Textbook 3, First Edition.
Rick Chapman and Richard Gasparovic.
© 2022 American Geophysical Union. Published 2022 by John Wiley & Sons, Inc.
Companion website: www.wiley.com/go/chapman/physicsofearthremotesensing

Index

Remote Sensing Physics: An Introduction to Observing Earth from Space, Advanced Textbook 3, First Edition.
Rick Chapman and Richard Gasparovic.
© 2022 American Geophysical Union. Published 2022 by John Wiley & Sons, Inc.
Companion website: www.wiley.com/go/chapman/physicsofearthremotesensing

Printed and bound by CPI Group (UK) Ltd, Croydon, CR0 4YY

16/04/2025

14658468-0004